MW00388598

Telecommunications Law in the Internet Age

The Morgan Kaufmann Series in Networking
Series Editor, David Clark, MIT

Telecommunications Law in the Internet Age
Sharon K. Black

Optical Networks: A Practical Perspective, 2e
Rajiv Ramaswami and Kumar N. Sivarajan

Internet QoS: Architectures and Mechanisms
Zheng Wang

TCP/IP Sockets in Java: Practical Guide for Programmers
Michael J. Donahoo and Kenneth L. Calvert

TCP/IP Sockets in C: Practical Guide for Programmers
Kenneth L. Calvert and Michael J. Donahoo

Multicast Communication: Protocols, Programming, and Applications
Ralph Wittmann and Martina Zitterbart

MPLS: Technology and Applications
Bruce Davie and Yakov Rekhter

High-Performance Communication Networks, 2e
Jean Walrand and Pravin Varaiya

Computer Networks: A Systems Approach, 2e
Larry L. Peterson and Bruce S. Davie

Internetworking Multimedia
Jon Crowcroft, Mark Handley, and Ian Wakeman

Understanding Networked Applications: A First Course
David G. Messerschmitt

Integrated Management of Networked Systems: Concepts, Architectures, and their Operational Application
Heinz-Gerd Hegering, Sebastian Abeck, and Bernhard Neumair

Virtual Private Networks: Making the Right Connection
Dennis Fowler

Networked Applications: A Guide to the New Computing Infrastructure
David G. Messerschmitt

Modern Cable Television Technology: Video, Voice, and Data Communications
Walter Ciciora, James Farmer, and David Large

Switching in IP Networks: IP Switching, Tag Switching, and Related Technologies
Bruce S. Davie, Paul Doolan, and Yakov Rekhter

Wide Area Network Design: Concepts and Tools for Optimization
Robert S. Cahn

Practical Computer Network Analysis and Design
James D. McCabe

Frame Relay Applications: Business and Technology Case Studies
James P. Cavanagh

For further information on these books and for a list of forthcoming titles, please visit our Web site at *http://www.mkp.com*.

Telecommunications Law in the Internet Age

Sharon K. Black, Attorney-at-Law

Law Offices of S. K. Black, Boulder, Colorado

AN IMPRINT OF ACADEMIC PRESS
A Harcourt Science and Technology Company
SAN FRANCISCO SAN DIEGO NEW YORK BOSTON
LONDON SYDNEY TOKYO

Editor	Rick Adams
Publishing Services Manager	Scott Norton
Senior Production Editor	Cheri Palmer
Assistant Acquisitions Editor	Karyn Johnson
Project Management	Dusty Friedman, The Book Company
Cover Design	Ross Carron Design
Cover Image	© Corbis Image/PictureQuest
Text Design	Andrew Ogus
Composition	Thompson Type
Copyeditor	Jane Loftus
Proofreader	Debra Gates
Indexer	Bill Meyers
Printer	Edwards Brothers

Morgan Kaufmann Publishers
340 Pine Street, Sixth Floor, San Francisco, CA 94104-3205, USA
http://www.mkp.com

ACADEMIC PRESS
A Harcourt Science and Technology Company
525 B Street, Suite 1900, San Diego, CA 92101-4495, USA
http://www.academicpress.com

Academic Press
Harcourt Place, 32 Jamestown Road, London, NW1 7BY, United Kingdom
http://www.academicpress.com

Printed in the United States of America

06 05 04 03 02 5 4 3 2 1

Library of Congress Control Number: 2001094370

ISBN 1-55860-546-0

This book is printed on acid-free paper.

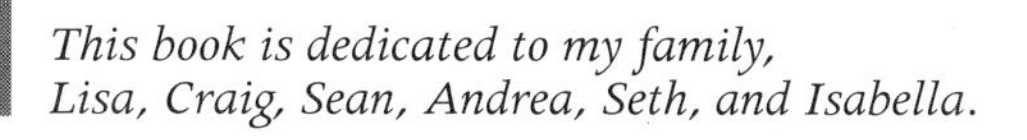

This book is dedicated to my family,
Lisa, Craig, Sean, Andrea, Seth, and Isabella.

Contents

Chapter 6
Participating in Global Telecommunications Trade: U.S. Import and Export Laws 197

Chapter 7
Licensing to Protect Telecommunications Intellectual Property 233

Chapter 10

Cyberlaw: Evolving Legal Issues with the Internet 389

Preface

OVERVIEW

The term "Internet Age" describes the remarkable state of nearly instantaneous worldwide voice, data, and video communications available today. With the "dial" of a phone or the click of a mouse, we can exchange information across distances and international borders in a manner never imagined by previous generations.

Communications in the Internet Age are unique from all other communications phenomena that have existed before. Hundreds of thousands of networks link together in such a manner that any person in the world with access to the Internet or a telephone can contact any other person in the world with access to the Internet or a telephone, regardless of their location, type of equipment being used, or standard controlling that equipment. Communications can be (1) one-to-one, one-to-many, or many-to-many; (2) interactive or noninteractive; and (3) local, long-distance, or international, depending on the communicator's preference.

These modern communications, however, also raise a number of legal issues including regulation, pricing, import/export, licensing, privacy, encryption, e-commerce, jurisdiction, taxation, intellectual property, merger/acquisition, and antitrust issues. Integral to the Internet Age are the laws that guide, direct, balance, and respond to these issues, and thus are a key part of each telecommunications product and service. It is important to understand these laws when communicating in this new environment. The purpose of this book is to provide that understanding.

APPROACH

To accomplish this purpose, this book addresses the primary opportunities, risks, and responsibilities integral to the new global telecommunications environment and outlines the key areas that users, corporate strategists, and network managers should be aware of in order to take full advantage of the opportunities. It discusses the legal issues raised by this environment and describes how each of those issues affects individuals and companies. In doing so, it highlights the three key aspects of these issues: (1) legal/regulatory, (2) international, and (3) social impact on users.

Since both telecommunications and the law are very dynamic disciplines, it is difficult to keep a book such as this current. Books on quickly changing topics continually run the risk of being out-of-date the moment they are printed. Therefore, to create a resource that has ongoing usefulness to readers, the structure of this book provides the background of each issue, the law to date governing each issue, and indications of where the law is likely to go as we move forward into even more fluid communications. With this foundation, readers can more fully understand the rationale behind the changes as they occur and thus more knowledgeably track the ongoing evolution of the law.

The goal of this book is first to avoid duplicating information presented in other telecommunications law books, but rather to add to that information. As such, this book moves beyond a description of the primary federal U.S. telecommunications law and addresses the state and international laws affecting the topics. Second, it also focuses less on telephone, broadcast, and cable law, and more on Internet law and the converged technologies of voice, data, and video communications systems. Third, while it addresses the era before 1996 to lay a foundation on certain issues, this book concentrates mainly on the changes occurring since passage of the Telecommunications Act of 1996.

CONTENT

To accomplish its purposes, after Chapter 1 introduces the book and describes the changes and current conditions creating "The Internet Age," this book is divided into three parts:

Part I: The New Competitive Telecommunications Environment

Part II: Embracing the Expanded Global Telecommunications Market

Part III: Legal Issues with Advanced Technologies

The first part, Part I, containing Chapters 2, 3 and 4, describes the key legal and regulatory events creating the new competitive telecommunications environment. Chapter 2 provides the historical background to understand more fully these events and their impact on market structure and opportunities. It provides a foundation on which the reader can more knowledgeably track the ongoing activities, including the massive changes brought about by the Telecommunications Act of 1996. Chapter 3 describes the key aspects of the U.S. Telecommunications Act of 1996, including the duties and obligations of various industry participants. Knowledge of details in the 1996 Act is critical to practitioners in this industry because it is the most extensive change occurring in U.S. telecommunications law in the 62 years since the Communications Act of 1934 and affects all aspects of the U.S. telecommunications industry. Chapter 4 discusses several ongoing issues resulting from the Telecommunications Act of 1996 that affect doing business in the new U.S. telecommunications environment and the impact of these issues on market structure, opportunities, economics, and pricing, and the various strategies among companies. These issues include local number portability, universal service, and access charge reform, including reciprocal compensation.

Part II, containing Chapters 5, 6, and 7, addresses the extremely important international aspects of the new telecommunications environment. Chapter 5 describes the World Trade Organization (WTO) and the significance of its telecommunications-related agreements to U.S. telecommunications companies and their international partners. The markets opened by the WTO dwarf the current U.S. market, and thus most of the telecommunications activity over the next 10–20 years will be in international trade. For this reason, Chapter 6 reviews current U.S. trade law, describing the export-import regulations critical to U.S. telecommunications firms seeking to provide products and services to the global telecommunications market. Chapter 7 suggests licensing options required to protect the intellectual property of telecommunications companies when sold abroad.

Part III, containing Chapters 8, 9, and 10, discusses several ongoing legal issues involved in the personal use of the new advanced technologies available to users today in the Internet Age. Chapter 8 explores the privacy concerns associated with that use and the relatively weak protections that the law provides. Chapter 9 describes the technological response to these privacy issues, encryption. Chapter 10 highlights 10 unresolved legal issues with the Internet. These issues include e-commerce, taxation, jurisdiction, First Amendment, tort and criminal issues. They are, perhaps, more accurately called "threshold issues" because they indicate the future direction of telecommunications law in the Internet Age.

INTENDED AUDIENCE

This book is intended, first, for any person using modern communications who is interested in knowing his or her rights, obligations, and legal protections concerning those communications. Second, it is designed to provide specific information useful to all telecommunications professionals interested in the legal issues surrounding the manufacture, provision, sale, and exchange of telecommunications products and services; the officers, corporate strategists, and network managers of companies using telecommunications products and services; and the attorneys and financial professionals assisting these companies or practicing in the area of telecommunications law. Third, it provides the background and current information required by the law schools, business schools, and telecommunications programs when training the above-mentioned professionals. Fourth, it offers information useful to the technologists and entrepreneurs embracing the many opportunities provided by this new environment.

ACKNOWLEDGMENTS

First, my sincerest thanks to my family: Lisa, Craig, Sean, Andrea, Seth, and Isabella, who have watched me work on "the book" for the past several years. They have been wonderfully patient and supportive as I worked around their activities and put numerous "should do's" off until this project was completed.

I also sincerely appreciate the loyalty and encouragement of my friends, for whom I was largely unavailable during the past few years. I thank them for continuing to believe in me and including me in plans even when I could not attend.

Concerning actual input to the book, special thanks go to Roger P. Newell and Mary Wand, whose work provided the basis for much of the information in Chapters 3 and 4, respectively. They are experts in their fields and were very generous to make their work available to others through this medium. I also want to thank Matthew Wenig for his significant contributions to Chapter 6.

Special acknowledgment goes to the reviewers of this book, whose contributions were invaluable. They include: Brent Alderfer, Former Commissioner, Colorado State Public Utilities Commission; Rachelle Chong, former FCC Commissioner; John Fike, Texas A&M University; Randy Hayes, University of Northern Iowa; David Loomis, Illinois State University; and Martin Weiss, University of Pittsburgh. I was honored that such luminaries in this field would take their valuable time to guide and comment on the information on these pages. Their input greatly enhanced the quality, readability, and general direction of the book.

I am also thankful for the wonderful efforts of the many students who assisted me with research for the book: Rick Lin, Randy Shefman, Andrew Lawrence, Christina Book, Christopher King, and Yeong Young Lee.

Heartfelt thanks also go to Jennifer Mann and Karyn Johnson at Morgan Kaufmann Publishers, who were both my editors and very good friends. They were patient with the delays that resulted from the demands of my clients and the changes required because of changes in the law. They were able to keep the process moving forward in a professional manner and have been marvels at coordinating things and encouraging everyone when the process seemed to bog down.

Additionally, I would like to thank Jim Cavanagh of The Consultant Registry in Atlanta, Georgia, who introduced me to Jennifer Mann and got me started on this project. He is an amazing person with incredible energy, insight, and industry knowledge, who has also learned the invaluable lesson of how to be a good friend.

This book also would not be possible without the foresight of Professors George C. Codding, Jr. and Frank Barnes at the University of Colorado (C.U.), who started the first telecommunications degree program in the United States in 1971. As the first student in the Interdisciplinary Telecommunications Program, I was encouraged by both professors to continue this work through engineering and the law. Both are visionaries who saw the need for this combination of disciplines and provided the educational model for the industry.

Finally, I would like to thank all of the wonderful people with whom I have worked over the years who have in their various ways have taught me a great deal about the fascination of telecommunications, the industry's technology and law, and the astounding impact each has on our daily lives.

CHAPTER 1 Introduction—The New Telecommunications Environment

". . . to provide for a pro-competitive, de-regulatory national policy framework designed to accelerate rapidly private sector deployment of advanced telecommunications and information technologies and services to all Americans by opening all telecommunications markets to competition . . ."

The goal of the U.S. Telecommunications Act of 1996[1]

TELECOMMUNICATIONS LAW HAS EXISTED SINCE SYSTEMS WERE FIRST developed to carry communications over distances. Early law and international agreements addressed the technical and legal issues associated with postal and semaphoring systems, telegraph, and radio communications systems, especially ship-to-ship and ship-to-shore communications. These issues included the (1) details of how one system interconnected with another system to complete the communications, (2) pricing of services, (3) security, (4) ownership, (5) access to the systems, (6) control of information, and (7) content of many communications, generally to exclude obscenity or to assist law enforcement and national defense agencies.

As the telephone, television, cable television (CATV), microwave systems, and satellites came into being, the law addressed the legal issues accompanying these new technologies. Governments allowed regulated monopolies to provide and control the communications systems and worked with international bodies, such as the International Telecommunications Union (ITU) in Geneva, Switzerland, to enable the exchange of international communications. Up to this point, however, countries and governments could still control communications much like goods passing across borders.

In the last 25 years, the telecommunications environment in the United States and around the world changed dramatically—creating what is called "the Internet Age" in communications. With it, a person with access to modern communication sytems can communicate nearly instantaneously with any other person in the world with access to modern voice, data, and/or video communications. While generally positive, this exchange of information impacts multiple areas of our lives and can no longer be controlled easily by governments or systems. This reality raises numerous new legal issues and expands the group of persons interested in telecommunications law from specialized attorneys and regulators to anyone using modern communications, especially the Internet. To assist you in more fully understanding these issues, the purpose of this chapter is to identify the technical and structural changes creating this expanded environment.

1.1 NEW TRANSPORT TECHNOLOGIES

Among the many important changes creating this expanded telecommunications environment are the technologies, such as digital, packet communications, wireless, global satellite systems, and the Internet, that have significantly increased the availability and transport capacity of the communications systems and drastically changed the economics of communications networks, which in turn lowered the cost of using the networks.

For the first time in history, the charge for most communications shifted from time and distance based to a more averaged or flat rate. This phenomenon is known as the *death of distance*. For example, with packet systems, each call is separated into smaller bundles or *packets,* addressed and sent over the cleanest, fastest, and least expensive path available. At the receiving end, the equipment collects and sequences the packets and presents them to the receiving party in a manner that is nearly indistinguishable in quality, but more efficient to transmit than other, more traditional methods. It also means that the cost of transporting each call must be averaged over all of the systems that carried portions of that call.

Similarly, with satellite systems, the cost of each call is primarily for the uplink and downlink. Thus, a call from California to New York or Paris costs basically the same as a call from California to Colorado. This trend in making distance much less of a factor in the cost of modern communications has spurred major changes in network design and pricing decisions throughout the telecommunications industry.

1.2 CONVERGED VOICE, DATA, VIDEO, AND GRAPHICS SYSTEMS

As a result of the new transport technologies, communications and computing in the United States began to converge to provide combined voice, data, video, and graphics systems to customers. In most countries of the world, communications services have always been combined under the Postal, Telephone, and Telegraph (PTT) department of each country's government. In the United States, however, separate, monopolistic companies traditionally focused on only one segment of the industry. In an effort to spread the information sources and balance ownership of the communications delivery systems, the U.S. Congress, the Federal Communications Commission (FCC), and courts agreed that the Bell Companies and a few independent telephone companies should provide voice or telephone communications, while Western Union should provide telegraph services. Similarly, the networks ABC, CBS, and NBC were the main television broadcasters, and IBM initially dominated the data industry. Reflecting these divisions in the communications industry, most U.S. corporations had separate voice and data departments, and U.S. law differentiated between *telecommunications* and *information* services.

This pattern continued in the United States until the Telecommunications Act of 1996 acknowledged this technical convergence and reversed the historical concept of telecommunications as a *natural monopoly* to encourage a far more liberal division of ownership among companies. This has created tremendous opportunities for new companies in the telecommunications industry and completely changed the structure of the industry. The history of this transition and ownership patterns is so important to our understanding of the current U.S. telecommunications industry that it is described in more detail in Chapter 2.

1.3 LEGAL CHANGES

Along with these technical changes, several legal events occurred in the United States and worldwide in the 1980s and 1990s that dramatically changed the telecommunications industry, products and services offered, and options for customers. These events included the following: (1) the breakup of the Bell System in 1984, (2) the opening of the Internet in the mid-1990s, (3) changes in the laws in nearly two-thirds of the states in the United States in 1995 to open local telecommunications markets to competition, (4) the enactment and implementation of the Telecommunications Act of 1996 in the United States, and (5) the signing of the World Trade Organization agreement by 69 countries in February 1997.

1.3.1 Breakup of the Bell System

The breakup of the Bell System was an out-of-court settlement of an antitrust case against the Bell System. Nonetheless, it dramatically changed the structure of the U.S. telecommunications industry, dividing it into regulated local markets and competitive long-distance markets. This action set the stage for the current competitive environment in the industry.

1.3.2 The Internet

The Internet began as the ARPANET, providing military communications under the Advanced Research Project Agency (ARPA). In 1979, the *packet* and *store-and-forward* concepts tested on ARPANET were used to form the NSFNET to provide a research computer network among selected universities under projects funded by the National Science Foundation (NSF). In 1992, the World Wide Web technology was released, connecting over 1 million host sites. In 1993, the White House "went online," and finally, on April 30, 1995, the NSFNET closed and the new Internet officially opened. This tremendous change in the concept and structure of communications reverberates still.

1.3.3 Changes in State Law

Also in the early 1990s, nearly a decade after the 1984 divestiture of the U.S. Bell System, state legislatures across the United States began studying the results of the divestiture and its implication for the future of telecommunications in their states. They noted that three things had occurred: (1) prices for long-distance service had dropped dramatically, (2) the providers of telecommunications services delivered new technology to customers much faster than before, and (3) customers had more options for service. The state governments speculated that since these results had occurred in the long-distance communications markets, they would also work in their local telephone markets if competition were introduced. Therefore, two-thirds of the state governments passed state laws opening their local markets. They specifically stated their goals in the various state laws so that if these results did not occur for some reason, they could reverse their decision for competition and return to a more regulated environment.

What is not always realized is that the majority of states introduced local competition in telecommunications before the U.S. federal law mandated it. This state action resulted in differences among the states that impacted nationwide products, causing some "churning" as states subsequently realigned their state laws to become more consistent with the federal Telecommunications Act of 1996.

1.3.4 The Telecommunications Act of 1996

The Telecommunications Act of 1996 is the most significant piece of telecommunications legislation in the United States in the 62 years since the passage of the Communications Act of 1934. It opened competition in the local telecommunications markets throughout the United States and mandated changes in the rights and responsibilities of all portions of the telecommunications industry, including: (1) equipment manufacturers, (2) existing and new service providers, (3) wired and wireless companies, and (4) all voice, data, video, and graphics technologies. The 1996 Act is of such tremendous importance to the telecommunications industry that two chapters, Chapters 3 and 4, of this book are dedicated to discussing it.

1.3.5 World Trade Organization Agreement of 1997

Also in the early 1990s, governments around the world began assessing the impact of the Internet and the other new technologies on their countries. Many governments, stating that they chose not to be left out of the Internet Age, signed a World Trade Organization (WTO) agreement in February 1997 in which they agreed to (1) continue the liberalization and privatization of their national telecommunications networks; (2) open their previously closed, national networks to foreign investment and operation; (3) streamline and publish the regulations by which foreign companies must operate in their countries; (4) simplify their import/export, construction, operations, and taxation procedures; and (5) not nationalize systems once they were installed and functional. Under the 1997 WTO agreement, each country establishes its own rules and levels of involvement, so the rules and agreements are not uniform worldwide. However, the 1997 WTO agreement has had such a significant impact on the future of global telecommunications that an entire chapter, Chapter 5, discusses this event.

1.4 INTERNATIONAL TELECOMMUNICATIONS EQUIPMENT MARKETS

The impact of the changes in the international telecommunications environment brought about the 1997 WTO agreement has been staggering. For example, a report by the International Telecommunications Union (ITU) in 1997 stated that nearly 60% of the world's population has never made a phone call.[2] In marketing terms, this means that, as robust as the telecommunications products and services markets have been in the United States, Europe, and other developed nations of the world, they represent only 40% of the total global market. The 1997 WTO agreement opened much of the other 60%.

At present, the only companies capable of providing the equipment and services needed for this new 60% of the market are the telecommunications companies of the United States, Europe, and Asia. This means that the primary markets for these companies are no longer their previous monopolistic territories, but rather the new international markets, a shift that has been acknowledged in the strategy of nearly every telecommunications company today. It also means that the amount of global telecommunications investment over the next 10–20 years is expected to eclipse all previous worldwide telecommunications investment. As a result, it will significantly impact the world's economy, but perhaps more importantly, it will continue to change the information available to the world's population, the manner in which people obtain that information, and from where they obtain it.

1.5 TECHNICAL STANDARDS

As telecommunications companies move out of their traditional monopoly markets and into the new international market, they do not want to retool their production systems to provide the products and services required for each segment within that market. Instead, the current trend in telecommunications equipment is to move away from proprietary technical standards and toward more open international standards. This trend also enables telecommunications companies to effectively partner or enter into joint ventures with other telecommunications companies and assists customers who prefer equipment that is readily connectable and serviceable by any provider.

This trend toward more fluid, international standards is best seen in Europe today. Just as the European Community (EC) developed the euro as a standard monetary currency, it is facilitating the migration of the separate national telecommunications equipment standards in Europe to more uniform European standards. This process, documented in the "European Community's 1990 Open Directive," recognizes the strengths of the telecommunications providers in each country and seeks to make those strengths available throughout Europe.

1.6 WHAT IS COMMUNICATED?

As a result of the new technologies, service providers, and technical standards being used, people have changed the way in which they communicate. Many rely daily on new telecommunications products and services, such as voice mail, email, e-commerce, data information, electronic news, and the Internet. People no longer just make phone calls, but instead transmit voice, video, and data information, and conduct on-line research and commerce, often transmitting personal information such as credit card numbers and health and insurance information.

1.7 WITH WHOM ARE WE COMMUNICATING?

In addition to *what* is communicated, the patterns of *with whom* people are communicating has changed. Communications are no longer just the one-to-one conversation of the telephone or the one-to-many communication of broadcast mediums, but the one-or-many to one-or-many interaction of the Internet.

1.8 WHERE ARE WE COMMUNICATING?

Additionally, communications have moved from predominately local and national levels to international levels. In previous generations, international communications were expensive and therefore rarely used, but today they are nearly daily events for anyone using the Internet. As people click into Web sites, they generally do not think about where the servers for those sites are located. However, many are, in fact, located outside the United States. In addition, all Internet communications are transmitted over the clearest and most available path for that communication. At times, this means that domestic communications may actually be transmitted outside the United States en route to its final domestic destination.

1.9 NEW LOCAL ACCESS OPTIONS

In response to these new communications patterns, telecommunications provider companies have begun offering expanded network access options for local customers. These options include technologies such as (1) digital subscriber lines (DSL) and Integrated Services Digital Networks (ISDN) over a twisted pair of copper lines, (2) two-way CATV, (3) two-way wireless, (4) two-way satellite, (5) expanded optic fiber, and (6) distribution of communications over existing electric utility lines. Each option has moved users from traditional low-speed, narrowband access to greatly improved, high-speed, broadband access that facilitates use of the Internet.

1.10 UNIVERSAL SERVICE FUND SUPPORT OF INTERNET CONNECTIONS

Historically, one important aspect of telecommunications law has been to ensure that communications users have access to low-cost, high-quality networks through federally mandated programs such as the Universal Service Fund (USF). In the past, this fund has paid for the installation of communications networks in high-cost, low-density, unprofitable, generally rural areas and thus transformed telephones from a luxury item to a standard household item that most people could afford. Today, such programs are extending access to the Internet by funding connections to schools and libraries. In mandating this assistance, Congress hoped to make Internet connection available to most people and thus transform modern Internet communications from a luxury item to a common household item. Congress also hoped that this would avoid the *Digital Divide* or a situation where only those households wealthy enough to own a personal computer and to afford an Internet service provider (ISP) would have access to the Internet. With the availability of voice over the Internet, this may become increasingly important to all telephone users.

1.11 STRUCTURAL CHANGES

With these technological and legal changes, the telecommunications industry has shifted from several regulated *natural monopolies* to an industry that is economically capable of supporting multiple service providers in each market. For this reason, Congress, in the Telecommunications Act of 1996, was able to open the telecommunications market to as many entrants as possible in an effort to reduce the price of the communications systems and services to their lowest, most efficient level. However, what appears to be happening is that through mergers and acquisitions, the industry is regrouping into a few, very large companies, which are now global, not national, companies.

For example, following the 1984 breakup of the Bell System in the United States, seven Regional Bell Operating Companies existed: (1) Ameritech; (2) Bell Atlantic; (3) Bell South; (4) NYNEX; (5) Pacific Telesis; (6) SBC (formerly Southwestern Bell Corp.); and (7) US WEST. By 1999, only four of the seven remained, and each had grown into a large, diversified, international company. The three largest long-distance providers, AT&T, Sprint, and MCIWorldcom followed a similar path. What this means for users is unclear at present, but the trend will definitely open extensive antitrust issues in the industry, as well as concerns for users and governments who acknowledge the power of communications in our daily lives.

CONCLUSION

While the final outcome of this evolution is unknown, these changes in the telecommunications industry over the past 25 years are enormously important because they created the environment that catapulted communications into the 21st century. They encompass the technologies, options, services, and industry structure available to us. With these changes, the world entered one of the most exciting and vital eras in its history.

As with any change, new legal issues also surfaced, including those concerning the following: (1) the reintroduction of competition into a previously monopolistic, regulated industry; (2) the new rights, responsibilities, and obligations brought about by the Telecommunications Act of 1996; (3) access cost issues; (4) the international opportunities and concerns brought about by the WTO agreement; (5) the increased importance of import/export procedures as companies move to embrace international opportunities; (6) heightened intellectual property concerns and the need for improved licensing; (7) heightened privacy, security, and law enforcement concerns; (8) encryption issues; and (9) the as yet unsettled concerns raised by e-commerce, jurisdiction, and taxation. The remaining chapters of this book explore these issues and the current telecommunications law in the Internet Age.

ENDNOTES

[1]The goal of the Telecommunications Act of 1996 as stated by Mr. Biley, Chairman of the Joint Committee, in its Report accompanying Senate Bill 652 (S.652) upon its submittal to Congress as the Telecommunications Act of 1996. House of Representatives, Report 104-458, 104th Congress, 2nd Session, 1996.

[2]International Telecommunications Union, "Worldwide Calling Patterns," Geneva (1997).

CHAPTER 2 Competition and Regulation—A Continuing Telecommunications Cycle

"To make available, so far as possible, to all the people of the United States a rapid, efficient, Nation-wide, and world-wide wire and radio communication service with adequate facilities at reasonable charges . . ."

Stated goal of the Communications Act of 1934[1]

THE GOAL FOR U.S. COMMUNICATIONS HAS NOT CHANGED SINCE 1934 and continues to guide U.S. telecommunications policy today. However, the means to accomplish that goal have moved from competition to regulation and now back to competition again. With the recent mergers and acquisitions in the telecommunications industry, the communications service providers seem to be regrouping into a few very large companies with such tremendous, even global, power that they likely will need to be regulated again in some manner in the future.

Therefore, understanding the progression of the competition–regulation cycle in the United States helps us to understand the current telecommunications industry, laws, and policies and to evaluate future options available to us. Awareness of the background of the various issues considered by the telecommunications industry and the rationales behind the solutions that were applied to each issue enables us to sort through which options succeeded and were retained, or failed and were rejected. This information provides a richer context in which to find appropriate solutions to current and future issues. The purpose of this chapter is to provide that information.

To accomplish this purpose, Section 2.1 describes why the United States and other countries regulate industries such as telecommunications. It notes the trend in many industries to start with several pioneering companies competing against one another, and then to have one or two companies take control of the industry, imposing high prices on restricted products to the detriment of consumers. Finally, the section discusses the efforts of the government to control these dominant companies through regulation and antitrust actions.

Section 2.2 expands on the specifics of this trend in the communications industry. It traces the path from vibrant competition early in the telegraph and telephone industries, through the rise of the American Telephone and Telegraph Company (AT&T) and the Bell System to dominance in Section 2.3, and the various efforts of the government to control this dominance in Section 2.4. The discussion describes the resulting separation of the voice, data, and video portions of the industry that still exists today and the cost controls that still define the industry. Section 2.5 addresses the technological developments that have occurred in the telecommunications industry and that converged in a manner that blurs the established boundaries between voice, data, and video services and products. These converged technologies provide both unique options and conflicts for regulators defining the issues being considered by the industry today. The description of events in each section is purposely brief in order to highlight trends, not to provide a full and complete history of the telecommunications industry. That task is left to the many excellent detailed histories documented in other books, some of which are cited in the endnotes.

2.1 COMPETITION VERSUS REGULATION—SEEKING A BALANCE

Competition is not a new concept in the telecommunications industry. The United States's free-market approach to commerce has always held that competition allows the market to most creatively meet the needs of customers, at the best prices and in the most efficient manner. Consistent with this philosophy, significant competition existed early in telecommunications history, when the telegraph, telephone, and radio were new industries. However, as with many industries, certain trends occurred as the technology evolved. The purpose of this section is to explore these trends and the United States's solution to them.

2.1.1 Rise of Trusts and Anticompetitive Behavior

Industries such as petroleum, railroad, steel, cotton, sugar, and whiskey all had robust competition early in their histories.[2] In each case, a few driven individuals emerged to lead their companies to dominance in their industries. In general, these individuals are referred to as the robber barons, because their business practices were anticompetitive. They (1) cut prices to drive competitors out of business; (2) then bought the failing competitors, forming "trusts," and (3) once in control, raised prices. These anticompetitive business practices resulted in huge profits for the controlling companies, but they also frequently ignored new inventions and technological improvements, diverted funds from product development, and gradually provided a more restricted range of products. As a result, the quantity, quality, and availability of affordable products to the public decreased, while the political power of the trusts increased, protecting their monopoly position. All of this led to public outcry against the dominant companies and bitter criticism of them in the newspapers and books of the time.

2.1.2 Antitrust Law and Regulation

In response to this public outcry, state and federal agencies began to investigate the trusts and used four basic approaches to control them: (1) definitional distinctions, (2) structural separation, (3) regulation of rates and operations, and (4) antitrust legislation.[3] Frequently, combinations of all four of these approaches were used.

Under the first approach, *definitional distinctions,* the government divided the market in question into segments and allocated each segment to specific companies to ensure competition. In the telecommunications industry, for example, the government split communications into voice, data, and video services, and assigned certain companies to each. The government often prohibited dominant companies from entering certain segments of the market to ensure the survival of their competitors.

With the second approach, *structural separations,* regulators required the dominant companies to conduct a portion of their business through subsidiaries. Each subsidiary was expected to be a completely separate company from the parent and be able to function without preference by or support from the parent company. This allowed the dominant companies to enter new markets, but removed the opportunity for them to subsidize their new efforts with profits from the parent company.

Under the third approach, *regulation of rates and operations,* Congress and the state legislatures delegated responsibility for the oversight of an industry to a commission or investigative body. In telecommunications, these include the Federal Communications Commission (FCC) and the state public utilities commissions (PUCs).

Finally, with the fourth approach, the U.S. Congress and state legislators made certain anticompetitive behavior illegal through *antitrust legislation* and provided legal remedies to the "victims" of the trusts. In most of these laws, the Department of Justice was appointed to be the primary enforcer of the law and the representative of the public's interest against the companies exhibiting anticompetitive behavior. Among the series of laws enacted to regulate antitrust behavior were the following: (1) the Interstate Commerce Act of 1887, forming the Interstate Commerce Commission (ICC); (2) the Sherman Antitrust Act of 1890; (3) state action; (4) the Mann-Elkins Act of 1910; (5) the Clayton Act of 1914, forming the Federal Trade Commission; (6) the Willis-Graham Act of 1921; (7) the Robinson-Patman Act of 1936; and (8) the Kefauver-Cellar Act of 1950. Each is described briefly below.

1887—The Interstate Commerce Act and the ICC

The Interstate Commerce Act of 1887 laid the foundation for antitrust law in the United States.[4] In it, Congress established the Interstate Commerce Commission (ICC) and gave it authority to (1) review the management of interstate carriers, (2) examine the carrier records and documents, (3) sue the companies in federal court when the investigations warranted such lawsuits, (4) summon witnesses, (5) regulate rates and trade practices of the companies, and (6) outlaw price fixing. *Price fixing, price setting,* or *pooling* are terms that refer to the special rates, rebates, and price discriminations that the trusts have used against persons, places, or commodities. The focus of the Interstate Commerce Commission was to ensure fairness, efficiency, and the protection of the "public good" in interstate transactions.

1890—The Sherman Antitrust Act

Within the first three years following the establishment of the Interstate Commerce Commission, Congress realized that the Interstate Commerce Act of 1887 did not adequately address numerous anticompetitive issues, and thus Congress added the Sherman Antitrust Act of 1890[5] to more forcefully restrain the activities of the trusts and holding companies. It outlawed the monopolization and constraint of interstate trade and prohibited any "unreasonable" interference with the competitive pricing or distribution systems of the open market. Congress intended the Sherman Act to ensure safe, economical, efficient, and adequate services to the public, and placed responsibility for enforcement of the Act with the federal government.

However, even with the passage of the Sherman Act by Congress and continued public outcry concerning the abusive activities of the large trusts in the United States, the Sherman Act was not actively enforced for at least a decade, until the early 1900s. The delay was due mainly to an economic recession, the development of the West, and the absorption of Alaska and Hawaii as territories, all three of which distracted the attention of the government. Responding to the public's frustration over the delay in antitrust regulation, Theodore Roosevelt campaigned on the issue during his bid for the U.S. presidency in 1900. After his election, he directed the Department of Justice to focus on enforcement of the Sherman Act. "Trust-busting" in the United States began in earnest and continued through the Theodore Roosevelt (1901–1909), William H. Taft (1909–1913), and Woodrow Wilson (1913–1921) administrations.

State Action and the Rise of State Regulatory Commissions

In general, the states welcomed the federal antitrust legislation and federal responsibility for antitrust enforcement because for more than 30 years, the state governments had been suing the dominant companies under common law, and the state courts had ruled against the companies, ordering them to dissolve. In response, the trusts simply moved their assets to another state or into holding companies in which they appeared to be divested, but actually continued to operate under consolidated management. Thus, while state action and antitrust legislation closed specific operations in that state, they could not effectively stem the growth or negative operations of the trusts overall. In most cases, the trusts continued to grow, merging their properties into the new organizations operating in other states.

An example of this fluidity of the trusts occurred in 1892 when the Supreme Court of Ohio dissolved the Standard Oil Trust because of the vastness of its holdings and public criticism of the Trust's business methods. The Court required Standard Oil to dissolve and the companies in the trust to operate as the separate entities they had been 10 years earlier, before the trust was formed in 1882.[6] However, this effort to curb the abuses of Standard Oil through *structural separation* affected only the company's activities within the state of Ohio. Seven years later, in 1899, Standard Oil left Ohio and moved to New Jersey. In New Jersey, management amended the charter of the Standard Oil Company of New Jersey to pull the nearly 20 companies affected in the dissolved trust in Ohio into the New Jersey company through stock transfers. The Standard Oil Company of New Jersey thus preserved its vast oil empire and became one of the largest and most powerful holding companies in the world. It continued as such for another 19 years, despite the Sherman Act.

Beyond the renewed interest in trust-busting at the federal level in the early 1900s, the states established regulatory commissions to monitor the activities of companies in intrastate commerce.[7] By 1915, most states were reviewing the activities and loosely regulating the rates of companies, primarily in the oil, water, transportation, and communications industries. By 1922, 40 of the 48 states had state regulatory commissions.[8]

1910—The Mann-Elkins Act

Acknowledging the increased size and anticompetitive practices of AT&T, Congress passed the Mann-Elkins Act of 1910[9], extending the authority of the Interstate Commerce Act and the Interstate Commerce Commission to the telecommunications industry. It heralded the first specific application of antitrust law to the telecommunications industry.

1914—The Clayton Act and the Federal Trade Commission

Even with these efforts, large, powerful companies continued to dominate major U.S. industries. In the presidential campaign of 1912, Woodrow Wilson called for the destruction of monopolies and trusts, and during his administration, Congress passed the Clayton Antitrust Act[10] in October 1914. It was another attempt to bolster the Sherman Act by banning price discrimination, anticompetitive mergers, interlocking directorates, and exclusive-dealing arrangements. It also permitted private individuals to file lawsuits when affected by a monopoly's violation of these provisions and created the Federal Trade Commission (FTC)[11] as the agency of the federal government designated to address these violations. In the Clayton Act, Congress empowered the FTC to conduct investigations and to issue cease and desist orders against companies found to be engaging in unfair or anticompetitive business practices. The FTC's mission was to protect consumers from these business practices.

1921—The Willis-Graham Act

In 1921, Congress passed the Willis-Graham Act,[12] in which it specifically affirmed the natural monopoly concept of utilities such as the telephone, telegraph, water, natural gas, and electric industries. Congress considered a natural monopoly to exist where (a) a massive outlay of capital was required to provide the service, (b) the average cost of the service declined with each additional customer, and thus (c) efficient, low-cost provision of service required a concentration of customers.[13] In such cases, Congress determined that it was not in the best interests of the public to duplicate systems or services in each geographic market or to encourage several providing companies to compete for customers in the same geographic area.[14] Instead, Congress declared that these companies should be regulated as public utilities.

1936—The Robinson-Patman Act

In 1936, Congress noted that some companies were still discriminating in the prices they offered to customers. To correct this, Congress passed the Robinson-Patman Act,[15] in which Congress strengthened the Clayton Act. The new legislation made it illegal for goods and services of like grade and quality to be sold at widely varying prices in order to reduce competition.

1950—The Kefauver-Cellar Act

In 1950, Congress passed the Kefauver-Celler Act[16] to tighten control of business mergers that either reduced competition or increased monopoly control of a company. Nonetheless, regulators also began to realize that regulation of the monopolies was not really as effective as they desired because the large companies always had more time, money, and professional resources than the regulators.

2.1.3 Impact of Technological Changes

While these legislative and regulatory efforts attempted to control companies, technological developments often redefined an industry and made previous regulations or restrictions obsolete or impractical. New technologies or technological development of existing services frequently introduced new options or products that effectively challenged the old companies and encouraged competition once again. In this process, it is important to note, however, that the law generally follows technological changes and does not lead them. Thus, the laws governing the changing industries usually evolve more slowly than the technological changes. In seeking the proper balance between competition and regulation, regulators have agreed to let the market regulate the industries.

2.2 EARLY COMPETITION IN THE COMMUNICATIONS INDUSTRY

The communications industry followed the same developmental path described above, with full competition that flourished for most of the first 50 years of the industry. The following sections describe this evolution from invention to industry development.

2.2.1 Telegraph

The telegraph, invented in 1837 and patented in 1840 by Samuel F.B. Morse, became instantly popular and created a vibrant industry because it provided a significantly faster means of communicating over long distances than any other method available at the time. The ability to send and receive messages quickly also caused most people to overlook the fact that telegraph service was not private. Senders gave their messages to telegraph operators to transmit, while operators at the receiving end decoded the messages, placed them in envelopes, and handed them to telegraph messengers to deliver to the recipients.[17]

Since messengers delivered the telegrams at the receiving end, the service was predominantly a long-distance service rather than a local one. However, some companies in time-dependent industries, such as the stock market and newspapers, installed their own private lines to the telegraph office and thus were able to send and receive directly both local and long-distance communications without the delay of telegram messengers.

In addition to not being private, telegraph communications were one-way or *simplex,* not *duplex* or bidirectional. Thus, even if both communicating parties were standing next to the telegraph operators at each end, only one could communicate at a time. In many ways, this is similar to email and *chat rooms* today, except no operator is required.

Congress subsidized the construction of the first telegraph line, connecting Baltimore and Washington, D.C. On May 24, 1844, Morse sent the first message, "What hath God wrought!," from the Supreme Court chamber in the Capitol

Building in Washington, D.C. to his assistant, Alfred Vail, in Baltimore, Maryland. Vail then sent the same message back to Morse from Baltimore to Washington, D.C. Later that same year, the telegraph was the first medium to convey the news, from Baltimore to Washington, D.C., that the Democratic National Convention had nominated James K. Polk for president. Following this event, the news-gathering agencies pooled their resources to build a network for telegraphed news among several newspapers, initiating the Associated Press.

However, the technology of telegraph was so simple that Morse could not defend his patent, and thus use of the technology expanded rapidly. By 1846, New York City was linked to Washington, D.C., and by 1851, over 50 telegraph companies existed in the United States. On April 4, 1856, 12 of the 50 companies combined to form the Western Union Telegraph Company, marking the beginning of unified service in the United States.

By 1861, the communications needs of the U.S. government during the Civil War accelerated the construction of transcontinental telegraph lines. On October 24, 1861, Western Union completed the first line from Nebraska to California, routed through Salt Lake City, Utah. This transcontinental telegraph service ended the Pony Express Service after it had operated only 19 months. After the Civil War, the telegraph played a crucial role in the development of the West, with the telegraph lines generally placed in the railroad rights of way. Today, the same railroad rights of way contain major optic fiber links, forming much of the backbone for modern U.S. communications.

2.2.2 International Telegraph Communications

Along with the domestic market, demand for international telegraph grew. After four unsuccessful attempts, the first undersea, transatlantic telegraph cable was successfully laid in 1866. Since the phone had not yet been invented, the cable carried only telegraph traffic.

2.2.3 Equipment Manufacturing

To meet this increasing domestic and international demand for the telegraph, the Western Electric Manufacturing Company (WEMC) was founded in 1869 by three men: Elisha Gray, an inventor; Enos Barton, a former Western Union telegrapher; and Anson Stager, a vice-president of the Western Union Telegraph Company. Initially, the company manufactured only telegraph equipment, but after its first seven years, Alexander Graham Bell invented the telephone in 1876, and the company began manufacturing telephone equipment. Six years later, in 1882, the Bell Company acquired the WEMC, using it to provide high-quality, reliable telephone equipment. The Western Electric Manufacturing Company then became known simply as Western Electric and became the equipment-manufacturing branch of the Bell System.

In 1925, the research portion of Western Electric, known as the Western Electric Research Force, was moved into a single research organization and renamed the Bell Telephone Laboratories (Bell Labs). As such, Western Electric and the American Bell Telephone Company jointly owned Bell Labs.[18] Bell Labs also became a key part of the Bell System and was responsible for developing most of the telephone standards in the United States.

2.2.4 Telephone

On February 14, 1876, Alexander Graham Bell, then 29 years old, applied for a patent for the "talking telegraph" that he had invented.[19] In his patent application, Bell wrote

> The advantages I claim to derive from the use of an undulatory current in place of a merely intermittent one are, first, that a very much larger number of signals can be transmitted simultaneously on the same circuit; second, that a closed circuit and single main battery may be used; third, that communication in both directions is established without the necessity of special induction-coils; fourth, that cable dispatches may be transmitted more rapidly than by means of an intermittent current or by the methods at present in use; for, as it is unnecessary to discharge the cable before a new signal can be made, the lagging of cable-signals is prevented; fifth, and that as the circuit is never broken a spark-arrester becomes unnecessary. . . .[20]

On March 7, 1876, the U.S. Patent Office granted to Bell the patent for the invention of the telephone. Upon receipt of the patent, Bell offered to sell it to the Western Union Company for $100,000. Western Union refused Bell's offer.[21]

1877—Formation of the Bell Telephone Company and Western Union's American Speaking Telephone Company

With the refusal of Western Union to purchase his patent, Bell formed the Bell Telephone Company, opening business in 1877. Unlike the telegraph, the telephone allowed users to speak directly with one another in real time, rather than one at a time through a telegraph operator. While the telephone was technically primitive at first and met with significant criticism, it rapidly became a popular technology, growing to over 11,000 telephones in its first year.[22] For customers using both telephone and telegraph service, each technology offered specific advantages. If customers desired a written communication or *record,* they used the telegraph. If customers desired privacy or real-time conversation, they used the telephone. As the two industries grew, telephones tended to become less expensive than the telegraph because experienced telegraph operators were not required for each message. [23] Instead, switchboard operators could connect numerous customers in much shorter times.

1878—Patent Case: Bell v. Western Union

With Bell's success, the Western Union Company immediately realized its mistake in not buying Bell's patent and began its own telephone subsidiary named the American Speaking Telephone Company. Under U.S. law, patents provide the owner of the patent exclusive use of the technology for a limited number of years. In Bell's time, the general rule was 17 years after receipt of the patent. After 1996, the general rule is 20 years after filing of the patent. However, extensions, challenges, and backlogs can change these terms. Thus, from 1876 to 1893, Bell held the exclusive rights to use his telephone patent. Since Western Union, therefore, could not use Bell's patented design, its subsidiary, the American Speaking Telephone Company, used a transmitter designed by Thomas Edison and a receiver developed by Elisha Gray.[24]

In response to Western Union's new subsidiary, the Bell Telephone Company filed suit against Western Union and its American Speaking Telephone Company, claiming that Edison's and Gray's equipment infringed on Bell's patents. This lawsuit, filed in 1878, was the first of many lawsuits between competitors in the telecommunications industry. While waiting for resolution of the court case against Western Union, Bell also expanded, forming a second company in 1878, the New England Telephone Company. Western Union immediately mirrored its competitor's expansion by forming its own second subsidiary called the Gold and Stock Telegraph Company.[25] Gradually, both Bell and Western Union added central offices for more efficient network deployment, hired telephone company switchboard operators, provided electrical power to the system, and upgraded telephones in their systems. All of these changes significantly improved the performance and economics of telephone over telegraph.

The patent suit was decided the next year, on November 10, 1879, with some *definitional distinctions* imposed by the court. As part of the settlement, Western Union agreed to sell its telephone interests and to stay out of the "voice" industry, while Bell agreed to sell its telegraph interests and to stay out of "areas already occupied by Western Union."[26] The telephone was thus defined as *voice* service and the telegraph as *record carriage* service. Eventually, record carriage grew to include telex and later *data* services.[27] Thus, this agreement initiated the current division between voice and data services that still exists in most U.S. corporations today and throughout U.S. telecommunications law. The agreement between the Bell and Western Union companies did not, however, immediately affect international communications because the transatlantic cables at the time were not capable of carrying voice communications. They carried only telegraph communications until 1956.[28] Voice communications across the Atlantic did not become available until 1927, when AT&T developed a high-frequency radio method that remained in use as the only international voice service until 1956. In 1956, AT&T installed improved cables capable of carrying both voice and data communications.[29]

2.3 THE RISE OF THE BELL SYSTEM DOMINANCE

By November 10, 1879, once Bell's patent case against Western Union was resolved and the Bell Telephone Company had verified control over the use of its telephone patent, the Bell Company began increasing its mergers and buyouts of other, smaller telephone companies. The largest of these was the merger of the Bell Telephone and New England Telephone Companies. In 1879, the two companies merged to form the National Bell Telephone Company. A year later, in 1880, the National Bell Telephone Company changed its name to the American Bell Telephone Company.

As noted earlier, in February 1882, the American Bell Telephone Company bought the Western Electric Company and made it the sole supplier of Bell equipment.[30] Prior to this acquisition, the equipment for Bell-licensed companies had been made by a variety of manufacturers, with varying results. The purchase of Western Electric Company gave the American Bell Company control over the design, production, and quality of its telephone equipment.

In 1884, American Bell installed the first long-distance telephone line in the United States, between Boston and New York. As it built its long-distance voice network, it frequently used telegraph facilities and rights of way to carry the voice signal. To facilitate its access to these telegraph resources, it purchased shares of Western Union. Thus, in February 1885, American Bell created the American Telephone and Telegraph (AT&T) Company as its long-distance subsidiary. AT&T's mission was to establish telephone service throughout the world by "wire, cable, and other appropriate means."

2.3.1 Bell's Patent Exclusivity and Return on Investment

With exclusive use of its telephone patent and strong leadership, the Bell System experienced tremendous growth. Between 1876 and 1893, Bell won over 600 patent infringement suits and became clearly the dominant telephone company in the United States. During this time, Bell's average return on investment was nearly 46% of each dollar invested, creating the capital Bell needed to expand, but providing little incentive to extend telephones into less profitable areas. In fact, during these years the number of phones in service increased only about 6.3% annually, and plant investment grew at only 8.2% annually.[31]

However, on March 7, 1893, Bell's patent expired, and the technology became publicly available. This meant that other companies had access to Bell technical designs and could develop telephone equipment without infringing on Bell's patent rights. Given the remarkable return on investment in the telephone industry, numerous new providers of telephone service entered the market. Competition began in earnest, and independent telephone companies immediately proliferated. They first opened business in rural areas not served by Bell and later entered cities in direct competition with the Bell System. In addition to private investment, the early independent telephone companies also included *cooperative* (customer-owned) and *farmer line* (do-it-yourself extensions of service) systems.[32]

2.3.2 AT&T Parent Over Bell System and Western Union

To fight competition and strengthen its own position, the American Bell Company reorganized in 1899. It inverted the positions of the parent and subsidiary companies, making AT&T the parent corporation over the Bell Companies. In 1907, Governor Charles Hughes (New York) and Senator Robert M. Lafollette (Wisconsin) endorsed the idea of state regulation requiring a single carrier to serve all who wanted communications service in a specific geographic area. This requirement opened the door two years later, in 1909, for AT&T to acquire control of the Western Union Telegraph Company in a stock purchase. This economically driven move reversed the division of voice and data that existed since the settlement of the 1896 Bell Company/Western Union patent case and recombined portions of the voice and data industries into one company.

2.3.3 AT&T's Refusal to Interconnect with Competitors

At this same time, AT&T increased its anticompetitive behavior. More troubling to customers than its size or its provision of both telephone and telegraph services in certain areas, AT&T refused to physically interconnect the wires of its Bell System facilities with the wires of its competitors, the independent telephone companies. In AT&T's 1910 annual report, Theodore Vail, AT&T's president, explained that interconnection with the independents "would force the Bell System to place at the disposal of and under the control of any opposition company . . . one of [our] circuits . . . and disconnect it, for the time being, from the circuits of the Bell System. . . . The fact that the opposition exchange could get such facilities would enhance its importance at the expense of the Bell System."[33] Many similar arguments are made today by CATV, copper, fiber, and wireless companies unwilling to share their resources with competitors.

Therefore, instead of interconnecting with its competitors, AT&T simply purchased them and added their facilities to the Bell System. When the independents would not sell, AT&T refused to interconnect with them. This refusal significantly limited the value of telephone service for the customers of the independent telephone companies because it meant that they could not call friends, family, and businesses that were customers of the Bell System. AT&T's practices thus increased public outrage against AT&T, but were very lucrative for the company.

2.4 REGULATION OF THE COMMUNICATIONS INDUSTRY

With the public outcry and the renewed antitrust activities of the Roosevelt and Taft administrations, Congress passed the Mann-Elkins Act of 1910, extending the Interstate Commerce Commission's authority to the telecommunications industry.

2.4.1 1912 Antitrust Case: Settled with 1913 Kingsbury Commitment

However, over the next two years, the new legislation did not slow AT&T's anticompetitive business practices, and thus, in 1912, a group of independent companies filed a complaint with the U.S. Department of Justice based on the Mann-Elkins Act. The independent telephone companies argued that AT&T's control of Western Union and its refusal to physically interconnect Bell's facilities with the facilities of the independents violated U.S. antitrust laws.[34] One of the leaders of the lawsuit was Clarence MacKay, President of the Postal Telegraph Company (Postal), AT&T/Western Union's greatest rival.

In January 1913, as the new Wilson administration took office, Attorney General George W. Wickersham advised AT&T that, "in his opinion, certain of its [AT&T's] currently planned acquisitions of independent companies in the Middle West were indeed in violation of the Sherman Antitrust Act."[35] That same month the Interstate Commerce Commission began an investigation of AT&T's activities. Following the ICC's initial findings, the Department of Justice began preparing an antitrust suit against AT&T. The suit was settled out of court with the Kingsbury Commitment.

On December 19, 1913, AT&T Vice President Nathan C. Kingsbury sent a letter to James McReynolds, Wickersham's successor as attorney general, agreeing to three specific actions.[36] First, he stated that "wishing to put their affairs beyond criticism and in compliance with your suggestions formulated as a result of a number of interviews between us during the last sixty days," AT&T agreed to dispose of its Western Union stock "in such a way that the control and management of [Western Union] will be entirely independent." Second, AT&T agreed to purchase no more independent telephone companies except with the approval of the Interstate Commerce Commission. Third, it agreed to make "arrangements . . . promptly under which all other telephone companies may secure for their subscribers tool service over the lines of the companies in the Bell System."[37]

The results of the Kingsbury Commitment were astonishing. First, within months AT&T sold its Western Union stock and, as a result, voice and data technologies once again operated as separate industries. Second, the independent telephone companies considered the Kingsbury Commitment a great victory because it ensured their survival and preserved competition in the telecommunications industry.[38] By 1920, the independents grew to over 8,500 companies. Third, the Interstate Commerce Commission established a *uniform system of accounts* in 1913 for telephone companies' in-state financial reports. Interstate telephone rates, on the other hand, remained virtually unregulated until the 1930s. Fourth, continuing its antitrust momentum, Congress enacted the Clayton Act of 1914.[39] In it, Congress created the Federal Trade Commission (FTC)[40] and empowered the FTC to conduct investigations of large companies. If the investigations found that companies were involved in unfair business practices, Congress empowered the FTC to issue cease-and-desist orders against the companies. Fifth, the concept of universal service evolved from the terms of the Kingsbury Commitment.

2.4.2 Universal Service

Given the market restrictions on the Bell System following the Kingsbury Commitment, Theodore Vail, President of AT&T, developed a marketing plan that he called *universal service*.[41] He stated that AT&T's goal was to provide basic telephone service at an affordable price in the U.S.[42] Vail frequently referred to the phone as a "necessity" and described universal service as a means to assist the public in acquiring that necessity.[43] He noted that

> If there were no Bell System, only dissociated individual companies or groups of companies, no line over a few hundred miles long would have been built, or if built it could not be operated. . . satisfactorily.[44]

Vail's concept of universal service caught the attention of the U.S. government because, as the rumblings of World War I flared in Europe, the U.S. government was concerned about the availability of communications facilities across the United States. For military, commercial, and political reasons, the U.S. government recognized that communications was the technology to unify the United States and it adopted Vail's concept of universal service as a national policy. In doing so, the government established a federal Universal Service Fund to pay for the installation of telephone facilities in high-cost, low-profit areas of the United States where privately funded companies could not afford to provide service. The Universal Service Fund was built and maintained by collecting a government-imposed monthly contribution from all existing telephone customers as part of their monthly phone bills. This practice continues today.

By January 1915, the first transcontinental phone line was constructed and, for the inaugural coast-to-coast long distance phone call, Alexander Graham Bell and Thomas Watson were invited out of retirement to make the call.[45] Many of the photos of Bell as an old man in a full white beard are from this era. As additional telephone lines were installed in previously unserved areas, it made little sense to regulators to duplicate equipment in any one area. This strengthened the concept of the telephone as a *natural monopoly* and convinced Congress that the public would not benefit from having competing telephone companies providing service in the same geographic area. Instead, Congress decided that each telephone company should be a regulated *public utility.*[46]

One might expect that Vail would oppose such regulation, arguing that it was unnecessary since so many independent telephone companies existed. Surprisingly, however, Vail did not oppose regulation of the telephone industry by the U.S. or state governments. In fact, as Congress considered placing AT&T under the control of the U.S. Postal Service, as other countries around the world had done with their Postal, Telephone, and Telegraph (PTT) agencies, Vail argued that such a move would truly make telephone an *unregulated monopoly*. Instead, he argued that continued private ownership of the monopoly industries, with government regulation, resulted in a situation more in the *public's interest*. In a 1915 speech to the National Association of Railway Commissioners, entitled "Some Observations on Western Tendencies," he stated

> Regulatory bodies, state and federal, should be thought of and should think of themselves as juries charged with "protecting the individual member of the public against corporate aggression or extortion, and the corporate member of the community against public extortion and aggression." Regulators should see it as their duty "to restrain and suppress . . . certain evils that have been ingrained in our commercial practices," and also "to restrain an indignant and excited public."[47]

He further stated that AT&T had

> no serious objection to regulation, so long as it was such . . . as to encourage the highest possible standard in plant, the utmost extension of facilities, the highest efficiency in service, and to that end should allow rates that will warrant the highest wages for the best service, some reward for high efficiency in administration, and such certainty of return on investment as will induce investors . . . to supply all the capital needed to meet the demands of the public.[48]

Nonetheless, for one year, from August 1, 1918 to August 1, 1919, a joint resolution of Congress placed the United States's telegraph and telephone systems under the control of the U.S. Post Office as an emergency measure. On July 31, 1918, President Wilson announced the change and the next day handed control of the Bell System's operations, assets, services, and holdings to Postmaster General A.S. Burleson. However, the arrangement did not work well, and after one year, Congress reversed its decision and returned the facilities to private ownership.

By 1919, *universal service* became an even more widely accepted policy of the U.S. government. As described by the U.S. House of Representatives in a report concerning the extension of the telephone system in the District of Columbia, "[Telephone] [s]ervice should be provided at reasonable cost, in fact at as low as efficient service permits, so that the largest number [of citizens] possible may use it"[49] Even the state governments supported the concept. For example, the Public Utilities Commission of South Dakota stated that, "It has been the general policy of this Commission to encourage the extension of [telephone] exchange service into every community and household of the state. . . ."[50] Similarly, the Oregon Public Service Commission stated that, "[a]ny attempt to make rates proportionate to cost would result in residence charges so high that few householders would use the telephone at all and this would result in greatly impairing its value to the business user."[51]

These statements reflect the general regulatory feeling in the United States at that time, and as a result, a system of non–cost-based, averaged rates developed to make telephone charges affordable to all. The system was justified because most people believed that enabling more households and businesses to have telephone service increased the value of all telephones and benefited all subscribers. This policy also resulted in what was called the *switchboard-to-switchboard* principle, which assigned (1) the total cost of all subscriber equipment, local distribution and exchange equipment to local exchange rates; and (2) the cost of long-distance switching equipment and lines to toll calls. In 1930, the U.S. Supreme Court rejected this prevalent concept and determined that proper rates should spread the costs to both local and long-distance service. However, the Bell System largely ignored this decision until the 1950s.[52]

Thus, over the next 70 years, from the 1920s to the 1990s, the government's policy of using universal service to assure "the availability of adequate telephone service to the widest practical number of users" continued through subsequent federal legislation. Funding for telephone service evolved into a complex system of cost-support mechanisms, funds, programs, averaged rates, shared costs, cross-subsidized services, and full area service obligations.[53] Congress and the state governments also provided various Rural Electrification Administration grants and low-interest loans to rural telephone companies and cooperatives to encourage development of telephone service in rural areas.[54] The House Committee report for telephone legislation recognized the Bell System as a national resource that provided a unique technology to the citizens of the United States and argued that "cream skimming" be made unlawful because it would "relegat[e] farmers in the less profitable areas to a perpetual telephone wilderness."[55]

2.4.3 1921 The Willis-Graham Act

Although the concept was already well established, Congress codified that utilities such as telephone, water, and electricity, were natural monopolies in the Willis-Graham Act of 1921.[56] This concept continued into the 1990's through the franchising of both telephone and cable television systems.

Most notably, the Willis-Graham Act exempted telephony from the portion of the Sherman Act that restricted the consolidation of competing companies. Instead, Congress permitted such consolidation where the economies of scale of the larger phone companies enabled universal service.[57] On this exclusion, the Bell System continued to grow throughout the 1920s and 1930s beyond the original geographic constraints imposed by the Kingsbury Commitment, merging into larger regional companies, with the addition of each new area approved by the Department of Justice and later the FCC. As part of this expansion, Bell Laboratories was formed in 1925.

2.4.4 1934 Communications Act

In 1934, with the continued expansion of domestic and international communications facilities, Congress decided to pull all U.S. communications law together into one statute and then to add any elements deemed to be necessary but absent. Prior to 1934, U.S. laws existed covering radio, telegraph, and telephone communications, but the laws were scattered, inconsistent, and somewhat out of date. With the Communications Act of 1934 (1934 Act),[58] Congress (1) consolidated those laws, (2) established the overarching goals for U.S. communications, and (3) created the FCC to regulate the operations and pricing of interstate and maritime communications. At the same time, the state commissions reviewed their responsibilities and expanded their procedures to include the added tasks of (1) rate analysis, (2) management of the state's universal service or *high-cost fund,* (3) distribution of funds among companies for the carriage of telephone traffic between companies based on separation agreements, and (4) dispute resolution between the utilities and the public. The chapters or sections within the 1934 Act are called *titles,* each addressing a specific aspect of the communications industry.

The 1934 Act remains the primary law in telecommunications today. In fact, the 1996 Act simply updates, not replaces, the 1934 Act, and its titles exist for convenience of the 1996 Act. They do not, however, replace the overriding industry identification titles of the 1934 Act. For example, Title I of the 1934 Act covers General Provisions, while Title II addresses the common carrier industry, including the telephone and telegraph companies and their services. Title III contains the laws affecting the radio or wireless industry, while Titles IV and V address "Procedures and Administrative Provisions" and "Penal Provisions—Forfeitures," respectively. In the 1940s, Title VI was added to cover cable television issues and Title VII to address "Miscellaneous Provisions." These titles provide the industry labels used today. For example, many discussions and documents in telecommunications refer to services and products as being Title II (Telephone/Telegraph), Title III (Wireless), and Title VI (Cable) elements. These titles in the 1934 Act should not, however, be confused with the titles of the 1996 Act, discussed in more detail in Chapter 3.[59]

2.4.5 1949 Antitrust Case: Settled with 1956 Consent Decree

Two years after the 1934 Act, in 1936, Congress passed the Robinson-Patman Act[60] to bolster the price discrimination section of the Clayton Act. The new Act made it unlawful for any seller engaged in commerce to discriminate, directly or indirectly, in the price charged to purchasers for commodities of like grade and quality, especially where the effect reduced competition.

Based on the Robinson-Patman Act, the FCC ordered an in-depth investigation of the Bell System's vertical structure in the late 1930s. Almost from its inception, AT&T had a vertical structure that was unique. With its products developed at Bell Labs, its equipment manufactured at Western Electric, its phone lines and equipment installed only by Bell employees, and the fact that the company leased, but never sold, its Bell-manufactured telephone equipment to consumers, the company had a tight vertical hierarchy that was very different from other utilities. For example, customers did not purchase their lamps and sinks from the electric and water utilities and nonutility electricians and plumbers could install the facilities. As a result of the investigation and on behalf of the *other equipment manufacturers* (OEMs), the Department of Justice began preparing another antitrust suit against AT&T. The Department of Justice sought to break up the Bell System, requiring it to sell Western Electric and open the product supply market to competition. However, the case was postponed with the outbreak of World War II.

In 1949, with World War II over, the Department of Justice reopened the antitrust case against the Bell System.[61] The suit continued for seven years before it was settled out of court in the 1956 Consent Decree. The 1956 Consent Decree[62] accomplished several things, but three specific results significantly impacted the industry. First, the 1956 Decree allowed AT&T to keep its vertical structure, so long as Western Electric manufactured only products used by the Bell operating companies.[63] Second, AT&T agreed that it would not collect royalties from the other equipment manufacturers on its then-existing patents and would license future patents to any applicant on a nondiscriminatory basis at reasonable royalties.[64] Third, and perhaps most importantly, the decree limited the services that the Bell System could provide. The Bell Operating Companies (BOCs) agreed that they would confine their activities, with certain exceptions, to providing only voice communications services.[65] This concession emphasized the Bell System's core business, but eventually barred the Bell System from the very lucrative data communications market. It also, once again, divided voice and data communications, causing business customers to separate the *voice communications departments* in their companies, from the *data* or *information-processing departments*. It also set the United States apart from the more converged communications provided by the Postal, Telephone, and Telegraph agencies in most other countries, an evolution pattern in communications that the United States is now adjusting.

The 1956 Decree remained in effect until 1984, when the *Modified Final Judgment* broke up the Bell System, dividing it horizontally into local and long-distance sectors. This action was called the Modified Final Judgment because it modified the 1956 Decree.

2.4.6 1974 Antitrust Case: Settled with 1982 Modified Final Judgment

On August 11, 1982, the district court for the District of Columbia entered a consent decree, known as the Modification of Final Judgment (MFJ)[66], that settled an eight-year antitrust case initiated by the Department of Justice against AT&T and the Bell System in 1974 on behalf of long-distance and other competitive carriers. The MFJ, effective January 1, 1984, was an out-of-court settlement that divested the Bell System into AT&T, a competitive long-distance carrier, and seven newly created Regional Bell Operating Companies (RBOCs), providing basic local exchange telephone service. In addition, Bell Labs and Western Electric, became competitive companies under AT&T.

The main benefit of the MFJ to AT&T was that it permitted AT&T to become a competitive company, able to enter the data communications and other markets that it had agreed to stay out of in the 1956 Decree. This enabled AT&T to redefine its services to more closely address customer requirements in the Information Age.

Its impact on the seven new RBOCs was that they became independent, regulated local telephone companies that maintained monopoly control of their service areas, but continued to be prohibited from (1) providing information services, (2) providing interLATA (LATA: local access and transport areas) services, and (3) manufacturing and selling telecommunications equipment, including customer premises equipment. Five years later, in 1987, the MFJ was relaxed slightly to allow the RBOCs to provide voice-messaging services and to transmit data information generated by others.[67] In 1991, the restriction on RBOC ownership of content-based information services was lifted.[68] The telecommunications industry structure created by the MFJ remained in effect until 1996, when the Telecommunications Act of 1996 was enacted and replaced many of the details of the MFJ.

2.5 TECHNOLOGICAL CONVERGENCE AND RENEWED COMPETITION

During the years that telegraph, telephone, and radio communications were divided into separate industries, regulators decided which companies could enter each market, which geographical areas each could serve, what products each could provide, and at what price. The impact of these constraints is still evident in the structure and terminology of the United States's telecommunications industry today.

However, technological developments challenged this traditional division of the communications industry and opened it to competition. This section discusses the development of five technologies: customer premises equipment, microwave, undersea cable, satellite, and computers. Each technology is capable of carrying voice and data traffic with equal ease, changing the concept of a natural monopoly and the historical division of technologies.[69] Both microwave and satellite are also wireless technologies capable of carrying communications traffic at lower cost than the traditional wire systems.

2.5.1 Customer Premises Equipment

First, since the U.S. government recognized the importance of high-quality telephone service in America throughout the telephone industry's history, it rejected anything that could reduce the performance of the network or unnecessarily raise its operating cost. Therefore, from the 1870s until the 1960s, the government supported a policy of absolute prohibition of any direct electrical connection of "foreign" equipment to the network. *Foreign* was defined as equipment not provided by the telephone company. This policy protected the U.S. public switched telephone network (PSTN) from interference, damage, or degradation of service caused by the attachment of *customer-provided equipment* (CPE). The carriers stated "it was not possible to ensure efficient telephone service to the public if equipment or devices supplied by parties with no legal responsibility for the quality of service on the network were attached to the network by customers."[70] Two court cases following World War II, however, caused the U.S. government to change its rigid *nonattachment policy*, challenged the vertical hierarchy of the Bell System, and opened the end-user equipment market to the competition that we know today.

1956—The Hush-A-Phone Decision

The first court case was the Hush-A-Phone decision, issued by a District of Columbia court in 1956.[71] The Hush-A-Phone was a small plastic nonelectrical cup that snapped over the mouthpiece of the telephone handset. Much like cupping one's hand around the mouthpiece, it enhanced the user's privacy when talking into the phone and removed ambient noise for the listener. AT&T reacted to the use of the Hush-A-Phone by threatening to cut off telephone service to any customer found to be using it. Hush-A-Phone backers responded by suing AT&T.

The court ruled against AT&T and permitted the attachment of the Hush-A-Phone and other similar nonelectrical devices that posed no harm to the telephone network. The court still supported the prohibition against *direct* electrical connections of foreign devices to the telephone system, but it believed that the phrase "attachment of foreign devices" could not be interpreted as broadly as AT&T had done in the Hush-A-Phone case.

1968—The Carterfone Decision

The second case, the *Carterfone Decision*,[72] further extended the movement toward using foreign or non–telephone company equipment on the telephone network. The Carterfone, developed and produced by Carter Electronics, a Texas firm, was a device used to connect radio calls to the wireline telephone network primarily for field personnel in the petroleum industry. Since oil exploration and drilling generally occurred in remote areas not served by wireline phone facilities, the petroleum workers used radios for communications.

However, when the workers needed to call someone who did not have a radio, such as their suppliers or families, they could not make a call on the telephone system from their radio.

The Carterfone provided a solution to this problem. It was a small box with two rubber *cups* spaced to accommodate the ear and mouthpiece of a standard telephone handset. The backside of the box connected to the radio system. To use it, a radio user called into the base station and asked the base station operator to dial a desired telephone number. When the called party answered, the attendant placed the handset of the wireline phone into the Carterfone box. The box managed the *push to talk* and *release to listen* functions of the radio. When the call was completed, the attendant removed the phone from the Carterfone box and disconnected the call.

Critical to the court and regulators, the Carterfone provided only an acoustical connection between the radio and the phone system. Like the Hush-A-Phone, it did not electrically connect to the phone system. However, AT&T again responded to this new device by threatening to disconnect the service of any customer who used a Carterfone. In protest, Carter sued AT&T and won.

After Carterfone, AT&T developed protective devices to interface between the network and foreign equipment. Nearly a decade later, in November 1975, the FCC established an extended registration program in its Report and Order in Docket 19528. Under the program, equipment manufacturers submitted technical information to the FCC to prove that the manufacturers' equipment could safely be connected directly to the network. If the customer-provided equipment was approved and registered with the FCC, carrier-provided protective coupling devices were no longer necessary. Otherwise, the equipment could be used with a registered protective coupling device.[73] This program marked the beginning of intense competition in the customer premise equipment and customer-owned telephone switch or private branch exchange markets.

2.5.2 Microwave

A second technology that became competitive after World War II, which significantly changed the methods and costs of distributing communications and blurred the divisions between voice, data, and video technologies, was the microwave industry. Microwave technology became instantly popular, especially for long-distance communications, because it offered companies significant savings over wired communications systems and the opportunity to carry combined voice and data traffic.[74]

Prior to World War II, the FCC assigned most of the microwave spectrum to the telecommunications common carriers, mainly AT&T. Spectrum for private microwave systems was leased only on a case-by-case basis to government or business users with "special communications needs."[75] Following World War II, however, the demand for microwave technology grew so quickly that the FCC began to relax its restrictions on the private companies. By the end of

1951, over 13,000 route miles of private microwave systems were in place or under construction.[76] The common carriers protested this action by the FCC, arguing that the additional users would congest the airwaves and result in degraded service. The private users, on the other hand, wanted to expand the service.

1959—The "Above-890 MHz" Decision

Microwave continued to grow over the next decade, and in 1959 the FCC opened the frequencies higher than 890 megahertz (MHz) for private systems and removed nearly all restrictions on the use of microwave for private systems. Again, the common carriers strongly opposed this action, but the FCC decided in favor of the extended use of the technology.[77]

1969—MCI and Switched Service

As a result of this extended use, in 1963 Microwave Communications Inc., now known as MCI, filed an application with the FCC to build a microwave system between Chicago and St. Louis on which it planned to provide point-to-point (nonswitched), private-line services to business customers as a common carrier. MCI argued that this was necessary to bring the benefits of the above-890 MHz decision to small businesses. Although the process took six years, the FCC finally approved the application in 1969, and competition in the telecommunications industry took a quantum leap forward.[78] The decision proved so popular that by the next year, 1970, the FCC had received more than 1,000 applications from companies wanting to provide point-to-point (nonswitched), private-line microwave service similar to MCI's.

1970—Specialized Common Carrier Decision of 1971

In 1971, the FCC responded to these multiple applications in its Specialized Common Carrier Decision.[79] In the decision, the FCC decreed that the private microwave companies be allowed to compete directly with the regulated common carriers in selling private-line network transmission services.[80] It was a momentous decision that opened competition in point-to-point (nonswitched), private-line services.

1978—MCI Telecommunications Corporation v FCC [*Execunet* I]

In 1974, MCI moved the industry ahead again when it filed an application to provide public switched long-distance service. The application was a revision of the tariff under which MCI furnished its nonswitched private-line interstate service, but in the revision, MCI proposed a class of *metered-use* services, including its *Execunet* service. With *Execunet,* a subscriber could use any telephone to reach any other telephone in a distant city served by MCI simply by dialing a local MCI number followed by an access code and the number in the distant city in a manner similar to using a calling card today. *Execunet* customers were billed for each call on a time and distance basis, subject to a monthly minimum.[81] Thus, the *Execunet* services[82] were designed to allow MCI to compete directly with AT&T.[83]

The FCC first denied MCI's *Execunet* application in both 1975[84] and 1976,[85] but MCI appealed the denial. In 1977, the District of Columbia Court of Appeals reversed the FCC's decision and remanded the case for a clearer explanation of why competition in long distance, specifically with Message Toll Service (MTS)[86] and Wide Area Telecommunications Service (WATS),[87] was not in the public interest.[88] In response, the FCC initiated a rule-making procedure to consider competition in long distance. The rule-making process eventually concluded in 1980 with a recommendation that competition should be instituted in long distance permitting companies to offer limited resale of switched services. This recommendation set the stage for the opening of competition in long distance in 1984.

2.5.3 Undersea Cables

A third technology to change the structure, economics, and dynamics of the telecommunications industry was undersea cables. For 90 years, from 1866 when the first undersea cable was laid until 1956, undersea cables carried only telegraph communications. Voice communications across the Atlantic began in 1927 using a high-frequency radio method developed by AT&T. This remained the sole method of international voice communications until 1956.[89] Both forms of international communications, undersea telegraph cable and high-frequency radio voice, were *point-to-point* connections. The *points* were *gateway* cities where the cables came ashore or the radio communications transmitters and receivers were located. The first three U.S. gateway cities were New York, Washington, D.C., and San Francisco, with additional gateway cities added later.[90] To transmit from *non–gateway* locations, customers used domestic providers to carry the traffic from the remote location to a gateway city, known as *backhauling*. The message was then *handed off* to an *international record carrier (IRC)* such as AT&T, Western Union, ITT WorldCom, RCA Globecom, and Houston International Teleport.

In 1942, the Board of War Communications prohibited radio voice communications as a security measure during World War II.[91] This pushed more traffic to the telegraph circuits. To more efficiently meet the increased wartime traffic on telegraph communications, Congress permitted the merger of the two primary telegraph companies in the United States, Western Union and Postal Telegraph and Cable System in 1943. The merger made the resulting Western Union Company the primary provider of domestic telegraph service in the United States, just as AT&T was the primary provider of domestic telephone service.[92] However, in approving the merger, the U.S. government required Western Union to remove itself from providing international telegraph service or international record carriage services, leaving this business for the other international record carriers.

In 1956, a decade after World War II ended, AT&T developed underwater repeaters reliable enough to carry voice communications over the primitive undersea cables. AT&T then partnered with the British government to share the high cost of laying the first transatlantic telecommunications (TAT-1) voice cable. The new cable connected to facilities at the gateway cities,[93] and the transatlantic voice-grade channels could be subdivided into smaller telegraph circuits. These two features removed the technological differences between international voice and record (telegraph) traffic over cable.[94] In addition, the significantly superior quality of the transmission on these channels caused the IRCs to lease circuits from AT&T rather than use the existing cable or high-frequency radio.

1955—AT&T's Dataphone Service

Since AT&T was already carrying voice and data (telegraph) service over the same international cables, in 1955 the FCC authorized AT&T to combine the services on its international undersea cable between California and Hawaii.[95] AT&T called this combined service its *Dataphone* service. When Hawaii became a state and shifted from an international to a domestic destination, the FCC permitted the Dataphone service to continue.[96]

1959—AT&T's Alternate Voice-Data Service

Four years later, in 1959, the FCC permitted AT&T to provide *alternate voice-data* service to the Air Force over shared circuits,[97] and the following year, in 1960, the FCC permitted the alternative voice-data service to become a public tariffed offering to any customer requiring both voice and data service between United States and Puerto Rico or Bermuda.[98]

1964—Restrictions of Combined Voice and Data

However, in 1964, the FCC stopped this progression toward unified voice and data when AT&T applied for authorization to lay a fourth transatlantic cable (TAT-4) and to offer alternative voice-data service to all customers.[99] The FCC wanted to protect the other international record carriers and reestablished the regulatory division between voice and data services, excepting only the services to Hawaii and those covered by certain defense contracts.[100] The FCC's plan did not work, however, because within 15 years, by 1980 AT&T provided 97% of all international message service for U.S. carriers.[101]

1981—The Record Carrier Competition Act

In response, Congress enacted the Record Carrier Competition Act of 1981,[102] in which it (1) introduced competition into the international communications industry by removing the previous division between domestic and international services and allowing each record carrier to provide both domestic and international record carriage services;[103] (2) deregulated the international record carriers; and (3) as part of the new competition, mandated that all record carriers interconnect with all other record carriers, so that subscribers of one record carrier could communicate with subscribers of all other record carriers. The interconnection was to be, "(I) equal in type and quality; and (II) made available at the same rates and upon the same terms and conditions as each carrier provided for itself."[104]

The result of the Record Carrier Competition Act of 1981 was a series of mergers in the international communications market. For example, Western Union immediately purchased ITT World Communications in 1981, followed by MCI's merger with the following: (1) Western Union International (Xerox) in 1982; (2) RCA Globecom in 1988; (3) Houston International Teleport in the early 1990s; and (4) World Communications, Inc. in the 1990s. These mergers significantly increased MCI's share of the international communications market, reaching 20.7% in 1992.[105] In contrast, Sprint served 7% and AT&T's share dropped from its previous high of 97% in 1980 to 68.7% in 1992. Other international communications companies serve the remainder of the market.

2.5.4. Satellites

A fourth challenge to the historical division between voice and data services and competition within the industry came with the development of satellite technology. Like customer-premises equipment, microwave, and undersea cables, satellites can carry both voice and data communications with equal capability and efficiency over long distances.[106] Satellites are unique, however, because the industry developed with tremendous technical, financial, and regulatory support from the U.S. government.

The concept of using satellites for communications was an old idea, delayed only until humans developed the capability to launch an object into space and place it in a sustaining orbit. In 1945, Arthur C. Clarke, a science writer and author of, among other works, *2001: A Space Odyssey,* calculated that an object orbiting the earth at 22,300 miles (35,800 kilometers) over the equator would move around Earth at basically the same speed as Earth rotates. Since the satellite would move in "sync" with Earth, it would appear to be *stationary* from any particular point on Earth. Thus, such satellites are called *geosynchronous* or *geostationary* (GEO) satellites. The band at the appropriate altitude over the equator for this phenomenon to occur is known as the geosynchronous orbit. Clarke also noted that the *footprint* or view of Earth from a satellite at that altitude is large enough that three satellites, positioned over the Atlantic, Pacific, and Indian Oceans, could provide communications services to all of the populated areas of Earth. The only areas not adequately covered would be the North and South Poles.

A decade later, in 1957, the Russians developed sufficient launch capability to place the world's first artificial satellite, *Sputnik,* in an orbit a few hundred miles above Earth's surface. In the midst of the Cold War, this accomplishment by the Soviets was an affront to the ongoing efforts of the U.S. military to develop satellite and launch technologies.[107] However, the following year, in 1958, the United States Air Force launched *Score,* generally considered to be the world's first communications satellite to successfully transmit telecommunications signals. Like *Sputnik, Score* was a low Earth orbit (LEO) satellite, placed several hundred miles above Earth, not in the geostationary orbit. In 1959, the International Telecommunications Union in Geneva, Switzerland, allocated radio spectrum to space use, and shortly after, in 1962, the United States launched *Telestar,* a more powerful communications satellite.

1962—Communications Satellite Act

Also in 1962, Congress passed the Communications Satellite Act[108] in which it opened commercial use of space for communications and created the Communications Satellite Corporation (Comsat). Organizationally, Comsat was a unique entity. It was a new, semiprivate corporation, half owned (50%) by the international communications carriers, including AT&T, and the other half (50%) owned by public investors.[109] As a semiprivate corporation, Comsat was required to report to the Congress and the president each year.[110]

Congress delegated responsibility for satellite regulation to the FCC in three ways. First, Congress declared Comsat to be a common carrier and thus included satellite service in Title II, the common carrier portion of the Communications Act of 1934 regulated by the FCC. Second, the FCC's authority over satellites derived from Title III of the 1934 Act, as amended, which gave the FCC broad power to regulate use of the radio-frequency spectrum for communications purposes.[111] Third, under its enabling statute, Congress delegated to the FCC responsibility for implementing all communications law, including the Communications Satellite Act. However, to keep a monopoly from developing in the satellite industry, the FCC regulated details in satellite communications such as Earth station ownership. For example, the FCC required that Comsat never own more than 50% of each satellite Earth station, while the other 50% be owned by the remaining international carriers.

1964—Intelsat Interim Agreement

In 1964, two years after the United States created Comsat, 14 governments met at an international conference in Washington, D.C. and signed an international interim agreement creating the International Telecommunications Satellite Consortium (Intelsat).[112] The original signatories were Australia, Canada, Denmark, France, Germany, Italy, Japan, the Netherlands, Norway, Spain, Switzerland, the United Kingdom, the United States, and the Vatican. The U.S. Congress designated Comsat as the United States's representative to Intelsat.

The Interim Agreement stated that Intelsat's purpose was to establish "a single, global commercial communications satellite system."[113] To accomplish this purpose, the signatories (1) established the Interim Communications Satellite Committee to coordinate the activities of the consortium; (2) acknowledged that Comsat, with more than a 50% investment share, was clearly the majority owner and thus had veto over any decision of the Interim Communications Satellite Committee; and (3) confirmed Comsat "as the manager in the design, development, construction, establishment, operation and maintenance of the space segment." Comsat thus dominated Intelsat's development and progress for its first decade.

The world's view of INTELSAT was also unique. First, as a consortium it was considered by the signatories to be an *international organization,* with all of the legal status, privileges, and immunities of an international organization, including having its agreements and documents be considered international treaties, supported by the state departments of the signatory countries. Second, as a commercial organization with independent status and the power to make contracts and to hold property, it was given the responsibility of providing public international telecommunications through communications satellites.

In June 1965, Intelsat launched the world's first international communications satellite known as *Early Bird* or *Intelsat I.*[114] Positioned over the Atlantic Ocean in a geostationary orbit, *Early Bird* linked first North America and Europe and later other countries on both sides of the Atlantic. Operated under Comsat's management, it was the first satellite to provide commercial telephone and television service.[115] Two years later, in 1967, the consortium placed *Intelsat II* over the Pacific, linking those areas to the Atlantic, and in 1969, it added *Intelsat III* over the Indian Ocean to complete full global coverage.[116]

Prescribed-Use Policy

Interestingly, when INTELSAT's service became available, AT&T and the other international carriers had adequate capacity and technological sophistication in their undersea cables to serve all customers at the time. Thus, neither the customers nor the 50% international carrier owners had much motivation to use satellites for their operations. However, in 1966, the U.S. government, in an effort to ensure the survivability of the new satellite technology, adopted a *prescribed-use policy.* This policy established *loading requirements* prohibiting the international communications carriers from using or constructing additional undersea cable circuits to carry communications traffic until it used an equal amount of satellite capacity.[117] In addition, in 1979, the FCC authorized the international carriers to accept messages for international communications directly at satellite Earth stations in order to reduce backhaul costs and thus facilitate satellite use.[118]

This prescribed-use policy had both positive and negative impacts. It made Comsat a true *carriers' carrier* in the United States by requiring that the retail common carriers, AT&T, and the other international carriers, buy half-circuits from Comsat.[119] It also established a system of mutual backup for terrestrial and satellite circuits, which has proven useful in ensuring continuous communications when either satellite or terrestrial circuits have failed. On the other hand, the policy removed the purchase and sale of international circuits from the competitive marketplace. Instead, the FCC established the rates for circuits based on a composite average of both cable and satellite circuit costs.[120] This removed accurate information about the true cost of satellite and cable technology and services, a concern that still exists today.

Open Skies Policy

Satellite technology was so inviting that in 1966, the broadcast company ABC filed a request with the FCC for authority to operate a satellite system to distribute television broadcast signals within the United States.[121] ABC suggested that the service could also be used for a variety of communications services, including domestic telephone and high-speed long distance. In the next few years, the FCC received several other applications for authority to operate domestic satellite systems. By 1970, after significant consideration, Congress decided that an open, competitive, domestic satellite environment would be in the public interest because it would (1) encourage innovation, (2) enable rapid development of the technology, and (3) meet consumers' needs in the most effective manner possible.[122] This decision is known as the *Open Skies Policy.* To implement it, Congress decided to allow limited open entry by all qualified U.S. entities with the financial and technical capability to launch and operate a domestic satellite system. Congress also decided that it would refrain from suggesting the technology to be used, but instead elected to let the market determine the best choice for each system. Unlike previous technologies, this meant that the competitive entrants were not protected from fair competition or guaranteed sufficient market to survive.[123]

In announcing the Open Skies Policy, the FCC requested proposals from potential providers. It received several applications and, after reviewing the proposals, the FCC approved eight systems in 1972. By December 1973, domestic satellite service began providing communications in the United States. The Open Skies Policy did not go unchallenged, however. Several groups raised objections, arguing environmental, First Amendment, and antitrust issues.[124] However, the courts upheld the FCC's decision, stating that the FCC had followed proper procedure in making public interest determinations. Over the next decade, the Open Skies Policy was deemed to be a success with very positive results because eight companies continued to provide domestic satellite service, and after 20 years, by 1992, over 22 provided service. These companies include the following: AT&T; Western Union; Hughes; GE Americom; Ecostar; Loral Space and Communications Ltd.; Orion Network Systems; Inc.; and GTE SpaceNet.[125] It had also developed numerous services such as cable television, broadcast printing, wide-area telephone and data services, and it had placed the United States in a leadership position in satellite technology.

Also in 1973, after years of negotiation, a Definitive Agreement[126] replaced Intelsat's Interim Agreement, providing new leadership and a new organizational structure. First, with the Open Skies Policy, Comsat resigned as Intelsat's manager, and the consortium developed a governing charter composed of two separate international agreements: (1) a Definitive Agreement among government signatories, and (2) an Operating Agreement signed by a single telecommunications entity in each member country, designated by the government. It established an *Assembly of Parties* as Intelsat's *principal organ,* and shifted responsibility for the planning, implementation, and operation of international satellites from the Interim Communications Satellite Committee to a *board of governors* composed of investors. Finally, a director general was named to manage day-to-day management operation of Intelsat. By 2001, Intelsat remains a multinational organization headquartered in Washington, D.C., with 136 members, all members of the International Telecommunication Union. Another 50 countries use the satellites, but are not members of Intelsat. Intelsat owns and operates 23 satellites, providing global, regional, and domestic telephone, television, teleconferencing, facsimile, data, and telex services. Portions of Intelsat are being privatized under the Open-Market Reorganization @Normal: for the Betterment of International Telecommunications Act (the Orbit Act) to reflect other industry moves toward more competitive markets. The outcome of these changes on the structure of the organization, however, is not yet known.[127]

2.5.5 Computers

The fifth technology to blur the technological distinctions between voice and data services and the structural divisions in the telecommunications industry was the development and propagation of computers in the 1960s and 1970s. Computers were integrated into the telephone industry in two critical ways. First, nearly as soon as computers became available, they were incorporated in the internal switching and signaling functions of the telephone industry, resulting in the digitization of voice traffic. The digitization of communications made voice, data, and video communications nearly identical. Second, as the computer industry evolved from *card-fed machines,*[128] to *terminals* connected to *mainframe computers* through telephone lines, to fully linked computer networks, customer demand for high-speed, high-capacity communications lines skyrocketed. These distributed data-processing networks, using common carrier facilities, resulted in data communications traffic that nearly eclipsed the voice portion of the carriers' business. As the FCC noted, "[I]t is clear that data processing cannot survive, much less develop further, except through reliance upon and use of communication facilities and services."[129] In these cases, where the data service was dependent on the use of the common carriers' transmission facilities, some persons argued that the FCC was statutorily required to regulate the services. The FCC, however, rejected this argument.

Instead, it favored competition, but was unclear how to separate the regulated and unregulated portions of communications services.

In 1966, the FCC began an examination of the increasing overlap of computers and communications and its related issues. The scope of the study was very broad, but, in addition to building a technical awareness of the merging of the two technologies, the FCC's primary concerns focused on three questions. First, should regulated common carriers, mainly the telephone companies, be permitted to provide data processing services? Second, if so, should the data processing services be regulated? Third, if regulated telephone companies were permitted to provide unregulated data services, how should the government prevent the companies from subsidizing their data services with revenues from their basic voice services, allowing them to price the services far below their competitors, and thus drive their competitors from the market?[130] Since the 1956 Consent Decree barred AT&T and its local Bell Company subsidiaries from offering data-processing services, these questions did not initially apply to the Bell System.

To assist it in analyzing these questions and their market implications, the FCC hired the Stanford Research Institute (SRI). In 1969, SRI delivered a five-volume report in which it recommended that regulated common carriers should be permitted to provide data services and that the data services should not be regulated. However, the scope of the report did not include suggestions concerning how the FCC and state regulators should distinguish between the unregulated data services and the regulated voice services.[131] That was left to the FCC.

Shortly after the SRI report, in 1970, the FCC opened the first of three investigative proceedings, known as the *First, Second,* and *Third Computer Inquiries,* or *Computer I, II,* and *III,* to further address the issues and options of voice and data services. The approach of the three proceedings was to define the relationships, possible efficiencies, and public benefits of voice and data services offered both separately and as combined services.[132] The results of these three proceedings have significantly impacted U.S. telecommunications policy and legislation since the early 1970s and formed the basis for competition in today's telecommunications environment.

1971—*Computer I*

In the *First Computer Inquiry,*[133] *or Computer I,* completed in 1970–1971, the FCC declared that regulated common carriers would be permitted to provide unregulated data processing services, but only through "*maximally*" separate subsidiaries from the carriers' voice services. Maximally separate subsidiaries must maintain separate officers, personnel, accounting books, credit, facilities, and equipment. In addition, the subsidiaries were required to treat affiliates and nonaffiliates equally, using nondiscriminatory tariffed rates and terms for their services. However, the regulated carriers could not buy, sell, or promote the data-processing services of their own affiliates,[134] nor make their own computer facilities available to other entities to create unregulated services. These *struc-*

tural separation requirements were intended to prevent (1) cross-subsidization of costs from the regulated company to the competitive subsidiary, (2) anticompetitive tying of products available only to the regulated companies, and (3) any misuse of the monopoly communications facilities to benefit only them. The FCC exempted the smaller independent telephone companies, those with combined operating annual revenues under $1 million dollars, from these restrictions. This exemption applied to over 300 independent companies.[135] The structural separation solution, of course, did not affect AT&T and its local Bell Company subsidiaries because they were barred from providing data services.

To distinguish between regulated *telecommunications services* and unregulated *data-processing services,* the FCC focused on the functions performed by computers in the two services. It decided that if the computer simply transmitted or *switched messages* without any manipulation of the content or use of the message, it was part of the carrier's traditional *message-switching* function and should be regulated under Title II of the Communications Act of 1934 covering Common Carrier Services.[136] On the other hand, if the computer were used for data-processing services, additional manipulation of the information occurred that put the process beyond simple *message switching.* In this case, the FCC decided that the service should not be regulated, even if the user's terminals were connected to central computers by common carrier lines and could not function without the carrier's facilities.[137]

This *definitional approach* used by the FCC established a distinction between telecommunications and data-processing services that still guides regulation today.[138] In addition, the FCC applied these two definitions to categorize the communications services that used computers into three broad groups: (1) message-switching services, (2) data-processing services, and (3) *hybrid services,* which included elements of both message-switched and data processing services. The FCC then answered the question of what should be regulated by stating that it would regulate the messaging services,[139] would not regulate the data-processing services,[140] and would decide the level of regulation or nonregulation of the hybrid services on a case-by-case basis. Its case-by-case decisions would depend on the predominance of telecommunications or data-processing attributes in each hybrid service.[141] In general, the FCC viewed the data-processing services as new services provided by unregulated data-processing companies. These companies operated in the highly competitive data-processing market, governed by market conditions and therefore did not require FCC regulation.[142] The FCC further stated that

> in a field as dynamic and innovative as this one, it is not possible to formulate a sufficient number of hypothetical situations which would provide meaningful guidelines. We are also troubled lest such formulation serve a purpose exactly contrary to that which is intended by inhibiting the ingenuity and responsiveness of the interested parties and causing them to limit or construct their services in accordance with hypothetical cases we have listed rather than the actual needs of users.[143]

In its Final Report and Order for *Computer I,* the FCC did not discuss the issue raised by some commentators that the FCC had no statutory authority to extend its common carrier regulatory jurisdiction to data-processing services. The FCC simply stated that, with its decision not to regulate data processing, it was neither "relevant nor necessary" to discuss the scope of its jurisdiction regarding data processing. However, as with the rest of *Computer I,* the Commission left the issue open by stating that "if there should develop significant changes in the structure of the data processing industry, or if abuses emerge[d] which [would] require the exercise of corrective action by the Commission, we shall not hesitate to re-examine the polices set forth herein."[144]

1973—*Computer I* Affirmed

In 1973, the FCC's rules resulting from its *Computer I* was largely affirmed on appeal in the Second Circuit Court.[145] The court reversed only two items, holding that (1) the regulated carriers could not be barred from purchasing data-processing services from their own unregulated affiliates, and (2) the FCC could not forbid the affiliates from using a telephone company's name or logo. The court reversed these two rules because it stated that the rules protected the data-processing industry, an industry that was "beyond [the FCC's] charge and which the Commission itself has announced it declines to regulate."[146]

1974—Antitrust Case Against the Bell System

In 1974, two years after the FCC completed *Computer I* and two years before it began *Computer II,* the Department of Justice initiated the antitrust case against the Bell System that was resolved eight years later in the MFJ of 1982. At the same time, while the FCC's *Computer I* ruling was upheld on appeal, significant changes in computer technology were occurring that made some of the terms and definitions in *Computer I* obsolete. These technological changes included the following: (1) the development of integrated circuits, (2) the trend toward *distributed processing* with remote terminals attached to large mainframe computers via telephone lines, and (3) the evolution of *smart customer premise equipment* including phones with computer intelligence built into them.[147] These changes made voice and data services so technologically intertwined that clear demarcations dissolved between them.[148] The changes also made the case-by-case determination of which hybrid services qualified as *unregulated data processing* rather than *regulated telecommunications (Title II) services* too time-consuming for the FCC.[149] In response, the FCC initiated its *Second Computer Inquiry* or *Computer II* in 1976.[150] It took four years to complete, with the Final Report and Order issued in 1980.

1976—*Computer II*

The purpose of *Computer II* was similar to that of *Computer I* in that it reviewed the voice and data services and the appropriate regulation or nonregulation of each. In *Computer II,* the FCC reconfirmed that common carriers should be allowed to provide data services and that those data services should not be regulated.[151] The FCC also required in *Computer II* that enhanced services be provided through separate subsidiaries. However, it focused on requiring separate subsidiaries only for AT&T as the dominant carrier, once AT&T was permitted to provide such services.[152] The FCC reasoned that the carriers other than AT&T would lack the national market power to cross-subsidize their voice and data services, would misuse transmission facilities to destroy their competitors, or negatively affect customers.[153] Hence, all other carriers would be permitted to provide such services directly through their normal channels, with no separate subsidiary requirement. In fact, the FCC's initial order in *Computer II* applied the separate subsidiary requirement to both AT&T and GTE. However, upon reconsideration, the restriction against GTE was removed because (1) GTE was dependent upon AT&T for the vast majority of its interstate transmission lines, and (2) only AT&T was considered large enough to significantly cross-subsidize or engage in anticompetitive conduct. Thus, *Computer II's* separate subsidiary requirement affected only AT&T.

Beyond the structural requirement for subsidiaries, and the definitional approach used in *Computer I,* the FCC focused more on a structural approach in *Computer II*. It distinguished between *basic services* and *enhanced services* and permitted carriers to provide both. Traditionally, the FCC had always described telephone services fairly broadly. However, in *Computer II,* it defined basic services very narrowly, limiting them mainly to traditional switching and transmission functions.[154] All other services were broadly defined as enhanced services and opened to the competitive marketplace.[155] The FCC also required all carriers to unbundle the basic components of their services from the enhanced and to offer the unbundled transmission capacity to other enhanced service providers on equal tariffed terms and conditions that they provided to themselves.[156] This had an enormous impact on the industry and still guides it today.

Specifically, the FCC defined basic services in *Computer II* as "common carrier offerings of transmission capacity for the movement of information," or "pure transmission capability over a communications path that is virtually transparent in terms of its interaction with customer supplied information."[157] The FCC noted that "data processing, computer memory or storage, and switching techniques can be components of a basic service if they are used solely to facilitate the movement of information."[158] "In a basic service, once information is given to the communications facility, its progress towards the destination is subject to only those delays caused by congestion within the network or transmission priorities given by the originator."[159] Basic services provide a "pure transmission capability over a communications path that is virtually transparent in terms of its interaction with customer-supplied information."[160]

In contrast, the FCC defined enhanced services as "services, offered over common carrier transmission facilities used in interstate communications, which employ computer processing applications that act on the format, content, code, protocol or similar aspects of the subscriber's transmitted information; or provide the subscriber with additional, different, or restructured information, or involve subscriber interaction with stored information."[161] This *Computer II* definition included both the data-processing services and hybrid forms of communications from *Computer I.*[162] In blending the two, the FCC acknowledged that "a substantial data processing component [exists] in all of these enhanced services, over which the agency had never imposed regulation."[163] The FCC further recognized that "some enhanced services are not dramatically dissimilar from basic services or dramatically different from communications as defined in the *Computer I,*" and the "plausible arguments can be tendered for drawing [the distinction] elsewhere."[164]

To ensure that the states did not regulate enhanced services and customer premises equipment, in a manner inconsistent with the federal regulators, the FCC invoked *ancillary jurisdiction* under Title I of the Communications Act of 1934. Section 152 of the 1934 Act gave the Commission jurisdiction over "all interstate and foreign communication by wire or radio," and Section 153 defined *communication by wire* as "the transmission of writing, signs, signals, pictures and sound of all kinds . . . incidental to such transmission."[165] Therefore, the FCC changed its position from *Computer I* by stating in *Computer II* that it had jurisdiction over enhanced services under the ancillary jurisdiction of Title I of the Communications Act of 1934, since enhanced services "constitute the electronic transmission of writing, signs, signals, pictures, etc., over the interstate telecommunications network."[166] The FCC also imposed certain separate subsidiary requirements where required to ensure wire communications services at reasonable rates.[167]

In response to the changes made in *Computer II* and the MFJ, AT&T formed American Bell Inc. in June 1982 as its enhanced service provider and AT&T Communications as its basic service provider. American Bell Inc.'s name was later changed to American Telephone and Telegraph Information Services (ATTIS). Initially ATTIS provided only enhanced services, but by January 1983 it also began selling new customer premises equipment.

1982—*Computer II* Affirmed

The FCC's *Computer II* decision was affirmed in its entirety two years later, in 1982, by a unanimous panel of the District of Columbia Circuit Court.[168] The same year, the Department of Justice and the Bell System signed the MFJ, completely changing the telecommunications industry structure. Although the FCC had moved away from definitional distinctions in *Computer II,* the MFJ used both the definitional distinctions between telecommunications and information services of *Computer I* and the structural distinctions between basic and enhanced services of *Computer II.* The MFJ opened the data market to AT&T, but prohibited the BOCs from providing information or interLATA services. Five years later, in 1987, however, the MFJ was changed to allow the Bell Operating

Companies to provide voice-messaging services and to transmit information services generated by others.[169] In 1991, the restriction on Bell Operating Company ownership of content-based information services was lifted.[170]

1986—*Computer III:* Reversed Definitional and Structural Approach

The FCC completed *Computer II* believing that (1) the regulatory distinctions between basic and enhanced services in *Computer II* were straightforward and definitive, and (2) the structural separation approach of requiring that certain activities of telephone companies be conducted through separate subsidiaries were fair, especially since both aspects of the Final Order had been affirmed by the court.

In the first few years following *Computer II,* however, AT&T requested a series of waivers from the requirements of the structural separation approach in *Computer II,* which revealed flaws with that approach.[171] For example, in 1981, AT&T requested a waiver permitting the Bell Operating Companies to provide certain *custom calling* services, such as call answering, advance calling, and remote access, directly to their customers instead of through a subsidiary.[172] The FCC refused the waiver because it believed the services could be provided through a subsidiary.[173] The delay in making this decision, however, withheld the introduction of these popular services from consumers for nearly a decade, negatively impacting the quantity and quality of services offered to the public.[174]

Another challenge to *Computer II* came with *code and protocol conversion.* By 1985, the FCC had eight protocol conversion petitions from the Bell Operating Companies to consider.[175] The FCC found that it could not keep up with the ad hoc nature of the waiver-making process and recognized that the "process is insufficient to provide the stability needed by users, carriers, value-added networks, and other enhanced service providers, whose business planning and decision-making are affected by our decisions."[176] Thus, the FCC announced a Notice of Proposed Rulemaking, called the *Third Computer Inquiry* or *Computer III,*[177] "to formulate general rules of future applicability to govern the treatment of protocol conversion and similar enhanced services."[178]

In *Computer III,* issued on June 16, 1986, the FCC acknowledged that the definitions and rules from *Computer I and II* were out of date and that the structural separation that each mandated decreased innovation and efficiency, resulting in higher costs to the public. As a result, the FCC reversed itself in *Computer III,* permitting telephone companies to provide enhanced services without separate subsidiaries.[179] However, the FCC also recognized that the telephone companies could behave in an anticompetitive manner, so it replaced the structural separation requirement with two nonstructural safeguards.[180] First, the FCC announced that it would develop cost allocation methods to minimize the Bell Operating Companies' ability to shift costs from their unregulated to regulated activities. Second, the FCC adopted regulations specifically designed to prevent the Bell Operating Companies from using their "substantial market power in providing network access" to discriminate against competing providers of enhanced services.

These antidiscrimination regulations contained three prongs. First, the FCC endorsed an open-network policy requiring the Bell Operating Companies to make the telephone networks as accessible to competitors as they are to the Bell Operating Companies themselves. Second, the FCC required each Bell Operating Company to notify its competitors in the enhanced services industry of changes in the network that may affect the provision of enhanced services so as to permit competitors to take advantage of the changes. Third, the FCC required each Bell Operating Company to provide its competitors with information about customer use of the telephone network so that the competitors may design their services to suit customer needs. [181]

The open-network policy in the first prong also contained two subcomponents: (1) Comparably Efficient Interconnection (CEI), requiring each Bell Operating Company to provide competitors with connections to the local exchange network that are equal to the connections available to the Bell Operating Company's own enhanced services; and (2) open network architecture, requiring each Bell Operating Company to incorporate CEI concepts into the overall design of its basic service network. It also imposed a customer proprietary network information requirement that if AT&T and the Bell Operating Companies had information about customer use of the basic network services, they must establish procedures to honor requests from customers that their information be withheld from enhanced services personnel and be released to other enhanced services vendors.

1991—*Computer III* Reversed

In 1990, the Ninth Circuit Court of Appeals reversed and remanded the *Computer III* decision, on the grounds that (1) the FCC's substitution of nonstructural safeguards for the federal structural separation requirement to which the Bell Operating Companies were subject was unlawfully arbitrary and capricious and (2) the FCC failed to carry its burden of showing that its preemption orders are necessary to avoid frustrating its regulatory goals.[182]

The FCC decided not to take the Ninth Circuit Court's decision to the Supreme Court, but instead, on remand, essentially reinstated the *Computer III* rules and increased it efforts to more clearly address the court's concerns.[183] As part of this, the FCC emphasized that the *Computer III* requirements for the Bell Operating Companies to (1) file reports concerning the installation, maintenance, and quality of basic services; (2) disclose network design information to enhanced service providers; and (3) continue to engage in open network architecture were valid and appropriate. It also argued that its four years of experience with cost-accounting rules was an effective safeguard to "protect ratepayers against cross-subsidization by the Bell Operating Companies, and is a realistic and reliable alternative to structural separation."[184] The FCC further argued that price cap regulation "serves as an effective complement to these cost accounting safeguards by reducing Bell Operating Companies incentives to cross-subsidize since carriers are not able automatically to recoup misallocated nonregulated costs by raising basic service rates." [185]

In 1994, the Ninth Circuit reviewed the FCC's approach and concluded that the measures taken to combat the carriers' incentive to cross-subsidize were adequate, but that since the new open network architecture rules had not been fully implemented, the FCC was still not justified in dropping its structural separation requirements.[186] The Court did approve, however, the FCC's position that it could preempt state rules that were inconsistent with the FCC's rules and that affected facilities and personnel used to provide the intrastate portion of jurisdictionally mixed enhanced services.

The issue of the Ninth Circuit's approval of the FCC's *Computer III* decision, however, became less important with the U.S. Congress's increased work on the Telecommunications Act of 1996.[187] While the 1996 Act retained the distinction between voice or *telecommunications* services and data or *information services,*[188] it also sought to move the telecommunications industry toward more competition and less regulation. The 1996 Act acknowledged that the technologies are converging and included options for future review of the definitions and their continued use.[189] Once enacted, the 1996 Act addressed most of the issues in *Computer III,* thus very few of the *Computer III* rules, even as modified by the Ninth Circuit, survive, except at the intraLATA level. Since intraLATA services generally are in-state, they are less affected by the 1996 Act. Thus, for example, a 1996 Report and Order stated that the pre-1996 Act *Computer III* rules still apply to the intraLATA provision of information services.[190] These rules require a *local exchange carrier* to either use a separate subsidiary to provide information services or file a comparably efficient interconnection report with the FCC. The details of the Telecommunications Act of 1996 are described in greater detail in the next chapter, Chapter 3.

CONCLUSION

The history of competition and regulation in the telecommunications industry has been a history of seeking the best method to achieve the goal of providing "rapid, efficient, nationwide and worldwide wire and radio communication service, with adequate facilities, at reasonable charges." We still seek that today.

While the goal for U.S. communications has not changed since 1934, the various phases, formats, solutions, and policies implemented to accomplish that goal have moved from competition to regulation, and now back to competition again. The methods used have provided low-cost, high-quality communications to most people in the U.S., and have adjusted with technological advances, but they have also created a history of reversals, confusing programs, and industry structural changes that are astounding. As we look ahead into the future, the ongoing mergers and acquisitions in the telecommunications industry seem to indicate that this process is far from finished.

ENDNOTES

[1]Communications Act of 1934, 47 U.S.C. § 151 (1934).

[2]Petroleum (Standard Oil Co.), railroad (J.J. Hill), steel (United States Steel Corporation), cotton (American Cotton Trust), sugar (Sugar Trust) and whiskey (Whiskey Trust).

[3]*See* M. Horwitz, The Transformation of American Law: 1780-1860, 109-39 (1977); Jones, *Historical Development of the Law of Business Competition, 36 Yale L.J. 207 (1926); May, Antitrust Practice and Procedure in the Formative Era: The Constitutional and Conceptual Reach of State Antitrust Law, 1880-1918,* 135 U. Pa. L. Rev. 495 (1987).

[4]49 U.S.C. § 10101 *et seq.*, enacted on February 4, 1887.

[5]15 U.S.C. §§ 1-7.

[6]State of Ohio v. Standard Oil, Ohio Sup. Ct., 49 Ohio St. 137, 30 N.E. 279 (1892).

[7]Some states call their commission by a name other than Public Utilities Commission (PUC). A current list of the state commissions, under their appropriate names, is available at *www.naruc.org.*

[8]John Brooks, Telephone: The First Hundred Years, 144 Harper & Row, New York (1975).

[9]Pub. L. No. 61-218, 36 Stat. 539 (1910). Commerce Court (Mann-Elkins) Act, ch. 309, §7, 36 Stat. 539 (1910). *See also* Essential Communications v. AT&T, 610 F.2d 1114 (3d Cir. 1979) (AT&T's obligation "to provide services at just and reasonable rates, without unjust discrimination or undue preference.")

[10]15 U.S.C. §§ 12-27.

[11]15 U.S.C. §§ 41-45.

[12]Willis-Graham Act of 1921,42 Stat. 27 (1921), current version at 47 U.S.C. § 221(a) (1976).

[13]*See* Posner, *Natural Monopoly and Its Regulation,* 21 Stan. L. Rev. 548 (1969).

[14]Telecommunications in Transition: The Status of Competition In the Telecommunications Industry, Report of the Majority Staff of the Subcommittee on Telecommunications, Consumer Protection, and Finance of the Committee on Energy and Commerce, U.S. House of Representatives (Nov. 3, 1981), at 1; and U.S. Dept. of Commerce, National Telecommunications and Information Administration (NTIA) Alternatives Report 8 (1987) [hereinafter NTIA Alternatives Report].

[15]15 U.S.C. § 13.

[16]15 U.S.C. §§ 18, 21.

[17]After the invention of the telephone, senders of telegrams often phoned the messages to the telegraph office to reach persons without phones.

[18]Western Electric Company (1976).

[19]His assistant, Watson, was only 21 years old.

[20]"Telephone Cases," 126 U.S. 1 and 7,8 (1888).

[21]R.F. Rey, Engineering and Operations in the Bell System, AT&T Bell Laboratories, Murray Hill, N.J., 1983, at 700. *See also,* Brooks, *supra* note 8, at 61.

[22]Brooks, *supra* note 8, at 69.

[23]George A. Codding, Jr., The International Telecommunications Union: An Experiment in International Cooperation, Ayer Press (1977).

[24]Rey, *supra* note 21, at 62.

[25]*Id.*

[26]Telephone Cases, *supra* note 20, at 1.

[27]Overseas Communications Services, 84 F.C.C.2d 622, 623 (1980).

[28]Telephone Cases, *supra* note 20, at 2c. *See also,* "Overseas Communications Services, *supra* note 27; Goldberg, *One-Hundred and Twenty Years of Intl. Communications,* 37 Fed. Comm. L.J. 131 (1985); FCC, Common Carrier Bureau, An Overview of In-

ternational Telecommunications: Industry Structure and Commission Policies 2 (July 1977) [hereinafter FCC Intl. Report].

[29]FCC Intl. Report *supra* note 28, at 2; Goldberg, *supra* note 28, at 133-134. (This technological development reopened the potential for merger of international voice and data (record carriage).)

[30]Rey, *supra* note 21, at 700.

[31]Warren Lavey, *The Public Policies that Changed the Telephone Industries into Regulated Monopolies: Lessons from 1915,* 39. Fed. Comm. L.J., 171, 178 (1987).

[32]*Id.* at 179 (could not compete with Bell in Bell-served areas).

[33]Brooks, *supra* note 8, at 135 (citing the AT&T Annual Report of 1910).

[34]*Id.*

[35]*Id.*

[36]Letter sent on December 19, 1913 by an AT&T Vice President Nathan C. Kingsbury to James C. McReynolds, U.S. Attorney General, Department of Justice [hereinafter Kingsbury Committment].

[37]Brooks, *supra* note 8, at 136.

[38]*Id.* at 160.

[39]15 U.S.C. §§ 12-27.

[40]15 U.S.C. §§ 41-45.

[41]*See* Milton Mueller, Universal Service, MIT Press (1997) for an excellent discussion of this topic.

[42]Rey, *supra* note 21, at 3.

[43]Brooks, *supra* note 8, at 144.

[44]*Id.*

[45]*Id.* at 139.

[46]Telecommunications in Transition: The Status of Competition In the Telecommunications Industry, Report of the Majority Staff of the Subcommittee on Telecommunications, Consumer Protection, and Finance of the Committee on Energy and Commerce, U.S. House of Representatives, 1, (Nov. 3, 1981), and NTIA Alternatives Report, *supra* note 14.

[47]Brooks, *supra* note 8, at 144.

[48]*Id.* at 143.

[49]Extension of the Telephone System in the District of Columbia, H.R. Rep. No. 379, 65th Cong., 2d Sess. (1918). (Efficient provision of service); and Lavey, *supra* note 31, at 172.

[50]*See* Northwestern Bell Tel. Co., 71 Pub. Util. Rep. (PUR) 1, 11 (SoD. Pub. Utils. Comm'n. 1947); and Lavey, *supra* note 31, at 174.

[51]Pacific Telephone and Telegraph Co., 1919D Pub. Util Rep. (PUR) 345, 370-371 (Ore. Pub. Serv. Comm'n 1919).

[52]Smith v. Illinois Bell Telephone Co., 282 U.S. 133, 51 S. Ct. 65, 75 L.Ed. 255 (1930).

[53]Lavey, *supra* note 31.

[54]7 U.S.C. § 921 (1982); Lavey, *supra* note 31, at 174.

[55]H.R. Rep. No. 246, 81st Cong., 1st Sess. 8 (1949). Lavey, *supra* note 31, at 174.

[56]Willis-Graham Act of 1921, 42 Stat. 27 (1921), current version at 47 U.S.C. § 221(a) (1976).

[57]Brooks, *supra* note 8, at 160.

[58]Communications Act of 1934, 43 Stat. 1064 (1934) current version at 47 U.S.C. §§ 151 *et. seq*. (1976).

[59]The titles of the 1996 Act include: Title I (Telecommunications Services); Title II (Broadcast Services); Title III (Cable Services); Title IV (Regulatory Reform); Title V (Obscenity and Violence); Title VI (Effect on Other Laws); and Title VI (Miscellaneous Provisions).

[60]15 U.S.C. § 13.

[61]United States v. Western Elec. Co., (1949).

[62]United States v. Western Elec. Co., 1956 Trade Cas. (CCH) 68,246 (D.N.J. 1956).

[63]However, as an exception, Western Electric was allowed to continue to make equipment for the certain government projects and in cases where no competitive products existed. For example, the Bell System continued to provide the "artificial larynx" for disabled persons because no competitive commercial supplier would undertake the task.

[64]Rey, *supra* note 21, at 695.

[65]*Id.* at 694.

[66]United States v. AT&T Western Elec. Co., 552 F. Supp. 131, 226-32 (D.D.C. 1982), *aff'd sub nom.* Maryland v. United States, 460 U.S. 1001 (1983); United States v. Western Elec. Co., 569 F. Supp. 1057 (D.D.C. 1983) (Plan of Reorganization), *aff'd sub nom.* California v. United States, 464 U.S. 1013 (1983); *see also* United States v. Western Elec. Co.,

No. 82-0192 (D.D.C. Apr. 11, 1996) (vacating the MFJ).

[67]*See* United States v. Western Elec. Co., 673 F. Supp. 525 (D.D.C. 1987), United States v. Western Elec. Co., 714 F. Supp. 1 (D.D.C. 1988).

[68]United States v. Western Elec. Co., 767 F. Supp. 308 (D.D.C. 1991), *stay vacated,* United States v. Western Elec. Co., 1991-1 Trade Cases (CCH) ¶69,610 (D.C.Cir. 1991).

[69]Glen O. Robinson, *The Titanic Remembered: AT&T and the Changing World of Telecommunications,"* 5 Yale J. Reg. 517, 520-32 (1988) (book review).

[70]Rey, *supra* note 21, at 693.

[71]Hush-A-Phone Corp. v. FCC, 238 F.2d 266 (D.C. Cir. 1956) (AT&T tariff prohibiting attachment of passive device to telephone handset held invalid as unreasonable restriction on customer telephone use.)

[72]*In re* Use of the Carterfone Device in Message Toll Telephone Service, 13 F.C.C.2d 420 (1968), 1968 FCC LEXIS 1269 (1968).

[73]Rey, *supra* note 21, at 464.

[74]Weisman, *Default Capacity Tariffs: Smoothing the Transitional Regulatory Asymmetries in the Telecommunications Market,* 5 Yale J. on Reg. 149 (1988) "Years of monopoly control resulted in revenues far above actual costs especially in long-distance services."

[75]Rey, *supra* note 21, at 693.

[76]*Id.*

[77]Allocation of Microwave Frequencies Above 890 Mhz, 27 F.C.C. 359 (1959), *aff'd on rehearing* 29 F.C.C. 825 (1960).

[78]Microwave Communications Inc., 18 F.C.C. 2d 953 (1969).

[79]Specialized Common Carriers, Notice of Inquiry, 24 F.C.C.2d 318 (1970); Specialized Common Carriers, First Report and Order, 29 F.C.C.2d 870 (1971), *aff'd sub nom.* Washington Utilities & Transportation Commission v. FCC, 513 F.2d 1142 (9th Cir. 1975), *cert. denied* 423 U.S. 836 (1975).

[80]Rey, *supra* note 21, at 693.

[81]MCI Telecommunictions Corp. v. FCC [*Execunet* I], 561 F.2d 365, 365 (D.C. Cir. 1977), *cert. denied,* 434 U.S. 1040 (1978).

[82]Newton's Telecom Dictionary states that MCI's "*Execunet*" service was named by MCI Vice President Karl Vorder-Bruegge, and made successful in Dallas, Texas by Gerry Taylor using a 104-port Action WATSBOX.

[83]*Dial-up long distance* is also known as *switched intercity service.*

[84]FCC letter to MCI dated July 2, 1975.

[85]MCI Telecommunications Corp., 60 F.C.C.2d 25 (1976).

[86]Message Toll Service (MTS) is a generic name for *pay-by-the-minute* dial-up (switched), long distance telephone service. It includes both conventional long distance, also called Direct Distance Dial (DDD) and measured WATS. Other, more infrequent definitions for MTS are Message Telephone Service or Message Telecommunications Service.

[87]Wide Area Telecommunications Service (WATS) is a volume-discounted, long-distance telephone service provided by both long-distance and local telephone companies, which allows companies and government agencies with high volumes of telephone calls to pay a bulk-price for telephone service. With WATS, customers pay a flat monthly rate, plus a usage rate. As the customer makes more minutes of calls, the cost per minute decreases through various tiered-prices. WATS is offered in three ways: (1) call direction (in-bound calls or out-bound calls from the customer's location); (2) coverage area (bands); or (3) geographic type of call (such as for intrastate, interstate, or international calls). Sometimes WATS has also been defined as Wide Area Transmission Service or Wide Area Transport Service, but these are less common.

[88]MCI Telecommunications Corp., 561 F.2d 365 (*supra* note 81); *See also* MCI Telecommunications Corp. v. FCC, 580 F.2d 590 (D.C. Cir.) (*Execunet* II), *cert. denied,* 439 U.S. 980 (1978).

[89]FCC Intl. Report, *supra* note 28, at 2; Goldberg, *supra* note 28, at 133-134.

[90]*See* Intl. Record Carriers Scope of Operation, 58 F.C.C.2d 250 (1976) (addition of Miami and New Orleans as gateway cities); and 76 F.C.C.2d 115 (1979) (addition of 21 gateway cities and discussion of history of gateway cities including FCC regulation).

[91]*See* Note, United States Regulation of Intl. Record Telecommunications, 3 B.U. Intl. L.J. 99, 104-105 (1985).

[92]Goldberg, *supra* note 89, at 137.

[93]*Id.* at 138; Overseas Communications Services, *supra* note 27, at 624.

[94]*Id.*

[95]AT&T, No. P-C-3630, Mimeo No. 22818 (FCC Sept. 7, 1955), *recon. denied,* 44 F.C.C. 602 (1955).

[96]*Dataphone,* 38 F.C.C. 1222 (1965).

[97]AT&T, 27 F.C.C. 113, 120, 121 (1959); *See also* Goldberg, *supra* note 28, at 139.

[98]AT&T, 28 F.C.C. 221 (1960) (Puerto Rico, considered an international point); AT&T, Nos. P-C-4612 and S-C-L-18, Mimeo Nos. 91693 and 91694 (FCC July 27, 1960).

[99]AT&T, 37 F.C.C. 1151, 1159 (1964).

[100]*Id.* at 1159-1162.

[101]WU-IRC Interconnection Arrangements, 89 F.C.C.2d §§ at 207 (1982).

[102]Record Carrier Competition Act of 1981, Pub. L. No. 97-130, §2, 95 Stat. 1687 (1981), codified 47 U.S.C. § 222(e) (1981).

[103]47 U.S.C. § 222(d) (1981).

[104]47 U.S.C. § 222(c) *See also* AT&T, 6 FCC Rcd. 115 at ¶ 20 (1990) ("sunset" of some interconnection provisions).

[105]Industry Analysis Division, Common Carrier Bureau, Federal Communications Commission, Trends in the International Communications Industry, at 35-38 (March 1994).

[106]Goldberg, *supra* note 28, at 141.

[107]Rodriquez, *International Telecommunications and Satellite Systems Intelsat and Separate Systems: Cold War Revisited,* 15 Int'l. Bus. Law 321 (1987).

[108]Communications Satellite Act of 1962, Pub. L. No. 87-624, §102, 6 Stat. 419 (1962) *codified as amended at 47* U.S.C. §§701-57 (1988). *See* FCC Intl. Report *supra* note 28, at 12.

[109]*See* Snow, *Competition by Private Carriers in International Commercial Satellite Traffic: Conceptual and Historical Background,* in Tracing New Orbits 33, 34 (D.A. Demac ed. 1986).

[110]Ownership and Operation of Initial Earth Stations in the United States, 5 F.C.C.2d 812, 819 (1966).

[111]47 U.S.C. § 301.

[112]Agreement Establishing Interim Arrangements for a Global Commercial Communications Satellite System, Aug. 20, 1964, 15 U.S.T. 1745, T.I.A.S. No. 5646.

[113]*Id.*

[114]Comsat Study—Implementation of Section 505 of the International Maritime Stellite Telecommunications Act, Final Report and Order, 77 F.C.C.2d 564, 595 (1980).

[115]Communications Satellite Corp. Decision in Docket 16070.

[116]*Id.*

[117]ITT Cable & Radio, Inc.—Puerto Rico, 5 F.C.C.2d 812, 832 (1966).

[118]International Record Carriers, Scope of Operations in the Continental United States, 76 F.C.C.2d 115 at Para. 95 (1979).

[119]Goldberg, *supra* note 28, at 141, 143, FCC Intl. Report, *supra* note 28, at 17-18.

[120]Aeronautical Radio, 77 F.C.C.2d 535, 552–555 (1980); Authorized User Policy, 90 F.C.C.2d 1394, 1424 (1982).

[121]Establishment of Domestic Communications Satellite Facilities by Nongovernmental Entities, Supplemental Notice of Inquiry in Docket 16495, 5 F.C.C.2d 354 (1966).

[122]Establishment of Domestic Communications Satellite Facilities by Nongovernmental Entities, Report and Order in Docket 16495, 22 F.C.C.2d 86, 88-90 (1970).

[123]Establishment of Domestic Communications Satellite Facilities by Nongovernmental Entities, Second Report and Order in Docket 16495, 35 F.C.C.2d 844, 850 (1972).

[124]Network v. FCC, 511 F.2d 786, 790-791 (D.C. Cir. 1975).

[125]Assignment of Orbital Locations to Space Stations in the Domestic Fixed Satellite Service, Memorandum Opinion and Order, 84 F.C.C. 2d 584, 587 (1981) (Orbital Locations).

[126]Agreement on the International Telecommunications Satellite Organization (INTELSAT), Aug. 20, 1971, entered into force on Feb. 12, 1973, T.I.A.S. No 7532, 23 U.S.T. 3813, Art. XV(b).

[127]Orbit Act of 2000, Pub. L. No. 106-180, (2000). *See also* 47 U.S.C. § 763a(1) (2000).

[128]These computers took up whole air-conditioned rooms, were attended by large staffs, and operated

on Grosch's Law: the efficiency of a computer increased with its size.

[129]Regulatory Policy Problems Presented by the Interdependence of Computer and Communication Facilities, Final Decision and Order, 28 F.C.C.2d 267, 269 (1971) [hereinafter *Computer I, Final Decision and Order*], *aff'd sub nom.* GTE Service Corp. v. FCC, 474 F.2d 724 (2nd Cir. 1973), *decision on remand,* 40 F.C.C. 2d 293 (1973). *See also, The FCC Computer Inquiry: Interfaces of Competitive and Regulated Markets,* 71 Mich. L. Rev. 172, 173-4 (1972); Irwin, *The Computer Utility: Competition or Regulation* 76 Yale L.J. 1299 (1967); Irwin, *Computers and Communications: The Economics of Interdependence,* 34 Law & Contemp. Probs. 360, (1969).

[130]Frieden, *The Computer Inquiries: Mapping the Communications/Information Processing Terrain,* 33 Fed. Comm. L.J. 55 (1981).

[131]Dunn, Stanford Research Institute Report Prepared for the FCC (1969). *See also,* Dunn, *Policy Issues Presented by the Interdependence of Computer and Communications Services,* 34 Law & Contemp. Prob. 369 (1969).

[132]Rey, *supra* note 21, at 695.

[133]*Computer I, Final Decision and Order supra* note 129.

[134]*Id.* at 269-270 (quoting *Computer I,* Tentative Decision, 28 F.C.C.2d at 302). *See* Note, FCC Review of Regulation Relating to Provision of Data Processing Services by Communications Common Carriers, 15 B.C. Indus. & Com. L.Rev. 162 (1973).

[135]*Id.* at 275.

[136]*Id.* at 267, 287.

[137]*Id.*

[138]*See* 47 C.F.R. § 64.702.

[139]Pure message-switched or telecommunications services include services such as: Message Telecommunications Service (MTS), Wide Area Telecommunications Service (WATS), video transmission, and private line services. All would be regulated by the FCC.

[140]Pure data processing services, including word processing and data base services would be unregulated.

[141]Hybrid services, including remote access data processing, and automated information services such as directory services.

[142]*Computer I, Final Decision and Order, supra* note 129, at 275-276.

[143]*Id.* at 278-279.

[144]*Id.* at 268.

[145]GTE Service Corp. v. FCC, 474 F.2d (*supra* note 129), at 731.

[146]*Id.* at 733.

[147]In 1977, the FCC ruled in its *Dataspeed 40/4* Order that AT&T could offer a data terminal under tariff, but stated that it would continue to review the decision as it proceeded through its *ComputerII.* See AT&T Revisions to Tariffs FCC No. 269 and 267 Relating to Dataspeed 40/4, 62 F.C.C. 2d 21 (1977).

[148]Amendment of Section 64.702 of the Commission's Rules and Regulations, Second Computer Inquiry, 77 FCC 2d 384, 394 [*hereinafter Computer II, Final Decision*], *modified on recon.,* 84 FCC 2d 50 (1980) (Computer II, Reconsideration Order), *further modified on recon.,* 88 FCC 2d 512 (1981), (Computer II, Further Reconsideration Order)*, aff'd sub nom., Computer and Communications Industry Ass'n v. FCC,* 693 F.2d 198 (D.C. Cir. 1982), cert. denied, 461 U.S. 9389 (1983); Aff'd on second further recon., FCC 84-190 (released May 4, 1984).

[149]Frieden, *supra* note 130, at The Effect of the *Second Computer Inquiry on Telecommunications and Data Processing,* 27 Wayne L.Rev. 1537 (1981); *Comment, Interdependence of Communications and Data Processing: An Alternative Proposal for the Second Computer Inquiry,* 73 Nw.U. L.J. 307, 308 (1978).

[150]*Computer II, Final Decision,* 77 FCC 2d (*supra* note 148).

[151]*Id.* at 435. "Any agency regulatory decision in this area must assess the merits—as we do in this order—of extending regulation to an activity simply because a part of it is subject to the agency's jurisdiction where such regulation would not be necessary to protect or promote some overall statutory purpose."

[152]*Id.*

[153]Barbara Esbin, Internet over Cable: Defining the Future in Terms of the Past, FCC Office of Plans and Policy (OPP), Working Paper No. 30 (August 1998) at 31.

[154]*Computer II, Final Decision,* supra note 148, at 428 and 430-435.

[155]*Id.* at 387.

[156]*Id.* at 475.

[157]*Id.* at 419-20.

[158]*Id.* Such basic services include abbreviated dialing, directory assistance, billing, and voice encryption services. *See also,* Esbin, *supra* note 153, at 29.

[159]*Computer II Final Decision, supra* note 148.

[160]*Id.*

[161]*Id.* at 387 and 47 U.S.C. § 64.702(a) *supra* note 138.

[162]Such services include voice store and forward services that emulate telephone answering machines, and mass calling services including dial-up stock quotations or sports services.

[163]*Computer II Final Decision, supra* note 148, at 435.

[164]*Id.* at 434.

[165]Computers and Communications Industry Assn. v. FCC, 693 F.2d at 205; and Comment, Storming the AT&T Fortress: Can the FCC Deregulate Competitive Common Carrier Services?, 32 Fed. Comm. L.J. 205, 217-219 (1980).

[166]*Computer II Final Decision, supra* note 148, at 435. The FCC also stated that it tried to distinguish between the two services "in a manner which distinguishes wholly traditional common carrier activities, regulatable under Title II of the Act, from (historically and functionally) competitive activities not congruent with the Act's traditional forms," in recognition of the policy "that substance not form govern the treatment of services within the Act's reach." "We have acted upon that belief by applying traditional Title II regulatory mechanisms to basic services and applying no direct regulatory mechanism for enhanced service.

[167]*Id.*

[168]Computer and Communications Industry Assn. v. FCC, 693 F.2d 198, 203 (D.C. Cir. 1982).

[169]*See* United States v. Western Elec. Co., 673 F. Supp. 525 (*supra* note 67), United States v. Western Elec. Co., 714 F. Supp. 1 (*supra* note 67).

[170]United States v. Western Elec. Co., 767 F. Supp. 308 (*supra* note 68).

[171]*Computer II Final Decision, supra* note 148, at 420-21.

[172]AT&T Petition for Waiver of §64.702 of the Commission's Rules and Regulations, 88 F.C.C.2d 1 (1981).

[173]*Id.* at 16.

[174]*See* Kellogg, p. 551. AT&T Petition for Waiver, 88 F.C.C. 2d at 4.

[175]Petitions for Waiver of §64.702 of the Commission's Rules, Memorandum Opinion and Order, 100 F.C.C.2d 1057 (1985).

[176]*Id.* at 1061-62.

[177]*Amendment of Section 64.702 of the Commission's Rules and Regulations (Third Computer Inquiry), Phase I, Report and Order,* 104 FCC 2d 958 (1986) (*Computer III Phase I Order*), *modified on recon.,* 2 FCC Rcd 3035 (1987), *further modified on recon.,* 3 FCC Rcd 1135 (1988), *second futher recon.,* 4 FCC Rcd. 5927 (1989), *Phase II, Peport and Order,* 2 FCC Rcd. 3072 (1987) (*Computer III Phase II Order*), *modified on recon.,* 3 FCC Rcd. 1150 (1988), *further modif. on recon.,* 4 FCC Rcd. 5927 (1989), rev'd in part sub nom., *California v. FCC,* 905 F.2d 1217 (9th Cir. 1990), *on remand,* 6 FCC Rcd. 7571 (1991), *vacated in part and remanded, California v. FCC,* 39 F.3d 919 (9th Cir. 1994). *See also Computer III Further Remand Proceedings: Bell Operating Company Provision of Enhanced Services, Notice of Proposed Rulemaking,* CC Docket No. 95-20, 10 FCC Rcd. 8360 (1995).

[178]Petitions for Waiver of §64.702 of the Commission's Rules, Memorandum Opinion and Order, 100 F.C.C.2d 1057, (*supra* note 175) at 1062.

[179]The PC (microcomputer) became an increasingly important desktop product and part of users' telecommunications equipment or customer premise equipment (CPE).

[180]California v. FCC [*Computer III*], 905 F.2d 1217, 1222 (9th Cir. 1990). On remand, 6 FCC Rcd.

7571 (1991), vacated in part and remanded, California v. FCC, 39 F.3d 919 (9th Cir. 1994). *See also, Computer III* Further Remand Proceedings: Bell Operating Company Provision of Enhanced Services, Notice of Proposed Rulemaking, CC Docket No. 95-20, 10 FCC Rcd. 8360 (1995).

[181]*Id.*

[182]*Id.*

[183]*See Computer III* Remond Proceedings, 57 Fed. Reg. 4.373 (1992).

[184]*Id.*

[185]*Id.*

[186]California v. FCC, 39 F.3d 919 (9th Cir. 1994) (*supra* note 177).

[187]See Implementation of the Telecommunications Act of 1996: Telecommunications Carriers' Use of Customer Proprietary Network Information and Other Customer Information, Implementation of the Non-Accounting Safeguards of Section 271 and 272 of the Communications Act of 1934, as Amended, CC Docket No. 96-115, CC Docket No 96-149, Second Report and Order and Further Notice of Proposed Rulemaking, FCC 98-27 (released Feb. 28, 1998) ("Use of CPNI") at ¶ 46 (summarizing Commission precedent as indicating that telecommunications services and information services are "separate, non-overlapping categories, so that information services do not constitute "telecommunications" within the meaning of the 1996 Act".)

[188]See H.R. Rep. No. 204, Part I, 104th Cong. 1st Sess. 125 (1995).

[189]H.R. Conf. Rep. No. 458, 104th Cong. 2d Sess. 116 (1996).

[190]First Report and Order, Further Notice of Proposed Rulemaking, In the Matter of Implementation of the Telecommunications Act of 1996, Telemessaging, Electronic Publishing, and Alarm Monitoring Services, CC Docket No. 96-152, 1996 LEXIS 7126.

CHAPTER 3 The Telecommunications Act of 1996

". . . to promote competition and reduce regulation in order to secure lower prices and higher quality services for American telecommunications consumers and encourage the rapid deployment of new telecommunications technologies."

Purpose of the Telecommunications Act of 1996.[1, 2]

THE TELECOMMUNICATIONS ACT OF 1996, PASSED BY CONGRESS ON February 1, 1996,[3] and signed into law on February 8, 1996 by President William J. Clinton, was the most extensive change in U.S. communications law in the 62 years since the Communications Act of 1934. It altered the structure and operation of the industry in at least four significant ways. First, it opened the local telephone markets to competition, completely reversing the historical concept of a *natural monopoly* in telephones. As such, it removed previous barriers to entry for multiple market participants.

Second, it removed divisions between technologies. For example, cable television (CATV) companies and electric and gas utilities can now provide telephone service. It also allowed other equipment manufacturers and entertainment companies to provide new services that are redefining what "basic" communications is and how it is delivered to customers. These changes, in turn, are redefining how people communicate.

Third, it affected all sectors of the telecommunications industry. Unlike the Kingsbury Commitment, the 1956 Decree and the 1982 Modified Final Judgment (MFJ),[4] which affected only AT&T and its subsidiaries in the Bell system, the 1996 Act affects all sectors of the telecommunications industry. This includes existing local telephone service providers; independent providers such as GTE, United, and Continental; small, independent telephone companies; long-distance carriers; equipment manufacturers; and new entrants to the market.

Fourth, it removed the restrictions imposed by prior consent decrees, including the 1956 Decree and the 1982 MFJ, that are inconsistent with the 1996 Act.[5] This means that the 1996 Act supersedes most of the provisions of these historical agreements with the new obligations of the 1996 Act.

Considering the breadth and depth of these changes, the 1996 Act will direct how the communications industry will be organized and will function for many years to come. Understanding the Act is critical when evaluating the opportunities available to companies in the new competitive telecommunications environment and the restrictions that define their market strategies. Therefore, the purpose of this chapter is to describe (1) the intent, general structure, and organization of the 1996 Act; (2) the new definitions and provisions it provides; (3) the overall duties and responsibilities it imposes on the various segments of the industry; and (4) challenges to the Act.

To accomplish this, Section 3.1 describes the purpose of the 1996 Act, while Section 3.2 outlines the Act's general structure and organization. Section 3.3 provides several key definitions, and Section 3.4 discusses several key provisions to open competition in the local telecommunications market. Section 3.5 describes the duties and obligations of the various categories of carriers identified in the 1996 Act, while Section 3.6 discusses four constitutional challenges to certain of these duties and obligations.

3.1 PURPOSE OF THE TELECOMMUNICATIONS ACT OF 1996

In general, the 1996 Act is part of an ongoing effort to protect the public from the potential negative domination of large, powerful companies and to encourage innovation. In the decade following the breakup of the Bell System in 1984,[6] the U.S. Congress and state legislatures noted that three positive results had occurred from competition in long distance. First, long-distance prices dropped significantly from nearly $3.00 per minute to less than 10¢ per minute. Second, new technology was introduced to consumers more quickly than had occurred previously; and third, these changes resulted in more equipment and service options for consumers. The state and federal legislators hoped to (1) obtain the same three results for consumers with competition in the local market, (2) bring communications law more into line with the technological developments that integrated voice and data services, and (3) thus place the U.S. communications industry in a better position for the Internet Age. To accomplish these goals, the Act introduced competition in the local exchange market.[7]

3.1.1 Federal Action

In detailing its objectives, Congress declared in an early draft of Senate Bill S. 652, printed on June 23, 1995,[8] that

(1) Competition, not regulation, is the best way to spur innovation and the development of new services. A competitive market place is the most efficient way to lower prices and increase value for consumers. In furthering the principle of open and full competition in all telecommunications markets, however, it must be recognized that some markets are more open than others.

(2) Local telephone service is predominantly a monopoly service. Although business customers in metropolitan areas may have alternative providers for exchange access service, consumers do not have a choice of local telephone service. Some States have begun to open local telephone markets to competition. A national policy framework is needed to accelerate the process.

(3) Because of their monopoly status, local telephone companies and the Bell operating companies have been prevented from competing in certain markets. It is time to eliminate these restrictions. Nonetheless, transition rules designed to open monopoly markets to competition must be in place before certain restrictions are lifted.

(4) Transition rules must be truly transitional, not protectionism for certain industry segments or artificial impediments to increased competition in all markets. Where possible, transition rules should create investment incentives through increased competition. Regulatory safeguards should be adopted only where competitive conditions would not prevent anticompetitive behavior.

(5) More competitive American telecommunications markets will promote United States technological advances, domestic job and investment opportunities, national competitiveness, sustained economic development, and improved quality of American life more effectively than regulation.

(6) Congress should establish clear statutory guidelines, standards, and time-frames to facilitate more effective communications competition and, by so doing, will reduce business and customer uncertainty, lessen regulatory processes, court appeals, and litigation, and thus encourage the business community to focus more on competing in the domestic and international communications marketplace.

(7) Where competitive markets are demonstrably inadequate to safeguard important public policy goals, such as the continued universal availability of telecommunications services at reasonable and affordable prices, particularly in rural America, Congress should establish workable regulatory procedures to advance those goals, provided that in any proceeding undertaken to ensure universal availability, regulators shall seek to choose the most procompetitive and least burdensome alternative.

(8) Competitive communications markets, safeguarded by effective Federal and State antitrust enforcement, and strong economic growth in the United States which such markets will foster are the most effective means of assuring that all segments of the American public command access to advanced telecommunications technologies.

(9) Achieving full and fair competition requires strict parity of marketplace opportunities and responsibilities on the part of incumbent telecommunications service providers as well as new entrants into the telecommunications marketplace, provided that any responsibilities placed on providers should be the minimum required to advance a clearly defined public policy goal.

(10) Congress should not cede its constitutional responsibility regarding interstate and foreign commerce in communications to the Judiciary through the establishment of procedures which will encourage or necessitate judicial interpretation or intervention into the communications marketplace.

(11) Ensuring that all Americans, regardless of where they may work, live, or visit, ultimately have comparable access to the full benefits of competitive communications markets requires Federal and State authorities to work together affirmatively to minimize and remove unnecessary institutional and regulatory barriers to new entry and competition.

(12) Effectively competitive communications markets will ensure customers the widest possible choice of services and equipment, tailored to individual desires and needs, and at prices they are willing to pay.

(13) Investment in and deployment of existing and future advanced, multipurpose technologies will best be fostered by minimizing government limitations on the commercial use of those technologies.

(14) The efficient development of competitive United States communications markets will be furthered by policies which aim at ensuring reciprocal opening of international investment opportunities.

These Congressional "findings" translated into the following explanations of Congress's goal for the 1996 Act:

To provide for a national policy framework designed to accelerate rapidly private sector deployment of advanced telecommunications and information technologies and services to all Americans by opening all telecommunications markets to competition, and for other purposes.

. . . to establish a national policy framework designed to accelerate rapidly the private sector deployment of advanced telecommunications and information technologies and services to all Americans by opening all telecommunications markets to competition, and to meet the following goals:

(1) To promote and encourage advanced telecommunications networks, capable of enabling users to originate and receive affordable, high-quality voice, data, image, graphic, and video telecommunications services.

(2) To improve international competitiveness markedly.

(3) To spur economic growth, create jobs, and increase productivity.

(4) To deliver a better quality of life through the preservation and advancement of universal service to allow the more efficient delivery of educational, health care, and other social services.

3.1.2 States' Action

While Congress was outlining the goals and directives of the federal Telecommunications Act of 1996, approximately two-thirds of the states' legislatures changed their state laws in 1995 to open their local markets to competition. Most of these laws allowed one year for the state public utilities commission (PUC), representatives of industry, public interest groups, and legislative committees to write the rules and regulations implementing the law. This meant that most of the states required local telephone competition to be in place and operating in their states before the federal Telecommunications Act of 1996 was enacted. This was difficult because the precise format, details, and dates of the federal implementation were unknown. For the states, this required meeting state-imposed statutory deadlines, while still tracking the federal activity to avoid as much inconsistency with the federal law as possible. It was also a time of cooperation within the states as the representatives of industry, public interest groups, staffs of the PUCs, and legislative committees met to draft the implementing rules in each state. In addition, the states often consulted with the staffs of other states to trade ideas, discuss processes, and determine the best approaches and solutions to common issues and problems. This level of cooperation among the states helped stabilize implementation ideas from one state to another and is the primary reason that some similarity exists among the state laws. This similarity is crucial for telecommunications companies seeking to offer products in multiple states.

3.2 STRUCTURE AND ORGANIZATION OF THE 1996 ACT—47 U.S.C. §§ 151 *et seq.*

The Telecommunications Act of 1996 is not a new law that replaces the Communications Act of 1934. Instead, it amends the 1934 Act by modifying certain portions of the older Act, and adding new sections to accommodate technological changes and the new competitive environment. Thus, to find the 1996 Act, one must go to the 1934 Act, found at 47 U.S.C. §§ 151 *et seq.*, and look for the 1996 updates.

When the 1996 updates are printed separately as the Telecommunications Act of 1996 Act, four general trends emerge. First, the sections of the Telecommunications Act are non-sequential, in that the sections start at Sections 1, 2, and 3, then jump to Sections 101 through 104, and then to section 151. The remaining sections move forward in a similar hopscotch manner throughout the Act, identifying only those portions of the 1934 Act that were changed or added as new sections.

Second, both the 1996 Act, when printed alone, and the 1934 Act, printed in its entirety, each have seven titles, or chapters, but they have different names and address different topics. For example, the seven titles in the 1996 Act are as follows:

Title I: Telecommunication Services
Title II: Broadcast Services
Title III: Cable Services
Title IV: Regulatory Reform
Title V: Obscenity and Violence
Title VI: Effect on Other Laws
Title VII: Miscellaneous Provisions

The seven titles in the 1934 Act are as follows:

Title I: General Provisions
Title II: Common Carrier (Wireline)
Title III: Radio (Wireless)
Title IV: Procedural and Administrative Provisions
Title V: Penal Provisions—Forfeitures
Title VI: Cable Communications
Title VII: Miscellaneous Provisions

Nonetheless, since the 1934 Act is the fuller law, the industry follows the 1934 titles when referring to Information and Enhanced Service Providers as Title I providers; to common carriers, incumbent local exchange carriers and competitive vocal exchange carriers, as Title II providers; wireless companies as Title III providers, and cable television companies as Title VI providers.

Because of this difference in the titles within the two Acts, a particular citation for information may not be the same in the two Acts. Where the location of an item differs, this book provides both citations so the reader can find the correct citation in either document.

Third, the 1996 Act provides mainly broad, overarching goals for U.S. communications in the new competitive environment. For example, Congress mandated in the 1996 Act that (1) telecommunications facilities must be interconnected at any "technically feasible point,"[9] (2) rates, terms, and conditions in the new competitive environment shall be "just, reasonable, and nondiscriminatory,"[10] and (3) the Federal Communication Commission (FCC) and each state commission "shall encourage the deployment on a reasonable and timely basis of advanced telecommunications capability to all Americans . . . in a manner consistent with the public interest."[11] Congress delegated to the FCC[12] and the states[13] the task of determining (1) where the "technically feasible point" is for each product or system;[14] (2) what "just, reasonable, and nondiscriminatory" mean; and (3) how to best implement these goals. While the state interpretations cannot conflict with the FCC's interpretation, they often vary widely among themselves and are usually more detailed than the federal interpretations. Where different states have different interpretations of what these mean and how to best achieve them, implementation of a nationwide product becomes difficult for companies.

Fourth, the 1996 Act begins with three introductory sections, Sections 1, 2, and 3. Section 1 states that the short title and citation for the 1996 Act is the Telecommunications Act of 1996, and the 1996 Act, "[e]xcept as otherwise expressly provided," amends the Communications Act of 1934. Section 2 then provides the table of contents, listing the seven titles and their subsections, and Section 3 provides new and modified definitions to the 1934 Act, many of which significantly alter the structure and direction of the new competitive environment. These are discussed in greater detail below.

3.3 DEFINITIONS—SECTION 3 OF THE TELECOMMUNICATIONS ACT OF 1996

One of the most important aspects of Section 3 of the 1996 Act is the modified and new definitions it provides to the industry.[15] Among the more surprising changes and additions are the definitions of telecommunications, carriers, and equipment. Since each is a common word, one would expect the definitions to be quite straightforward. However, each has a peculiar twist that has a great impact on the new competitive industry. Each term also has subsections, for example, (1) *telecommunications* includes definitions of telecommunications service, telecommunications carriers, and the telecommunications industry; (2) *carrier* includes definitions of local exchange carriers, including both incumbent and competitive local exchange carriers, Bell Operating Companies, and the two exceptions provided for mobile service providers and small, rural telephone companies; and (3) *equipment* includes definitions of telecommunications equipment and customer-provided equipment. The details of each are discussed below.

3.3.1 Three-Pronged Definition of Telecommunications

The 1934 Act did not contain a definition for the word *telecommunications*. Instead, it defined only the word *communications*. The 1996 Act, however, updated this by providing definitions for telecommunications, telecommunications service, and telecommunications carrier, creating what is known as the three-pronged definition of telecommunications.[16] In addition, it provides a definition of the telecommunications industry.[17] The four definitions concerning telecommunications follow.

Telecommunications, § 3(a)(2)(48)

The 1996 Act defines the term *telecommunications* as "the transmission, between or among points specified by the user, of information of the user's choosing, without change in the form or content of the information as sent and received."[18]

The three key elements in this definition, (1) between or among points specified by the user; (2) carrying information of the user's choosing, and (3) without change in the form or content of the information as sent and received, seem quite familiar. We recognize that users specify the points of communications when they dial phone numbers or address emails. Similarly, we recognize that the information in the communication is unchanged in form or content when phone conversations include mispronounced words and emails include misspelled words. However, during implementation of the 1996 Act, it was noted that modern communications generally convert most communications from analog to digital as they pass through various equipment in the network and that this was a change in the form of the information. It was subsequently decided that since these digitization changes are made only to facilitate the transmission process and are transparent to the users, they are not to be considered changes in the form or content of the information as sent and received. Instead, Congress defined telecommunications as the transmission of messages "passed through" just as the user communicated it, with its content unedited by a third party.

Telecommunications Service, § 3(a)(2)(51)

The term *telecommunications service* is defined in the Telecommunications Act of 1996 as "the offering of telecommunications for a fee directly to the public, or to such classes of users as to be effectively available directly to the public, regardless of the facilities used."[19]

The three key phrases in this definition are the offering of telecommunications (1) for a fee, (2) to the public, and (3) regardless of the facilities used. These highlight three specific characteristics of a telecommunications service not specified previously.

Telecommunications Carrier, § 3(a)(2)(49)

The term *telecommunications carrier,* is defined in the 1996 Act, as "*any* provider of telecommunications services, except . . . aggregators of telecommunications services (as defined in Section 226). A telecommunications carrier shall be treated as a common carrier under this Act only to the extent that it is engaged in providing telecommunications services, except that the Commission shall determine whether the provision of fixed and mobile satellite service shall be treated as common carriage."[20]

This definition has three surprising portions. First, it defines a telecommunications carrier as any provider of telecommunications services. Since two of the elements of a telecommunications service are *for a fee* and *to the public,* this definition means that a telecommunications carrier is anyone who offers a telecommunications service for a fee. This is a much larger group than we have previously considered to be telecommunications carriers. It can include such groups as (1) business centers in airports or shopping centers that send or receive faxes and other communications for a fee; (2) colleges that add the price of a phone in the fee for a dorm room; and (3) office-building owners who add fees for telecommunications facilities in the rent. Traditionally, these groups have not considered themselves to be telecommunications carriers and thus have applied for exemptions from the FCC or the state legislators or PUCs. Some exemptions have been granted, others have not. The second surprising portion of this definition is that the only exemption provided in the 1996 Act is for "aggregators," such as hotels, where the price of the phone is included in the room rental cost. The difference between considering dorms and office buildings as "carriers," and hotels as "not carriers" appears to be in the length of time of a tenant's stay. Shorter or temporary stays seem to be excluded, although no clear distinction is provided in the statute or legislative history. The third surprising portion of this definition is that it states that the FCC will make a later determination about whether a satellite communications system is to be considered a common carrier. No explanation concerning this deferral was provided in either the legislation or the Joint Explanatory Statement of the Conference Committee.

Telecommunications Industry, § 714(k)(3)

The term *telecommunications industry,* as defined in the 1996 Act, means "communications businesses using regulated or unregulated facilities or services and includes broadcasting, telecommunications, cable, computer, data transmission, software, programming, advanced messaging, and electronics businesses."[21] This definition indicates Congress's partial acknowledgement of the convergence of technologies and the variety of potential new providers of telecommunications services in the new competitive environment.

3.3.2 Definitions of Carriers

With the extended definition of telecommunications carriers, the 1996 Act further divides the term *carriers* into four specific categories including the following: (1) local exchange carriers, (2) incumbent local exchange carriers, (3) competitive local exchange carriers, and (4) Bell Operating Companies, as well as two important exclusions, (5) wireless companies, and (6) small, rural telephone companies. All six categories are described in more detail below.

Local Exchange Carriers, § 3(a)(2)(44)

The 1996 Act defines the term *local exchange carrier* (LEC) as "any person that is engaged in the provision of telephone exchange service or exchange access. Such term does not include a person insofar as such person is engaged in the provision of a commercial mobile service under Section 332(c), except to the extent that the Commission finds that such service should be included in the definition of such term."[22]

As one would expect, this definition includes anyone who provides local telephone exchange service to the public for a fee. Surprisingly, however, the definition includes two additional elements. First, it includes any company providing exchange access. *Exchange access* is defined in the 1996 Act as "the offering of access to telephone exchange services or facilities for the purpose of the origination or termination of telephone toll services."[23] Hence, a local exchange carrier is not just a local service provider, but *any* service provider that originates or terminates toll traffic. With the introduction of technologies such as satellite, fiber, microwave, cable, and Internet service providers (ISPs), toll service could be originated and terminated without the use of the local telephone company's facilities.

Second, the definition of local exchange carriers specifically excludes *commercial mobile service providers*. These wireless companies provide radio communications service between mobile stations and fixed, land stations, or among mobile stations,[24] and include services such as cellular telephone, personal communications systems, and enhanced specialized mobile radio (ESMR) services. This exclusion means that these wireless companies are not considered local exchange carriers, even if no wire facilities exist in the area and the wireless companies provide the only telephone service available. This exclusion also means that the mobile service providers are not responsible for most of the duties and obligations of local exchange carriers, but they are considered new entrants and thus can receive all of the benefits and privileges provided to new entrants in the 1996 Act.

Exclusion—Mobile Service Providers

Congress exempted the commercial mobile service providers for several reasons. First, wireless service had always been viewed as a supplemental service to basic landline telephone service. For this reason, it had never been regulated as a local telephone company. Second, the advanced wireless systems tend to provide combined voice, data, and video services, each addressed differently by the 1996 Act. Third, Congress wanted competition to start immediately after the 1996 Act was signed. It recognized that the migration from monopoly to competition could take years if duplicate wired systems had to be installed. Instead, a much quicker method to provide a duplicate local infrastructure was through wireless systems. Congress realized that it was the easiest, fastest technology to bring about competition and thus to encourage the expansion of wireless facilities. Congress (1) made certain federal land available for wireless antenna sites, (2) reduced the time required for antenna site approval, (3) created an appeal process if a site request were denied, and (4) placed the research and responsibility for setting radio frequency emission standards with the FCC.[25]

Incumbent Local Exchange Carriers, § 251(h)(1)

The 1996 Act defines an *incumbent local exchange carrier* (ILEC) as, "with respect to an area, the local exchange carrier that

(A) provided telephone exchange service in the area on the date of enactment of the Telecommunications Act of 1996; and

(B) (i) was deemed to be a member of the National Exchange Carrier Association [pursuant to section 69.601(b) of the Commission's regulations (47 C.F.R. 69.601(b))] on such date of enactment; or

(B) (ii) is a person or entity that, on or after such date of enactment, became a successor or assign of a member described in clause (i); . . .[26]

Beyond recognizing that the telephone companies providing local service on February 8, 1996 are incumbent carriers, this definition adds three future elements to the definition of an incumbent local exchange carrier. It states that the FCC may consider a local exchange carrier to be an incumbent local exchange carrier if

(A) such carrier occupies a position in the telephone exchange service market that is comparable to the position occupied by the existing ILEC;

(B) such carrier has substantially replaced an incumbent local exchange carrier described in paragraph (1); and

(C) such treatment is consistent with the public interest, convenience, and necessity and the purposes of this section.[27]

The qualification that an incumbent local exchange carrier (ILEC) must be deemed to be a member of the National Exchange Carrier Association (NECA),[28] excludes small rural telephone companies that may have been providing the telephone service in an area for many years prior to 1996 and are technically incumbent local exchange carriers, but whom Congress is excluding from the responsibilities of incumbent local exchange carriers because of their small size. Congress did not want to place undue burden or cost on the subscribers of companies who may not experience competition.

Exclusion—Rural Telephone Companies, § 3(a)(2)(47)

A number of the smaller, rural carriers feared that the duties and responsibilities placed on incumbent local exchange carriers by the 1996 Act could bankrupt them. Therefore, to relieve the small carriers' concerns, Congress exempted rural telephone companies in the 1996 Act, defining a *rural telephone company* as a local exchange carrier[29] that

(A) provides common carrier service to any local exchange carrier study area that does not include either—

(i) any incorporated place of 10,000 inhabitants or more, or any part thereof, based on the most recently available population statistics of the Bureau of the Census; or

(ii) any territory, incorporated or unincorporated, included in an urbanized area, as defined by the Bureau of the Census as of August 10, 1993;

(B) provides telephone exchange service, including exchange access, to fewer than 50,000 access lines;

(C) provides telephone exchange service to any local exchange carrier study area with fewer than 100,000 access lines; or

(D) has less than 15 percent of its access lines in communities of more than 50,000 on the date of enactment of the Telecommunications Act of 1996.

While this definition and exclusion are important to the smaller carriers, Congress also stated that if a carrier grew beyond these qualifications, the carrier loses this rural telephone company exclusion and must meet the obligations of the incumbent local exchange carriers.

Congress exempted small, rural telephone companies from the definition of a local exchange carrier, and therefore from the duties and obligations imposed on local exchange carriers (by Sections 251 and 252)[30] because it was anticipated that few competitive companies would likely enter this market. Therefore, preparing for competition would be an unnecessary expense for the customers. However, in case new opportunities, such as demand for high-speed Internet connections, changed the market dynamics, Congress stated that this exemption lasted only until the rural markets received a request from a competitive local exchange carrier to interconnect their systems. At that point, the smaller, rural systems would be required to interconnect in a nondiscriminatory manner. Section 251(f)(2) also states that local exchange carriers with less that 2% of the nation's lines may petition the state for suspension or modifications of any of the requirements of Section 251(c) (local exchange carrier and incumbent local exchange carriers duties). Both allow the states to provide some flexibility in the application of the Act. The carriers requesting the exemption bear the burden of proof in their request.

Competitive Local Exchange Carriers

Surprisingly, the 1996 Act does not contain a definition of the term *competitive local exchange carrier.* It is, however, the term used by the industry to identify the new entrants providing switched services in local markets. In addition, throughout the 1996 Act, Congress reiterates that various providers are free to enter the opened local exchange markets as new competitive carriers. These include the following: (1) long-distance service providers or interexchange carriers; (2) cable television companies; (3) wireless carriers; (4) competitive access providers, including microwave and fiber-based carriers; (5) public utilities such as the gas and/or oil and electricity providers; and (6) entrepreneurs.

Bell Operating Companies, § 3(a)(2)(35)

The 1996 Act defines the term *Bell Operating Company* (BOC) as

(A) *any of the following companies: Bell Telephone Company of Nevada, Illinois Bell Telephone Company, Indiana Bell Telephone Company, Incorporated, Michigan Bell Telephone Company, New England Telephone and Telegraph Company, New Jersey Bell Telephone Company, New York Telephone Company, US West Communications Company, South Central Bell Telephone Company, Southern Bell Telephone and Telegraph Company, Southwestern Bell Telephone Company, The Bell Telephone Company of Pennsylvania, The Chesapeake and Potomac Telephone Company, The Chesapeake and Potomac Telephone Company of Maryland, The Chesapeake and Potomac Telephone Company of Virginia, The Chesapeake and Potomac Telephone Company of West Virginia, The Diamond State Telephone Company, The Ohio Bell Telephone Company, The Pacific Telephone and Telegraph Company, or Wisconsin Telephone Company; and*

(B) *includes any successor or assign of any such company that provides wireline telephone exchange service; but*

(C) *does not not include an affiliate of any such company, other than an affiliate described in subparagraph (A) or (B).*[31]

The first part of this definition, the list of 20 companies, is important because the 1996 Act placed the heaviest obligations on the Bell Operating Companies. These duties and responsibilities are described in Section 3.5.4, but it is interesting to note that many large, non–Bell companies providing local service, such as GTE, Sprint, and AT&T, were not given the same level of obligations. It is also intriguing that in some cases whole Regional Bell Operating Companies, such as US WEST and Southwestern Bell Co. (SBC), were included in the list, while others, such as Ameritech, BellSouth, Bell Atlantic, and NYNEX were named only as individual companies. No explanation for this was given in the legislative history.

The second part of this definition, the inclusion of any successor, applies directly to the companies, such as (1) US West, bought by Qwest; (2) Ameritech, Pacific Bell, and Southern New England Telephone, merged with SBC; and (3) NYNEX merged with Bell Atlantic. It guaranteed that these companies could not escape the heavy responsibilities of being a BOC by changing ownership, name, or stock allocations.

The third part of this definition, the exclusion of affiliates, links directly back to the issues of *Computer I, II,* and *III* discussed in Chapter 2.

3.3.3 Definitions of Equipment

The 1996 Act further provides two clarifying definitions regarding telecommunications equipment. The first identifies the equipment used by the carriers to provide services, and the second identifies the equipment used by customers, on their premises, to access the services.

Telecommunications Equipment, § 3(a)(2)(50)

The 1996 Act states that "the term *telecommunications equipment* means equipment, other than customer premises equipment, used by a carrier to provide telecommunications services, and includes software integral to such equipment (including upgrades)."[32]

The two key elements in this definition are the distinction between telecommunications equipment and customer premises equipment and the inclusion of software, with upgrades, integral to the equipment. The first element underscores the shift from the vertical structure of the Bell Companies, which provided end-to-end communications. The second element recognizes the importance of computers and software in today's telecommunications industry.

Customer Premises Equipment, § 3(a)(2)(38)

On the other hand, the 1996 Act defines the term *customer premises equipment* as "equipment employed on the premises of a person (other than a carrier) to originate, route, or terminate telecommunications."[33] A major part of the new competitive market opens opportunities in the customer premises equipment market and thus is important to all new service providers, facilities entrants, and customers seeking new applications of telecommunications.

3.4 PROVISIONS TO OPEN THE COMPETITIVE MARKET

Since one of the stated purposes of the 1996 Act was to open the local telecommunications market to competition, Section 251 provides three ways for competitive companies to enter the market, a clear process and timetable for implementation of the provisions of the 1996 Act, and the requirements for negotiating and writing interconnection agreements. A discussion of each follows.

3.4.1 Three Ways to Enter the New Competitive Market, 251 (c)(2-4)

The 1996 Act provided three basic ways for companies to enter the new competitive telecommunications market: facilities-based systems, unbundled access, and resale networks. Specifically, the 1996 Act requires an incumbent local exchange carrier (1) to interconnect its facilities with the facilities of a requesting new entrant in the incumbent local exchange carrier's local market in order to facilitate local telephone services over both facilities-based systems;[34] (2) to provide competing telecommunications carriers with access to individual elements of the incumbent local exchange carrier's own network on an unbundled basis (unbundled access);[35] and (3) to sell to its competing telecommunications carriers, at wholesale rates, any telecommunications service that the incumbent local exchange carrier provides to its customers at retail rates, in order to allow the competing carriers to resell the services (resale).[36] The Eighth Circuit Court of Appeals refers to these three duties as "the local competition provisions."[37]

Facilities-Based Interconnection, § 251(c)(2)

To accomplish facilities-based interconnection, the 1996 Act requires the incumbent local exchange carriers to

> *provide, for the facilities and equipment of any requesting telecommunications carrier, interconnection with the local exchange carrier's network-*
>
> *(A) for the transmission and routing of telephone exchange service and exchange access;*
>
> *(B) at any technically feasible point within the carrier's network;*
>
> *(C) that is at least equal in quality to that provided by the local exchange carrier to itself or to any subsidiary, affiliate, or any other party to which the carrier provides interconnection; and*
>
> *(D) on rates, terms, and conditions that are just, reasonable, and nondiscriminatory, in accordance with the terms and conditions of the agreement and the requirements of this section and section 252.*[38]

These requirements seek to ensure the interconnection of separate, facilities-based systems for transparent exchange of communications traffic between carriers so that all customers can seamlessly call any other person no matter which company serves each customer.

Unbundled Access, § 251(c)(3)

The 1996 Act defines *unbundled access* as

> *the duty to provide, to any requesting telecommunications carrier for the provision of a telecommunications service, nondiscriminatory access to network elements on an unbundled basis at any technically feasible point on rates, terms, and conditions that are just, reasonable, and nondiscriminatory in accordance with the terms and conditions of the agreement and the requirements of this Section and Section 252. An incumbent local exchange carrier shall provide such unbundled network elements in a manner that allows requesting carriers to combine such elements in order to provide such telecommunications service.*[39]

This option allows companies, such as cable television, wireless providers, and electric/gas utilities, who already have customers, installation and maintenance crews, and billing systems, to lease the network elements they are missing, such as a voice switches, or electronic ordering systems, to complete their local telephone system. The 1996 Act, thus mandates that the traditional vertical hierarchy of the telephone industry be divided into parts, known as *unbundled network elements,* so that competitive new entrants can choose what they need to provide competitive local telecommunications service.

Resale, § 251(c)(4)

The 1996 Act defines the third form of entry, *resale,* as

> *the duty of ILECs—*
>
> *(A) to offer for resale at wholesale rates any telecommunications service that the carrier provides at retail to subscribers who are not telecommunications carriers; and*
>
> *(B) not to prohibit, and not to impose unreasonable or discriminatory conditions or limitations on, the resale of such telecommunications service, except that a State commission may, consistent with regulations prescribed by the Commission under this section prohibit a reseller that obtains at wholesale rates a telecommunications service that is available at retail only to a category of subscribers from offering such service to a different category of subscribers.*[40]

Congress realized that when a new carrier entered a local market, it would take time for that new carrier to build its own facilities in order to serve customers and would be a significant investment for those companies. Both factors meant a delay in the start of competition and thus a delay in realizing the benefits of competition for consumers. Instead, Congress wanted competition in the telecommunications industry to start the day the 1996 Act was signed and thus made provisions for new entrants to resell the incumbent local exchange carriers services. By purchasing at wholesale and reselling at retail, the new entrants would be able to make a profit. Telecommunications services are available for resale in accordance with the requirements of 47 U.S.C. §§ 251(b)(1), 251(c)(4), 252(d)(3), and 271(c)(2)(B)(xiv).

3.4.2 Implementation of the 1996 Act, § 251(d)(1)

Congress required in the 1996 Act that "within 6 months after the date of enactment of the Telecommunications Act of 1996, the commission [FCC] shall complete all actions necessary to establish regulations to implement the requirements of this section." This meant that since the Act was signed on February 8, 1996, the FCC was required to complete the implementing rules for the Act by August 8, 1996. It also meant that the states already offering local competition must wait until August, 1996 before the federal government's final directions on the 1996 Act were available.

Notice of Proposed Rulemaking

To complete the rules needed to implement the 1996 Act, the FCC first drafted a set of more than 80 new rules addressing (1) interconnection, (2) unbundling,

(3) resale, (4) universal service, (5) access charges, (6) local number portability, (7) interconnection agreements, (8) the procedures by which companies could enter the new markets, and (9) new ownership rules. Second, before a government agency, such as the FCC, can impose new rules on an industry, the federal Administrative Procedures Act requires that the agency publish a draft of the rules and request public comment on them. This process is known as issuing a Notice of Proposed Rulemaking (NPRM) and is meant to ensure due process. Thus, in May 1996 the FCC published its proposed rules and requested comments on them from all interested persons.[41]

In any proposed rulemaking process, two rounds of public comment are provided for. In the first round, parties comment on the rules, and in the second round, parties review the responses of all other parties and comment on which they agree or disagree with and why. Over 450 parties responded to the two rounds in the May 1996 NPRM concerning the 1996 Act, including the governments of all 50 states; the U.S. territories, including Guam, the Virgin Islands, and Puerto Rico; the major telephone carriers, industry associations, and consumer groups including the offices of consumer council of most states; and associations representing retired persons and disabled.

During June and July 1996, both sets of public responses were considered by the FCC and many comments were incorporated into the FCC's final set of rules, issued on August 8, 1996, in the FCC's *First Report and Order, Implementation of the Local Competition Provisions in the Telecommunications Act of 1996*.[42] The FCC's rules are now codified throughout various sections of Title 47, Code of Federal Regulations,[43] and require uniform, nondiscriminatory interconnection, unbundled network elements, reciprocal rates for transport and termination, collocation of equipment, and rates based on forward-looking cost methodologies.

Initial Opposition to the Rules

Despite the broad-based public review of the FCC's proposed implementation rules in the two rounds of the NPRM, parties in nearly every jurisdiction of the United States filed court cases opposing the rules within the first few days and weeks following the FCC's release of its First Report and Order. These motions to stay the First Report and Order, were filed mainly by incumbent LECs and state utility commissions protesting pricing.

Although most of the petitioners requested the court to stay the entire First Report and Order, their specific attacks focused primarily on the FCC's rules regarding pricing.[44] The petitioners opposed the rules regarding both: 1) the prices that the incumbent LECs could charge their new competitors for interconnection, unbundled access, and resale; and 2) the prices for the transport and termination of local telecommunications traffic.[45] "Transport and termination of telecommunications" is the process whereby a call that is initiated by a customer of a telecommunications carrier is routed to a customer of a different telecommunications carrier and completed by that carrier. The telecommunications carrier that "terminates" or completes the call to its customer typically charges the other telecommunications carrier for the cost of terminating the call. The Act imposes a duty on all local exchange carriers, incumbents and new entrants, to establish reciprocal compensation arrangements for such transport and termination of phone calls.[46] The petitioners argued that the FCC exceeded its jurisdiction in establishing prices for what are essentially local intrastate telecommunications and that the pricing rules violate the terms of the Act.[47]

Eighth Circuit Court Review

When multiple cases are filed on an issue in several jurisdictions, multiple courts do not hear the cases because it would be inefficient and could lead to inconsistent decisions. Instead, the cases are consolidated in one court, pursuant to Rule 24 of the *Rules of Procedure of the Judicial Panel on Multidistrict Litigation.*[48] In the case of the complaints against the FCC's *First Report and Order Implementing the 1996 Act,* the motions were consolidated at the U.S. Eighth Circuit Court of Appeals in St. Louis, Missouri, by the September 11, 1996 order of the Judicial Panel on Multidistrict Litigation, Docket No. RTC-31. This was surprising to many people, because it was one of the few times a court outside of Washington, D.C. reviewed a telecommunications law and its implementing rules that would affect all aspects of the industry throughout the United States.

The Eighth Circuit Court spent most of September 1996 reviewing the FCC's findings and rules in the First Report and Order, and found most of the rules to be "good law." On October 3, 1996, in *Iowa Utilities Board v. FCC,*[49] the court stayed, temporarily, only two sections: (1) the operation and effect of the pricing provisions, and (2) the *pick and choose* option.[50] *Pick and choose* is a portion of the 1996 Act that allows a CLEC negotiating an interconnection agreement with an incumbent local exchange carrier to pick and choose provisions from other interconnection agreements previously negotiated by that incumbent local exchange carrier with other competitive local exchange carriers. The intent of the provision is to level the negotiating "playing field" so that all competitive local exchange carriers can obtain similar provisions from the incumbent local exchange carrier. Several applications to vacate these stays by the Eighth Circuit Court were sent to the U.S. Supreme Court, but all were denied.[51] During the remainder of 1996 and early 1997, the Eighth Circuit Court reviewed these two sections more fully and issued its decision on July 18, 1997 in *Iowa Utilities Board v. FCC [II].*[52] The Eighth Circuit Court's decision was amended upon rehearing on October 14, 1997.

3.4.3 Process for Writing Interconnection Agreements, § 252

Whichever of the three ways a company selects to enter a telecommunications market, facilities based, unbundled, or resale, the primary method of doing business in the new environment is through written contracts between two (or more) telecommunications carriers, called *interconnection agreements.* These contracts are very detailed documents that include all aspects of the business relationship between the two carriers, including the technical interconnections between their equipment and facilities, how they will exchange communications traffic, and how and at what rate the two companies will financially compensate each other for the traffic. In the case of equipment or service failure, the interconnection agreements identify which carrier will respond, how quickly, and in what manner. The contract also describes coverage of obligations such as local number portability and universal service, as well as which jurisdiction and method of resolution will be used if a contract dispute arises. Examples of interconnection agreements can be found on the Internet Web pages of most state PUCs.

Section 252 of the 1996 Act establishes the process, timing, and constraints by which these interconnection agreements are to be developed.[53] First, in each case, the interconnection agreements must be filed in each state in which the two carriers will do business. This is true even if the two negotiating carriers plan to offer the same services in several states. Second, if the parties cannot independently negotiate an agreement, the 1996 Act provides for mediated or arbitrated agreements.

Third, to begin the process of writing an interconnection agreement, the competing carrier sends a formal, written *Request to Negotiate* to the incumbent carrier. The ILEC must then notify, in writing, the utility commission of each state affected by the request. This is because the 1996 Act sets a very tight, nine-month calendar in which an interconnection agreement must be completed and signed. Congress included this deadline in the 1996 Act to keep the negotiation process from stalling, thwarting, or slowing the introduction of competition in the local market. The start date of the nine-month calendar begins with the incumbent carrier's acknowledgment of the receipt of the competing carrier's CLEC's *Request to Negotiate.*

Negotiated Agreements, § 251(c)(1), § 252(a)(1)

During the process of drafting an interconnection agreement, Congress's first preference is for the carriers to negotiate voluntary agreements among themselves. Section 252 of the 1996 Act states that each party, the incumbent local exchange carrier and the competing local exchange carrier, has a duty to negotiate, in good faith, the terms and conditions of an agreement that accomplishes the Act's goals.[54] The carriers may meet and negotiate the details of their agreement and, when completed, submit the agreement to the state utility commission for approval. Again, the state commission then has three months to review such agreements and to approve or deny them.

Mediated Agreements, § 252(a)(2)

However, if at any point during the negotiation process, the negotiations are not going well, either party may request that the state utility commission or its designees, usually its staff or an outside contractor hired by the commission, attend the negotiations and participate in the process as a *mediator.* In a mediated agreement, the mediator simply facilitates the process, while the decision making remains with the parties. When a mediated agreement is signed, it must also be submitted to the state utility commission for approval. The commission has three months in which to approve or deny the agreement.

Arbitrated Agreements, § 252(b)

If the mediation process does not improve the situation, either party may request that the state utility commission arbitrate the interconnection agreement. However, this request may only be made during a window from 135 to 160 days (inclusive) into the negotiations, calculated from the date of the CLEC's *Request to Negotiate*. This window is approximately four and one-half months into the mandated nine-month limit. The main difference between mediation and arbitration is that with arbitration the decision moves from the parties to the arbitrator. Thus, while the mediator serves as a facilitator of the process, the arbitrator becomes the judge. Additionally, pursuant to Section 252, the arbitrator's decision is *binding* on both parties, although Section 252(b) limits the arbitrator's authority to resolving only the *open issues* as specifically stated in the petition requesting arbitration. Once an arbitrated interconnection agreement is completed and presented to the state utility commission for approval, the state commission has only 30 days in which to review and approve or deny the agreement. This is because with arbitration, Congress assumed that the state commission would be intimately familiar with the details of the agreement and may have mandated many of them. Thus Congress felt that the commission needed time only to ensure that its directions were clearly included in the written contract, and a shorter review period was in the best interest of the public.

3.4.4 Post-Approval Issues

Once an interconnection agreement has been approved by the state utility commission, it must be made publicly available within 10 days. Usually, this is done by posting it on the state PUC's Internet Web site and providing an archived paper copy at the state PUC's offices. After that, however, issues such as tariffs and problem resolution may continue to arise. Generally, the 1996 Act does not address these, but the states have provided for them as follows.

Tariffs

A carrier's tariff is more than just a price list. It contains detailed descriptions of each service offered, complete with technical specifications, product guarantees, and quality of service parameters. It also details the respective obligations of the carrier and the customer and the resolution procedures to be used when problems arise. As such, it is a written contract between the telecommunications provider and its customers.

While the 1996 Act explicitly removed rate-of-return regulation, many states still require carriers doing business in their states to file tariffs. Unlike in the previous regulated environment, the states do not hold rate hearings to approve or deny the tariffs, but many require companies to file tariffs in order to have a publicly available, written statement from each carrier outlining its obligations and agreements with its customers. Some states require all companies, both incumbent local exchange carriers and competitive local exchange carriers, to file tariffs, and other states require only the incumbent local exchange carriers to file tariffs.

Problem Resolution

As with other contracts, problems can arise after the interconnection agreements are filed with the state. When this occurs, the parties can, of course, sue one another and seek judicial relief. However, most states realized that this could thwart or restrict competition, and thus they provided for mediation and arbitration to resolve disputes.[55]

3.5 DUTIES AND OBLIGATIONS OF CARRIERS

Title 1 of the 1996 Act, "Telecommunications Services,"[56] established two new sections to be added to the 1934 Act, specifically in Title 2, "Common Carriers."[57] The two new sections are "Part II - Development of Competitive Markets" and "Part III - Special Provisions Concerning Bell Operating Companies." The significance of these two new sections is that they delineate the rights and obligations of each participant in the new competitive local market. As such, these new laws establish the parameters in which the market participants must operate, the opportunities open to them, and the requirements that each competitor must meet.

The new Part II starts at Section 251, and includes Sections 251 through 261.[58] The new Part III starts at Section 271 and includes Sections 271 through 276.[59] Section 251 lists the obligations and duties of the new companies competitive local exchange carriers, while Section 271 contains the Competitive Checklist that the Bell Operating Companies must pass before they can provide ancillary services beyond basic local and long-distance telephone service. Sections 272–276, generally known as the special provisions or BOC safeguards, establish the parameters around four new ancillary services, including (1) equipment manufacturing, (2) electronic publishing, (3) alarm monitoring, and (4) payphone services. To clarify the rights and responsibilities of

TABLE 3.1 CHART OF CARRIERS' DUTIES
WHO MUST DO WHAT UNDER THE ACT (PART 1)

Duty	Statute	All Carriers	LECs	Incumbent LECs	BOCs
Interconnection	251 (a) (1)	A-1			B-1
Persons with Disabilities	251 (a) (2)	A-2			
Network Coordination	251 (a) (2)	A-3			
Resale	251 (b) (1)	—	L-1	I-4	B-14
Number of Portability	251 (b) (2)	—	L-2		
Dialing Parity	251 (b) (3)	—	L-3		B-12
Poles, Ducts, Conduits, and Rights-of-Way	251 (b) (4)	—	L-4		B-3
Reciprocal Compensation	251 (b) (5)	—	L-5		B-13
Duty to Negotiate	251 (c) (1)	—	—	I-1	
Interconnection Plus	251 (c) (2)	—	—	I-2	B-1
Network Elements	251 (c) (3)	—	—	I-3	B-2
Resale at Wholesale Rates	251 (c) (4)	—	—	I-4	
Notice of Changes	251 (c) (5)	—	—	I-5	
Collocation	251 (c) (6)	—	—	I-6	

Source: Used by permission of Telecommunications Research Associates.
Note: LECs, local exchange carriers; BOCs, Bell Operating Companies.

market participation, Sections 251 and 271 of the 1996 Act group the competitive market participants into the following four categories: (1) All telecommunications carriers; (2) all local exchange carriers; (3) all incumbent local exchange carriers, and (4) the Bell Operating Companies. Tables 3.1 and 3.2 summarize the rights and obligations of each group.

3.5.1 All Telecommunications Carriers, § 251(a)

The category of *telecommunications carriers* as defined in the 1996 Act[60] includes all incumbent local exchange carriers, all new entrants, and all long-distance service providers. For specific items, the category also includes

TABLE 3.2 CHART OF CARRIERS' DUTIES
WHO MUST DO WHAT UNDER THE ACT (PART 2)

Duty	Statute	All Carriers	LECs	Incumbent LECs	BOCs
Interconnection	271 (c) (2) (B) (i)	A-1			B-1
Network Elements	271 (c) (2) (B) (ii)	—	—	I-3	B-2
Poles, Ducts, Conduits, Rights-of-Way	271 (c) (2) (B) (iii)	—	L-4		B-3
Access to Local Loop	271 (c) (2) (B) (iv)	—	—	—	B-4
Trunk-Side Local Transport	271 (c) (2) (B) (v)	—	—	—	B-5
Local Switching	271 (c) (2) (B) (vi)	—	—	—	B-6
911/Directory Assistance/Operator	271 (c) (2) (B) (vii)	—	—	—	B-7
White Pages Listings	271 (c) (2) (B) (viii)	—	—	—	B-8
Telephone Numbers	271 (c) (2) (B) (ix)	—	—	—	B-9
Databases for Call Handling	271 (c) (2) (B) (x)	—	—	—	B-10
Interim Number Portability	271 (c) (2) (B) (xi)	—	—	—	B-11
Info re Local Dialing Parity	271 (c) (2) (B) (xii)	—	L-3		B-12
Reciprocal Compensation	271 (c) (2) (B) (xiii)	—	L-5		B-13
Resale at Wholesale Rates	271 (c) (2) (B) (xiv)	—	—	I-4	B-14

Source: Used by permission of Telecommunications Research Associates.
Note: LECs, local exchange carriers; BOCs, Bell Operating Systems.

equipment providers, wireless carriers, and satellite service providers, but generally these three groups are exempt from any specific obligations. The 1996 Act identifies only three duties that all telecommunications carriers must meet: interconnection, consistency with U.S. laws regarding persons with disabilities, and coordinated network planning. Each is described further below.

All Carriers' Duty 1: Interconnection, § 251(a)(1)

First, the 1996 Act requires that all telecommunications carriers "interconnect directly or indirectly with the facilities and equipment of other telecommunications carriers."[61] This requirement dramatically changed U.S. communications law by making each carrier's duty to interconnect immediate with no required finding of need. For 62 years, from 1934 to 1996, U.S. communications law supported the concept of a *natural monopoly* in each market and resisted the addition of any other company's equipment to the telephone network, arguing that it might degrade the performance of the network. For this reason, the 1934 Act required a lengthy and complicated process to prove that each interconnection was "in the public necessity, interest, and convenience."[62] Typically the process took several years, was rarely granted, and thus was generally avoided. The 1996 Act reversed that process and made the *duty to interconnect* explicit, immediate and applicable to all telecommunications market participants.[63]

All Carriers' Duty 2: Persons with Disabilities, § 251(a)(2)

Second, the 1996 Act requires that the products and services of all telecommunications carriers be consistent with the Americans With Disabilities Act of 1990.[64] It states that all telecommunications carriers have the duty "not to install network features, functions or capabilities that do not comply with the guidelines and standards established pursuant to Section 255." Section 255 requires that manufacturers of telecommunications equipment, manufacturers of customer premises equipment, and providers of telecommunications services, shall ensure that their equipment or services are "accessible to and usable by individuals with disabilities, if readily achievable."[65]

Beyond placing payphones at wheelchair level, redesigning phone booths, placing larger numbers on telephones, and volume controls on handsets, this requirement helps to bring the benefits of digital communications to persons with disabilities to provide them with fuller access to enhanced communications. For example, it could make it possible for visually impaired persons to have Internet information, books, and other research and written communications be "read" to them through a *codec*. The word *codec,* is a shortened form of coder/decoder, a device that converts analog to digital communications and back again. It can convert digital or computerized information into analog form so the human ear can understand it. At present these systems often sound like very mechanical voices, but technology is improving. Codecs can also convert spoken or oral communications into written communications for hearing-impaired persons. This has exciting potential for the future, but also places significant responsibilities on the manufacturers of telecommunications equipment and providers of telecommunications services. Most surprising of all, the federal agency given the responsibility for developing the guidelines in this area is the U.S. Architecture and Transportation Board, a group previously not prominent in telecommunications equipment and services, but now responsible for the final approval of all new telecommunications equipment and services.

All Carriers' Duty 3: Coordinated Network Planning, § 251(a)(2)

Third, the 1996 Act requires all telecommunications carriers to not "install network features, functions or capabilities that do not comply with the guidelines and standards established pursuant to Section 256." Section 256 requires the FCC to establish procedures for FCC oversight of coordinated network planning for effective and efficient interconnection of public telecommunications networks.[66]

Historically, the individual Bell Operating Companies in the United States coordinated their network and equipment planning with one another to seamlessly transfer calls across the country. While they were separate companies, they were also part of the Bell System, and it increased the value of the phones to their customers to be able to call anyone. The companies also had monopoly control of a geographic area, so even interconnection with independent companies following the 1913 Kingsbury Commitment did not mean market loss for them. The 1996 Act continues this tradition of seamless communications by requiring all local exchange carriers to interconnect with one another even in a competitive environment. Taking this one step further, the 1996 Act requires all telecommunications carriers to coordinate the planning and system design of public networks to allow a free flow of calls from one company to another in a nondiscriminatory manner.[67]

3.5.2 Duties of Local Exchange Carriers, § 251(b)

Within the larger group of *all carriers* is a subset of companies called the *local exchange carriers*. The 1996 Act defines local exchange carriers as any person who provides telephone exchange service or exchange access, including the (1) incumbent local exchange carriers, (2) new competitive local exchange carriers, (3) Bell Operating Companies, and (4) non-Bell Operating Companies or independent telephone companies. It does not include, however, two very important groups within the industry, the commercial mobile service providers (wireless), operating pursuant to 47 U.S.C. 332(c), and small, primarily rural telecom providers,[68] except to the extent that the FCC finds that their services should be included in certain cases.

In addition to the three general duties of *all carriers* described in the section above, the 1996 Act adds five additional duties that the local exchange carriers must meet. In its Joint Explanatory Statement, Congress stated that these five *additional* duties make sense only in the context of a specific request from another telecommunications carrier or any other person who actually seeks to connect with or provide services using the local exchange carrier's network.[69] The five duties are presented below.

Local Exchange Carriers' Duty 1: Resale, § 251(b)(1)

First, the 1996 Act, requires all local exchange carriers "not to prohibit, and not to impose unreasonable or discriminatory conditions or limitations on the resale of its telecommunications services."[70] This requirement is significant for three reasons. First, it protects the resale option of market entry discussed earlier in Section 3.4.1. Second, the requirement does not apply to wireless systems, because they are not considered local exchange carriers. Third, it does not require that the resale occur at a wholesale price, as is the case with resale by the Bell Operating Companies, discussed later in Section 3.5.4.

Initially, companies even as large as AT&T announced that it would use resale contracts to enter at least 30% of the local markets by the end of 1996.[71] However, after the first several such resale agreements, competitive local exchange carriers became frustrated with the level of service provided under the resale agreements and decided instead to purchase local facilities on which they could enter the market. AT&T, for example, purchased companies such as TCG and TCI and began to integrate all of its facilities to provide services in the local market.

Local Exchange Carriers' Duty 2: Number Portability, § 251(b)(2)

Second, the 1996 Act requires all local exchange carriers "to provide, to the extent technically feasible, (local) number portability" in accordance with FCC requirements.[72] The Act defines *number portability* as "the ability of users of telecommunications services to retain, at the same location, existing telecommunications numbers without impairment of quality, reliability, or convenience when switching from one telecommunications carrier to another."[73]

This duty is important because it allows customers to keep their phone number if they change local service providers. It is an enormously active part of the post-1996 competitive environment that eventually will impact every phone call in the United States. Likely also, it will create a third portion of the industry to manage phone number assignments, and ultimately will continue to raise serious cost issues because the database of "ported" telephone numbers must be continually maintained and updated. The process and issues surrounding *local number portability* are discussed in greater detail in Chapter Four. It is also an obligation that applies to the wireless industry. Commercial mobile providers, including cellular, personal communications systems and certain specialized mobile radio carriers, will also participate in the long-term solution so that customers may also keep their wireless telephone numbers when they switch providers. Currently, the date for this to occur is November 24, 2002, but that date may change.

Local Exchange Carriers' Duty 3: Dialing Parity, § 251(b)(3)

Third, the 1996 Act requires all local exchange carriers "to provide dialing parity to competing providers of telephone exchange service and telephone toll service, and to permit all such providers to have nondiscriminatory access to telephone numbers, operator services, directory assistance, and directory listing, with no unreasonable dialing delays."[74] It defines *dialing parity* as enabling a person that is not an affiliate of a local exchange carrier to "provide telecommunications services in such a manner that customers have the ability to route automatically, without the use of any access code, their telecommunications to the telecommunications services provider of the customer's designation from among 2 or more telecommunications services providers (including such local exchange carrier)."[75]

Since the introduction of competition in the long-distance industry in 1984, many customers have grown accustomed to (1) dialing codes to access their long-distance service provider, (2) then dialing the phone number they are calling, and (3) finally entering their personal identification number. This process often requires 25 or more dialed digits. Recognizing this, Congress stated in the 1996 Act that parity must exist in local dialing.[76] Thus, if local calls can be made by dialing seven digits, then the customers of the competitive local exchange carriers must also be able to make local calls by dialing seven digits. Similarly, if the region has adopted a 10-digit local dialing requirement, then the customers of both incumbent local exchange carriers and competitive local exchange carriers must be able to make local calls with 10 digits. Congress did not want competition to be discouraged by the evolution of a situation similar to the "special access code," which required some customers to dial more digits to make a call in the long-distance market. Congress, instead, mandated *parity* in the number of digits dialed by customers of both the competitive local exchange carriers and incumbent local exchange carriers to place local calls.

Additional Services

In addition, section 251 (b)(3) of the 1996 Act requires that the incumbent local exchange carrier in each region must also provide to the competitive exchange carriers in that region, nondiscriminatory access to (1) telephone numbers to assign to customers, (2) operator services, (3) directory assistance, and (4) white-page directory listings. In this manner, Congress determined that if five telephone companies provided local telecommunications services in a city, residents would not have to have five different telephone books. Instead, the incumbent local exchange carrier is required to print, and distribute for free, the white pages. The yellow pages can be marketed and printed for profit in whatever manner the incumbent local exchange carrier selects. Similarly, operator services and directory assistance must be provided as separate *network elements* by the incumbent local exchange carriers, so that the competitive local exchange carriers can choose to purchase them from the incumbent local exchange carrier or provide their own service.

Network Elements

The 1996 Act defines a *network element* as "a facility or equipment used in the provision of a telecommunications service. Such term also includes features, functions, and capabilities that are provided by means of such facility or equipment, including subscriber numbers, databases, signaling systems, and information sufficient for billing and collection or used in the transmission, routing, or other provision of a telecommunications service."[77]

The FCC's implementing rules identify the following eight network elements:[78]

(1) Local loop
(2) Local switching
(3) Tandem switching
(4) Interoffice transmission
(5) Databases and signaling systems
(6) Operation support system
(7) Operator service
(8) Directory assistance

The *local loop* is the transmission link between the distribution frame at the central office and the network interface at the customer premises. The incumbent local exchange carrier must include conditioning requested by the customer, although unbundling of subloop elements, between concentrators and local drops, are not required by the 1996 Act or the FCC at this time.

Local switching is the connection between the local loop and trunk. It includes all lineside and trunkside functions, plus all vertical functions including dial tone, 911, and Custom Local Area Signaling Services. *Tandem switching* is the connection between trunks, including all related switch functions. *Interoffice transmission* is the link between the central office or wire center of the ILEC and CLEC, with all relevant network functions.

Databases and signaling systems are the functional and Advanced Information Network (AIN) databases that allow access to and work with the Signaling System 7 (SS7). They include the carriers' *Operation Support System* (OSS) or electronic service-ordering systems. At present, OSS is a key reason that many of the incumbent local exchange systems have not passed their section 271 checklist and was an area of tremendous activity from 1997 through 2001.

Operator service and *directory assistance* are two services that may be confusing to customers if widely duplicated. Thus, if a competitive local exchange carrier obtains these services from an incumbent local exchange carrier under a resale agreement, the incumbent local exchange carrier must provide each service as a separate element and rebrand if technically possible, or if not possbile, use no brand.

Before 1996, when a customer dialed *O* for the operator, or 411 for directory assistance, the customer heard something like "Thank you for calling US WEST. What city please?" After 1996, to meet the *nondiscriminatory* requirement in operator and directory assistance services, the ILEC operators answering calls for several companies either had to identify each company by name or use no company name or *brand* at all. In early experiments, it became too confusing for the operators to identify each company correctly. When they did not, it was confusing to customers who thought they were dialing the operator of one company, only to have a second company answer the call. For these reasons, most ILECs have chosen to use no brand. Instead, callers now hear something generic such as, "What city please?" with no identification of the carrier providing the operator service.

Local Exchange Carriers' Duty 4: Access to Rights-of-Way, § 251(b)(4)

Fourth, the 1996 Act requires all local exchange carriers "to afford access to the poles, ducts, conduits, and rights-of-way of such carrier to competing providers of telecommunications services on rates, terms, and conditions that are consistent with Section 224."[79] Section 244 states that, "All utilities must allow new entrant telecom providers to use their rights-of-way." Section 703 extends these duties to all public utilities including gas, electric, and water providers.

This requirement is critical because, as seen in the development of telegraph, telephone, cable television, and fiber optics systems today, rights-of-way are crucial to telecommunications distribution systems. Competition cannot evolve unless competing carriers have affordable access to equal rights-of-way. The four main points of concern with this requirement include the following: (1) the states are to determine the *reasonableness* of the rates on a case-by-case basis; (2) the owner of the right-of-way, pole, or duct may not reserve space for its own future needs; but (3) a CLEC may reserve space based on a long-range growth plan; and (4) other than section 252 of the 1996 Act, no guidance exists about how to resolve contention among companies for the limited space available.

Local Exchange Carriers' Duty 5: Reciprocal Compensation, § 251(b)(5)

Fifth, the 1996 Act requires all local exchange carriers "to establish reciprocal compensation arrangements for the transport and termination of telecommunications."[80] Initially, this requirement that local carriers compensate one another for communications traffic exchanged between them seemed quite reasonable. However, the interpretation and implementation of this requirement has become one of the most litigated issues in the post-1996 telecommunications industry.

Reciprocal compensation mandates that payments to a terminating carrier must be equal regardless of the direction of the flow of traffic or the type of carrier receiving the traffic. It requires that the fees paid be nondiscriminatory, or equal among carriers, and be applied only to traffic that is originated and terminated in a local area. The states are to define what is meant by "local area." This has raised several issues. First, in most cases, the difference between local and long-distance traffic is readily defined, but for commercial mobile radio wireless service carriers, reciprocal compensation applies to <u>all</u> traffic that originates and terminates in a single metropolitan trading area. Metropolitan trading areas are the license area of personal communications service carriers and only 51 MTAs exist in the United States. Therefore, what is considered local for wireless carriers covers a much larger area than for wireline communications. A second issue concerns traffic to and from ISPs. Payments to terminate this traffic are disproportionate among companies, raising issues about whether Internet-bound traffic is local or not. The details of the litigation analyzing this are discussed in Chapter 4.

3.5.3 Duties of Incumbent Local Exchange Carriers, § 251(c)

As defined in Section 3.3.2, an incumbent local exchange carrier (ILEC) is any local carrier that (1) provided telephone exchange service on February 8, 1996; (2) is large enough to be a member of the National Exchange Carrier Association (not a small, rural carrier); or (3) acquired sufficient market share to match or absorb the previous incumbent carrier's serving area. Therefore, the term incumbent local exchange carrier includes both the Bell Companies and the non-Bell companies such as GTE and dozens of other larger independent telephone companies that were providing telephone service on February 8, 1996. Since the incumbent local exchange carriers are also part of the *all carriers* and *local exchange carriers* categories discussed previously, they are responsible for the 8 duties required of those two groups. In addition, the 1996 Act imposes the following six specific obligations on the incumbent carriers, making a total of 14 duties required of all incumbent local exchange carriers.

Incumbent Local Exchange Carriers' Duty 1: Negotiate, § 251(c)(1)

First, the 1996 Act requires all ILECs "to negotiate in good faith" interconnection agreements with other competing carriers. The interconnection agreements must include specific acknowledgements of and accommodations for all of the duties of the LECs [local exchange carriers] and ILECs [incumbent local exchange carriers], including the three duties of telecommunications carriers, all five additional duties of LECs, and all six duties of ILECs."[81] The requesting telecommunications carrier also has a duty to negotiate the terms and conditions of the agreement in good faith. Section 252 addresses the details of the negotiation process and alternatives, such as arbitration, if either party delays, obscures, or thwarts the negotiations in some manner.

Incumbent Local Exchange Carriers' Duty 2: Interconnect, § 251(c)(2)

Second, the 1996 Act requires all incumbent local exchange carriers to

> *interconnect the facilities and equipment of any requesting telecommunications carrier, with the local exchange carrier's network—*
>
> *(A) for the transmission and routing of local telephone traffic and access to long distance service;*

(B) *at any technically feasible point within the carrier's network;*

(C) *with quality that is at least equal to that which the local exchange carrier (LEC) provides to itself, a subsidiary, affiliate, or any other party; and*

(D) *on rates, terms, and conditions that are just, reasonable, and nondiscriminatory.*[82]

This duty is more detailed than "Duty 1: Interconnection" of *all telecommunications carriers* and thus is less arguable.

Incumbent Local Exchange Carriers' Duty 3: Unbundle Access to Network Elements, § 251(c)(3)

Third, the 1996 Act mandates that an incumbent local exchange carrier shall

> *provide, to any requesting telecommunications carrier for the provision of a telecommunications service, nondiscriminatory access to network elements on an unbundled basis at any technically feasible point on rates, terms, and conditions that are just, reasonable, and nondiscriminatory in accordance with the terms and conditions of the agreement and the requirements of sections [251] and 252. An incumbent local exchange carrier shall provide such unbundled network elements in a manner that allows requesting carriers to combine such elements in order to provide such telecommunications service.*[83]

While the concept and purpose of *unbundling* and *network elements* were discussed in Section 3.4.1, this portion of the 1996 Act places specific responsibility on the incumbent local exchange carriers to unbundle the local network into elements and make them available to the new entrants "in a manner that allows requesting carriers to combine the elements in order to provide telecommunications service." Second, the law requires the incumbent local exchange carriers to provide these network elements at the same quality as they provide for their own customers, with no limitations on their use. Third, the incumbent local exchange carriers must provide the network elements upon request, not in a delayed manner. Fourth, the elements must be provided at *any technically feasible point* and at *rates, terms, and conditions that are just and reasonable.* Different states have divided the network into as many as 22 different elements, and the rates charged vary widely.

Incumbent Local Exchange Carriers' Duty 4: Resale, § 251(c)(4)

Fourth, the 1996 Act mandates that an incumbent local exchange carrier must "offer for resale any telecommunications service that the carrier provides at retail to subscribers who are not telecommunications carriers."[84] The Act also requires that this be done at wholesale rates and in a manner that does "not prohibit or impose unreasonable or discriminatory conditions or limitations on the resale of such telecommunications service." However, "a State commission may, consistent with regulations prescribed by the Commission [under the 1996 Act], prohibit a reseller that obtains at wholesale rates a telecommunications service that is available at retail only to a category of subscribers from offering such service to a different category of subscribers."

Congress included this requirement to allow new entrants to enter the local telecommunications market earlier than they could if they had to first build and install their own facilities. The new entrant can obtain full service, or any portion needed, from the incumbent local exchange carriers at wholesale rates and sell it to customers at retail rates. Each state determines the difference between the retail price of the service and the wholesale price to the competitive local exchange carrier.

The only guidance the 1996 Act provides to the states concerning setting the wholesale rate is found in Section 252(d)(3) of the 1996 Act. It states that "The wholesale rates for services to be resold shall be the retail rates minus the costs of marketing, billing, collection, and other costs that will be avoided" by the incumbent local exchange carrier since it no longer serves those customers. The issue that exists for the state commissions is the difference between *avoided* and *avoidable*. Many incumbent carriers argue that very few costs are avoided because even if they lose some customers, they must still have a billing system and run marketing campaigns. Beyond the cost of the stamps and envelopes to those few customers, all other costs remain despite the number of customers the incumbent local exchange carrier has or loses to other carriers. This argument is countered by the competitive local exchange carriers who argue that more costs could be avoided if the incumbent local exchange carriers were more serious about avoiding costs.

Incumbent Local Exchange Carriers' Duty 5: Notice of Changes, § 251(c)(5)

Fifth, the 1996 Act states that incumbent local exchange carriers have a duty "to provide reasonable public notice of changes in the information necessary for the use or interoperability of the ILEC's facilities and networks." The purpose of this duty is to prevent any surprises from the ILEC that would restrict interconnection and interoperability from occurring.

Incumbent Local Exchange Carriers' Duty 6: Collocation, § 251(c)(6)

Sixth, the 1996 Act states that incumbent local exchange carriers have a duty "to provide (for) physical collocation of equipment necessary for interconnection or access to unbundled network elements: (1) on rates, terms, and conditions that are just, reasonable, and nondiscriminatory; (2) at the premises of the local exchange carrier; and (3) the carrier may provide for virtual collocation only if the local exchange carrier demonstrates to the State commission that physical collocation is not practical for technical reasons or because of space limitations."

Webster's Dictionary defines *collocation* as residing side by side. *Physical collocation* means that both the incumbent local exchange carriers' (ILEC's) and competitive local exchange carriers' equipment are located in the ILEC's switch room. *Virtual collocation* means that two companies' equipment are connected through high-capacity trunks, but are actually located often miles apart. The 1996 Act's requirement for physical collocation was a surprise to many of the people working in the industry because most observers were expecting that Congress would require virtual collocation. The advantage to new entrants of physical collocation is that their equipment is in the same environment as the incumbent local exchange carrier's equipment. The disadvantage is that it complicates the ability of the competitive carriers to get to their equipment without an escort from the incumbent carrier. Both generally are items discussed in negotiating interconnection agreements.

3.5.4 Duties of Bell Operating Companies, § 271

Section 271 of the 1996 Act places 14 additional duties on the 20 Bell Operating Companies (BOCs) listed at 47 U.S.C. 153(4).[85] The definition of a Bell Operating Company also includes any successor or assignee of the above companies that provides wireline telephone exchange service. The definition excludes any affiliate of the above companies, other than an affiliate providing local telephone service. The 1996 Act defines an *affiliate* as "a person that (directly or indirectly) owns or controls, is owned or controlled by, or is under common ownership or control with, another person. For purposes of this paragraph, the term 'own' means to own an equity interest (or the equivalent thereof) of more than 10 percent."[86]

Since the Bell Operating Companies are also part of all carriers, local exchange carriers, and incumbent local exchange carrier categories discussed previously, they are responsible for the 14 duties required of those three groups. The additional 14 obligations required in Section 271 therefore make a total of 28 duties required of the Bell Operating Systems. The 14 extra duties for the Bell Operating Companies, are known as the *Fourteen-Point Checklist* to "pass their [section] 271 [requirements]."

Two important points should be noted concerning Section 271. First, its requirements do not affect GTE, Continental, United, or any of the hundreds of other non-Bell incumbent local exchange carriers. Second, Section 271 contains four subsections covering the following: (1) the markets and services that the Bell Operating Companies can provide immediately, (2) the markets and services that the Bell Operating Companies cannot enter until after it passes its Section 271 requirements, (3) the 14 points, and (4) the process to be followed by each company to pass their 271 requirements. Each is described next.

In-Region versus Out-of-Region Services, § 271(a)

Section 271(a) of the 1996 Act divides the new competitive U.S. telecommunications market into two groups: those that the Bell Operating Companies can enter immediately after February 8, 1996, and those in which the Bell Operating Companies must have FCC approval of the companies' Section 271 applications before they can enter.[87] These two markets are defined primarily as *in-region* and *out-of-region* markets, respectively.

In-region markets are the markets in which the Bell Operating Company was the previous monopoly holder[88] and in which it is still considered the dominant provider. In this market, the Bell Operating Company has the "home team advantages" of an installed base of customers, name recognition, and established billing systems to the majority of customers. In contrast, in the out-of-region markets, the Bell Operating Company is simply another competitive local exchange carrier (CLEC) that must attract customers from the established incumbent local exchange carrier (ILEC). First, when providing services out-of-region, the Bell Operating Company is a competitive local exchange carrier CLEC. For this reason, the 1996 Act allows the Bell Operating Companies, or their affiliates, to provide out-of-region services immediately after February 8, 1996[89] so long as the Bell Operating Companies do so through separate subsidiaries and subject to section (j) of the FCC's rules. In-region services are more regulated and cannot be openly provided until after the Bell Operating Company "passes its 271."

Among the regulated services are interLATA long-distance services. The 1996 Act defines the term *interLATA service* as "telecommunications between a point located in a 'local access and transport area' and a point located outside such area."[90] It also defines the term *local access and transport area* (*LATA*) as

> *a contiguous geographic area: (A) established before the date of enactment of the Telecommunications Act of 1996 by a Bell operating company such that no exchange area includes points within more than 1 metropolitan statistical area, consolidated metropolitan statistical area, or State, except as expressly permitted under the AT&T Consent Decree; or (B) established or modified by a Bell operating company after such date of enactment and approved by the Commission.*[91]

In long-distance services, however, the 1996 Act primarily distinguishes between long-distance services based on their point of origination. For example, the 1996 Act restricts a Bell operating company from providing interLATA long-distance services *originating* in any of its in-region states until it has "passed its 271" and obtained FCC authorization to do so. The only services excepted from this requirement are those services that are previously authorized, as defined in Section 271(f), or are *incidental* to the provision of another service, as defined in Section 271(g).[92] On the other hand, if the long-distance service originates out of region, the BOC may provide it immediately.

Incidental InterLATA Services, § 271(b)(3)

Similarly, the 1996 Act requires no authorization for a

> *Bell operating company (BOC), or any affiliate, to provide incidental InterLATA services*[93] *originating in any State (in-region and out-of-region) after February 8, 1996, including:*

(A) audio, video, or other programming services—such as cable television.

(B) customer interactive services providing the ability to select or respond to the audio, video or other programming services.

(C) alarm monitoring services, if the BOC was already providing them before February 8, 1996. The service must be continued, however, through a separate subsidiary.

(D) two-way interactive video services or Internet services over dedicated facilities to or for elementary and secondary schools as defined in Section 1996 Act 254(h)(5);

(E) commercial mobile services, including long-distance calls to and from the BOCs' cellular and PCS customers,

(F) services that permit a customer, located in one LATA, to retrieve stored information from, or file information in, information storage facilities operated by the BOC but located in another LATA;

(G) signaling information used in connection with the provision of telephone exchange services or exchange access by a local exchange carrier; or

(H) network control signaling information to, and receipt of such signaling information from, common carriers offering interLATA services at any location within the area in which such Bell operating company provides telephone exchange services or exchange access."[94]

Special Provisions, §§ 272–276

In addition to the InterLATA opportunities described in Section 271, Sections 272 to 276 of the 1996 Act outline several *special provisions* and additional opportunities open to the BOCs in the new competitive environment once they obtain approval of their separate 271 applications. The sections include:

272: Separate affiliate, safeguards

273: Manufacturing by Bell Operating Companies

274: Electronic publishing by Bell Operating Companies

275: Alarm-monitoring services

276: Provision of payphone service

Fourteen-Point Checklist, § 271(c)(2)(B)

The 1996 Act, Section § 271(c)(2)(B), is titled the "Competitive Checklist," but, as discussed earlier, it is also known as the "271 Fourteen-Point Checklist."[95] Its purpose is to answer the question of When does sufficient competition exist in a market to allow the Bell Operating Company in that market to be able to enter the restricted lines of businesses of out-of-region long distance, manufacturing of equipment or other services? Is the answer 10%, 40%, or 50%. By how many companies? The passage of § 271 is the benchmark adopted to answer these important questions, and thus it is critical to both the incumbent local exchange carriers and the competitive exchange carriers.

The checklist identifies fourteen items, that the Bell Operating Company must convince the FCC that it has met or accomplished in each state in which it provides local service as the incumbent local exchange carrier. The burden of proof rests on each Bell Operating Company to prove that it has adequately enabled competition in that local market. Table 3.2 on page 77 provides a summary of the 14 points. As can be noted from Table 3.2, the first two of the 14 points repeat two of the duties placed on all incumbent local exchange carriers: (1) proof that the Bell Operating Company has provided nondiscriminatory access for its competitors to its network elements, and (2) proof that it adequately interconnects with all competitors who request interconnection. The next 8 points, points 3 through 11, list the eight network elements identified in the FCC's *First Report and Order.* The Bell Operating Company (BOC) must show that it has made each element available in a nondiscriminatory manner and at a *reasonable price.* The last 3 of the 14 points, numbers 12 through 14, check that the requirements for interconnection, reciprocal compensation, and resale at a wholesale rate have been adequately met. The 14 points are as follows:

> *Access or interconnection provided or generally offered by a Bell operating company [BOC] to other telecommunications carriers must include each of the following requirements:*
>
> 1. *Interconnection: The BOC must show that it has adequately interconnected with the facilities and equipment and meets the requirements of sections 251(c)(2) and 252(d)(1).*
> 2. *Nondiscriminatory access to network elements: The BOC must show that it is offering nondiscriminatory access to unbundled network elements (UNEs) in accordance with the requirements of sections 251(c)(3) and 252(d)(1).*
> 3. *Nondiscriminatory access to the poles, ducts, conduits, and rights-of-way owned or controlled by the Bell Operating Company: The BOC must show that it is providing nondiscriminatory access to its poles, ducts, conduits and rights-of-way at just and reasonable rates in accordance with the requirements of section 224.*
> 4. *Local loop transmission from the central office to the customer's premises: The BOC must show that it has unbundled local loop transmission from local switching or other services, and is providing it as an unbundled service to its competitors on a nondiscriminatory basis at a fair and reasonable price.*
> 5. *Local transport from the trunk side of a wireline local exchange carrier switch: The BOC must show that it has unbundled from switching or other services.*
> 6. *Local switching: The BOC must show that it has unbundled local switching from transport, local loop transmission, or other services.*

7. *Nondiscriminatory access to—The BOC must show that it has unbundled 911 and E911 services; directory assistance services; and operator call completion services.*
8. *White pages: The BOC must show that it has unbundled directory listings for customers of the other carrier's telephone exchange service.*
9. *Until the date by which telecommunications numbering administration guidelines, plan, or rules are established, nondiscriminatory access to telephone numbers for assignment to the other carrier's telephone exchange service customers. After that date, compliance with such guidelines, plan, or rules.*
10. *Nondiscriminatory access to databases and associated signaling necessary for call routing and completion.*
11. *Until the date by which the Commission issues regulations pursuant to section 251 to require number portability, interim telecommunications number portability through remote call forwarding, direct inward dialing trunks, or other comparable arrangements, with as little impairment of functioning, quality, reliability, and convenience as possible. After that date, full compliance with such regulations.*
12. *Nondiscriminatory access to such services or information as are necessary to allow the requesting carrier to implement local dialing parity in accordance with the requirements of Subsection 251(b)(3).*
13. *Reciprocal compensation arrangements in accordance with the requirements of Subsection 252(d)(2).*
14. *Resale at wholesale rates. Telecommunications services are available for resale in accordance with the requirements of Subsections 251(c)(4) and 252(d)(3).*

The "271" Approval Process, § 271(d)

Although the Bell Operating Company must pass its 14-Point Checklist in each state, it is not the state PUC that makes the determination. Instead, the written application is approved by the FCC. However, the FCC does not make a unilateral decision. Instead it "consults" with two groups before it makes its decision: the PUC in the state concerned with the Bell Operating Company's application, and the attorney general at the U.S. Department of Justice. The question asked of the Department of Justice is primarily whether the Bell Operating Company is under investigation for corporate misdeeds, such as tax evasion, fraud, or sexual discrimination, and thus whether it is a good corporate citizen before the Bell Operating Company can be cleared to enter into other markets. The law requires that the Department of Justice respond within 30 days, which is reasonable given that the Department of Justice can simply check its records for past and current proceedings.

The question asked of the state PUC, on the other hand, is whether the Bell Operating Company has adequately met the 14 points in *that* state. Pursuant to state and federal Administrative Procedure Acts (APAs),[96] the PUC cannot unilaterally answer this question. Instead it must hold hearings to consider comments on the issue from the public, affected competitors such as AT&T, MCI, Sprint, and other competitive local exchange carriers, and any other party with standing. However, a public hearing requires at least 20 days notice before the hearing can begin, and for some reason, the 1996 Act only provided 20 days for the state to respond to the FCC's request for comment. Recognizing this conflict in timing, the telecommunications industry came up with a unique solution. The National Association of Regulatory Utility Commissioners (NARUC) drafted a "best practices letter" to be sent to the corporate executive officer of each affected Bell Operating Company, suggesting that it provide a 90-day *prior notice* to its state PUC *before* the Bell Operating Company made application to the FCC. This would allow the PUC time to hold the hearings and determine a response before the Bell Operating Company applied for "271" clearance from the FCC. All Bell Operating Companies readily agreed to do this since it was in their best interests. Without such correction in the timetable, the Bell Operating Companies would have great difficulty "passing their 271" at the FCC.

Once the FCC has the proper information, it then considers the responses and must make its determination on each Bell Operating Company's application within 90 days. Within 10 days following its approval, the FCC must publish a brief description of the determination in the Federal Register.[97] Once a Bell Operating Company passes its 271 requirements, it can become a full competitor in all markets, just as other companies.

3.6 CHALLENGES TO THE CONSTITUTIONALITY OF SECTION 271 REQUIREMENTS

Frustrated by the difficulty of clearing the Section 271 hurdle, the Bell Operating Companies challenged the constitutionality of Section 271. The process involved several steps.[98] As the Fifth Circuit Court of Appeals described it:

> On April 11, 1997, plaintiff SBC Communications, which is of course one of the RBOCs (Regional Bell Operating Companies), applied to the FCC pursuant to § 271 to have the long distance line-of-business restriction lifted for its local service area of Oklahoma. The FCC determined that the statutory criteria had not been met, and therefore denied the application on June 26, 1997. SBC appealed the ruling to the D.C. Circuit, where it was affirmed on March 20, 1998.[99]
>
> Without waiting for the outcome of that appeal, however, on July 2, 1997, SBC and its subsidiaries filed suit against the United States and the FCC in the Federal District Court for the Northern District of Texas, alleging that all of the Special Provisions were facially unconstitutional under

> the Bill of Attainder and Equal Protection Clauses [of the U.S. Constitution] and that § 274 violated the Free Speech Clause as well. Several long distance companies, including MCI Telecommunications Corp., Sprint Communications Company, and AT&T, the BOCs' erstwhile parent, intervened on the government's side in the dispute, and two other RBOCs, U.S. West Communications and Bell Atlantic Corp., intervened on SBC's. Bell Atlantic added a slightly more nuanced separation of powers challenge to SBC's other constitutional complaints.[100]
>
> On December 31, 1997, ruling on cross-motions for summary judgment, District Judge Kendall held that the Special Provisions constituted an unconstitutional bill of attainder and that they were severable from the rest of the Act. He therefore granted SBC's motion and declared the challenged sections void."[101]

The United States, the FCC, and the defendant-intervenors appealed to the United States Court of Appeals for the Fifth Circuit. The Fifth Circuit Court of Appeals reviewed the constitutionality of the federal statute *de novo* and, on September 4, 1998, reversed the lower court, holding that the special provisions are constitutional.[102] Since *bill of attainder* is a less familiar term, it warrants additional explanation.

3.6.1 Bill of Attainder

> Article I, sec. 9, cl. 3 of the United States Constitution provides that "[n]o Bill of Attainder or ex post facto law shall be passed [by Congress]. Article I, sec. 10, cl. 1 contains a parallel provision applicable to the states."

The term, *bill of attainder,* is not a commonly used or understood phrase, so as

> the Supreme Court has often clarified, [i]n forbidding bills of attainder, the draftsmen of the Constitution sought to prohibit the ancient practice of the Parliament in England of punishing without trial 'specifically designated persons or groups.'[103] Consistent with this characterization, the Court has generally defined a bill of attainder as 'a law that legislatively determines guilt and inflicts punishment upon an identified individual without provision of the protections of a judicial trial.'[104] Where, as here [in the Special Provisions in the 1996 Act], the liability in question clearly attaches by operation of the legislative act alone, the constitutional test may be summarized in the following two-pronged test: First, has the legislature acted with specificity? Second, has it imposed punishment?[105]

Concerning the first prong, *specificity,* the U.S. Fifth Circuit Court of Appeals, stated that

> Notwithstanding beguiling arguments that support the district court's holding, at bottom, we simply cannot find a constitutional violation in this case. Even assuming that the Bill of Attainder Clause applies to [specific]

corporations, and even assuming that the Special Provisions are sufficient to meet the specificity prong of the test, there simply cannot be a bill of attainder unless it is also the case that the Special Provisions impose punishment on the BOCs."[106]

Concerning the second prong, *punishment,* the Fifth Circuit discussed the various arguments and held that

For all of the foregoing reasons, we find that the Special Provisions ultimately are nonpunitive as an historical, functional, and motivational matter. They are therefore not an unconstitutional and odious bill of attainder as that term has been defined by the Supreme Court. To the extent that the district court concluded otherwise, it was in error, and its decision on that point is accordingly reversed."[107]

As noted above, however, SBC and the other appellees also urge three additional constitutional arguments as alternate bases for affirming the judgment of the district court. Having found the Special Provisions not to constitute a bill of attainder, we must obviously consider these alternate theories. We do so only briefly, however, as they are far less substantial."[108]

3.6.2 Separation of Powers

First, the appellees contend that the Special Provisions violate separation of powers because they address themselves to a particular judicial consent decree—the MFJ—in such a way as to alter the result. They rely on the well accepted rule that it violates separation of powers principles for Congress to reopen any adjudication that represents the "'final word of the judicial department' on a case. Yet under Pennsylvania v. Wheeling and Belmont Bridge Co., 59 U.S. (18 How.) 421, 15 L.Ed. 435 (1855), it has long been clear that Congress may change the law underlying ongoing equitable relief, even if, as in Wheeling itself, the change is specifically targeted at and limited in applicability to a particular injunction, and even if the change results in the necessary lifting of that injunction. . . . In the light of [numerous precedents], we simply cannot see a separation-of-powers problem based on the Special Provisions' interference with the MFJ in this case.[109]

Therefore, the "Special provisions of Telecommunications Act of 1996 imposing line-of-business restrictions on named Bell operating companies (BOCs) do not violate the constitutional requirement of separation of powers by replacing restrictions imposed by judicial consent decree."[110]

3.6.3 Equal Protection Clause

The appellees next argued that the Special Provisions violate the Equal Protection Clause by discriminating against the BOCs by name. Under City of

> New Orleans v. Dukes, *427 U.S. 297, 96 S.Ct. 2513, 49 L.Ed.2d 511 (1976), however, specification of named parties in economic regulation is clearly permissible for equal protection purposes so long as the regulation is rationally related to a legitimate governmental interest and does not trammel fundamental personal rights or draw upon inherently suspect distinctions such as race, religion, or alianage. As should be manifest from the entire history of this area of the law, regulation of an LEC's conduct in the local telephone service market neither restricts fundamental individual rights nor lacks rational relation to the government's legitimate interest in ensuring greater competition in all telecommunications markets. Furthermore, the specification of the BOCs in the Special Provisions at issue here was not based on invidious criteria like race, religion, or alienage. As such, the Special provisions are not inconsistent with the Equal Protection Clause.*[111]

3.6.4 Bell Operating Companies' Right to Free Speech

> Finally, the appellees urge that, even if the other Special Provisions are allowed to stand, § 274 must go as it impermissibly infringes the Bell Operating Companies right to free speech. The D.C. Circuit Court recently rejected an identical challenge to § 274 by another regional Bell Operating Company, however, see BellSouth, 144 F.3d at 67-71, and we can find no reason to disagree with the result and analysis. Because § 274 does not in any way differentiate speech on the basis of content, its speech restricting provisions are subject only to (at most) intermediate scrutiny review under *Turner Broadcasting System, Inc. v. FCC,* 512 U.S. 622, 642, 114 S.Ct. 2445, 129 L.Ed.2d 497 (1994) (Turner I). Under that standard, a restriction will be upheld

if it advances important governmental interests unrelated to the suppression of free speech and does not burden substantially more speech than necessary to further those interest. Obviously the competition-embracing interests discussed above are manifestly sufficient to meet the first hurdle. Furthermore, because § 274 merely imposes a structural separation requirement on speech activities, not an absolute bar, its restrictions are practically de minimis in this necessarily corporate context, and certainly do not burden substantially more speech than necessary to accomplish its legitimate goals. For these reasons, the contention that § 274 violates the BOCs right to free speech is entirely lacking in merit.[112]

CONCLUSION

The details, duties, and obligations incorporated in the Telecommunications Act of 1996 create the most extensive change in U.S. communications law since

1934. While the Act is constantly being challenged and refined, it is important for participants in the telecommunications industry to understand the Act in order to more fully evaluate the opportunities and restrictions that it presents to them. This understanding also assists participants in analyzing the ongoing issues as discussed in Chapter Four.

ENDNOTES

[1]The author sincerely thanks Roger Newell, J.D., a colleague at Telecommunications Research Associates (TRA), for Mr. Newell's and TRA's permission to use in this chapter the materials developed by Mr. Newell summarizing the duties and obligations of telecommunications carriers. Mr. Newell can be reached at *www.rnewell@tra.com*. TRA can be reached at *www.tra.com* or 1-800-872-4736 in St. Marys, Kansas.

[2]Telecommunications Act of 1996, Pub. L. No. 104-104, purpose statement, 110 Stat. 56, 56 (1996) [hereinafter Telecommunications Act of 1996].

[3]The Telecommunications Act of 1996 was sent to Congress for enactment on January 3, 1996 as S. 652. Its final form was a compromise created by the Conference Committee from versions drafted by the House of Representatives (HR) in HR Report 104-458, 104th Congress, 2d Session, and the Senate in Senate Report S. 652, 104th, 2d Session, passed on July 19, 1995, and printed on June 23, 1995 [hereinafter S.652].

[4]The MFJ refers to the order entered August 24, 1982, in the antitrust action *United States v. Western Electric,* Civil Action No. 82-0192, in the United States District Court for the District of Columbia and implemented on Janurary 1, 1984 [hereinafter MFJ]. The term "AT&T Consent Decree" was substituted for "Modification of Final Judgment" in the Joint Explanatory Statement "in order to characterize more accurately the intent of the Senate bill and House-amendment with respect to the supersession issues addressed in title VI. In Section 3(a)(34) of the 1996 Act, the term "AT&T Consent Decree" includes any judgment or order with respect to such action entered on or after August 24, 1982" [hereinafter AT&T Consent Decree].

[5]Telecommunications Act of 1996, *supra* note 2, at § 601(a).

[6]MFJ, *supra* note 4.

[7]Iowa Utilities Bd. v. FCC, 109 F.3d 418, 421 (8th Cir. 1996) [hereinafter Iowa I].

[8]S. 652, *supra* note 3, at § 5.

[9]47 U.S.C. § 251(c)(2)(B).

[10]47 U.S.C. § 251(c)(2)(D), 252 and 271.

[11]Telecommunications Act of 1996, *supra* note 2, at § 706(a).

[12]*See, for example,* 47 U.S.C. § 251(d)(1) (1996).

[13]*See, for example,* 47 U.S.C. §§ 251(d)(3), 251(f), and 252 (1996).

[14]47 U.S. C. §§ 251(d)(1), 251(d)(3), 251(f), and 252 (1996).

[15]The changes are codified at 47 U.S.C. § 153 (1996).

[16]Kevin Werbach, Counsel for New Technology Policy, *The Digital Tornado: The Internet and Telecommunications Policy,* FCC, Office of Plans and Policy, Working Paper No. 29, March 1997, at 30.

[17]Telecommunications Act of 1996, *supra* note 2, at § 714.

[18]*Id.* at § 3(a)(2)(48); codified at 47 U.S.C. § 153(43) (1996).

[19]*Id.* at § 3(a)(2)(51), codified at 47 U.S.C. § 153(46) (1996).

[20]*Id.* at § 3(a)(2)(49), codified at 47 U.S.C. § 153(44) (1996).

[21]*Id.* at § 714(k)(3).

[22]*Id.* at § 3(a)(2)(44), codified at 47 U.S.C. § 153(26) (1996).

[23]*Id.* at § 3(a)(2)(40), codified at 47 U.S.C. § 153(16) (1996).

[24]*Id.* at § 3(a)(2)(44), codified at 47 U.S.C. §§ 153 (26), (27), and (33) (1996).

[25]*Id.* at § 704, codified at 47 U.S.C. § 332(c) (1996).

[26]*Id.* at § 251(h)(1) (1996).

[27]47 U.S.C. § 251(h)(2) (1996).

[28]47 U.S.C. § 251(h)(1)(B)(i) (1996).

[29]Telecommunications Act of 1996, *supra* note 2, at § 3(a)(2)(47), codified at 47 U.S.C. § 153(37) (1996).

[30]*Id.* at § 251(f)(1), codified at 47 U.S.C. § 251(f)(1) (1996).

[31]*Id.* at § 3(a)(2)(35), codified at 47 U.S.C. § 153(4) (1996).

[32]*Id.* at § 3(a)(2)(50), codified at 47 U.S.C. § 153(45) (1996).

[33]*Id.* at § 3(a)(2)(38), codified at 47 U.S.C. § 153(14) (1996).

[34]47 U.S.C. § 251(c)(2) (1996).

[35]47 U.S.C. § 251(c)(3) (1996).

[36]47 U.S.C. § 251(c)(4). *See also,* Iowa Utilities Board v. FCC, 120 F.3d 753, 753 (1997) [hereinafter Iowa II].

[37]*Id.* at note 3.

[38]Telecommunications Act of 1996, *supra* note 2, at § 251(c)(2), codified at 47 U.S.C. § 251(c)(2) (1996).

[39]47 U.S.C. § 251(c)(3) (1996).

[40]47 U.S.C. § 251(c)(4) (1996).

[41]Notice of Proposed Rulemaking (NPRM) Concerning Implemention of the Local Competition Provisions in the Telecommunications Act of 1996, CC Docket No. 96-98 (May 1996). [hereinafter May 1996 FCC NPRM to implement 1996 Act].

[42]First Report and Order, Implementation of the Local Competition Provisions in the Telecommunications Act of 1996, CC Docket No. 96-98 (Aug. 8, 1996) [hereinafter August 8, 1996, FCC Order implementing 1996 Act].

[43]Iowa II, *supra* note 36, at note 6.

[44]*Id.* at 755-756.

[45]*Id.*

[46]*Id.* at note 7 citing § 251(b)(5) of the 1996 Act.

[47]*Id.* at 754-755.

[48]*See* 28 U.S.C. § 2112(a)(3) (1994).

[49]Iowa I, *supra* note 7, motion to vacate stay denied, 117 S. Ct. 429 (1996).

[50]47 U.S.C. § 252(i); 47 C.F.R. § 51.809.

[51]Federal Communications Commission and the United States, applicants, v. Iowa Utilities Board, et al., No. A-299, Nov. 12, 1996 [Former decision, 117 S.Ct. 378]; and the Association for Local Telecommunications Services, et al., applicants, v. Iowa Utilities Board, et al., No. A-300, Nov. 12, 1996 (Former decision, 117 S.Ct. 379).

[52]Iowa II, *supra* note 36.

[53]Telecommunications Act of 1996, *supra* note 2, at § 252, codified at 47 U.S.C. § 252 (1996).

[54]*Id.* at §§ 251(c) (1), 252(a)(1), codified at 47 U.S.C. §§ 251(c) (1), 252(a)(1) (1996).

[55]47 U.S.C. § 252 (1996).

[56]Telecommunications Act of 1996, *supra* note 2, at Title 1, Subtitles A and B.

[57]47 U.S.C. §§ 202-276 entitled Title II-Common Carriers.

[58]47 U.S.C. §§ 251-261.

[59]Telecommunications Act of 1996, *supra* note 2, at §§ 271-276, codified at 47 U.S.C. §§ 271-276.

[60]47 U.S.C. § 153(44) (1996).

[61]Telecommunications Act of 1996, *supra* note 2, at § 251(a)(1), codified at 47 U.S.C. § 251(a)(1) (1996).

[62]Communications Act of 1934, 47 U.S.C. § 201(a).

[63]Telecommunications Act of 1996, *supra* note 2, at § 251(a)(1), codified at 47 U.S.C. 251(a)(1) (1996).

[64]*Id.* at § 251(a)(2), citing § 255, codified at 47 U.S.C. § 251(a)(2) (1996).

[65]*Id.* at § 255, codified at 47 U.S.C. § 255 (1996).

[66]*Id.* at § 251(a)(2), citing § 256, codified at 47 U.S.C. § 251(a)(2) (1996).

[67]*Id.* at § 256.

[68]47 U.S.C. § 251(f) (1996).

[69]Joint Explanatory Statement of the Committee of Conference, Conference Agreement on New Section 251—Interconnection (1996).

[70]Telecommunications Act of 1996, *supra* note 2, at § 251(b)(1), codified at 47 U.S.C. § 251(b)(1) (1996).

[71] *AT&T Competitive Strategy,* Wall St. J., June 12, 1996.

[72] Telecommunications Act of 1996, *supra* note 2, at § 251(b)(2); codified at 47 U.S.C. § 251(b)(2) (1996).

[73] *Id.* at § 3(46); codified at 47 U.S.C. § 153(30) (1996).

[74] *Id.* at § 251(b)(3); codified at 47 U.S.C. § 251(b)(3) (1996).

[75] *Id.* at § 3 (39); 47 U.S.C. § 153(15) (1996).

[76] 47 U.S.C. § 251(b)(3) (1996).

[77] Telecommunications Act of 1996, *supra* note 2, at § 3 (45); 47 U.S.C. § 153(29) (1996).

[78] *In re* Implementation of the Local Competition Provisions in the Telecommunications Act of 1996, First Report and Order, FCC 96-325, 11 FCC Rcd 15499, 1996 FCC LEXIS 4312 (1996).

[79] Telecommunications Act of 1996, *supra* note 2, at § 251(b)(4); codified at 47 U.S.C. § 251(b)(4) (1996).

[80] *Id.* at § 251(b)(5); codified at 47 U.S.C. § 251(b)(5) (1996).

[81] *Id.* at § 251(c)(1); codified at 47 U.S.C. § 251(c)(1) (1996).

[82] *Id.* at § 251(c)(2); codified at 47 U.S.C. § 251(c)(2) (1996).

[83] *Id.* at § 251(c)(3); codified at 47 U.S.C. § 251(c)(3) (1996).

[84] *Id.* at § 251(c)(4); codified at 47 U.S.C. § 251(c)(4) (1996).

[85] *Id.* at § 3(35); codified at 47 U.S.C. § 135 (4) (1996).

[86] *Id.* at § 3(33); codified at 47 U.S.C. § 135 (1) (1996).

[87] 47 U.S.C. § 271(d)(3) (1996).

[88] 47 U.S.C. § 271(i)(1) (1996).

[89] 47 U.S.C. § 271(b)(2) (1996).

[90] Telecommunications Act of 1996, *supra* note 2, at § 3(42); codified at 47 U.S.C. § 135 (21) (1996).

[91] *Id.* at § 3(43); codified at 47 U.S.C. § 135 (25) (1996).

[92] Joint Explanatory Statement of the Committee of Conference, Conference Agreement on New Section 271 (1996).

[93] 47 U.S.C. § 271(g) (1996).

[94] 47 U.S.C. § 271(b)(3) (1996).

[95] Telecommunications Act of 1996, *supra* note 2, at § 271(c)(2)(B); codified at 47 U.S.C. § 271(c)(2)(B) (1996).

[96] Administrative Procedure Act, 60 Stat. 237, (5 U.S.C.).

[97] 47 U.S.C. § 271(d)(5) (1996).

[98] Eric M. Swedenburg, *Promoting Competition in the TC Market: Why the FCC Should Adopt a Less Stringent Approach to 271,* Cornell L. Rev., V. 8, at 1418, 1420 (1999).

[99] SBC Communications, Inc. v. FCC, 154 F.3d 226, 232-233 (5th Cir. 1998), describing SBC Communications, Inc. v. FCC, 138 F.3d 410 (D.C. Cir. 1998).

[100] *Id.* 154 F.3d 226, 233 (5th Cir. 1998).

[101] *Id.*

[102] United States v. Bailey, 115 F.3d 1222, 1225 (5th Cir. 1997), *de novo* review: as cited in SBC Communications, Inc. v. FCC, *supra* note 99, at 233.

[103] Selective Service v. Minnesota Public Interest Research Group, 468 U.S. 841, 847 (1984) (quoting United States v. Brown, 381 U.S. 437, 447 (1965)).

[104] *Id.,* quoting Nixon v. Administrator of General Services, 433 U.S. 425, 468 (1977).

[105] SBC Communications, Inc. v. FCC, *supra* note 99, at 233.

[106] *Id.* at 234, citing Plaut v. Spendthrift Farm, Inc., 514 U.S. 211 (1995).

[107] *Id.* at 244.

[108] *Id.*

[109] *Id.*

[110] *Id.* at 227.

[111] *Id.* at 246.

[112] *Id.* at 247.

CHAPTER 4 Outstanding Issues from the Telecommunications Act of 1996

AS WITH ANY NEW LEGISLATION OR MAJOR CHANGE IN EXISTING legislation, a number of issues remain unresolved years after the enactment of the Telecommunications Act of 1996. Among these are four main issues that continue to dominate the Federal Communications Commission's (FCC's) calendar and various court dockets: (1) local number portability, (2) universal service, (3) access charges, and (4) reciprocal compensation. The resolutions of these four issues will significantly impact the future of the telecommunications industry in the United States and, to some extent, around the world. They also affect each user of telecommunications service because these issues appear as line items on each user's monthly telephone bill.

The purpose of this chapter is to identify the primary concerns and unresolved arguments in each of these four areas in order to enable the reader to track the development of these issues as they occur and to appreciate the importance of their final resolutions. To do this, Section 4.1 discusses local number portability, Section 4.2 outlines the current issues in universal service, and Sections 4.3 and 4.4 describe the arguments in access charges and reciprocal compensation.

4.1 LOCAL NUMBER PORTABILITY

In drafting the Telecommunications Act of 1996, Congress recognized that a person's telephone number is an important piece of personal information that is disruptive and often costly when changed. For this reason, Congress acknowledged that if individuals or companies were required to change their phone numbers whenever they changed local telephone service providers, they probably would not change providers as readily. Therefore, to encourage competition, Congress mandated in the Telecommunications Act of 1996 that people be allowed to take or *port* their local telephone numbers with them when they changed local telephone service providers. In Section 251(b)(2) of the Telecommunications Act of 1996, Congress declared that all local exchange carriers including incumbent local exchange carriers, competitive local exchange carriers, and wireless providers have a "duty to provide, to the extent technically feasible, number portability" in accordance with FCC requirements.[1]

The deadline for compliance by landline companies in the 100 largest telephone markets in the United States originally was December 31, 1998 and in markets under the 100 largest within six months after another carrier made a request. Until that time, the carriers must forward all calls on the ported numbers to the new carrier. This is referred to as *interim number portability.* After the compliance dates, *permanent number* portability will "dip" into a database to find the correct carrier. Due to implementation problems, however, the deadlines were extended somewhat. Wireless phones, including cellular, personal communications systems and certain specialized mobile radio, are required to provide long-term number portability beginning November 24, 2002, and to be completed nationwide within 18 months.

Technology To Be Used

With these deadlines, Congress did not specify a technology or method to be used for local number portability, but instead, asked the FCC to set only minimum performance requirements and left the detailed decisions about implementation to the telecommunications industry. To facilitate this process, Congress established a Federal-State Joint Board (Joint Board) to be responsible for the process of implementing local number portability. The federal portion of the Joint Board consisted of staff from the FCC, while the state portion consisted of staff from whichever state public utility commissions (PUCs) could afford to participate. Since the requirements of the 1996 Act were unfunded mandates on the states, not all states could afford to participate. Therefore, representatives from industry comprised the majority of the Joint Board.

Immediately after the 1996 Act was signed, the Local Number Portability Joint Board began meeting. The LNP Joint Board met for several weeks to review the technological options for local number portability, selected the preferred method (originating and next-to-last "database dip"), issued a request for proposals (RFP) to build the database, received and evaluated responses to the RFP, and selected two winners to develop the required database: (1) Ross Perot's EDS and (2) Lockheed-Martin. On August 18, 1997, Congress approved the LNP Joint Board's two selections and work on the database began. EDS has since dropped out of the process, but work continued under Lockheed-Martin.

In general, local number portability of landline numbers is possible by referencing the first six digits of a standard ten-digit telephone number. This is known as a *six-digit look-up.*

Ongoing Issues

Several issues concerning local number portability exist. First, the cost of developing and implementing number portability originally was shared by all carriers in proportion to each carrier's revenue, traffic, number of lines, and number of subscribers. Second, if a user has several lines, the charge is paid on each line. However, many consumers do not have competitive options for telephone service in their markets and therefore cannot port their numbers. These consumers question why they must pay this charge, and several consumer advocacy groups have taken up their cause. Third, once fully implemented, the database used for local number portability must be maintained and updated. These charges will continue to appear on consumers' monthly bills. Fourth, the technical processes that make local number portability available, including a centralized database of ported numbers, create a new portion of the telecommunications industry that has never existed previously, but touches every telephone call made in the United States. This places great importance on the function, accuracy, and control of the database. Fifth, copies of the database must be located in telephone switching equipment throughout the United States, even in privately owned switches, of which millions exist in the United States. Thus, the implementation and maintenance of local number portability involves nearly every telephone switch in the United States and hundreds of thousands of hardware and software personnel. Sixth, part of the good news regarding local number portability is that it will position the U.S. telecommunications industry to provide one number, or *personal communications service* (PCS) as the term is used worldwide. This is expected to help relieve the telephone number shortage arising and places the U.S. telecommunications industry in a position to provide the up-to-date services and technologies required in an expanded global, converged voice/data/video communications environment. For these reasons, local number portability has been the source of tremendous activity in the United States and an area whose final resolution and ongoing presence will touch every telephone user in the United States for as far into the future as we can see.

4.1.1 Three Types of Number Portability

Actually, three types of number portability exist: provider portability, service portability, and geographic portability. *Provider portability* is the ability of end users to retain the same telephone numbers when changing from one telephone company, or service provider, to another. *Service portability,* is the ability of end users to retain the same telephone numbers when they use a service, such as 800 numbers, or change from one service, such as basic telephone or "plain old telephone service" (POTS), to another service, such as Integrated Services Digital Network (ISDN). *Geographic portability,* sometimes also called location portability, is the ability of end users to retain the same telephone numbers when they move from one location to another, either within the same city or to another city. Historically, customers have usually been able to keep their number if they moved within the area served by the same central office, but could not keep their number if they moved to an area served by a different central office. In all three cases, number portability requires service without change or reduction in service quality, reliability, or convenience.

It is important to note, however, that the only type of number portability required in the Telecommunications Act of 1996 is the first type, *provider portability*. Thus, users are able to port or take their local telephone numbers when they switch from one telecommunications carrier to another, but only while remaining at the same location. The 1996 Act does not require that users be able to keep their telephone number when they move from one location to another or change service. However, some telephone service providers do make that available, as with nationwide, toll-free wireless services.

4.1.2 The North American Numbering Plan

Number portability is possible technically because each telephone number identifies both the specific customer being called and the terminating telephone company's equipment receiving the call, known as the *terminating switch*. In the United States, telephone numbers are based on the 10-digit North American Numbering Plan developed by AT&T in the late 1940s and implemented by AT&T in the early 1950s. As is familiar to most people, the first 3 digits of the 10 are the area code, more formally known as the Numbering Plan Area (NPA) code. The term *area codes* implies location or a particular geographic area. Thus 202 identifies calls bound for the Washington, D.C. area. However, area codes such as 800 indentify "called party pays" services; 900 identifies "Pay per Call" services, and "500" identifies "Follow Me" services.

The second 3 digits of the 10-digit U.S. telephone number identify the switch receiving the call, known as the *terminating switch*. These are known as the *exchange* or *NXX* code. Since the exchange code routes calls to a specific central office switch, owned by a specific service provider and historically connected to a specific customer's location identified by the last 4 of the 10 digits of the telephone number, exchange codes have been inherently nonportable. To port numbers, calls were forwarded or redirected to a second or third switch. Modern switches, using Signaling System Seven (SS7) technology with databases, however, offer some options to facilitate all three types of number portability.

AT&T managed the North American Numbering Plan (NANP) until 1984 when, with the divestiture of the Bell System, Bellcore (now called Telcordia) was given the responsibility of administrating the NANP.[2] In 1993, Bellcore advised the FCC that it wished to relinquish this responsibility pending industry and/or regulatory resolution of the administrator issue. Telcordia stated that the increasing number of new entrants into the telecommunications market made continuation of the present form of administration untenable.

4.1.3 Pre-1996 Number Portability in the United States

Number portability was not a new concept created by Congress in 1996 solely to encourage competition. In fact, following AT&T's implementation of SS7 technology in the late 1970s and early 1980s, the telecommunications industry began exploring issues and options for number portability in order to provide services such as 800 service, commercial mobile radio service (CMRS), 500 and 900 service portability.

1986—Full 800 Service Portability

In 1986, the FCC formally opened Docket No. 86-10 on 800 Access and issued a Notice of Proposed Rulemaking (NPRM) on the service.[3] Work on the docket continued until 1991 when the FCC mandated number portability for 800 numbers to allow customers to retain their 800 number even if they moved and thus changed the termination location for the calls, changed 800 service providers, or obtained 800 service from multiple providers.[4] Pursuant to the FCC's Order, local telephone companies implemented the 800 Database Access System in 1993.

1991—Partial Wireless, 500, and 900 Service Portability

At the same time, partial number portability systems were also implemented in roaming on wireless phones (commercial mobile radio service), personal communications service "500, Follow Me," and "900, Caller Pays" services. These three services allowed customers to receive calls at other than their billing address.

4.1.4 Interim Methods

Two primary interim methods of number portability evolved after 1996: (1) remote call forwarding and (2) flexible direct inward dialing (Flex-DID). Three additional derivations of these two evolved: (3) enhanced remote call forwarding (ERCF); (4) route index/ portability hub; and (5) hub routing with advanced intelligent network (AIN) capacity. All five methods are transparent to the calling party. The details and differences of each of these five are important to attorneys and others interested in telecommunication law because they (1) are used in legal arguments concerning LNP; (2) are critical in contracts and interconnection agreements concerning LNP; and (3) impact the quality and cost of the resulting services.

Remote Call Forwarding

Remote call forwarding (RCF) requires carrier A (the customer's original carrier and operator of the central office switch designated in the NXX code in the customer's telephone number) to receive the call, translate the dialed number into a new number with the NXX of carrier B, and transfer it to carrier B, the customer's new local service provider. Carrier B then completes the routing of the call to its customer.

While the remote call forwarding method works fairly well and is used in most states, several concerns or limitations exist with it that drive the industry to find a longer-term solution. First, since the remote call forwarding method involves a translation from the dialed telephone number to a new number, this method requires the use of two, 10-digit telephone numbers. The United States is running out of telephone numbers, and the use of remote call forwarding for number portability exacerbates this problem.[5] Second, because remote call forwarding places a second call to a transparent telephone number, it may degrade transmission quality and generally will *not* support several custom local area signaling services (CLASS), such as caller ID. Third, remote call forwarding is capable of handling only a limited number of calls to customers of the same competing service provider at any one time, and thus may create call blockages. Fourth, since calls to customers who have selected new service providers must still be routed initially to the customers' original service provider, the incumbent local exchange carrier is always involved in the routing of those customers' calls. This is an inefficient use of the networks and may preclude the development of efficient competing networks. However, the incumbent local exchange carrier has little incentive to provide efficient routing services to their competitors. Additionally, since all terminating interstate calls pass over the incumbent local exchange carrier's network, the local carrier and not the competing local service provider recovers the interstate access charges from the IXCs under the existing access charge regime.

Flexible Direct Inward Dialing

Flexible Direct Inward Dialing (Flex-DID) is similar to remote call forwarding in that the original service provider (carrier A) and the new service provider (carrier B) perform the same functions, and all calls must be routed to carrier A. Unlike remote call forward, however, carrier A does not translate the dialed number, but instead routes calls to the ported number over a dedicated facility to carrier B's switch.

Flex-DID has many of the same limitations as remote call forwarding, including degradation of transmission quality and the inability to support certain custom local signaling service features such as caller ID. While Flex-DID can process more simultaneous calls to a competing service provider than remote call forwarding, there is still a significant limitation on how many calls it can process at any one time. Flex-DID is used in various states.

Derivations of the Two Methods

Several derivations of remote call forwarding and Flex-DID exist. The three most common are (1) enhanced remote call forwarding, (2) route index/portability hub, and (3) hub routing with advanced intelligent networks (AIN). Like remote call forwarding and Flex-DID, all three route incoming calls to carrier A first, resulting in service degradation and the loss of custom local signaling service and functionality. All three differ from remote call forwarding and Flex-DID in that they use the local carrier's tandem switches to aggregate calls to carrier B before the calls are routed. Use of the tandem switches is usually more efficient because it alleviates the need for direct connections between every end office and the switches owned by the various carriers B.

1. Enhanced remote call forwarding. With enhanced remote call forwarding (ERCF), calls are first routed to carrier A and then assigned an ERCF translation consisting of the same number preceded by a 10XXX prefix. The XXX is carrier B's ID code. The resulting 12- to 15-digit number is then sent to a tandem switch that recognizes the 5-digit prefix, strips it off, and routes the call to carrier B.
2. Route index/portability hub. Under route index/portability hub (RI/PH), carrier A inserts carrier B's ID code, called an iXX prefix, at the front of the telephone number. The resulting 10- to 13-digit number is then sent to the tandem switch that recognizes the prefix, strips it off, and routes the call to carrier B.
3. Hub routing with AIN. Under hub routing with AIN (HR/AIN), carrier A's switch queries or "dips into" a remote database containing routing information. It then adds this routing information to the number and sends the call to a tandem switch, which then routes it to carrier B's switch. This method requires that carrier A's switch be equipped with the ability to query a database and that the database be up-to-date.

4.1.5 Long-Term Database-Dip Methods

All long-term or permanent LNP solutions suggested involve querying or "dipping" into a database containing information about the ported telephone numbers before the number is processed. Four different versions of database dip were suggested including the following: (1) database dip at the originating end (by carrier A); (2) database dip by the first carrier to reach the portability island; (3) database dip by next-to-last carrier; and (4) database dip at the terminating end (by carrier B). All require use of database technologies, Signaling System 7 (SS7), and Advanced Intelligent Networks (AIN). This is a problem in rural areas that do not have SS7 and AIN. Additionally, determination of whether the called number has been ported or not must be made for every call. On the other hand, the four methods differ in the number of switches that must be capable of making database queries, the ability to route calls in the most efficient manner, and the ability to avoid duplicate database dips. The details are explained in the following sections.

4.1.6 Early Number Portability Regulation

July 13, 1995—NPRM on Telephone Number Portability, CC Docket No. 95-116, RM 8535

In 1994 and 1995, the FCC began preparing for the anticipated directives from Congress to provide number portability to general telephone users and service consumers. As a result, on July 13, 1995, the FCC opened a new docket on number portability (Docket No. 95-116), and issued a "Notice of Proposed Rulemaking (NPRM) In the Matter of Telephone Number Portability."[6] In the NPRM, the FCC noted the current thinking in the industry stating that (1) number portability is of benefit to consumers, (2) it contributes to the development of competition among alternative providers of local telephone and other telecommunications services, and (3) the FCC should assume a leadership role in developing a *national* number portability policy. With the release of this document, the FCC sought public and industry comments on the topic of number portability and the appropriate FCC role.[7]

July 13, 1995—NPRM on Administration of the NANP

On the same day, the FCC also released its "Report and Order in the Matter of Administration of the North American Numbering Plan" (NANP) [CC Docket No. 92-237], in which it recommended that the two issues of administration of the NANP, and administration of a long-term database solution to number portability may be merged into one national solution.[8] See Appendix A for a list of the documents and court cases generated as part of the search for that solution.

4.1.7 Ongoing Issues

While the technical details of both interim and permanent local number portability (LNP) are being resolved, a number of issues remain open. For most telephone numbers (both landline and wireless), the open issues include:

1. What additional monetary and nonmonetary costs are associated with further implementation and ongoing maintenance of LNP?
2. What are the feasibility, issues, limitations, and costs of the continued transition to a permanent number portability environment?

3. What assurances can protect 911 and other critical calls to and from ported numbers?
4. What is the fairest way to recover costs between carriers? At one time, NYNEX and other carriers charged the competitive local exchange carriers a monthly, per line fee that placed responsibility for the costs on the parties who directly benefit from the number portability. However, this deters consumers from changing providers. Therefore, what alternative cost recovery methods are available?

For nongeographic telephone numbers such as 800, 900, and 500 services, the open issues include:

1. Will service portability allow customers to respond more readily to service and price differences among service providers, promoting competition?
2. What are the monetary and nonmonetary costs associated with implementing LNP?
3. What are the feasibility, issues, limitations, and costs for these services.

For long-term considerations, the open issues include:

1. Who will be the administrator and/or provider of the service management systems administering the database?
2. Who are the candidates for a suggested neutral third-party number administrator?
3. How will the owner or operator of the LNP database be selected?
4. How will the costs of maintaining the LNP database be recovered?
5. By what date will the administrative duties transfer?
6. Will LNP result in lower prices and enhanced service?

Initially, the costs to implement LNP capability have been covered by the industry members, but in May 1999,[9] many telephone service providers received permission from their state commissions to begin passing the cost on to their customers. Therefore, in most states, a local number portability charge now appears on telephone customers' monthly telephone bills. The amount of the charge varies widely across the United States, ranging from 24¢ to over $1.18. Generally, the charge for business customers is unspecified, but appears as a fee. While wireless companies must comply by November 24, 2002, no word has been received about how or to what extent wireless customers will be charged for local number portability.

While an expensive and unsettled process, the purpose of local number portability is to contribute to the development of competition among alternative providers of local telephone service and thus to offer U.S. telephone users with greater choice in telecommunications services and providers. It also helps position the United States for evolving uses of technology, such as one-number technology that will relieve the pressure for additional telephone numbers. However, a tremendous amount of work is still being invested in the issue and likely will continue for some time.

4.2 UNIVERSAL SERVICE

4.2.1 The Early Concept

As discussed in Chapter 2, *universal service* originally was a concept voiced by many people early in the 1900s as a concept to provide affordable telephone service to most persons in the United States. It was a marketing concept developed by Theodore Vail, Chairman and CEO of the Bell System in the early 1900s. His plan was to price residential telephone service so low that nearly anyone in the United States could afford it. The U.S. government adopted the concept around World War I, as a method of unifying the United States's economy, military defense, and democratic participation. In doing so, the U.S. government established a federal Universal Service Fund to provide the money needed to develop the U.S. telephone infrastructure. It allowed the pricing of telephone service to be based not on *actual costs,* but rather on *averaged costs* throughout the system to ensure "the availability of adequate telephone service to the widest practical number of users."[10] This policy provided telephone service to all areas of the United States, "whether or not the installation of any particular phone was profitable."[11] It thus ensured telephone service to *high-cost* rural areas that likely would not have been served under a normal business case. This *non–cost-based* pricing was successful in that today most Americans think of telephone service "as available as tap water."[12] It is also the reason why business telephone lines typically have cost twice as much per month as residential lines when the two are essentially the same. However, it was also believed that once nearly every location in the United States was wired, the monthly universal service charges would disappear. Instead, they are increasing.

4.2.2 Universal Service in the 1996 Act

While the concept of universal service is somewhat inconsistent with a competitive environment, Congress retained and even expanded it as an important public policy in the Telecommunications Act of 1996. This is because Congress recognized that access to telephone service is "crucial to full participation in our society and economy which are increasingly dependent upon the rapid exchange of information."[13] As such, the entire Section 254 of the Telecommunications Act of 1996 mandates universal service. In it Congress established several new programs including the "Education or E-rate" to connect schools and libraries to the Internet and other advanced telecommunications services and funds for rural health telecommunications services. It also expanded the concept of making phone service *affordable.* Congress created a second Federal-State Joint Board to focus on issues surrounding universal service and delegated to the FCC the task of implementing and overseeing the programs. In doing so, Congress directed the FCC to take into account the extent to which the services (1) are essential to education, public health, or public safety; (2) have been subscribed to by most residential users; (3) are being deployed in public networks; and (4) are consistent with the public interest, convenience, and necessity.[14]

Hence, universal service affects every person in the United States who has a telephone or uses the library. All telephone customers pay a monthly universal service fund fee, at both the federal and state levels, which is collected by the telephone company and paid to the federal and state treasuries. It is then distributed to companies investing in high-cost, low-profit areas. Numerous issues, however, still surround the implementation of these programs. The main issues today include: (1) the E-rate and rural health-care programs; (2) the implementation of the Universal Service Joint Board's recommendations; (3) pricing of implemented services; (4) who must *pay* into the universal service funds (USF); and (5) who may *draw from* the funds. Each of these is described briefly in the following five sections.

E-Rate and Rural Health-care Programs

In addition to the traditional universal service funds to install telephone facilities in high-cost, usually rural areas, Congress set aside specific amounts to provide funds to wire schools, libraries, and rural health-care providers to the Internet. The purpose of the program is to avoid a "digital divide." Initially, Congress allocated $2.25 billion to wire classrooms and libraries to the Internet. The eligible K through 12 schools and libraries pay a discounted fee for the service, and the difference is reimbursed from the fund.[15] Additional funds have been tagged for Internet access to rural health-care providers.[16]

Recognizing that the program could be under-funded and avoiding unnecessary subsidies to areas able to pay the total amount, the FCC adopted rules of priority that give preference to the most economically disadvantaged applicants. For schools, this is to be determined by the number of students on a subsidized lunch program. The neediness of libraries is harder to determine because they tend to serve several economic and cultural sections of cities.

To receive funds, the school, library, or health-care facility must be qualified by a recognized group in each state. Usually this is a coalition of a state's Department of Education and Board of Libraries. Since its inception, the E-rate program has aided many schools and libraries in obtaining needed telecommunications services, but its implementation has also experienced significant confusion and setbacks. One of the biggest blows was that in June 1998, the FCC reformed the E-rate and reduced the amount going to the E-rate by almost half,[17] from $1.28 billion to $1.28 billion. [18]

The reform also changed the E-rate funding year for the schools and libraries from a calendar year (January 1–December 31) to a school year (July 1–June 30) to synchronize the E-rate programs with the budget and planning schedules of the recipient schools and libraries.

Federal-State Joint Board for Universal Service

In the 1996 Act, Congress instituted a Federal-State Joint Board to review the issues involved with universal service. Congress delegated to the Universal Service

Joint Board the task of developing recommendations for implementing the goals of the Telecommunications Act of 1996 and required that the Joint Board submit its recommendations and a timetable for implementing its recommendations within nine months of the date of enactment, or by November 8, 1996. Congress also required the FCC to initiate a proceeding to implement the Joint Board's recommendations and complete the proceedings by May 8, 1997, 15 months after the date of enactment.

November 8, 1996—Universal Service Joint Board's Recommendations

The Universal Service Joint Board met its deadline and submitted its recommendations on November 8, 1996.[19] In its recommendations, the Joint Board defined eight supportable services deemed to be of value to society and required that to be eligible for universal service monies, a carrier must provide and advertise all eight required services and be facilities-based, unbundled, or a combination of both. Carriers offering their services entirely by resale of another carrier's services are not eligible for universal service support because they do not build infrastructure.[20] The eight services were (1) voice-grade access to the public switched network, with the ability to place and receive calls; (2) touch-tone or dual-tone multi-frequency signaling; (3) single-party service; (4) access to emergency services including 911 and Enhanced-911; (5) access to operator services; (6) access to long-distance services; (7) access to directory assistance; and (8) access to Lifeline and Link-Up services for low-income customers, with provisions for toll blocking and other limitations on services.

If the carrier is unable to offer single-party, Enhanced-911, or toll limitation services, they are able to receive universal service support for their high-cost areas for a specific length of time while they upgrade their networks to provide these services.[21] BellSouth requested to the FCC to clarify that service for the first residential line should be fully supported by universal service and that service to single-connection businesses should be made available at a reduced rate. BellSouth also made recommendations concerning the following: (1) the affordability of service, (2) the eligibility of carriers to receive universal service support, (3) the problems with financing universal service for high-cost areas and low-income customers, and (4) support for schools, libraries, and rural health care programs.

May 8, 1997—FCC's Implementation of Joint Board's Recommendations

The FCC, required to implement the Universal Service Joint Board's recommendations by May 8, 1997, or within 15 months after the date the 1996 Act was signed, issued its Report and Order to Congress on Universal Service.[22] In it the FCC acknowledged that competition in rural areas is unlikely, because the areas offer little potential for profit. The 1996 Act grants significant power and discretion to the states to determine how competition should evolve in the states.[23]

Pricing of Implemented Services

Since its inception, universal service justified the non–cost-based, averaged pricing of basic telephone service. The Telecommunications Act of 1996, however, changed that and required cost-based pricing using the *forward-looking total long-run cost* of each element be used. The Act identified this as the *total element long-run incremental cost* (TELRIC). Since, historically, telephone service was averaged and cross-subsidized, the industry had little real knowledge about these costs. In the interim, the industry requested that the Congress and FCC accept figures on total telephone *services* rather than on each *element* of a service. This was accepted and is known as *total service long-run incremental costs* (TSLRIC).

After the Telecommunications Act of 1996, two computerized models were developed to begin to compile cost information. The two were the Benchmark Cost Model, supported by the regional Bell Operating Companies, and the Hatfield Model, supported by AT&T and MCI. The FCC, state commissions, and courts all began to use these models, at least to verify cost statements, and the models have evolved into important tools for understanding the costs of providing telecommunications services. What is not clear is whether both models will survive in the future, how one or both will be maintained and updated, or what role they will play in the telecommunications industry, universal service, and other programs.

Who Pays into the Universal Service Fund?

Little disagreement exists in the telecommunications industry that the services supported by the universal service funds are socially valuable. The larger, ongoing issue is "What is the fairest way to pay for them?" While all telephone service customers pay a universal service fee each month on each line, the amount of the fee is based on revenue paid to the carrier. Each carrier must collect the funds and deliver them to the state and federal treasuries. The concern is that if the contributions to the fund are based on revenues, it could place a disproportionate burden on certain carriers or types of carriers.

Section 254(b)(4) of the Telecommunications Act of 1996, notes that, "All providers of telecommunications services should make an equitable and nondiscriminatory contribution to the preservation and use of universal service," but Section 254(d) also states that, "Every telecommunications carrier that provides interstate telecommunications services shall contribute" . . . to universal service. For the long-distance carriers and others, this confusion and concern keeps the issue of who must pay into the funds open.[24]

Who Draws from the Universal Service Fund?

Only carriers designated as eligible telecommunications carriers by the state public utilities commissions (PUCs) may receive universal service support. To be eligible, the carrier must meet two basic requirements. First, it must offer services that are supported by the universal service fund throughout the carrier's service area. Second, the carrier must advertise the availability of those services and charge for them throughout its service area.[25]

In general, the Telecommunications Act of 1996 reaffirmed the concept of universal service as one of making telecommunications services accessible and affordable to all members of society that desire it, including those in rural areas and those with lower or fixed incomes. While this is a benevolent concept, it has not been totally successful in the past and will require continually increasing funds in the future. In addition, some opponents view the universal charges as a tax or continuing subsidy that has not been approved by voters. Others support it, arguing "universal service is appropriate because it enables all Americans to communicate with each other 'at reasonable rates' and 'on equal footing.'"[26] Acknowledging both arguments, Congress made universal service a huge part of the Telecommunications Act of 1996, affirmed continuing monthly charges to each telecommunications user, and extended the concept from simply infrastructure development to providing affordable access to the Internet. This clearly changed the tone of the concept and opened the discussion that has continued since 1996.

4.3 ACCESS

Access is the third of the ongoing issues dominating the telecommunications calendar since the enactment of the Telecommunications Act of 1996. It has been one of the most heavily litigated issues in telecommunications since 1996, and the results have taken a number of unexpected turns that have completely changed the pricing structures of most telecommunications products and services. The issue of "reciprocal compensation" has driven numerous mergers and acquisitions in the past five years and is singularly responsible for the financial success and/or failure of multiple telecommunications companies. It has therefore directed the marketing strategies of many telecommunications companies since 1997–1998.

The term *access,* in the telecommunications industry, refers to a customer's ability to dial into the telephone network to make a call, send a fax, or transmit data or video information. Most recently, it also includes the ability of Internet users around the world to dial into or access the Internet. For telecommunications equipment and service providers, access includes the requirements and technical standards associated with interconnecting systems to enable the transfer of calls from one carrier to another. As such, the pricing, regulation, terms, and conditions of access will affect the growth of the national information infrastructure and to some extent, the global information infrastructure, by determining how widespread and affordable access will be.

As discussed in Section 4.2 on universal service, the United States has long recognized the importance of widespread access to the network. With the transition from monopoly to competition in the same markets, the issues surrounding access have become more complicated. The purpose of this section is to describe the background and key events leading to the current discussion continuing today.

4.3.1 FCC's 1983 Access Charge Rules

In 1983, while preparing for the breakup or *divestiture* of the Bell System and the separation of long-distance service from local service on January 1, 1984, the FCC realized that certain new payments must be developed to compensate for traffic among the new types of carriers. These areas were: (1) payment from long-distance to local carriers, (2) payment from wireless to landline carriers, and (3) exemption of payment for enhanced service providers. Each is discussed in the following three sections.

Payment from Long Distance to Local for Access

Whenever a long-distance call would be made in the post-1984 environment, the FCC realized that at least three different telecommunications carriers would be involved in the call: (1) the local carrier where the call originated, (2) the long-distance carrier or carriers, and (3) the local carrier where the call terminated. These are known as the *originating, transport,* and *terminating* carriers, respectively. Most important to this issue is that at the originating and terminating ends, some portion of the local telephone company's lines, trunks, switches, and software would be required to connect the call, and this use of the local carrier's equipment and services is known as *access service.*

Additionally, the FCC realized that while all three companies were involved in the call, the customer making the call would pay only the long-distance carrier for it. Therefore, unless the long-distance carrier reimbursed the local carriers for their assistance with the call, the local carrier had no way to recover its costs associated with that call. To avoid this inequity, the FCC issued its *1983 Access Charge Order* (1983 Access Order) requiring that long-distance carriers reimburse the originating and terminating carriers for their assistance in establishing long-distance calls by paying per-minute access charges on all long-distance calls. The rules implementing this order, known as the FCC's *1983 Access Charge Rules* (1983 Access Rules), were placed in Part 69 of the FCC's full set of rules.[27]

The FCC further decided that the access charge reimbursement should be assigned to the *common line element* and recovered, at least in part, through a uniform national per-minute carrier-to-carrier access charge known as the *carrier common line charge* paid by the long-distance carriers to the local carriers. This access charge is also known by other names including the *carrier's carrier charge,* and the *carrier line charge,* all recovering common line costs. Since access costs are not dependent on the distance of the call but only the connection costs, the FCC decided that the amount of the charge should be based on the *fully distributed cost* of the network.

However, to avoid paying these access charges during the first five years after they were assessed, the long-distance companies tried bypassing the local exchange. To discourage this practice, in 1988 the FCC reduced the per-minute carrier common line charge to the long-distance carriers and spread the costs of local access over all subscribers capable of making a long-distance call. This new charge was a *subscriber line charge* placed on every customers' phone bill regardless of whether long-distance calls were actually made on the line or not.[28] Today, both the common carrier line charge, and the subscriber line charge are passed on to the customer. Thus, the long-distance carriers pay approximately 2.5¢ to 3¢ per minute for local access on long-distance calls,[29] and the monthly telephone bills of most U.S. telephone customers include most, but not necessarily all, of the following items:

1. Federal Access Charge: typically $3.50 per residential line per month and $4.50 per business line per month.
2. Carrier Line Charge: Typically 85¢ per residential line per month, 44¢ per business Centrex line, and $3.50 per business single line, multiple line, and basic rate interface, but varies among states and over time.
3. Subscriber or Customer Line Charge: approximately $9.50 per month, but varies with each state.
4. Primary Interexchange Carrier Charge: varies state to state.

Payment from Wireless-to-Landline Telephone for Access

The second portion of the telecommunications industry in which the 1983 Access Charge Rules applied between 1984 and 1996 was in wireless-to-landline communications. Since the purpose of the FCC's 1983 Access Charge Rules was to compensate all communications carriers involved in completing each call, wireless companies were required to reimburse landline carriers for each call made by a wireless customer to a landline customer. Subsequently, this cost was passed to the wireless customer. On the other hand, if a landline customer made a call to a wireless phone, the caller paid for the landline carrier's costs as part of his or her flat monthly rate. The costs incurred by the wireless company, however, were paid by the wireless customer as part of the wireless service. Either way, the wireless customer paid to both send and receive wireless calls.

Exemption of Payment for Enhanced Service Providers

One exception to access charges was for *enhanced service providers* (ESPs). Since *Computer I* and *II*, and other 1970–1980 decisions discussed in Chapter 2, the telecommunications industry in the United States was divided into *voice* and *data* networks.[30] Voice networks were regulated and required to pay access charges. Data networks, on the other hand, were not regulated and therefore not required to pay access charges.[31] Consistent with their name, the ESPs provide customers with *enhanced [data] services*, generally using the local public switched telephone network (PSTN) to reach those customers. In 1983, very few Internet service providers (ISPs) existed, but the few that did were officially defined as being part of the larger group of enhanced service providers.

Besides being *data* providers, the ESPs/ISPs were granted an access-charge exemption because the FCC decided that ESPs were *business customers* or *end users,* not *carriers.* In categorizing the ESPs/ISPs as end users, the FCC noted that, like other business customers, the ESPs/ISPs paid both local business rates for the use of their local access lines and interstate subscriber line charges for their switched access connections to the local exchange carrier's facilities.[32] For these reasons, the FCC determined that (1) the ESPs/ISPs, were end users, not carriers, and thus a carrier-to-carrier charge was not appropriate; (2) the FCC did not have jurisdiction over noncarrier end users; (3) the ESPs/ISPs were exempt from Title II regulation; and (4) the ESPs/ISPs were therefore exempt from paying the per-minute carrier-to-carrier access charges that the long-distance carriers were required to pay to the local telephone carriers for access to the public switched telephone network.[33]

4.3.2 Initial Challenge to the FCC's Exemption of ISPs—ACTA's Internet Phone Petition To Designate ISPs as Long-Distance Providers

The FCC's 1983 definition of ISPs as ESPs, and thus their exemption from the access charges, remained unchallenged in the telecommunications industry for 12 years, until technological developments made *Voice over Internet Protocol* (VoIP) possible. The first challenge to the FCC's classification of ISPs came in 1995 from the long-distance carriers or interexchange carriers. On March 4, 1995, the America's Carriers Telecommunications Association (ACTA), an organization of approximately 130 long-distance service and equipment providers, filed a petition with the FCC for a Declaratory Ruling, Special Relief, and for Institution of Rulemaking Proceedings.[34] This filing, also known as the *Internet Phone Petition,* initiated the FCC's rulemaking proceedings identified as *Rulemaking No. 8775 (RM No. 8775).*

In its petition, ACTA made two arguments, among others. First, ACTA argued that providers of Internet telephony software are telecommunications carriers, not ESPs, and therefore should be regulated and taxed as common carriers. Second, if VoIP were not regulated and taxed, the nation's telecommunications infrastructure would not have sufficient funds to support itself and universal service.

Opponents to ACTA's petition argued that it is specifically because voice calls on the Internet are transmitted as digitized packets over high-speed networks that it is not feasible to distinguish the voice calls from the data transmissions. This inability to distinguish between the two makes it difficult, costly, and not in the public's interest to regulate or tax voice calls on the Internet.

4.3.3 Impact of the Telecommunications Act of 1996

Therefore, from 1984 to 1996, the FCC's 1983 Access Charge Rules charged for voice communications, but not for data communications (ISPs or VoIP). The Telecommunications Act of 1996, however, introduced several changes that significantly affected the area of access charges especially concerning long-distance carriers, wireless companies, and competitive local exchange carriers. The FCC implemented these changes in its rules and Interconnection Order issued on August 8, 1996.[35] The following three sections describe these changes.

Impact on Long-Distance Access Charges

First, the Telecommunications Act of 1996 removed the structural barriers that had prohibited the long-distance carriers, also known as interexchange carriers, from providing local retail telephone services. This allowed the interchange carriers to become local exchange carriers and to keep any originating or terminating access charges rather than pay them to another local company. This significantly increased the long-distance carriers' revenues. Thus, if the long-distance carrier provided end-to-end service, including the origination, long-distance transport, and termination service, it could pocket the entire amount generated by the call and was not required to share its revenue with any other company. As another revenue option, the long-distance carriers could pass all or a portion of their local access charge savings on to their customers in the form of lower prices for long-distance calls. These lower prices attracted more customers and thus increased the long-distance carriers' total revenues. Both revenue options provided by access charge savings have lead to the significant decrease in long-distance prices that has occurred since 1996.

The incumbent local Bell Operating Companies, on the other hand, could not respond to this pricing strategy by becoming long-distance service providers until they passed the 14 points required of them in Section 271 of the 1996 Act. In addition, to prevent *predatory pricing,* the FCC imposed *access charge imputation* policies to ensure that the Bell Operating Companies charged themselves access charges at the same rate as they charged their competitors when supplying intrastate long distance services.[36] These imputation policies established *price floors* for end-user services that continue to be enforced today where systems are regulated.[37]

Impact on Wireless Access Charges—Exemption for Wireless

Second, as noted in Chapter 3, the Telecommunications Act of 1996 specifically excluded wireless communications providers from the definition of local exchange carriers.[38] This meant that, among other things, wireless carriers are not required to pay access charges. To underscore this intent, Congress further added in Section 705 of the 1996 Act entitled "Mobile Services Direct Access to Long Distance Carriers," now codified at 47 U.S.C. 332(c)(8), "Mobile Services Access" that

> A person engaged in the provision of commercial mobile services, insofar as such person is so engaged, *shall not be required to provide equal access to common carriers for the provision of telephone toll services.* If the Commission determines that subscribers to such services are denied access to the provider of telephone toll services of the subscribers' choice, and that such denial is contrary to the public interest, convenience, and necessity, then the Commission shall prescribe regulations to afford subscribers unblocked access to the provider of telephone toll services of the subscribers' choice through the use of a carrier identification code assigned to such provider or other mechanism. The requirements for unblocking shall not apply to mobile satellite services unless the Commission finds it to be in the public interest to apply such requirements to such services. (Emphasis added.)

Thus, after 1996, wireless telecommunications providers no longer had to pay landline companies access fees for both incoming and outgoing calls. Instead, wireless phone users need pay only for making (originating) calls, but not receiving them. To date, the only companies reflecting this change are those companies offering "the first incoming minute for free," but the future direction in pricing is likely to be for *call origination only,* as is charged in Europe, Asia, and other parts of the world.

Continued Exemption for ESPs/ISPs

Third, the exemption for ESPs/ISPs continued through the years. By 2001, however, with the increasing use and importance of ISPs, the definition of ISPs as unregulated ESPs, and thus their exemption from paying access charges, was being questioned. Since ESPs are not required to pay access charges, they avoid a significant part of the cost of providing modern services and encouraged Voice over Internet Protocol (VoIP).

4.4 RECIPROCAL COMPENSATION—PAYMENT FOR TRANSPORT AND TERMINATION OF COMMUNICATIONS

In drafting the Telecommunications Act of 1996, Congress decided that, just as the telecommunications industry's structural changes in 1984 (which created long-distance and local companies) required intercarrier access charges to compensate all carriers involved in completing a call, the structural changes in 1996 (which created incumbent local exchange carriers [ILECs] and competitive local exchange carriers [CLECs]), required some form of local access charges for the same reason.

Congress realized that if two telephone companies, companies A and B, provided local telephone service in the same town, and a customer of company A called a customer of company B, four steps must occur to complete the call. First, company A must *originate* the call on its local network. Second, company A must *hand-off* the call from its network to the network of company B. Third, company B must *transport* the call, or physically move it from the point of interconnection between the networks of the two companies to company B's switching facility. Fourth, company B must deliver or *terminate* the call from B's switching facility to the individual customer's line. In these steps, costs are incurred by both companies A and B. Company A's costs for originating and handing off the call are recovered when A's customer pays for the call. Company B's costs for transporting and terminating the call, however, are not recovered unless company A reimburses company B for B's costs.[39]

Congress called this post-1996, local version of the Access Rules, *reciprocal compensation* because Congress sought to provide for the "mutual and reciprocal recovery by each carrier of costs associated with the transport and termination on each carrier's network facilities of calls that originate on the network facilities of the other."[40] Congress also realized that, absent some legal obligation for each carrier to pay the other local carriers for their assistance in connecting calls, local exchange carriers (LECs) in a competitive environment could cause another local carrier to incur huge unrecovered costs. Thus, Congress mandated in the 1996 Act that all local exchange carriers, both ILECs and CLECs, have a "duty to establish reciprocal compensation arrangements for the transport and termination of telecommunications" with the other carriers in same local market.[41]

Congress further stated in the 1996 Act that these arrangements were to be (1) negotiated between the various LECs, (2) documented in the written contracts between the carriers, known as interconnection agreements, and (3) approved by the state commissions.[42] As discussed in Chapter 3, if the carriers could not reach an agreement, the state commissions pursuant to Section 252 would arbitrate agreement, the terms, and conditions of reciprocal compensation for them.

Third, Congress outlined the standards that the state commissions are required to follow in resolving any open issues and imposing conditions upon the parties to the agreement. These included that the state PUCs would do the following (1) ensure that the terms and conditions of all arbitrated interconnection agreements meet the requirements of Section 251, including Section 251(b)(5) [reciprocal compensation] and Section 251(d) [implementation]; (2) establish rates for interconnection, services, or network elements, and (3) provide a schedule for implementation of the terms and conditions.

Among the terms and conditions for reciprocal compensation, Congress provided the following (1) the traffic transported and terminated by each carrier was to be measured in *minutes of use*[43] rather than on a flat-rate or per-call basis; (2) the costs for transport and termination were to be determined "on the basis of a reasonable approximation of the additional costs of" these activities for each company;[44] (3) the *pricing* for the transport and termination of traffic was to be based on *reciprocal compensation;*[45] and (4) the method of payment to be used could be either direct cash payment for the transport and termination of traffic or other options, including arrangements such as *bill-and-keep,* which recovers costs through the offsetting of reciprocal obligations, also known as the *in-kind exchange of traffic without cash payment.*[46] However, in effect, it is also an agreement by each company to (1) be paid only for the origination of traffic and (2) to waive its right to payment for transport and termination costs. Thus, the term *reciprocal compensation* describes a range of nondiscriminatory *intercarrier compensation mechanisms* or methods for carriers to pay one another for assisting each other with local transport and termination of each others' local telecommunications traffic.[47]

Whichever method is used, reciprocal compensation differs from the local access charges paid by long-distance interexchange carriers to the local companies in two important ways. First, the access charges involve three companies, while reciprocal compensation, in either bill-and-keep and direct payment of minutes terminated involves only two companies. Second, access charges pay for traffic in both directions—both originated and terminated traffic—but exempts Internet-bound traffic. In contrast, the bill-and-keep method pays for only the origination of traffic, except for Internet-bound traffic, and the *direct cash payment* method pays only for the *minutes terminated,* including Internet-bound traffic.

Bill-and-Keep Method—Payment for Traffic Originated Only

If company A transports and terminates nearly the same amount of traffic for company B, as B transports and terminates for A, then their costs on behalf of one another are an even exchange. In such cases, it is easier for each company to simply bill its own customers and keep the revenue from their originated calls than to exchange cash payments with the other company for terminating the calls. Another way the companies viewed *bill-and-keep* is that they pay *in kind* for their costs to transport and terminate the other company's traffic.

Direct Cash-Payment Method—Payment for Traffic Terminated Only

Where the two companies have very different amounts of traffic, or rarely exchange traffic, Congress permitted the companies to use a straightforward cash-payment method of reciprocal compensation. In such cases, each company writes a check to the other company for the minutes of its traffic that the second company transported and terminated on behalf of the first company's customers. Since direct cash payment is based on minutes of use (MOU) or traffic terminated, it is also known as *minutes terminated.*

4.4.1 Initial Reaction of Incumbent Local Exchange Carriers and Competitive Local Exchange Carriers to Reciprocal Compensation

Initially, both the ILECs and CLECs viewed these reciprocal compensation requirements favorably because the concept of reciprocal compensation provided a fair and reasonable mechanism for the carriers to recover their costs incurred in terminating traffic for other carriers.

Incumbent Local Exchange Carriers' Reaction

The ILECs, in particular, found the concept of reciprocal compensation attractive because they anticipated that for many years they would remain the primary service provider in most markets, and thus they would terminate the majority of the traffic. This meant they would receive significant reimbursement for the largest number of minutes terminated[48] or, if the bill-and-keep method were used, costs would be equalized between companies. This was a huge win for the ILECs and it appeared as if the reciprocal compensation provisions in the 1996 Act would significantly increase the ILECs' revenues for years to come.

Competitive Local Exchange Carriers' Reaction—ISP Minutes Terminated Strategy

Surprisingly, however, the CLECs also favored reciprocal compensation, even though, with normal telephone traffic, they realized that they would have to pay the ILECs to terminate most of their customers' calls on the ILEC's network. This is because the ILECs were the "known" telephone companies and thus had most consumers' confidence. They also had all existing customers in their billing systems and knew the call volumes, equipment preferences, and extra services purchased by each customer. On this basis, the CLECs knew that even if they offered extraordinary services and pricing to attract customers, until they reached a more equal number of customers, they would be sending many more calls to the ILECs' network than they would receive from the ILECs.

Instead, the CLECs choose a strategy of focusing on business customers who had more incoming calls than outgoing. This call pattern would help minimize the CLECs' reciprocal compensation payments. In particular, they targeted ISPs as customers. The exponential growth of the Internet since 1996 created business opportunities that most participants in the industry had not envisioned in 1996 or during most of the ILECs' reciprocal compensation negotiations. The CLECs, however, realized that in a *minutes-terminated* environment, they needed only one customer in order to be profitable. That customer needed to be an *Internet Service Provider* because many Internet users call into their ISPs over "dial-up" local telephone lines provided by their ILECs and spend hours on the Internet each day.[49] In contrast, the ISPs rarely call back to those Internet users, resulting in far more *minutes terminated* at the ISP's local exchange carrier than at the ILEC's terminating location.

With ISPs as customers, the CLECs received significant reciprocal compensation payments, calculated in minutes terminated. On the other hand, they had little, if any, obligation to pay compensation in return. Thus, like the ILECs, the CLECs anticipated that they would receive the majority of the terminated minutes. Rather than viewing reciprocal compensation as a loss, many CLECs viewed it as a significant market opportunity. They knew that, with the minutes-terminated approach, any telecommunications carrier providing service to an ISP could be profitable in a reciprocal compensation environment.

The CLECs strategy worked. In the first few months, the number of minutes of traffic terminated by the CLECs for their ISP customers generated millions of dollars of reciprocal compensation payments from the ILECs to the CLECs.

4.4.2 Incumbent Local Exchange Carriers' Reaction to Competitive Local Exchange Carriers' ISP/Minutes-Terminated Strategy

Clearly the ILECs were not happy about having to pay their competitors to compete against them. Contrary to the ILECs' initial expectations that reciprocal compensation would work to their financial benefit, they frequently found that they were paying out far more to the CLECs in reciprocal compensation for ISP-bound traffic than the CLECs paid to them. In some cases, the ILECs complained that the payments to the CLECs were so large that they were paying more in reciprocal compensation for ISP-bound traffic than they were receiving in revenue for the local calls from their own customers.[50]

Incumbent Local Exchange Carriers Reaction 1: Refused to Pay the Reciprocal Compensation Charges

Immediately upon realizing the impact of the CLECs' minutes-terminated strategy to reciprocal compensation, the ILECs refused to pay reciprocal compensation minutes for Internet-bound traffic, especially to CLECs with extraordinarily high ratios of terminating versus originating traffic. Instead, the ILECs held the monies owed to the CLECs in an escrow or reserve account. This was a unilateral decision on the part of the ILECs, without the approval of the state public utilities commissions.

The result of the ILECs' refusal to pay its reciprocal compensation obligations, honor its contracts with the CLECs, and/or agree to new agreements, created a groundswell of litigation. The CLECs filed complaints with their state commissions to enforce the existing reciprocal compensation agreements. The CLECs seeking to renew existing reciprocal compensation provisions in agreements that were about to expire or to negotiate new agreements were forced to arbitrate pursuant to Section 252 of the 1996 Act. Nearly every state commission reviewing a reciprocal compensation complaint or arbitrating a new agreement, or every court reviewing a state commission decision concerning reciprocal compensation, has upheld the reciprocal compensation process and agreements.[51]

The ILECs argued to the state commissions and courts that they should not have to pay the CLECs for terminated minutes to carry Internet traffic because first, the extra Internet traffic on their networks increased the call duration time from an average of 4 minutes for a voice call to over 30 minutes for an Internet-bound call. Second, they argued the longer call times were degrading service for all customers. Third, this added traffic required additional equipment, which they must purchase, install, and maintain. Fourth, the ILECs argued that reciprocal compensation is for local traffic, and Internet-bound traffic is not local. Fifth, the ILECs argued that the exemption for ESPs and ISPs from paying local access charges is unfair because it forces local voice users to subsidize Internet users.

Instead, the ILECs requested first that Internet users pay some form of access charge to use the local exchange facilities. Bell Atlantic suggested a usage charge of 1¢ per minute for Internet access, compared with the 2.5¢ per minute then charged for long-distance access.[52] BellSouth announced its intention to assess access charges on VoIP providers even though, to the network, VoIP calls are data, are difficult to distinguish from other data, and therefore cost more to bill.[53] Second, the ILECs requested that ISP-bound traffic be excluded from the reciprocal compensation computations and that their interconnection agreements be modified to correct their contractual obligations under reciprocal compensation.

The CLECs responded that they were incurring costs to terminate this traffic originated by the ILECs' customers, costs that the ILEC avoided, and for which the CLECs should be compensated. After review of the situation, all of the state commissions and three federal courts agreed with the CLECs and ordered the ILECs to pay the agreed upon reciprocal compensation charges, including for Internet-bound traffic.

These actions resulted in the most significant amount of litigation and affected the financial viability of all telecommunications companies as well as the price of all telecommunications products. The debate continues. A brief review of the key events and arguments in this area is presented to assist readers in following this very important issue as it continues to unfold.

FCC's Notice of Proposed Rulemaking on Reform of Interstate Access Charges and Notice of Inquiry

On December 24, 1996, when the first instances of the strange twist in reciprocal compensation were raised, the FCC issued both a Notice of Inquiry (NOI) and a Notice of Proposed Rulemaking (NPRM) on Reform of Interstate Access Charges[54] to (1) address the primary access issues; (2) determine the impact of access rates, terms and conditions on the cost of Internet; (3) investigate what options exist to provide more efficient and effective data services to customers; and (4) determine whether the FCC should encourage higher-speed data communications-access facilities for customers.[55]

Incumbent Local Exchange Carriers' Reaction 2: Argued Increased Traffic and Degradation of Service from ISP-Bound Traffic

Throughout 1996 and 1997, the ILECs tracked the increased traffic on their networks from Internet-bound traffic and documented that it caused blockages in their facilities. The ILECs continued to argue that these blockages resulted in decreased service for all of their customers and required them to install additional equipment to accommodate the traffic. The ILECs reported to the state commissions, courts, and the FCC that the majority of this traffic was Internet-bound due to the increased use of the local telephone network by customers accessing the Internet through ISPs and the increased hold times on those calls.

In February 1997, for example, Bell Atlantic reported that its voice traffic averaged nine minutes per hour per line or 500 call seconds per hour (5 CCS), but that Internet traffic calls averaged 3,000 call seconds per hour or 30 CCS.[56] At 30 CCS, if 15% of the households in Bell Atlantic's territory accessed the Internet, Bell Atlantic reported that this doubled the previous traffic levels on its networks.[57] Similarly, Pacific Telesis (PacTel) reported that as of February 1997, its average voice call lasted approximately 4.5 minutes, while 30% of its dial-up Internet connections lasted 3 hours and 7.5% lasted more than 24 hours. PacTel argued that this unusually high volume of Internet-bound traffic in the Silicon Valley area resulted in blockages for approximately 16% of all local telephone calls.[58] In addition, the ILECs anticipated that the problem would get worse. PacTel reported in 1996 that 2.3 million households accessed the Internet through PacTel facilities, representing 27% of the total residential telephone traffic. However, within five years, by 2001, the company anticipated that number would increase to over 4.7 million on-line households, representing 50% of the local residential traffic.

In addition to severely degrading the performance of their networks and service to customers, the ILECs reported that the increase in Internet/ISP-bound traffic required them to invest significant additional funds to upgrade their networks to accommodate the increased traffic load. PacTel reported that it likely would be required to invest over $150 million in selected areas over the next five years to accommodate the growth of Internet traffic.[59] Bell Atlantic anticipated that its additional capital expenses would reach approximately $2,400 per subscriber line for peak periods of use compared with $245 per line historically used by Bell Atlantic as the basis for its line rates. In costs to customers, Bell Atlantic calculated that this would mean an increase in the charge for a standard subscriber line from $17 per month to approximately $75 per line.[60] Clearly, this new line cost was not acceptable to customers, regulators, or the carriers, and thus in 1997, the ILECs asked Congress[61] and the FCC to allow them to charge the ISPs either a special access charge or a per minute usage charge to cover these expenses.[62]

Competitive Local Exchange Carriers' Response

The CLECs and other small ISPs disagreed with the arguments presented by the ILECs *and* countered with the following arguments before the state commissions, courts, and FCC:

1. It is not apparent that the additional costs to the LECs are only from the Internet traffic;

2. Unlike long distance, where users pay the long-distance providers, Internet users are customers of the ILECs and have already paid for flat-rate, unlimited use of local lines;
3. The ILECs are being compensated for the Internet traffic in other ways, such as second lines for Internet access;
4. The ESPs are business customers, not competitive service providers;
5. Access charges would significantly alter the cost of the Internet and thus impact its growth and use; and
6. Alternate access technologies exist that could be used to carry additional Internet traffic.

About this time, other groups formed and submitted comments in the FCC's proceedings. The groups included the Digital Affordable Telecommunications Access (DATA) Coalition, (composed of IBM, Microsoft, Netscape, Compaq, Electronic Data Systems, and the Business Software Alliance); the Open Net Coalition[63], (representing independent ISPs); the Internet Access Coalition (IAC); and groups representing Internet users. These groups strongly opposed the arguments made by the ILECs and supported the CLEC's/ISPs' arguments.

1. Increased traffic is not just from the Internet. For example, these groups joined the CLECs and ISPs in pointing out that the increased traffic on the ILECs' networks is not solely due to Internet-bound traffic. Instead, they argued that the increased use of long distance, cellular telephones, and business communications networks also contribute to the network load, especially given the fact that the cost of these services dropped over the past five years. Thus, if additional costs are incurred by the local exchange carriers as a result of network congestion, costs should be recovered from all users generating that cost, not just from one group. If only Internet users are charged for the increased traffic, it would place an unfair "modem tax" on those users to upgrade the local voice network for other services.

2. Voice customers do not subsidize Internet use. The CLECs, ISPs and groups noted that, in terms of alleged subsidies, Internet users are also voice customers of the LECs. Since the LECs' charge their customers one flat, monthly rate for unlimited use of the local telephone lines, the customers have been led to believe that they have already paid for those lines and may appropriately use them for any call, including to transport Internet-bound traffic to their ISPs. The customers then pay an additional fee to their ISP to connect them to the Internet. Thus, these customers believe that if a local network rate increase is appropriate, and unlimited flat-rate service is to continue, the rate increase must be spread among all flat-rate monthly local line users.

3. ILECs are already compensated for additional Internet traffic. The CLECs and interested groups argued that the ILECs are already being compensated in other ways for any additional traffic on the local network from Internet-bound traffic. For example, many Internet users pay the ILECs monthly for the second telephone lines for Internet access. The CLECs/groups noted that the ILECs fail to include these substantial revenues in their arguments. To document the amount of this additional revenue, the DATA Coalition commissioned a study that concluded that in 1997 the ILECs collected $1.4 billion in revenues solely from the installation and use of these second home telephone lines. The study also showed that the phone companies invested only a fraction of those revenues to enhance their networks to support the additional usage. Thus, during that time the ILECs' profits grew by approximately 8–9%.

4. ESPs are business customers, not competitive providers. Opponents of an additional access charge noted that the ESPs/ISPs have been classified as business customers of the LECs who pay the standard business rates under the LECs' existing tariffs for their local lines. Once at the ISP, the traffic is then aggregated and the ISPs lease dedicated lines from both local and long-distance carriers to connect their customers to Internet sites around the United States and the world. The service that ESPs/ISPs provide, therefore, does not make uncompensated use of either the local or long-distance networks nor compete with the LECs as service providers. Thus, an additional carrier-to-carrier access charge, such as the long-distance carriers pay, is not reasonable or appropriate.

5. High access costs negatively affect Internet use. The CLECs and groups pointed out that a study by the Organization for Economic Cooperation and Development (OECD), released in March 1997, showed that the price of access to the Internet directly affected the volume of Internet usage and thus Internet growth. The study revealed that the United States, Finland, Iceland, and Sweden have the lowest Internet access costs in the world (approximately $20 per month) and the highest volume of Internet use. Austria, Ireland, and Mexico plus other countries have the highest Internet access costs ($90 per month) and the lowest volume of Internet use. Since the access issue is not unique to the United States, many countries are watching the United States to discern the best Internet access and reciprocal compensation pricing policies.[64]

6. New, alternative access technologies should be encouraged. The groups encouraged the regulators to consider that, instead of a new charge or rate structure adjustment, the local access network needed to be upgraded and diversified through products such as Digital Subscriber Lines (DSL) services, planned since 1992, wireless service, CATV, and electric utility access. The groups advocated that Congress and state, federal, and international regulators encourage private investment in separate local networks to accommodate the growing Internet traffic. This would provide higher speeds and alleviate any local voice network congestion. Instead of higher access fees, which would only discourage Internet use, these groups suggested a technical solution that would upgrade the availability of local networks, redirect extra traffic loads, handle the load differently, and thus eliminate the risk of system overload.

Incumbent Local Exchange Carrier's Reaction 3: Argued That ISP-Bound Calls Are Not Local

While regulators were evaluating the merits of a specific Internet-access fee, the ILECs added another argument, noting that since the ISPs connect their customers to interstate and international websites, Internet-bound calls are not local, but rather are interstate. Since Section 251 requires reciprocal compensation only for *local* traffic, this would mean that all *interstate, Internet-bound* traffic would not qualify for reciprocal compensation. To further support their arguments, the ILECs also pointed out that the FCC, in its August 1996 Interconnection Order, specifically stated that the reciprocal compensation provisions of Section 251(b)(5) should apply only to traffic that originates and *terminates within a local area,* as defined by each state commission.[65] (Emphasis added.) Thus, the ILECs noted that they should not owe any monies for ISP-bound traffic unless a CLEC could prove that the Internet traffic is local, and the definition of what constitutes a local call became the center of the reciprocal compensation controversy. This would also mean that the wording of existing interconnection agreements, approved by the state commissions, must be interpreted to automatically exclude payments for local ISP-bound calls.

Competitive Local Exchange Carriers' Response: The Definition of Local Calls

In response to the ILECs' argument that ISP-bound traffic is not *local,* the CLECs pointed out that the general practice in the telecommunications industry is to categorize telephone calls as local or long-distance, based on the area code and first three digits of the seven-digit local number of both the calling and called parties' telephone numbers, not their physical location.[66] Many ISPs, especially the larger ones, have enough customers to economically justify a physical presence in each local calling area with local telephone numbers assigned to them. Thus, when CLEC customers in that local calling area dial their ISP's local telephone number, it is, in fact, a local call. The call is physically routed to the ISP's location within the same local calling area where the ISP's equipment is located, and the call is then connected to the Internet.

Smaller ISPs, on the other hand, and those serving more rural areas, do not have a sufficient volume of traffic in each local calling area to justify maintaining a physical presence there. Instead, those ISPs allow their customers to access the Internet using either a local or toll-free number. Either way, the calls generally are routed to a more centralized location closer to its local carrier's switch, but usually outside the local calling area. From this more centralized location, the call is connected to the Internet.

The ISPs using local numbers receive a block of local telephone numbers from their LECs in each local calling area where the ISP's customers are located. Some states refer to this practice as the assignment of *virtual local numbers*.[67] The users of these virtual local numbers appear as local callers to the system and receive the same benefits as the customers of the larger ISPs. The practice of using virtual local numbers, and thus rating a call as *local* while routing it to a location outside of the local calling area, is permitted because it is not unique to ISP traffic. Instead, it is nearly identical to traditional *foreign exchange* service offered by most ILECs. For this reason, the ILECs initially treated Internet-bound traffic as local for purposes of paying reciprocal compensation to the ISP's local exchange carrier before the reciprocal compensation issue became so complex.

Most parties believe that the practice of using virtual local numbers provides affordable access to the Internet for small users and rural customers and thus greatly contributed to the growth of the Internet. Without this practice, persons in more remote areas could only access the Internet by paying toll charges. The attorney general of New York argued, "by entering the market for ISP-bound traffic, CLECs have contributed to the greater availability of Internet access to end-users."[68] From this perspective, abandoning reciprocal compensation for ISP-bound traffic could have the detrimental effect of limiting consumer choice in securing Internet access and increasing the price of such access.[69]

State Commission/Court Response: The Two-Call Theory

In considering these arguments, the state commissions and courts determined that all calls to the Internet actually consist of two separate calls. The first is between the calling party and the ISP's location. The second is between the ISP and a site or sites on the Internet. The state commissions and courts determined that the first call originates with the calling party and terminates at the ISP or its "virtual" local presence. This call is a basic telecommunications service, and if telephone numbers assigned to the calling party and the ISP are in the same local calling area, the call is considered to be local.

The second call, on the other hand, originates at the ISP and terminates at an Internet access point. It uses the ISP's own equipment, generally operating over telecommunications facilities leased from a telecommunications carrier, both to *deliver* information to the Internet and *retrieve* information from the Internet on behalf of its end-user customer, the calling party. The two-call theory holds that this second call is an information service, not a telecommunications service and therefore has no role in determining the jurisdiction of the call or whether it is local or long distance.

In this two-call theory, the key distinction between telecommunications (voice) and information (data) services rests on the functional nature of the end-user offering.[70] Telecommunications, as defined by Congress in the Telecommunications Act of 1996, is "the transmission, between or among points specified by the user, of information of the user's choosing, without change in the form of content of the information as sent and received."[71] An information service, on the other hand, is "the offering of a capability for generating, acquiring, storing, transforming, processing, retrieving, utilizing, or making available information via telecommunications . . ."[72] Thus, "[I]f the user can receive nothing more than pure transmission, the service is telecommunications service. If the user can receive enhanced functionality, such as manipulation of information and interaction with stored data, the service is an information service."[73]

With the two-call theory, the state commissions and courts determined that only the first call should be considered when deciding whether an Internet-bound call is a local or long-distance call for purposes of reciprocal compensation. In doing so, the state commissions and courts generally agreed with the CLECs and rejected the ILECs' argument that Internet-bound calls are totally interstate.

Incumbent Local Exchange Carriers' Reaction 4: Challenged Contractual Obligations in the Interconnection Agreements

Having little success arguing that ISP-bound traffic is not local, the ILECs changed their focus and argued that the details of the written contracts between the companies, known as their interconnection agreements, were flawed and thus could not be honored. The ILECs asked the courts and commissions to review the language, terms, and conditions of the contracts and to find that ISP-bound traffic did not qualify for reciprocal compensation under the terms of the existing agreements. They also challenged many of the reciprocal compensation agreements imposed on them by the states in arbitration.

Competitive Local Exchange Carriers' Response

In response, the CLECs argued first that the ILECs have clear contractual obligations to make the payments for all qualifying traffic in their interconnection agreements. The CLECs noted that the obligation to pay reciprocal compensation, as outlined in the Telecommunications Act of 1996, is for transport and termination services that one carrier provides to another carrier to complete a call made by a customer. They point out that the definition is not dependent on any particular type, level, or destination of traffic flow, but instead, is a contract between companies about compensating each for the transport and termination assistance in completing calls. The carriers signed contracts agreeing on the compensation for this exchange. Second the CLECs noted that the language in the FCCs rules makes no distinction about such traffic. Third, the CLECs argued that any shift away from a reciprocal compensation payment structure toward a system of negotiated rates would significantly alter the cost of access to the Internet and therefore impact the growth and use of the Internet.

State Commission/Court Response to Contractual Argument

In response to the ILECs' request to review the language of the carrier-to-carrier contracts, the state commissions and courts found that under traditional contract law analysis, the absence of language specifically excluding ISP-bound traffic from the definition of local traffic indicated that the parties had not intended to distinguish ISP-bound traffic from all other local traffic at the time the agreement was negotiated. The state commissions and courts therefore, rejected arguments by the ILECs that their existing agreements did not require reciprocal compensation payments for ISP-bound traffic, and ordered the ILECs to pay their reciprocal compensation obligations to the CLECs.[74]

For example, the Public Utilities Commission of Ohio concluded that *"Ameritech has not cited to, and we have not found, one sentence in the involved interconnection agreements that addressed ISP-bound traffic. Nor have there been any allegations that the parties ever agreed on treating ISP traffic different from all other locally dialed traffic. In the absence of such an agreement, the Commission was left with attempting to determine the parties' intent at the time the interconnection agreements were entered into."*[75] Accordingly, finding no intent to exclude ISP-bound traffic, the Ohio commission required Ameritech Ohio to pay reciprocal compensation to the complainants.

Thus, in the majority of these cases, the courts and state commissions found in favor of including Internet-bound traffic in reciprocal compensation. The ILECs then challenged these decisions in federal court.

Incumbent Local Exchange Carriers' Reaction 5: Refused to Negotiate New Agreements with Current Reciprocal Compensation Clauses

At the same time, when asked to renew existing interconnection agreements or to negotiate new agreements, the ILECs stated that they would agree to reciprocal compensation only if the definition of local traffic expressly excluded ISP or Internet-bound traffic. Based on earlier arguments, the state commissions simply stated that the terms would then be arbitrated and imposed on the ILECs.

Incumbent Local Exchange Carriers' Reaction 6: Created ISPs To Attract Terminated Minutes

Losing on all sides, the ILECs decided to become ISPs and attract their own minutes-terminated revenue or at least avoid paying it to another LEC. This was immediately complex because the ILECs then became both wholesalers and retailers of ISP services. They provide access facilities to other ISPs, including their competitor ISPs, and full Internet access services to their customers. The structure and price of each ILEC's ISP service offerings significantly affects the rate, terms, and conditions available to all ISP customers in that market.

Second, the move raised questions about the validity of the ILECs previous arguments. For example, the ILECs vigorously complained that Internet-bound traffic and the increased "hold times" overloaded their networks and required them to purchase additional equipment at an accelerated rate, but those same companies then began aggressively seeking customers and placing even heavier Internet traffic on their networks. In doing so, they became very quiet about the original argument of the impact of Internet-bound traffic on the local network. The ILECs had also vigorously complained about the economic burdens that reciprocal compensation placed on them and the harm that the use of virtual local numbers caused in the access pricing.[76] However, the ILECs are doing both now.

4.4.3 State Regulatory Concerns Regarding Competitive Local Exchange Carriers' ISP Strategy

The CLECs ISP-strategy did not pass unscrutinized, however. Both federal and state regulators were concerned the CLECs' focus on attracting ISPs as customers and using virtual local numbers, while not prohibited by any state or federal law, was questioned by some as perhaps "gaming" the system. The New York Public Service Commission, for example, noted,

> it is alleged that some CLECs are nothing more than ISPs that have adopted the trappings of CLECs solely to receive a reciprocal compensation revenue stream. Even in less extreme situations, it is argued that some CLECs are serving a niche market that is made lucrative by a perverse regulatory anomaly rather than the underlying economics of the situation[77]

or providing the benefits of competitive services to the public.

However, after hearing the various reciprocal compensation arguments, the state commissions and courts generally upheld the ILECs' obligations to pay reciprocal compensation to the CLECs. They did so because of the following four items. First, the commissions and courts noted that since no one envisioned the Internet having the significant impact on traffic carried by the telephone companies that it has had, the early interconnection agreements between companies were silent as to whether traffic terminating at ISPs should or should not qualify for reciprocal compensation. Second, the commissions and courts believed they must therefore turn to the specific wording of the law. Section 251(b)(5) of the Telecommunications Act of 1996 requires each carrier that originates local traffic to compensate any other carriers that terminate that traffic on their behalf for the costs of such termination. Third, the commissions and courts interpreted this to mean that, in a competitive local environment, all carriers involved in the completion of a revenue-generating call must be compensated. Fourth, the commissions and courts stated that an explicit exclusion exists for ISPs, including Internet-bound traffic passing through the ISPs.[78]

December 1996—FCC Notice of Proposed Rulemaking and Notice of Inquiry

In December 1996, the FCC took several steps concerning the reciprocal compensation issue. First, it initiated a Notice of Proposed Rulemaking and Notice of Inquiry to study the issue. Second, the FCC applied a $2.75 per line fee for businesses that have more than one telephone line, but did not apply any charges targeted to data network users. Third, the FCC adopted measures to keep basic local calling rates low and to facilitate the expansion of Internet access for schools and libraries, as directed in Section 254 of the Telecommunications Act of 1996.[79]

Four months later, in March 1997, Kevin Werbach, Counsel for New Technology Policy in the FCC's Office of Plans and Policy issued a widely quoted paper entitled "Digital Tornado: The Internet and Telecommunications Policy,"[80] in which he stated, among other things, that many of the developing Internet applications do not readily fit into existing regulatory categories. Instead of focusing on specific applications or technologies, such as voice over the Internet Protocol (VoIP) or wireless, the paper encouraged the FCC to create appropriate incentives to encourage the industry to build high-capacity, high-performance digital networks.

May 1997—FCC First Report and Order on Access Charge Reform

With the information and comments gained from the Notice of Proposed Rulemaking and Notice of Inquiry, the FCC issued its First Report and Order on access charge reform on May 16, 1997 entitled: "In the Matter of Access Charge Reform, Price Cap Performance Review for Local Exchange Carriers, Transport Rate Structure and Pricing, and End User Common Carrier Line Charges, First Access Charge Reform Report and Order."[81] In the document, the FCC

1. issued the second and third portions of its regulatory reforms implementing the Telecommunications Act of 1996;
2. adopted a market-based, competitive approach to encourage access charges to move toward a "true cost" level;
3. approved a move from usage-based to flat-rate common line charges (CLCs) because the FCC believed that the change would remove implicit subsidies and thus more accurately reflect the cost for local exchange carriers (LECs) to reach individual customers;
4. intended that the access charge reforms would decrease traditional access charges by approximately $1.7 billion in 1998 and $18.5 billion over the next five years;
5. increased the X-factor, to pass the productivity gains to customers;
6. required the LECs to exclude certain noncapitalized costs they had incurred in providing equal access to customers because they had already been recovered;
7. evaluated the ILECs' argument that the ISPs should have to pay access charges because the increased use of the Internet was placing a strain on the resources of the public switched telephone network;

8. retained the exemption for ESPs/ISPs from paying interstate access charge;
9. agreed that increased charges should be associated with the actual costs of each use but also stated at paragraph 345 that "many of the characteristics of ISP traffic are shared by other classes of business end-users";
10. stated that such charges must also provide appropriate incentives and obligations on network providers to ensure that high-quality, innovative data facilities are developed at reasonable prices and subject to fair terms of access and capable of facilitating expansion;
11. thus, even though some Internet access would be subsidized by the new universal service changes, the FCC believed that access-charge exemption for ESPs/ISPs fostered the early stages of Internet and data use;
12. avoided placing a disproportionate portion of true network costs on any one group of users—for both voice and data-access services; and
13. agreed to continue tracking the issue and to revisit it if necessary.

October, 1997— FCC Second Report and Order on Access Charge Reform

Five months later, on October 9, 1997, the FCC issued its Second Access Charge Reform Report and Order, entitled "In the Matter of Access Charge Reform; Price Cap Performance Review for Local Exchange Carriers; Transport Rate Structure, Second Access Charge Reform Report and Order."[82] It updated certain aspects of the FCC's comments, but did not substantively change its opinions or approach concerning the access reform and reciprocal compensation issues.

April 1998—FCC Report to Congress (CC Docket No. 96-45, FCC 98-67)

On April 10, 1998, the FCC filed a report with Congress concerning the findings of its studies and its resulting views on Internet access charges and universal service contributions. In the report entitled "In the Matter of Federal-State Joint Board on Universal Service (Report to Congress),"[83] the FCC reaffirmed its regulatory distinction between regulated voice services and unregulated data services. In doing so, the FCC reiterated that telecommunications service providers (or voice service providers) are subject to all of the following, while information service providers, (or data service providers), are not.

1. They provide transmission capability and capacity.
2. They receive access charges paid by long-distance carriers.
3. Their rates are regulated.
4. They have universal service obligations.

Where information service providers use their own facilities to provide content (information services), the FCC confirmed that it does not currently require them to contribute to the universal service fund, but reserves the right to do so in the future and on a continuing case-by-case review.[84]

August 1998—FCC Interconnection Memorandum Opinion and Order and Notice of Proposed Rulemaking

Six months later, the FCC ruled in the debate by refining the conditions around bill-and-keep and imposing *conditional bill-and-keep.* In its August 6, 1998 Interconnection Memorandum and Order,[85] the FCC ordered that the state commissions could impose bill-and-keep arrangements in arbitrations only if (1) "the volume of terminating traffic that originates on one network and terminates on another network is approximately equal to the volume of terminating traffic flowing in the opposite direction, and is expected to remain so;"[86] and (2) if the arrangement is mutually acceptable to both companies. Any bill-and-keep arrangement approved by a state must, however, include provisions that impose compensation obligations if the traffic becomes significantly out of balance according to a threshold adopted by the state or permits one party to request direct reciprocal compensation payment upon a showing that the traffic flows are inconsistent with the state-adopted threshold.[87] Any direct payments for this traffic between companies are to be measured in the number of *minutes of traffic terminated* by each carrier.

The FCC preferred *conditional* bill-and-keep over *unconditional* bill-and-keep as an intercarrier compensation mechanism under Section 251(b)(5) of the 1996 Act because the FCC believed that

> . . . carriers incur costs in terminating traffic that are not *de minimis,* and consequently, bill-and-keep arrangements that lack any provisions for compensation do not provide for the recovery of costs. In addition, as long as the cost of terminating traffic is positive, bill-and-keep arrangements are not economically efficient because they distort carriers' incentives, encouraging them to overuse competing carriers' termination facilities by seeking customers that primarily originate traffic. On the other hand, when states impose symmetrical rates for the termination of traffic, payments from one carrier to the other can be expected to be offset by payments in the opposite direction when traffic from one network to the other is approximately balanced with the traffic flowing in the opposite direction.[88]

As a result of the FCC's decision, the carrier-to-carrier compensation arrangements in the subsequent interconnection agreements, whether negotiated or arbitrated, took one of two forms: unconditional payment of reciprocal compensation or conditional bill-and-keep where the traffic balances fell within a specified threshold. In California, for example, the thresholds set forth in the interconnection agreements varied, but usually were in the range of 5–15% of the total traffic volumes. That is, if the threshold was set at 5%, the interconnection agreement required bill-and-keep if the traffic between the two carriers was reasonably in balance, but permitted a party to petition for reciprocal compensation if its ratio of terminating to originating traffic increased beyond a ratio of 55:45.

October 1998—FCC Memorandum Opinion and Order in the Matter of GTE Telephone Operating Cos.

On October 30, 1998, the FCC issued its "Memorandum Opinion and Order In the Matter of GTE Telephone Operating Cos."[89] In it, the FCC ordered that high-speed Internet services are *interstate* communications services and thus within the FCC's jurisdiction. While it chose to not press its jurisdiction, this decision was a significant departure from earlier decisions about ISP-bound calls being *local* and the Order set off a new round of inquiry, discussion, and court cases discussed in the following sections.

4.4.4 Challenges to FCC's Authority and Rules Implementing the Telecommunications Act of 1996

Throughout 1996, 1997, and 1998, as the FCC was investigating the issues surrounding access and preparing its two Access Charge Reports and Orders, numerous groups challenged the FCC's authority in this area. Nearly all challenges went to an appeals court and two went all the way to the Supreme Court. The key cases during this era include three decisions from the U.S. Court of Appeals for the Eighth Circuit: *Iowa Utilities Board v. FCC; Southwestern Bell Telephone Co. v. FCC;* and *AT&T Corp. v. Iowa Utilities Board.* Another key decision was from the Fifth Circuit, *SBC Communications Inc. v. FCC.* The third case, *AT&T Corp. v. Iowa Utilities Board* was heard by the Supreme Court, which ruled in favor of the FCC's authority. With the Fifth Circuit Court's case concerning SBC v. FCC, the Supreme Court denied *certiorari,* allowing the Fifth Circuit Court's decision to stand. The impact of each case is outlined below.

Case 1: *Iowa Utilities Bd. v. FCC,* 120 F.3d 753 [Case No. 96-3321, et. al.] (8th Cir. decided July 18, 1997)

During the controversy surrounding reciprocal compensation, the Iowa Utilities Board challenged the FCC's authority concerning the CLECs' access to ILEC facilities at the state level. The FCC disagreed with the Iowa Utilities Board's arguments and dismissed the challenge. The Iowa Utilities Board appealed to the U.S. Court of Appeals for the Eighth Circuit. In its decision, the Eighth Circuit Court's three-judge panel[90]

1. affirmed the FCC's ability to regulate local interconnection pricing and terms;
2. affirmed in part and vacated in part, the FCC's rules imposing reciprocal compensation obligation for terminating access for local traffic as set forth in the FCC's August 8, 1996, rules implementing the Telecommunications Act of 1996;[91]

3. Stated, however, that the Telecommunications Act of 1996 left most decisions concerning competition to states, and therefore prevented the FCC from intruding on state regulation;
4. overruled two of the FCC's key provisions implementing the Telecommunications Act of 1996 due to vagueness,[92] pricing and "pick and choose";
5. overruled the FCC's "pick and choose" provisions, which allowed any CLEC to pick and choose items from the interconnection agreements previously negotiated by the ILECs with other providers because it was inconsistent with private negotiations; and
6. decided that the FCC's rules for pricing of LECs' services were inconsistent with the directives of the 1996 Act.

Case 2: *Southwestern Bell Telephone Co. v. FCC,* 153 F.3d 523 (8th Cir. 1998)

In the second case, the Southwestern Bell Telephone Co. challenged the FCC's May 16, 1997, access charge reform Order, especially its continued exemption of ISPs from paying access charges. In reviewing the various arguments, the Eighth Circuit Court's three-judge panel unanimously upheld the FCC's access-charge reform order, decided that the FCC made a reasonable choice from the available policy alternatives it considered,[93] and affirmed the FCC's decision to continue the ISPs exemption from paying access charges.[94]

Case 3: *AT&T Corp. v. Iowa Utilities Bd,* Sup. Ct. Case No. 97-826 (decided January 25, 1999)

The same year, the Eighth Circuit Court also heard *AT&T Corp. v. Iowa Utilities Board,* Case No 97-826, *et. al.,* (January 27, 1998), in which the AT&T Corporation protested several rulings by the Iowa Utilities Board regarding the Telecommunications Act of 1996 that AT&T considered to be unusually heavy-handed. The Eighth Circuit Court upheld the Iowa board's decisions, and AT&T appealed the Eighth Circuit Court's decision to the Supreme Court. In its decision, released January 25, 1999, the Supreme Court upheld the FCC's authority to establish the local access rules, and reversed portions of the Eighth Circuit Court's decisions supporting the authority of the Iowa Board.[95]

Case 4: *SBC Communications Inc. v. FCC,* N.D.Tex. Case No. 7:97-CV-163-X (5th Cir. decided December 31, 1997)

An ILEC, SBC Communications, Inc., challenged the FCC[96] on various issues concerning competition.[97] In considering the case, Federal Judge Joseph Kendall reviewed various provisions of the Telecommunications Act of 1996, including the Act's restriction prohibiting the Regional Bell Operating Companies (RBOCs) from providing long-distance telephone services until the RBOC demonstrates that *sufficient competition* exists in the RBOC's local market. Judge Kendall ruled that this prohibition was unconstitutional.[98]

Case 5: *SBC Comm. Inc. v. FCC,* Case No. 98-10140 (5th Cir. decided September 4, 1998)

However, the FCC immediately appealed Judge Kendall's decision. The U.S. Court of Appeals for the Fifth Circuit heard the case, and in its decision released nine months later on September 4, 1998, sided with the FCC, reversed the Texas trial court's (Judge Kendall's) decision, and upheld the 1996 Act's long-distance prohibition.

Case 6: *SBC Comm. Inc. v. FCC,* Sup. Ct. Case No. 98-652, (Cert. denied January 21, 1999)

SBC then took the issue to the Supreme Court, but the Supreme Court declined to hear the case and let the Fifth Circuit Court decision stand.

4.4.5 February 1999—FCC's Declaratory Ruling

After the Supreme Court's affirmation of the FCC's authority in this area, the FCC issued its February 1999 Declaratory Ruling and Notice of Proposed Rulemaking entitled "In the Matter of Implementation of the Local Competition Provisions in the Telecommunications Act of 1996; Inter-Carrier Compensation for ISP-Bound Traffic, 14 FCCR 3689 at paragraphs 1034–1035 (February, 1999) (hereinafter the "Declaratory Ruling"). This Declaratory Ruling sought to clarify whether reciprocal compensation must be paid for ISP/Internet-bound traffic. While the FCC's Declaratory Ruling did not resolve the matter, the FCC did make the following six key decisions concerning reciprocal compensation.

Most ISP-Bound Traffic Is Interstate and the FCC Has Jurisdiction over Such Traffic

First, consistent with its earlier decisions, the FCC reiterated the following: (1) Internet-bound traffic is largely interstate, not local, and the FCC has jurisdiction over interstate traffic; (2) even if only 10% of the traffic were interstate, such as to a Web site or email, the FCC argued *de minimus* and considers it as 100%; (3) since reciprocal compensation payments apply only to local traffic,[99] ISP-bound traffic would not qualify; (4) however, ISPs are information service providers, exempted from paying intercarrier long distance charges; and (5) since ISP-bound traffic appears to be primarily interstate, and considering its earlier information service provider decisions, the FCC asserted jurisdiction over ISP traffic.

Not All ISP-Bound Traffic Is Interstate

Second, however, the FCC also recognized that ISP-bound traffic is not all interstate and some is under the jurisdiction of the states.

The Jurisdictional Nature of ISP-Bound Traffic Is To Be Considered Separately from the Regulatory Treatment

Third, the FCC distinguished the jurisdictional nature of ISP-bound traffic from the regulatory treatment of those calls. That is, while for jurisdictional purposes traffic to the Internet appears to be largely interstate and under the jurisdiction of the FCC, the FCC elected to discharge its regulatory obligations by treating ISP-bound traffic as though it were local.[100] Thus, the FCC also ruled that state regulatory commissions may treat such traffic as local for regulatory purposes, and, absent a federal rule to the contrary, states may enforce existing interconnection agreements and require reciprocal compensation for ISP-bound traffic in arbitrated agreements. The FCC stated that

> [W]e conclude that ISP-bound traffic is jurisdictionally mixed and appears to be largely interstate. This conclusion, however, does not in itself determine whether reciprocal compensation is due in any particular instance . . . In the absence, to date, of a federal rule regarding the appropriate intercarrier compensation for this traffic, *we therefore conclude that parties should be bound by their existing interconnection agreements, as interpreted by the state commissions.*[101] (Emphasis added.)

and

> A state commission's decision to impose reciprocal compensation obligations in an arbitration proceeding—or [in] a subsequent state commission decision that those obligations encompass ISP-bound traffic—does not conflict with any Commission rule regarding ISP-bound traffic.[102]

States Are Free to Regulate Reciprocal Compensation

Fourth, the FCC determined that no federal rule *prohibited* treating ISP-bound traffic as local for reciprocal compensation purposes, and thus state commissions were free to require carriers to include requirements for reciprocal compensation in the interconnection agreements they approved pursuant to their arbitrations under Section 252 of the Telecommunications Act of 1996.

The FCC rejects the "Two-call Theory" and Adopts an End-to-End Analysis

Fifth, while the FCC found that requiring reciprocal compensation for ISP-bound traffic did not violate any federal rule, the FCC rejected in its Declaratory Ruling the two-call theory as a basis for requiring payment of reciprocal compensation. The FCC held that rather than breaking a call to the Internet into component parts, as the two-call theory does, the call must be analyzed on an end-to-end basis. Thus, the FCC concluded that calls to the Internet do not terminate at the ISP, as required in the two call theory, but instead continue to their ultimate destination on the Internet Web site.[103] The FCC also recognized that the two-call theory was the basis for some states' decisions requiring reciprocal compensation for ISP-bound traffic and indicated that those states may decide to reexamine those decisions in light of the Declaratory Ruling.[104]

The FCC rejected the two-call theory as inconsistent with a long line of cases holding that a communication should not be divided into separate components for purposes of determining jurisdiction, relying instead on what it has termed *end-to-end analysis* for determining the jurisdiction of calls to enhanced service providers.[105] Under end-to-end analysis, the call is deemed to originate at the user's site and terminate at the Internet Web site, not the ISP location, and according to the FCC is likely to be interstate or international. The FCC's rejection of the two-call theory for Internet-bound traffic, however, remains subject to judicial review and further reconsideration at this time.[106]

Those courts and state commissions that relied on the two-call theory distinguished the FCC's cases requiring that the jurisdiction of calls to information service providers must be determined on an end-to-end basis on the grounds that those decisions had not considered calls to the Internet and that the Internet was fundamentally different from the other information services it had considered. As one court noted, "[i]ndeed, the FCC appears to define the very nature of Internet connections differently from interstate long-distance calls . . . For example, the FCC concluded that Internet access consists of more than one component, . . . a network transmission component, which is the connection over a local exchange network from a subscriber to an Internet Service Provider, in addition to the underlying information service."[107] This observation was prescient, as discussed further below. Therefore, while the two-call theory has been rejected by the FCC, it remains a potential argument in favor of paying reciprocal compensation for ISP-bound traffic given the United States Court of Appeals for the D.C. Circuit Court's finding that the FCC failed to adequately explain why the end-to-end theory determines whether ISP-bound traffic is local or interstate for purposes of Section 251(b)(5).

Outlined Factors for States' Reciprocal Compensation Decisions

Sixth, the FCC's Declaratory Ruling also provided states with a list of factors to consider when determining whether reciprocal compensation for ISP-bound traffic was reasonable, including

> For example, it may be appropriate for state commissions to consider such factors as: [1] whether incumbent LECs serving ESPs (including ISPs) have done so out of intrastate or interstate tariffs; [2] whether revenues associated with these services were counted as intrastate or interstate revenues; [3] whether there is evidence that incumbent LECs or CLECs made any effort to meter this traffic or other[wise] segregate it from local traffic, particularly for the purpose of billing one another for reciprocal compensation; [4] whether, in jurisdictions where incumbent LECs bill their end-users by message units, the incumbent LEC has included calls to ISPs in local telephone charges; and, [5] whether, if ISP traffic is not treated as local and subject to reciprocal compensation, incumbent LECs and CLECs would be compensated for this traffic."[108]

Thus, as a practical matter, aside from the elimination of the two-call theory, the FCC's February 1999 Declaratory Ruling did not change or clarify federal policy toward reciprocal compensation or reduce the amount of litigation. Instead, it simply concluded that the existing interconnection agreements, as interpreted by state commissions, are still binding until the FCC adopts a federal rule to the contrary[109] and thus, did not change the outcome of the litigation involving ILEC challenges to reciprocal compensation. Second, the FCC concluded that the states are free to treat Internet-bound traffic as local for ratemaking purposes.

FCC's Notice of Proposed Rulemaking

With its February 1999 Declaratory Ruling, the FCC initiated a Notice of Proposed Rulemaking (NPRM) entitled "In the Matter of Implementation of the Local Competition Provisions in the Telecommunications Act of 1996; Inter-Carrier Compensation for ISP-Bound Traffic, 14 FCCR 3689 (Feb. 1999)." In the NPRM, the FCC proposed rules on several items concerning *intercarrier compensation* and raised questions on other items on which it requested public comment. It avoided the term *reciprocal compensation* because that had come to mean a specific method of compensation between telecommunications carriers. Three areas that the FCC included in its queries were: jurisdiction, alternate methods for negotiations, and proposed inter-carrier compensation schemes.

Federal and State Jurisdiction

First, in its February 1999 Notice of Proposed Rulemaking, the FCC requested comment on whether the rules it adopts concerning intercarrier compensation should (1) apply only to interstate traffic, leaving the states the discretion to adopt separate rules for intrastate ISP traffic (mixed regulation); or (2) whether the FCC should adopt federal intercarrier compensation rules that would preempt the states' jurisdiction in this area (all federal regulation). A third option not considered by the FCC but raised in the comments was whether the issue should remain under state discretion with separate state rules (all state regulation). In raising this question, the FCC acknowledged that, for state and federal rules to coexist in a mixed-regulation situation, it would be necessary to identify local ISP-bound traffic versus interstate ISP-bound traffic and that technical and practical considerations may preclude doing so.[110]

Alternate Methods for Negotiations

Second, the FCC requested comment on two alternative proposals to govern the negotiations concerning ISP-bound traffic. Under its first proposal, parties would continue to use the existing method in which the parties undertake private negotiations on all issues under Section 251 of the Act, and if the parties fail to reach an agreement, any unresolved issues are arbitrated pursuant to Section 252.[111] This came to known as the mixed-jurisdiction method, and as a practical matter, commentors noted that it was under this procedure that reciprocal compensation became an issue.

Under the FCC's second alternative, the FCC would adopt a set of federal rules governing the intercarrier reciprocal compensation for ISP-bound traffic. Upon adoption of the rules, parties would negotiate the rates, terms, and conditions for ISP-bound traffic while, at the same time, separately negotiate the broader elements of an interconnection agreement within the framework of Section 251. Failure to reach a negotiated agreement for the ISP-bound traffic would be resolved by an FCC arbitration process separate from the Section 252 process.[112] This came to be known as the *federal-only* method.

With both proposals, the FCC stated that the reciprocal compensation negotiation process should be driven by market forces, rather than set by regulation. The FCC acknowledged that the "actual amounts, need for, and direction of intercarrier compensation might reasonably vary depending on the underlying commercial relationships with the end-user, who ultimately pays for transmission between its location and the ISP."[113] It believed that market-based negotiation would most likely reveal these variations and lead to a more effective outcome than regulation. Moreover, the FCC suggested that the current per minute-of-use (*minutes terminated*) pricing may not reflect how costs actually are incurred and noted that other pricing structures may result from more open, flexible negotiations that would better reflect the actual costs of terminating the traffic. By recognizing the different economic characteristics, the FCC suggested that voluntary agreements between the parties may be easier to reach.[114]

Proposed Compensation Schemes

Third, in its February 1999 Notice of Proposed Rulemaking, the FCC invited parties to submit proposals for alternative intercarrier compensation schemes that would advance its policy goals in this area of ensuring the broadest possible entry of efficient new competitors, eliminating the incentives for inefficient entry and irrational pricing schemes, and providing consumers the benefits of competition and emerging technologies.[115]

Reactions to the FCC's February 1999 Declaratory Ruling and Notice of Proposed Rulemaking

Not surprisingly, both the ILECs and CLECs read the FCC's February 1999 Declaratory Ruling as supporting their own positions, and both groups have quoted from it when arguing issues before the state commissions and federal courts. The ILECs, for example, emphasized the FCC's finding that ISP-bound traffic is essentially interstate in nature, not local, and therefore does not qualify for reciprocal compensation under the 1996 Act.[116] In contrast, the CLECs have emphasized the portion of the Declaratory Ruling that stated that until the FCC decides otherwise, there is no federal rule prohibiting payment of reciprocal compensation for ISP-bound traffic; thus, the FCC grants to the states the flexibility to treat the traffic as local for rate-making purposes. Both groups acknowledge that the real issue at stake is how the intercarrier compensation will be determined and at what levels. The state commissions, cities and counties, and state and federal courts, however, also had a range of reactions. Some of the more prominent reactions are discussed next.

State Commissions

Since more than one-half of the states had already ruled that the reciprocal compensation payments be made for Internet-bound traffic, the Declaratory Ruling required that these states revisit their decisions just to verify that they were in compliance. For states such as California, which relied on the two-call theory,[117] greater revisions had to be made. Since that time, California has continued to find in favor of reciprocal compensation, both as a general policy and in arbitrated interconnection agreements by applying the FCC's factors rather than relying exclusively on the two-call theory. Other state commissions have also continued to require reciprocal compensation provisions for ISP-bound traffic terminating at CLECs serving the ISPs.[118]

June 4, 1999—AT&T Corp. v. City of Portland and Multnomah County, *U.S. District Court in Oregon (Portland I)* [119]

Several cities and counties, however, have also addressed the issue of intercarrier compensation and attempted to impose equal access obligations on the cable system operators as telecommunications carriers and ISPs under their jurisdiction. For example, after AT&T purchased TCI and went to the City of Portland and Multnomah County in Oregon requesting the transfer of TCI's CATV franchise license to AT&T, both the city and county conditioned their approvals on AT&T's willingness to provide all ISPs with open access to customers over AT&T's cable facilities. AT&T refused and took the matter to federal district court in the case of *AT&T Corp. v. City of Portland and Multnomah County,*[120] which came to be known as *Portland I*. On June 4, 1999, the court concluded that the city and county had the authority to place such conditions on the transfer of a cable television franchise. Thus, the Court ruled for the city and county and dismissed AT&T's objections.[121] The court did not, however, address the merits or costs of open-access policies.

August 9, 1999—AT&T et al. v. City of Portland, *Appeal No. 99-35609 (Ninth Circuit Court of Appeals) (Portland II)*[122]

AT&T appealed to the Ninth Circuit Court of Appeals in *AT&T et al. v. City of Portland,*[123] a case that came to be known as *Portland II*.[124] It was joined by other cable television franchise holders from other cities and counties. The Ninth Circuit Court of Appeals heard a series of oral arguments. On one side were the CATV franchise holders and large ISPs, including AT&T, which also purchased Media One and had undertaken a joint venture with Time Warner/AOL, each with their own arguments, goals, and concerns. On the other side of the issue were various cities, counties, small ISPs, and open access groups, such as the Open Net Coalition,[125] advocating the enforcement of the sections of the Telecommunications Act of 1996 that required each carrier to provide open access for others. The arguments filed by the parties to date are available at the following Internet locations:

District Court Opinion:
www.techlawjournal.com/courts/portland/19990604op.htm

AT&T Appeal Brief:
www.techlawjournal.com/courts/portland/19990809.htm

Portland's Opposition Brief:
www.techlawjournal.com/courts/portland/19990907port.htm

FCC's Amicus Quriae Brief:
www.techlawjournal.com/courts/portland/19990916fcc.htm

On December 6, 1999, in response to some of the criticism raised by the open-access advocates against AT&T in Oregon, AT&T agreed to provide access through its cable systems to MindSpring Enterprises, Inc., an ISP. This action established a model for AT&T's possible relationships with nonaffiliated ISPs, which was highlighted by the FCC.[126] However, this issue is still open and deeply impacted by the outcome of the intercarrier compensation issue.

March 24, 2000—U.S. Court of Appeals for the District of Columbia Circuit Vacates the FCC's Declaratory Ruling

On March 24, 2000, the United States Court of Appeals for the District of Columbia Circuit, in a case known as *Bell Atlantic Telephone Companies v. FCC* issued a remarkable decision in which it vacated the FCC's February 1999 Declaratory Ruling and remanded it back to the FCC "for want of reasoned decision making."[127] Among other things, the D.C. Circuit Court held that (1) the traditional end-to-end analysis used by the FCC to determine whether a call is within its interstate jurisdiction does not automatically extend to determining whether ISP-bound traffic is local or interstate, and (2) the FCC failed to adequately explain why end-to-end analysis was appropriate.

In questioning the FCC's decision to apply end-to-end analysis to ISP-bound traffic, the D.C. Circuit Court noted a number of reasons why calls to ISPs do not clearly fit within the FCC's traditional end-to-end analysis. First, the Court pointed out that "the issues at the heart of this case is whether a call to an ISP is local or long-distance. Neither category fits clearly. Calls to ISPs are not quite local, because there is some communication taking place between the ISP and out-of-state websites. But, they are not quite long-distance, because the subsequent communication is not really a continuation, in the conventional sense, of the initial call to the ISP."[128]

Second, the Court found that "extension of the end-to-end analysis from jurisdictional purposes to the present context yields intuitively backwards results." That is, "calls that are jurisdictionally intrastate [local] will be subject to the federal reciprocal compensation requirements, while calls that are interstate are not subject to federal regulation but instead are left to potential state regulation . . . [t]he inconsistency is not necessarily fatal, . . . [b]ut it reveals that arguments supporting use of the end-to-end analysis in the jurisdictional analysis are not obviously transferable" for determining whether reciprocal compensation should be paid for ISP-bound traffic.[129]

Third, the court found that the FCC failed to adequately explain why calls to ISPs, which were then routed to the Internet, were analogous to the earlier cases that determined jurisdiction based on a single, continuous communication. Rather than consider ISP-bound traffic to the Internet as a single, continuous communication, the court found that the FCC failed to explain why the ISP is not "simply a communications-intensive business end user selling a product [Internet access] to other consumer and business end users."[130] Simply because the ISP may originate further telecommunications does not imply that original telecommunications did not terminate at the ISP, or why end-to-end analysis, which may be appropriate for jurisdictional purposes, works for reciprocal compensation.[131] In remanding the FCC's February 1999 Declaratory Ruling, the D.C. Circuit Court also opened the possibility that the FCC should reconsider its earlier decisions that reject the two-call theory and that the states have the power to require reciprocal compensation in the absence of a federal law to the contrary and to enforce reciprocal compensation provisions of existing agreements. As of October 2000, however, the issues raised by the D.C. Circuit Court had not been fully resolved.

The Reciprocal Compensation Adjustment Act (HR 4445)

In the interim, the ILECs turned their attention to Congress in an attempt to amend the Telecommunications Act of 1996. The Reciprocal Compensation Adjustment Act (HR 4445), introduced in the summer of 2000, sought to amend the 1996 Act concerning reciprocal compensation, although the details have changed frequently. In its various forms the bill proposed prohibiting minutes-terminated reciprocal compensation for all traffic, limiting the prohibition to Internet-bound traffic only, or exempting voice and wireless traffic from the prohibition. As of late October, 2000, no bill had passed the House.

Colorado PUC Ruling

At the state level, also during the summer of 2000, the ILECs appeared to have some success in their battle against uneven reciprocal compensation payments. In a rare victory, the Colorado PUC, in a Section 252 arbitration of an interconnection agreement between US WEST and Sprint Communications Company, refused to require US West to pay reciprocal compensation for ISP-bound traffic. In the major policy shift, the Colorado PUC concluded that "the originator of the Internet-bound call [acts] primarily as a customer of the ISP, not as a customer of US West. Both US West and Sprint are providing access-like functions to transmit the call to the Internet, similar to what their role would be in providing access to an IXC to transmit an interstate call."[132] Under these circumstances, reciprocal compensation would "bestow upon Sprint an unwarranted property right, the exercise of which would result in decidedly one-sided compensation."[133] The Colorado Commission recognized that this logic lead naturally to the conclusion that the ISP should pay access charges to both carriers and recover these costs from its end users, a result expressly prohibited by the FCC's regulations. Recognizing the limitations imposed by federal law, the Colorado PUC reverted back to a bill-and-keep type of mechanism for intercarrier compensation.[134] The issue of reciprocal compensation continues to be discussed at all levels—local, state, national and international.

CONCLUSION

While the Telecommunications Act of 1996 brought enormous changes to the telecommunications industry, a number of critical issues remain unresolved years after the 1996 Act was passed. These issues surrounding the four areas of local number portability, universal service, access charges, and reciprocal compensation dominate the current regulatory and legal calendar, as well as the efforts of many persons in the current telecommunications industry.

Local number portability, while an expensive and unsettled process, should position the United States for evolving uses of technology, such as one-number technology, which will relieve the pressure for additional telephone numbers. However, a tremendous amount of work is still being invested by the industry to make this happen and likely will continue for some time.

Universal service, while a well-established concept, moved from a means of providing infrastructure for U.S. residents to a means of providing affordable and available service. The impact of this massive change is still being discovered, but it suggests more rapid availability of modern telecommunications to users and a means to reduce the "digital divide" between the haves and have nots in the Internet Age.

Access is a critical factor. If excellent communications exist, but are not accessible, their presence is moot. While technology provides numerous exciting options for access, reciprocal compensation among carriers for assisting each other with the completion of calls has become one of the most complex and litigated issues in telecommunications since 1996.

The outcome of the current court cases and regulatory decisions in each of these areas will dictate the availability and quality of local, long-distance, and Internet communications in the future, how much access will cost customers, and which companies will provide it. The resolution of these issues will determine numerous areas of telecommunications law, the strategies of nearly all telecommunications companies, the pricing of most telecommunications products, and the monthly service bills of all telecommunications users. For these reasons, the progress of these issues and their eventual resolution are worth the attention of anyone involved in or using telecommunications.

ENDNOTES

[1]47 U.S.C.§ 251(b)(2) (1996).

[2]Plan of Reorganization, U.S. v. Western Electric Co., 569 F.Supp. 1957 (D.D.C. 1983).

[3]See Provision of Access for 800 Service (800 Access), Notice of Proposed Rulemaking, CC Docket No. 86-10, 102 FDD 2d 1387, 1388-89, para. 3-5 (1986).

[4]800 Access, Memorandum Opinion and Order on Reconsideration and Second Supplemental Notice of Proposed Rulemaking, CC Docket No. 86-10, 6 FCC Rcd. 5421 (1991); Order, 7 FCC Rcd. 8616 (1992).

[5]*See e.g. Illinois Commerce Commission Adopts Plan to Split Area Codes in Chicago Suburbs,* Communications Daily, at 7–8 (March 22, 1995); *North American Numbering Plan Manager Sees Companies "In Denial" on Changes,* Communications Daily, at 6 (March 24, 1995).

[6]FCC Notice of Proposed Rulemaking (NPRM) In the Matter of Telephone Number Portability, CC Docket No. 95-116, RM 8535, issued July 13, 1995; 10 FCC Rcd. 12350 (1995).

[7]*Id.*

[8]FCC Report and Order in the Matter of Administration of the North American Numbering Plan (NANP), CC Docket No. 92-237, adopted and released July 13, 1995; 11 FCC Rcd. 2588 (1995).

[9]Associated Press, March 4, 1999, at 1.

[10]7 U.S.C. § 921.

[11]H.R. Rep. No. 246, 81st Cong. 1st Sess. 8 (1949).

[12]Lavey, *The Public Policies That Changed the Telephone Industry into Regulated Monopolies: Lessons From Around 1915,* 39 Fed. Communications L.J. 171 (1987).

[13]MTS and WATS Market Structure, 50 Fed. Reg. 939, 941 (1985).

[14]Telecommunications Act of 1996, Pub. L. No. 104-104, §101, 110 Stat. 56 (1996) [hereinafter Telecommunications Act of 1996].

[15]47 U.S.C. § 214(e)(1) (1996).

[16]*Id.*

[17]In the Matter of Federal-State Joint Board on Universal Service, Fifth Order on Reconsideration and Fourth Report and Order, CC Docket No. 96-45, [FCC 98-67 —] (released June 23, 1998) [hereinafter June 1998 Universal Service Order].

[18]47 U.S.C. § 214(e)(1) (1996).

[19]In the Matter of Federal-State Joint Board on Universal Service, CC Docket No. 96-45, 12 FCC Rcd. 87; 1996 FCC LEXIS 6259, 5 Comm. Reg. (P&F) I (released Nov. 8, 1996) [hereinafter November 1996 Universal Service Order].

[20]*Battle Over Universal Service Order Continues,* Communications Today (Aug. 20, 1997).

[21]47 U.S.C. § 214(e)(1) (1996).

[22]In the Matter of Implementation of Section 254(k) of the Communications Act of 1934, as amended, (Continued) Docket No. 97-163; 1997 FCC LEXIS 2396, 6 Comm. Reg. (P&F) 1428 (released May 8, 1997).

[23]Telecommunications Act of 1996, *supra* note 14, at § 253(f).

[24]*Id.* at § 254.

[25]*Id.* at § 214(e).

[26]See Pressler and Schieffer, *A Proposal for Universal Telecommunications Service,* 40 Fed. Communications L.J. 351, 353 (1988); and Congressional Budget Office, U.S. Congress, *The Changing Telephone Industry: Access Charges, Universal Service,*

and Local Rates, at 27 (June 1984). *See also,* Kahn and Shew, *Current Issues in Telecommunications Regulation: Pricing,* 4 Yale J. on Reg. 191, 192 (1987), citing Posner, *Taxation by Regulation,* 2 Bell J. Econ. & Mgmt. Sci. 22 (1971).

[27]MTS and WATS Market Structure, Third Report and Order, 93 FCC.2d 241 (1983), *aff'd in principal part and remanded in part* National Ass'n of Regulatory Utility Commissioners v. FCC, 737 F.2d 1095 (D.C. Cir. 1984), *cert. denied* 469 U.S. 1227 (1985) [contains *FCC 1983 Access Charge Order*]. Implementing rules located at 47 C.F.R. § 69.1 *et seq.* (Part 69).

[28]See MTS and WATS Market Structure, Amendment of part 67 of the Commission's Rules and Establishment of a Joint Board, Report and Order, 3 FCC Rcd. 4543 (1988).

[29]Thomas E. Weber, *Baby Bells vs. the World: A Fight for Internet Fees,* Wall St. J., Feb. 27, 1997, at B6.

[30]Second Computer Inquiry [Computer II], 77 FCC.2d 384 (1980) (initial FCC preemption of state authority over enhanced services), upheld in Computer & Communications Industry Ass'n v. FCC, 693 F.2d 198 (D.C. Cir. 1982), *cert. denied,* 461 U.S. 938 (1983).

[31]California v. FCC, 905 F.2d 1217, 1223 (9th Cir. 1990) (California I) (Services involving data processing are "enhanced" services and not subject to FCC regulation. In contrast, common carriers provide "basic" communications services and are subject to FCC regulation).

[32]Amendments of Part 69 of the Commission's Rules Relating to Enhanced Service Providers, Order, 3 FCC Rcd. 2631 (para. 2n.8) (1988), (ESP Exemption Order).

[33]MTS and WATS Market Structure, Docket No. 78-72, Memorandum Opinion and Order, 97 FCC 2d 682, 711-22 (1983) (Access Charge Reconsideration Order) (referring to origination and termination of interstate communications by enhanced service providers (ESPs) as "leaky" private branch exchange (PBX) scenario). *See also,* Amendments of Part 69 of the Commission's Rules Relating to Enhanced Service Providers, CC Docket No. 87-215, Order, 3 FCC Rcd. 2631 (1988) (ESP Exemption Order).] *See also,* MTS and WATS Market Structure, (para. 4); 47 C.F.R. § 69.2(m)(1996) (*End User* means any customer of an interstate or foreign telecommunications service that is not a carrier.)

[34]ACTA Petition for Declaratory Ruling, Special Relief, and Institution of Rulemaking (RM No. 8775) (Mar. 4, 1995) [hereinafter Internet Phone Petition].

[35]In the Matter of Implementation of the Local Competition Provisions of the Telecommunications Act of 1996, CC Docket No. 96-98, FCC No. 96-325 (rel. August 8, 1996) at ¶ ¶ 1034-1035; II FCC Rcd 15499, 1601 3 (¶¶ 1034-1035) [hereinafter August 8, 1996, FCC Rules Implementing 1996 Act].

[36]*See, e.g.,* Application Access Charges to the Origination and Termination of Interstate, IntraLATA Services and Corridor Services, 57 Rad. Reg 2d (P&F) ¶ ¶ 9-11 (April 12, 1985); IntraLATA Exchange Telecommunications markets, Col. PUC, 113 P.U.R. 4th 272 (May 23,1990); Alternative Regulatory Frameworks for Local Exchange Carriers, 41 CPUC 89, 125 P.U.R. 4th 260 (July 24, 1991); IntraLATA Competition, Kansas State Corp. Commission, 143 P.U.R. 4th 542 (April 30, 1993).

[37]Larson and Parsons, *Telecommunications Regulation, Imputation Policies and Competition,* 16 Hast. Comm/Ent L.J. 1, 3-6 (1993).

[38]Telecommunications Act of 1996, *supra* note 14, at § 3(44), codified at 47 U.S.C. § 153(26) (1996).

[39]*See generally,* Illinois Bell Tel. Co. v. WorldCom Techs., 157 F.3d 500, 501 (7th Cir. 1998).

[40]47 USC § 252(d)(2)(A) (1996).

[41]Telecommunications Act of 1996, *supra* note 14, at § 251(b)(5).

[42]47 U.S.C. § 251 (1996).

[43]*See,* US WEST Communications, Inc. v. Colorado Public Utilities Commission, et. al., Dist. Ct., City & Co. of Denver, Case No. 96-CV-2566, at 10. (Oct. 27, 1997).

[44]47 U.S.C. § 252(d)(2)(A)(ii) (1996).

[45]47 U.S.C. § 252(d)(2)(B)(i) (1996).

[46]47 U.S.C. § 252(d)(2)(B)(i) (1996).

[47]Telecommunications Act of 1996, *supra* note 14, Joint Explanatory Statement at 7.

[48]Opinion and Order Concerning Reciprocal Compensation, State of New York Public Service Com-

mission, Opinion No. 99-10, August 26, 1999, mimeo at 2 [hereinafter New York Opinion].

[49]Dial-up lines are distinguished from xDSL services and other dedicated facilities such as T-1 and higher-speed lines, which, because they are in essence private lines not local exchange lines. Therefore they do not qualify for reciprocal compensation.

[50]Southwestern Bell has stated that a reciprocal compensation requirement could require it to pay as much as $421 monthly in termination fees for calls to the Internet initiated by its customers and terminated by a CLEC even though it receives only about $12 per month from its basic residential customers. Southwestern Bell Telephone Company v. Public Utility Commission of Texas, et al., 1998 U.S. Dist. LEXIS 12938, *14 (W.D. Texas, 1998).

[51] *Ruling of Illinois Commission on Ameritech vs. Focal,* Wall St. J., July 22, 1998; NewsEdge Corp, PR Newswire, File: p0722112.301.

[52]Thomas E. Weber, *Baby Bells vs. the World: A Fight for Internet Fees,* Wall St. J., Feb. 27, 1997, at B6.

[53]Jon Cornetto, *BellSouth to Charge for Voice-over-IP Services,* Infoworld Elec., Sept. 9, 1998.

[54]Notice of Proposed Rulemaking (NPRM) on Reform of Interstate Access Charges, CC Docket No. 96-262, FCC 96-488 (released Dec. 24, 1996), available at *www.fcc.gov.*

[55]Notice of Inquiry (NOI), CC Docket No. 96-262, FCC 96-488 (released Dec. 24, 1996), available at *www.fcc.gov.* (FCC initiated a study to examine whether FCC should encourage higher speed data communications to and from consumers and best methods to do so).

[56]"CCS" stands for "centum call seconds" or "hundreds of call seconds per hour." It is a measure of traffic used in the telecommunications industry. One hour of telephone traffic is equal to 36 CCS (60 seconds/minute × 60 minutes/hour = 3600 seconds/hour divided by 100 = 36 CCS).

[57]Report of Bell Atlantic on Internet Traffic, Feb. 12, 1997, at § 3, available at *www.ba.com/ea/fcc/report.htm.*

[58]*PacTel Reports On-Line Traffic Swamping Local Exchange,* Dow Jones Front Page, America Online, Mar. 25, 1997.

[59]*Id.*

[60]*Id.* at § 4.

[61]*Senate Panel to Discuss Issue of Internet Access Charges,* Dow Jones Front Page, America Online, Feb. 24, 1997.

[62]*Id.* at § 5.

[63]*www.opennetcoalition.org.*

[64]*See, www.oecd.org;* and Douglas Lavin and Jennifer Schenker, *High Access Costs for Net Depress Usage, Study Says,* Wall St. J., Mar. 14, 1997, at B3C.

[65]FCC Interconnection Memorandum Opinion and Order and NPRM, FCC 98-188, August 6, 1998 at ¶¶ 1034-1035; FCC Launches Inquiry, Proposes Actions to Promote the Deployment of Advanced Telecommunications Services By All Providers, 1998 FCC LEXIS 4109 (1998) (570 pages) [hereinafter FCC August 6, 1998 Interconnection Order].

[66]The term *NPA-NXX* represents the first six digits of a long-distance telephone number. A Numbering Plan Area (NPA) is more commonly known as an "area code." NXX is also known as a "prefix," or the first three digits of a local telephone number. Historically, most NPA-NXX have been associated with a geographic area, but as people have wireless roaming and free nationwide calling, this is changing. Also, 500, 800, and 900 services have always been geographically independent. The standard call rating system used in the United States determines whether a call as toll or local according to the distance between the geographic areas associated with the NPA-NXXs of the originating and terminating telephone numbers.

[67]New York Opinion, *supra* note 48, at 47.

[68]*Id.* at 46.

[69]*Id.*

[70]Federal State Joint Board on Universal Service, Report to Congress, CC Docket No. 96-45, FCC 98-67 (April 10, 1998), at ¶¶ 59 and 86 [hereinafter FCC April 10, 1998 Report to Congress].

[71]47 USC § 153(43) (1996).

[72]47 USC § 153(20) (1996).

[73]August 8, 1996 FCC Rules Implementing 1996 Act, *supra* note 35, at ¶ 59.

[74]In fact, many of the early agreements did not mention traffic terminating to ISPs because it was not something specifically contemplated by either party at the time the agreement was negotiated.

[75]New York Opinion, *supra* note 48, at 11-12.

[76]In addition to the obligation to pay reciprocal compensation, the incumbents also argue that virtual local numbers deprive them of toll and access charge revenues.

[77]New York Opinion, *supra* note 67, at 8.

[78]*Ruling of Illinois Commission on Ameritech vs. Focal, supra* note 51.

[79]On February 24, 1997 the U.S. Senate Commerce Committee also initiated a study on the issue of reciprocal compensation and began gathering information on the issues.

[80]Kevin Werbach, Counsel for New Technology Policy, *Digital Tornado: The Internet and Telecommunications Policy,* FCC, Office of Policy and Plans (OPP), Working Paper No. 29, March 1997. Available at *www.fcc.gov/Bureaus/OPP/working_papers/oppwp29pdf.*

[81]In the Matter of Access Charge Reform, Price Cap Performance Review for Local Exchange Carriers, Transport Rate Structure and Pricing, and End User Common Carrier Line Charges, First Access Charge Reform Report and Order, CC Docket Nos. 96-262, 94-1, 91-213, 95-72; 12 FCC Rcd. 15982 (May 16, 1997) [hereinafter First Access Charge Order]. *See also, FCC Approves Historic Universal Service and Access Charge Reforms,* Communications Daily (May 8, 1997).

[82]In the Matter of Access Charge Reform; Price Cap Performance Review for Local Exchange Carriers; Transport Rate Structure, Second Access Charge Reform Report and Order, FCC 97-368; 12 FCC Rcd. 16606 (1997) [hereinafter Second Access Charge Order].

[83]FCC April 10, 1998 Report to Congress, *supra* note 70.

[84]Telecommunications Act of 1996, *supra* note 14.

[85]FCC August 6, 1998 Interconnection Order, *supra* note 65.

[86]*Id.* at ¶ 1111.

[87]*Id.* at ¶ 1113.

[88]*Id.* at ¶ 1112.

[89]FCC Memorandum Opinion and Order in the Matter of GTE Telephone Operating Cos., CC Docket No. 98-79, FCC 98-292 (October 30, 1998). *See also, GTE Service is Ruled Interstate, Suggesting Likely Boon for Bells,* Wall St. J., Nov. 2, 1998, at B6.

[90]*See also,* MFS Intelenet v. Bell Atlantic, Case No. 8731, Order No. 74557, 1998 Md. PSC LEXIS 43 (September 3, 1998) (petition of Bell Atlantic Corp. for relief from barriers to deployment of Advanced Telecommunications Services et. al, CC Docket No. 98-11. 570 pages).

[91]August 8, 1996 FCC Rules Implementing 1996 Act, *supra* note 35, at ¶ 1034; Inter-Carrier Compensation for ISP-Bound Traffic, 14 FCC Rcd. 3689 (1999).

[92]*See* Stephanie Mehta, *Bells Win Ruling by Court Curbing FCC's Regulation,* Wall St. J., Jan. 23, 1998 at B6.

[93]Southwestern Bell Telephone Co. v. FCC, 153 F.3d 523, 537 (8th Cir. 1998).

[94]*Id.* at 544.

[95]AT&T Corp. v. Iowa Utilities Board, Sup. Ct. Case No 97-826, decided Jan. 25, 1999. *See also,* Stephanie Mehta and Edward Felsenthal, *Supreme Court Restores Federal Rules Aimed at Opening Local Phone Markets,* Wall St. J., Jan. 26, 1999 at A2.

[96]SBC Comm. Inc. v. FCC, (N.D. Tex., Case No. 7:97-CV-163-X, (5th Cir. Dec. 31, 1997).

[97]*Id.*

[98]*U.S. Judge Voids Key Provision of Telecom Law,* AOL Business Center, Jan. 1, 1998.

[99]Telecommunications Act of 1996, *supra* note 14, at § 251(b)(5) or 47 CFR § 51.70.i(e).

[100]*"In the Matter of Implementation of the Local Competition Provisions in the Telecommunications Act of 1996; Inter-Carrier Compensation for ISP-Bound Traffic,* 14 FCC Rcd. 3689, Feb. 1999 at ¶ ¶ 1034-1035 [hereinafter *"Declaratory Ruling"*].

[101]Id. at ¶ 1.

[102]*Id.* at ¶ 26.

[103]*Id.* at ¶ ¶ 15-16.

[104]*Id.* at ¶ 27.

[105]*See, e.g.* United States v. AT&T Co., 57 F. Supp. 451 (S.D.N.Y., 1944); National Association of Reg-

ulatory Utility Commissioners v. Federal Communications Commission, 746 F.2d 1492 (D.C. Cir. 1984); Petition for Emergency Relief and Declaratory Ruling Filed by the BellSouth Corporation, 7 FCC Rcd. 1619 (1992).

[106]Bell Atlantic Telephone Companies v. Federal Communications Commission, Case No. 99-1094, (March 24, 2000), (the United States Court of Appeals for the D.C. Circuit held that the FCC had not sufficiently explained its rejection of the two-call theory. Vacated the Declaratory Ruling, and remanded the matter back to the FCC for further reconsideration.)

[107]Southwestern Bell Telephone Company v. Public Utility Commission of Texas, 1998 U.S. Dist. LEXIS 12938, *31 (1998).

[108]Declaratory Ruling, *supra* note 100, at ¶ 24.

[109]Southwestern Bell Telephone Co., v. Brooks Fiber Communications of Oklahoma, Inc. et al., Case No. 98-CV-468-K(J), Order filed September 29, 1999 (ND OK).

[110]Declaratory Ruling, *supra* note 100, at ¶ 36.

[111]*Id.* at ¶ 30.

[112]*Id.* at ¶ 31.

[113]*Id.* at ¶ 29.

[114]*Id.*

[115]*Id.* at ¶ 33.

[116]*See, for example,* Michigan Bell Telephone Co. v. MFS Intelenet of Michigan, Inc. et al., 1999 U.S. Dist. LEXIS 12093, *6-7 (1999) ("Plaintiff's primary argument is that its agreements with the Defendant carriers are to be construed in accordance with federal law, and that the [Declaratory Ruling] establishes as federal law that ISP bound traffic is not local"); Nevada Bell's Motion for Summary Judgment, Nevada Bell v. Sheldrew et al. Case No. CV-N-99-00492-HDM (RAM).

[117]California noted that its decision finding that, as a general matter, reciprocal compensation is appropriate for traffic terminating to ISPs was based on a number of policies, not just the two-call theory, and reaffirmed its earlier decision. *See* California Public Utilities Decision No. 99-07-047 (1999).

[118]New York Opinion, *supra* note 48.

[119]AT&T Corp. v. City of Portland and Multnomah County, Case No. CV 99-65-PA (U.S. Dist. Ct. Or., June 4, 1999) [hereinafter Portland I].

[120]*Id.*

[121]*Id.*

[122]AT&T et al. v. City of Portland, Appeal No. 99-35609 (9th Cir. Aug. 9, 1999).

[123]*Id.*

[124]Corey Grice, *California ruling may affect ISP fees,* CNET News.Com, June 7, 1999 at *www.News.com/News/Item/0,4,37427,00.htmlt.ne.fd.gif.k.*

[125]Open Net Coalition Home Page, at *www.opennetcoalition.org.*

[126]Letter from AT&T to FCC Chairman William Kennard (Dec. 6, 1999) available at *www.techlawjournal.com/broadband/19991206let.htm.*

[127]Bell Atlantic Telephone Companies v. Federal Communications Commission, Case No. 99-1094 (2000).

[128]*Id.* at 5 and 6.

[129]*Id.* at 6.

[130]*Id.* at 7.

[131]*Id.* at 8.

[132]*Re* Sprint Communications Company, L.P., Docket No. 00B-011T, Decision No. C00-479, § C.g (May 5, 2000).

[133]*Id.* at § C.j.

[134]*Id.* at § C.l.

CHAPTER 5 The World Trade Organization and Its Telecommunications-Related Agreements

THE CREATION OF THE WORLD TRADE ORGANIZATION (WTO) IN 1994[1] likely will prove to be one of the most significant events in telecommunications history. Its agreements affecting telecommunications equipment and services have completely changed the structure of worldwide telecommunications and opened previously closed markets to new players. The WTO agreements provide the trade rules and regulations that will enable modern telecommunications to be used throughout the world, establishing the infrastructure and services required for the global, converged, broadband Internet Age.

As a result of the WTO agreements, worldwide sales of telecommunications equipment and services are expected to increase significantly over the next 10 years.[2] For U.S. telecommunications companies, the U.S. Department of Commerce stated in June 1994[3] that the WTO and its General Agreement on the Trade in Services (GATS)[4]

1. opened the telecommunications markets in over 75 countries, representing 90% of the 1994 worldwide telecommunications revenues. This significantly expanded trade opportunities for all segments of the U.S. telecommunications industry;
2. considerably reduced foreign tariffs on telecommunications services and equipment, resulting in substantially lower costs for U.S. exporters and international telecommunications customers;
3. streamlined and standardized customs procedures resulting in clearer, simpler, more uniform customs procedures and approvals processes throughout the world. This enables companies to import and export supplies, equipment, products, and services quicker, more easily, and at less cost;
4. significantly improved international protection of intellectual property rights. For software and hardware, in particular, the WTO member countries have agreed to compulsory licensing for patents and tougher measures against design piracy;

5. improved worldwide standards–setting procedures;
6. required regular publication of WTO member–country statutes and regulations affecting telecommunications with prompt updates as changes occur; and
7. recognized legally enforceable rights for telecommunications companies doing business internationally, strengthened by a new dispute-resolution process managed by the WTO that covers access and use of telecommunications services by other nations.

These measures significantly improved the conditions for telecommunications companies that want to participate in the global telecommunications market. The WTO and its agreements will impact all aspects of the telecommunications industry, all persons working in the industry and every person using telecommunications systems to communicate. However, it is a little-known area that is important for telecommunications users and U.S. companies doing business abroad to understand. The purpose of this chapter is to provide that understanding.

To do so, Section 5.1 describes the background of world trade law, policy, and organizations before the 1970s. Section 5.2 describes the General Agreement on Tariffs and Trade (GATT), with special emphasis on the nearly 25 years of effort to include services in world trade law and international negotiations, and the resulting agreements known as the Uruguay Round, that established the new world of global telecommunications. Sections 5.3 and 5.4 discuss the impact of the North American Free Trade Agreement (NAFTA) and the Global Information Infrastructure (GII) on international telecommunications. Section 5.5 describes the conclusion of the Uruguay Round and its multiple accomplishments, including the 1994 creation of the WTO, the General Agreement on Trade in Services (GATS), and Annexes on Telecommunications. They are the primary organizations and agreements guiding international trade today. Section 5.6 outlines the results of the 1996 Negotiations on Basic Telecommunications, while Section 5.7 describes its impact on the 1996 Singapore Ministerial Conference. Section 5.8 addresses the 1997 WTO Agreement on Basic Telecommunications, while Section 5.9 describes its inclusion in the GATS as its Fourth Protocol. Appendices, D through F list the legal instruments of the WTO, its members, and the commitments made by the various participating countries in the world. These summaries provide a guide for telecommunications companies seeking to enter these markets.

5.1 HISTORICAL INTERNATIONAL TRADE—GOODS, NOT SERVICES

For thousands of years, trade between different tribes or nations was limited almost exclusively to the exchange of tangible goods. Trade in services, as we know it today, was largely nonexistent. Services consisted mainly of the skills and information of the traders and the individuals traveling with them.

Even as late as the 1970s, the world's trade in goods still significantly exceeded the international trade in services. However, as computers and advanced communications propagated around the world, the trade in services grew by approximately 20% each year. In today's postindustrial economy, the importance of immediate communications, information management, the rise of international travel, and global business has caused an explosion in the international trade in services. Since 1996, this explosion has increased to nearly 50% of the developing countries' Gross Domestic Product (GDP), and nearly 70% of the GDP in developed countries.[5]

5.1.1 International Trade Law

The oldest and most established form of international law involves the sale of tangible goods, encompassing both private contract law and public trade law. In private contract law, the buyer and seller agree on the terms of the transaction and make a legally binding commitment to one another. This type of international transaction law has evolved over thousands of years, addressing differences in language, customs, monetary and banking systems, technical standards, customer preferences, customer service standards, jurisdictional issues, transportation arrangements, export and import laws, and problem resolution. Nonetheless, trade is still difficult because the parties deal with each other over a distance and often without an ongoing relationship or history of cooperation.

In public trade law, the international sale of goods raises several additional complexities beyond the issues noted above. International trade is so important to national interests that documents such as the U.S. Declaration of Independence discuss it, wars have been fought over it, and whole economies depend on it. While the laws of two or more countries must merge to address the transactions, each country's law will reflect its unique interests.

5.1.2 U.S. Trade Policy

Perhaps the most obvious or direct manner in which these national interests have emerged is in *tariffs,* or the additional charge, fee, or *duty* added to foreign products when they are imported into a country. The purpose of tariffs is to make imported products more expensive than domestic products, in order to protect domestic companies. For example, from colonial days to the Civil War, the northern states of the United States wanted high protective tariffs in the form of import duties to discourage the import of European manufactured goods that competed with the North's manufactured products. The southern states, on the other hand, wanted lower import and export costs for all goods, because they imported finished goods and exported raw cotton for use in the finished goods. The outcome of the Civil War resulted in highly protective U.S. tariffs that remained until after World War I. Throughout this time, U.S. tariffs generally exceeded the tariffs of most of the other developed countries.[6]

After World War I, the United States had one of the strongest economies in the world and began to reverse its trade policy. An important step in the U.S.'s tariff reduction movement was the Tariff Act of 1922. The reduction was quickly reversed, however, when the Great Depression began. With fears about exporting jobs and *dumping* of cheap imports on the U.S. economy, Congress enacted the Smoot–Hawley Tariff Act of 1930, which established the highest tariffs in U.S. history. Other nations retaliated by imposing their own high tariffs and blocking U.S. imports, with negative results for the United States. During the period from 1929 to 1933, U.S. exports fell nearly 70%.

Tariffs Reduced

The Trade Agreements Act of 1934 sought to change this. The Act, drafted in large part by Secretary of State Cordell Hull, delegated to the President of the United States authority to negotiate tariff reduction agreements up to 50% of the 1930 tariffs without further legislative action. This removed trade agreements from the treaty process where only the Senate approved them and gave power to all of Congress to give advance approval to agreements that met its terms. The lowering of tariffs proved effective and within five years, by 1939, U.S. exports had increased 50 to 60% over their 1934 levels. This increase occurred despite the ongoing worldwide depression. Learning from this experience, the United States today supports the concept of lower tariffs and retains its policy of presidential tariff negotiation within preset congressional parameters.

1944—International Monetary Fund, World Bank, and International Trade Organization

In 1944, an international monetary conference was held in Bretton Woods, New Hampshire, to develop a plan for rebuilding the world economy after World War II. At the meeting, the participants created both the International Monetary Fund (IMF) and the World Bank and proposed an International Trade Organization (ITO). The governments of the participating countries approved the IMF and the World Bank, but did not ratify the ITO.[7]

1947—Marshall Plan, United Nations, and GATT

After World War II, to assist in postwar recovery and reestablish world trade, the United States exported goods, some services, and capital, plus established collective security programs such as the Marshall Plan and the United Nations (UN). In 1947, more than 23 countries began meeting to develop a consensus on international trade,[8] and by 1948, the participants completed the "General

Agreement on Tariffs and Trade", which became the principal agreement guiding the global trade in goods for the next 47 years.

1993—World Trade Organization

The General Agreement on Tariffs and Trade was so successful that, in December 1993, the member countries voted to extend and formalize the GATT into the World Trade Organization. They signed an agreement on April 15, 1994, known as the Marrakesh Agreement Establishing the World Trade Organization. It was a part of the Final Act Embodying the Results of the Uruguay Round of Multilateral Trade Negotiations. On January 1, 1995, when the World Trade Organization agreement actually went into effect, the General Agreement on Tariffs and Trade became part of the newly created WTO. The agreement was absorbed into and replaced by the WTO, but continues to be the dominant agreement guiding the world's flow of products between countries.

At the same time, the member countries of GATT also signed the General Agreement on Trade in Services (GATS), with an Annex on Telecommunications that eventually resulted in: a Reference Paper prepared by the 1996 Negotiating Group on Basic Telecommunications (NGBT)'s and the 1997 WTO Agreement on Basic Telecommunications. The following sections review the basic philosophy, evolution, terminology, and structure of the GATT, WTO, GATS, and various telecommunications agreements and how they affect U.S. telecommunications companies.

5.2 1948—THE GENERAL AGREEMENT ON TARIFFS AND TRADE

The General Agreement on Tariffs and Trade (GATT) provided a set of rules guiding the international trade of goods that was adopted in 1948 and has been adhered to by most of the countries in the world since then.[9] The countries participating in the GATT are known as *member countries,* and over the years, the GATT's membership has grown from 23 countries in 1949 to over 100 countries in 1994.

5.2.1 Structure and Operation of GATT

To participate in GATT, each member country was represented by a *minister* from that country, and each country had one vote, regardless of its size or level of development. For the 47 years from 1947 through 1994, a simple majority vote of the ministers made all important, governing decisions in GATT at the GATT's annual business meetings.

Between these meetings, different groups operated within the GATT to carry out its tasks, including: (1) the GATT Secretariat, (2) the Council of Representatives, (3) tariff conferences, and (4) rounds of negotiations. The GATT Secretariat, located in Geneva, Switzerland, consisted of an administrative staff and international trade specialists who attended to the day–to–day operations of the GATT. The Council of Representatives, elected by the GATT members, addressed any urgent issues that arose between the annual business meetings. To deal specifically with tariff issues, tariff conferences were held periodically, usually at GATT headquarters in Geneva. Many of these groups continue, in some form, under the WTO.

Among the most important GATT meetings were periodic meetings called Negotiating Rounds that frequently continued over several years. In these rounds, the member countries through their ministers: updated the GATT, removed additional trade barriers, and reviewed complaints concerning violations of GATT rules. Between 1947 and 1994, the GATT has held eight rounds of negotiations as shown in Table 5.1.[10]

These rounds of negotiations were extremely important and effective. Each of the eight rounds ended successfully with trade agreements that were later ratified and implemented by the member countries. The first round, in 1947, rescued international trade recovery efforts by developing the GATT after the world's failed attempt to create an international trade organization. It resulted in 45,000 tariff concessions affecting $10 billion of trade, approximately one-fifth of the world's total at the time. The next six rounds focused on further tariff reductions. They were highly successful in that, as a result of these rounds, the average tariff levels around the world fell, the world's economies grew, and the standard of living in member countries improved dramatically.

The eighth round, the Uruguay Round, was the most extensive negotiating round in the GATT's history to that time and added significantly to the GATT's previous success. It began in September 1986 at Punta Del Este, Uruguay, with a GATT Ministerial Declaration, known as the *Uruguay Declaration,* and formally concluded in Marrakesh, Morocco on April 15, 1994, with *The Final Act Embodying the Results of the Uruguay Round of Multilateral Trade Negotiations,* signed by the GATT ministers. For this reason, the results of the Uruguay Round are frequently referred to as the Marrakesh Agreement or the *Marrakesh Declaration,* depending on whether the signed international agreement or major ministerial declaration resulting from the negotiations is being referred to.

The Uruguay Round established the WTO and new sets of agreements covering goods, services, intellectual property, and a dispute settlement. It obtained over 22,500 pages of specific commitments from 120 countries, stating what they would and would not agree to in international trade. The results provide a significantly higher degree of market security for traders and investors, freer, more robust world trade through negotiation, more predictable trade policies, and a better allocation of each nation's resources. In 1994, economists estimated that by the year 2005, the results of the Uruguay Round would contribute between $184 billion and $510 billion to the world's economy.

TABLE 5.1 GATT NEGOTIATING ROUNDS

Round	Name	Location	Dates	Objective
1	Geneva Round	Geneva, Switzerland	1947	Adoption of GATT by 23 countries
2	Annecy Round	Annecy, France	1949	Tariff reduction
3	Torquay Round	Torquay, England	1951	Tariff reduction
4	Geneva Round	Geneva, Switzerland	1956	Tariff reduction
5	Dillon Round	Geneva, Switzerland	1960–62	Tariff reduction
6	Kennedy Round	Geneva, Switzerland	1962–67	Tariff reduction GATT Anti-Dumping Agreement GATT negotiation rules
7	Tokyo Round	Tokyo, Japan	1973–79	Overall tariff reduction to an average level of 35% and 5–8% among developed nations.
8	Uruguay Round Concluded as Marrakesh Agreement	Punta del Este, Uruguay Final Act signed in Marrakesh, Morroco	1986–94	Inclusion of Services trade Establishment of the WTO Extension and update GATT Inclusion of Intellectual property Enhancement of Dispute resolution

5.2.2 1970s and 1980s—Movement to Include Trade in Services in GATT

For its first 46 years, from 1948 through 1994, the GATT addressed only the international trade in *goods*. In the 1970's, however, recognizing the increasing world trade in *services,* U.S. service providers lobbied the U.S. government to include trade in services in the GATT.[11] As a result, the United States required the U.S. Trade Representative's (USTR's) annual report to Congress to track both the growth and importance of the trade in services and include regular updates on foreign barriers to such trade.[12] It suggested that similar reports be developed within GATT and other international groups, and advocated that trade in services be a specific objective in future GATT negotiating rounds.

As a result, several reports emerged over the next several years, including a 1980–81 GATT produced report entitled *The Report of the Consultative Group of Eighteen to the Council of Representatives,*[13] which linked the trade in certain services to the trade in goods.[14] A second report, compiled by the Trade Committee of the Organization for Economic Cooperation and Development

(OECD Trade Committee), showed that trade in services was significant, but could be further stimulated by multilateral negotiations on trade barriers concerning trade in services.[15] Based on this second report, 25 members of the OECD, mainly from the developed nations, pushed to include services in the Uruguay Round to start in 1986.

A year later, at the 1982 GATT annual meeting, the United States formally proposed that trade in services be added to GATT.[16] The United States also suggested that each GATT member prepare a national study on its domestic service sectors. This suggestion was reinforced by the GATT Ministerial Declaration of 1982, with an additional recommendation that the parties share the results of their studies and determine whether to take action on their findings.[17] Some countries were opposed because they did not want to liberalize their markets as quickly as others wanted them opened.

The United States's proposal included one provision specifically addressing telecommunications services. In it, the United States acknowledged the unique dual role of telecommunications within the larger realm of services, noting that telecommunications is both a separate economic activity or service in its own right—and a backbone or transport for other services. As such, the United States pointed out, that telecommunications services should receive special attention in GATT discussions on services.

Within three years, by 1985, 14 countries had submitted reports to the GATT showing "that [in 1979] the growth rate of service industries outpaced the growth rate of all other sectors in both developed and developing countries, and that in 1980, forty-eight percent of all workers in the reporting countries were employed in service industries."[18] The reports of the United States and United Kingdom specifically called for negotiations for services similar to those for goods in order to encourage international cooperation in the services markets.[19]

5.2.3 September 20, 1986—Opening of the GATT's Uruguay Round

On September 20, 1986, a GATT Ministerial Declaration, the *Uruguay Declaration,* opened the Uruguay Round of GATT trade negotiations in Punta del Este, Uruguay.[20] In it, the ministers stated that the negotiations over the next several years were to include negotiations on both trade in *goods* and trade in *services*[21] and announced the formation of two separate groups to lead the negotiations in each area: a Group of Negotiations on Goods (GNG) and a Group of Negotiations on Services (GNS-Group). Both groups were to report to the ministerial–level Trade Negotiations Committee (TNC).[22]

1986 Ministerial Declaration on Services (MDS)

The Uruguay Declaration's section outlining the proposed negotiations on services was called the 1986 Ministerial Declaration on Services (MDS). In it, the ministers listed the duties of the Group of Negotiations on Services (GNS-Group), the topics to be addressed by the group, and the overall objectives for the negotiations, formally referred to as the General Negotiation on Services (GNS).

Concerning the duties of the Group of Negotiations on Services, the ministers stated that the group was to oversee the negotiations on trade in services, determine the order in which the issues would be addressed, and set the timeline for the negotiations. At each stage, the group was responsible for ensuring that the items negotiated met the objectives given to it by the ministers, and making periodic reports to the *Ministerial Trade Negotiations Committee* on the progress of the negotiations.[23] The topics identified by the ministers to be negotiated were very broad and included all areas and types of services.

The objectives for the Uruguay Round's General Negotiation on Services, as stated in the Ministerial Declaration on Services were

> to establish a multilateral framework of principles and rules for trade in services including an elaboration of possible disciplines for individual sectors, with a view to expansion of such trade under conditions of transparency and progressive liberalization and as a means of promoting economic growth of all trading partners and the development of developing countries. Such framework shall respect the policy objectives of national laws and regulations applying to services and shall take into account the work of relevant international organizations.[24]

5.2.4 GATT Principles

These objectives specifically mandated that the framework for trade in services to be established by the Group of Negotiations on Service include the principles and rules developed by the GATT for the trade in goods such as: transparency; progressive liberalization; promotion of the economic growth of all trading partners; inclusion of elements for the development of developing countries; respect for the policy objectives of the member countries, such as universal service; and consideration of the work of other international organizations. These principles also encompass the GATT principles of most-favored nation (MFN) or nondiscriminatory treatment; the national treatment of foreign service providers; and market access, or the rights of service providers when providing services in a country. Each of these GATT principles is important to U.S. telecommunications companies seeking to understand the international telecommunications trade environment, requirements expected of them, and concessions made to them. The principles are described in the following nine sections.

Transparency

First, the principle of *transparency* was a long–term, core principle of the GATT. As applied to services, it requires each participating country to publish all international agreements concerning trade in services to which the country is a party, to publish all domestic statutes and regulations that affect the provision of services in that country, and to notify the Council for Trade in Services when any new measures are adopted that significantly affect trade in services. The only exception is information that is confidential.[25]

Progressive Liberalization

Second, the *liberalization* of trade in services seeks to reduce the regulations and barriers restricting worldwide trade in services. Over their 47 year experience, the GATT members had learned that successful liberalization includes lowered tariffs and streamlined customs procedures. It increases fairness and equity; accelerates technological innovation; creates more diversity in products; and therefore, enhances consumer choice. Liberalization of a country's facilities also allows more cooperation between countries and facilitates competition. As such, it reduces anticompetitive behavior among countries and the negative impact of such behavior.

The GATT Ministers also recognized that liberalization of services could not happen overnight, but rather must be phased in. Hence the Ministerial Declarations on Services stated that the ministers' objective was progressive liberalization of regulations on trade in services, and generally, countries are permitted five years or more to implement their changes before many obligations apply to them.[26] The ministers also allowed the use of both multilateral and plurilateral agreements. *Multilateral agreements* were negotiated by the members of GATT, adopted by majority vote, and applied to the entire GATT membership. *Plurilateral agreements,* on the other hand, were negotiated between fewer countries and applied only to those countries. Thus, their application was more selective or conditional.

Promote the Economic Growth of All Trading Partners

Third, the GATT's definition of *services* included the opportunity for a country's service providers to establish operations in foreign markets and to have foreign companies provide services in its markets. Thus the movement of services, consumers, technological know-how, labor, and capital across borders were included.

Enable the Development of Developing Countries

Fourth, the GATT ministers recognized that the less-developed countries (LDCs) have needs that are significantly different from those of the more developed countries. These include access to information and distribution channels, markets for the LDCs' exports, and technology transfer. The Group of Negotiations on Services (GNS-Group) sought to address these in the negotiations and to include accommodations for them in the agreed upon framework of principles and rules for trade in services (services framework).

Coordinate with the Work of Other International Bodies

Fifth, as the Group of Negotiations on Services (GNS) considered how to relate their negotiations concerning trade in services to the work of other international bodies, the GNS realized that most international organizations and agreements do not try to liberalize trade in the sense of the GATT members' objectives.[27] For example, the OECD and the U.N. Conference on Trade and Development (UNCTAD) had arrangements for some services, but they mainly facilitated the exchange of information and established technical standards. While these helped to reduce nontariff trade barriers, many international agreements contained exceptions for services considered important to the GATT members.[28]

Most-Favored Nation (MFN) Treatment

Sixth, the term *most-favored nation* (MFN) *treatment* is somewhat confusing because it sounds as if the services of one country should be "favored" over the services of another country. However, in the GATT's multilateral agreements, the concept of MFN requires that the signatory countries treat the services and service suppliers of other countries equally. It requires that "no country's services and providers be treated more or less favorably than the similar services and service suppliers of any other country" and that this nondiscriminatory treatment be "accorded immediately and unconditionally" upon signing the agreement. In plurilateral agreements, MFN applies only if the country participated in the negotiation and thus it applies only on a conditional basis.

Given this concept of nondiscriminatory treatment, the GATT specified that the GATT's trade rules must be applied equally and fairly to all member nations. For example, a member granting a trade advantage to one country must grant the same advantage to all other member countries. The concept is also the reason the GATT grants each country one vote regardless of its size or level of development.

Negotiated Exceptions

Seventh, while MFN and nondiscriminatory treatment were key features of the GATT, the members also realized that no agreements can be absolute, so they allowed *negotiated exceptions* to specific rules if a country needed certain specific exceptions but was generally willing to adhere to most of the GATT rules. For example, the GATT members allowed the developed countries in the world to place lower trade barriers on items imported from developing countries than on similar items from developed countries. In addition, two or more countries may negotiate separate trade agreements that are more favorable than the GATT provisions, as was done under the North American Free Trade Agreement (NAFTA).

National Treatment of Foreign Service Providers

Eighth, the principle of national treatment of foreign service providers required that members treat service providers from other member countries no less favorably than it treats its own domestic service providers. For example, service providers from other countries must be afforded the same access to public switched telephone networks as domestic providers, regardless of whether the domestic providers are public or private.[29]

Market Access

Ninth, the principles of most-favored nation and national treatment stated that if a country allows other countries to enter its borders to operate in its markets, it must do so on a nondiscriminatory basis. In contrast, the principle of *market access* requires that a country take action to reduce its tariffs, quotas, and other restrictions to allow the greatest possible access for other countries to its market. The first two concepts are conditional, the third is absolute.

5.2.5 Drafts of the Services Framework

The framework of principles and rules for trade in services (services framework) established by the Group of Negotiations on Services is significant and worth some time for persons in the telecommunications industry to understand because it became the General Agreement on Trade in Services (GATS) adopted by the GATT member countries in 1994. It also provided the key principles and served as the basis for all major telecommunications legislation and international agreements since 1986, including NAFTA in 1992, the description of the Global Information Infrastructure (GII) in 1994, the U.S. Telecommunications Act of 1996, and the 1997 WTO Agreement on Basic Telecommunications, also known as the 1997 GATS.

The development process of the Group of Negotiations on Services, however, was not a direct path to its goal. Instead, the process required numerous reports and four separate drafts of the framework document to address the many issues facing the GATT members, before it was complete. In actuality, all four drafts were eventually rejected, but each helped to refine the GATT members' thinking about the issues and to shape the final compromise that resulted in the GATS. Some discussion of the reports and four drafts therefore is useful.

1988—Report to the Trade Negotiations Committee (TNC)

In 1986, the Group of Negotiations on Services began its assigned task of establishing a services framework by considering various issues, objectives and formats for the framework. Two years later, in 1988, the group made a Report to the Ministerial Trade Negotiations Committee, suggesting, first, that the agreement resulting from the framework negotiations should be similar to the GATT in format, but should be a separate agreement from the GATT. Hence, the concept of a GATS was advanced. Second, the group emphasized that while *crossover between concessions* made by the participating countries should be encouraged, no cross–retaliation should be permitted between trade in goods and trade in services. Third, the Group stated that the format of the resulting agreement should be as broad as possible.

1989—Examination of Specific Sectors

In 1989, the Group of Negotiations on Services held discussions concerning which services should be included. These were known as the Examination of Specific Sectors. The major issue in telecommunications was whether the negotiations should include both basic and enhanced services. In many countries, *enhanced services* were, for the most part, already open to competition, while *basic services* were generally still government–owned or provided by heavily regulated companies.

1990—July Draft

The first of the four drafts of the services framework was released in July 1990 by the Group of Negotiations on Services. It was similar in structure to the GATT and provided several suggestions for three issues: (1) national treatment, (2) government procurement, and (3) most-favored nation.

Concerning national treatment, the July Draft stated that all sectors of a member-country's services should be included in that member's *Schedule of Specific Commitments* and should therefore be subject to the concept of national treatment unless specifically exempted by a member-country in its schedule.[30]

The one exemption that the July Draft suggested should be allowed to all members was that of *government procurement* of services. The Group of Negotiations on Services agreed that GATT members should be able to exempt services procured by their governments for their own use from their obligations, including MFN requirements. It noted that the members differed only on how long the exemption should last and when the exemptions should be phased out.

Concerning the most-favored nation clause, the United States wanted to withdraw it because the United States had more open markets than other countries and was concerned about "free–riders," or countries that could enter the U.S. market without opening their own markets. Other parties did not want the MFN principle to apply to basic telecommunications. Still others wanted the MFN concept to apply in order to avoid turning the service agreements into a series of reciprocity agreements in specific sectors of each market. The July Draft of the framework for a trade in services agreement suggested a solution to these concerns by recommending that MFN be included only if the GATS signatories provided sufficient market access and application of the concept of national treatment in all sectors. To track this, the compromise suggested that the

GATT members be required to state in their schedules what they would and would not provide concerning market access and national treatment. The July Draft continued to be discussed until December 1990.

1990—December Draft

In December 1990, a second draft of the services framework, called the December Draft, suggested a solution to the continuing MFN problem by recommending both an annex on enhanced telecommunications services and an annex on continued negotiations on basic telecommunications.

While this was more acceptable to the participants, they wanted more time to review and discuss the suggestion. Therefore, the December 1990 Draft also was not accepted, and the negotiations continued for another year, until December 1991.

1991—Dunkel Draft

In December 1991, the third draft of the framework agreement, known as the Dunkel Draft, released the first consensus on the MFN issue.[31] In the Dunkel Draft, the member countries agreed that the MFN concept should be applied to all service sectors and member exceptions or reservations regarding such applications should only be taken under the specific conditions listed in an agreed upon Annex for Article II-MFN Exemptions.[32] However, the members did not resolve the *basic versus enhanced services* issue in telecommunications and thus the Dunkel draft was not accepted. As the negotiations on this issue continued, they were interrupted at various times by (1) a clash between the United States and the European Union over agricultural issues, (2) the United States's proposal to take an MFN exemption on basic telecommunications, (3) the European Union's proposal to take an MFN exemption to audiovisual communications, and (4) the lack of quality of service offered by the developing countries in their proposals.

1992—November Draft

Just before the GATT's 1992 annual meeting, the United States and European Union reached an agreement on the agriculture issue, but the issues in telecommunications continued. Therefore, at the 1992 annual meeting, the United States suggested a two-year extension of negotiations on basic telecommunications, hoping that the developing countries would raise the quality of their offers on services and open their markets to more reciprocal services and opportunities. The suggestion was accepted, and two years later, in 1994, a final framework was accepted as the General Agreement on Trade in Services, part of the results of the Uruguay Round of negotiations. It was announced in the Marrakesh Agreement of 1994.

5.3 DECEMBER 1992—NORTH AMERICAN FREE TRADE AGREEMENT NAFTA

During the GATT negotiations on services in 1988, the United States and Canada signed a separate bilateral agreement on many of the issues being discussed at GATT. The document was titled the U.S.–Canada Free Trade Agreement of 1988.[33] Four years later, in December 1992, Mexico joined the United States and Canada, expanding the agreement into the North American Free Trade Agreement (NAFTA) of 1992. NAFTA formed the largest free-trade bloc in the world[34] and incorporated many of the concepts, principles, wording, and elements of the GATS negotiations and discussions. NAFTA focused on removing specific problems faced by the three countries in trying to do business together and allowed only one year, until January 1, 1994, for the three countries to eliminate most trade and investment barriers among them.

A large part of NAFTA focused on telecommunications issues. The discussions of the United States, Canada, and Mexico on telecommunications in NAFTA impacted their positions on similar discussions at the GATT negotiations. The main objective of the three NAFTA countries in telecommunications was "to liberalize, as much as possible, trade in telecommunications equipment and services among the three countries."[35] Several provisions in NAFTA that impacted other international telecommunications agreements are discussed below.

5.3.1 Telecommunications Equipment and Service

First, in NAFTA, the three governments immediately eliminated import/export tariffs for 80% of telecommunications equipment. The remaining 20% was to be eliminated over the next five years.[36] Not all tariffs were removed as of mid-2001, but progress continues to be made in this area.

Second, concerning telecommunications services, Chapter 11 of NAFTA immediately removed all restrictions in Mexico on foreign investment and local presence of enhanced or value-added network (VAN) service providers.[37] Chapter 13 specifically required the three countries to guarantee[38] access to, interconnection with, and use of the public network on a reasonable and nondiscriminatory basis to VANs providers and private corporate networks[39] and freedom of movement of information to residents of the other NAFTA nations.[40] In addition, NAFTA allowed shared use of leased circuits among different users with a common business interest. This was a significant change from the previous agreements among many countries that allowed only single use of leased lines.[41]

Third, NAFTA allowed the three members to restrict network access in order to protect the technical integrity of the network, ensure universal service, and preserve the telecommunications companies' public service responsibilities.

5.3.2 Licensing

Fourth, NAFTA restricted the three governments from imposing regulations that unreasonably hinder or complicate the licensing application process. The governments must not discriminate against or hamper the entrance or existence of enhanced service providers into the market or in any other way use licensing as a nontariff barrier to trade. NAFTA requires that the information requested on the license applications be limited only to information concerning financial solvency and conformity with equipment standards. It also requires that the license applications be processed expeditiously and in a nondiscriminatory manner. Chapter 18 of NAFTA obligates Canada, Mexico, and the United States to ensure that their rules and regulations concerning licensing be applied only after individuals are given reasonable notice and an opportunity to present evidence in support of their positions.[42]

5.3.3 Technical Standards

Fifth, in establishing technical standards, NAFTA follows the U.S. policy that anything can be attached to the public network so long as it does not harm the network, is compatible with other users, and ensures physical safety for network personnel and users.[43] Thus, NAFTA disallows mandatory performance criteria[44] and prevents member countries from using technical standards as nontariff barriers to trade. It established a Telecommunications Standards Subcommittee to meet regularly to develop a "fully integrated North American telecommunications market,"[45] and requires the countries to accept test data from any qualified tester. The subcommittee and the member countries also seek to promote such standards on a global basis.[46]

5.3.4 Basic versus Enhanced Telecommunications Services

Sixth, in 1992, when NAFTA was signed, Canada, Mexico, and the United States still had monopolies providing *basic* telecommunications services and infrastructure. At Canada's insistence, NAFTA specifically excluded basic services telecommunications from its obligations. However, NAFTA required further discussions on the liberalization of trade in all telecommunications services and explicitly left open the possibility of future agreements on basic telecommunications services.[47]

For *enhanced* and value-added network services, however, NAFTA prohibited anticompetitive behavior. The three-member governments agreed to incorporate procompetitive measures in their domestic rules and regulations, including (1) accounting requirements, (2) structural separation (subsidiaries), (3) equal access, and (4) timely disclosures of technical changes.[48]

5.3.5 Antitrust Issues

Seventh, NAFTA required harmony among the antitrust and competition laws of the three member countries. Thus, NAFTA prohibits price fixing, cartels, and mergers that substantially lessen competition. It requires prenotification and review of acquisitions that exceed a preset threshold and provides measures for enforcement.[49]

5.3.6 Intellectual Property Issues

Eighth, Chapter 17 of NAFTA provides an elaborate framework for the protection and enforcement of intellectual property rights among the three countries. It requires Canada, Mexico, and the United States to offer equal levels of protection and enforcement of intellectual property rights and establishes strong enforcement remedies for violations, including damages, injunctive relief, and due process.[50]

5.3.7 Dispute Resolution

Ninth, Chapter 29 of NAFTA established the dispute resolution mechanism to be used when questions of interpretation or application arose among the three signatory countries.[51] It allowed private parties in each country to request that their trade representative investigate and take action on their behalf. For private parties in the United States, Title 19, section 2418 of the U.S. Code addresses the responsibilities and procedure to be used by the U.S. Trade Representative.[52]

Since its creation in 1992, the NAFTA has been significant to the U.S. telecommunications industry, and in March/April 1994, the NAFTA incorporated some of the results and principles of the GATT Uruguay Round in its agreement. The revised agreement was expected to do the following: (1) affect more than 360 million consumers; (2) create a $6 billion Mexican telecommunications market; (3) allow U.S. telecommunications equipment manufacturers to compete with European manufacturers, who had historically dominated the Mexican market; and (4) eliminate many investment restrictions on telecommunications companies in Mexico. The increased trade with and investment opportunities in Mexico were expected to bolster the Mexican economy and increase telecommunications capacity, especially for enhanced services and products.[53] After April 1994, NAFTA has used the WTO's Annexes on Telecommunications as its model, but NAFTA goes far beyond the WTO agreements.

Hence both the WTO agreements and NAFTA work together to enable solid U.S. participation in the new global telecommunications opportunities.

5.4 MARCH 1994—THE GLOBAL INFORMATION INFRASTRUCTURE (GII) CONCEPT IS INTRODUCED

As worldwide discussion on trade in all services and goods, not just telecommunications, came to fruition in the final weeks of the WTO's Uruguay Round, a World Telecommunications Development Conference was held in Buenos Aires, Argentina, in March 1994. At the conference, on March 21, 1994, Vice President Al Gore introduced a U.S. initiative to foster the development of a Global Information Infrastructure (GII). He described the Global Information Infrastructure (GII) as "a network of networks, to increase the ability to communicate around the globe."[54] He also stated that with "private investment, competition, open access to the network, appropriate and flexible regulation and universal service, the United States hopes to foster the growth of a communications infrastructure that is truly global in its reach and diverse in its service."[55]

Larry Irving, Assistant Secretary for Communications and Information, U.S. Department of Commerce, also pointed out that access to an advanced global communications infrastructure could increase the efficiency of existing businesses, create new businesses, provide developing countries with state-of-the-art communications systems, promote sustainable economic development and contribute to the quality of life.[56] These concepts echoed the reasons most of the countries of the world participated in the concluding activities of the Uruguay Round of GATT and signed the resulting documents.

5.5 APRIL 15, 1994—URUGUAY ROUND CONCLUDED, WTO CREATED IN MARRAKESH, MOROCCO

The negotiations comprising the Uruguay Round of GATT concluded on December 15, 1993, when the ministers voted on the final agreements, decisions, and declarations. Over the next four months, the legal documents were prepared for signatures, and the ministers met in Marrakesh, Morocco, on April 15, 1994, to sign The Final Act Embodying the Results of the Uruguay Round of Multilateral Trade Negotiations (Final Act).[57] The legal documents embodying the results of the Uruguay Round were published in a 34-volume set containing (1) the Final Act, (2) the Marrakesh Agreement Establishing the WTO, (3) various ministerial decisions, (4) individual agreements, (5) schedules from the member countries of their specific commitments on services, (6) tariff schedules from the member countries for trade in goods, and (7) multilateral agreements. The contents of each volume are noted in Table 5.2 on the next page.[58]

The Uruguay Round covered all areas of international trade, not just telecommunications. However, as with the GATT, many of the general agreements affect trade in telecommunications equipment and services, including provisions for finance and investment, customs procedures, intellectual property protections, dispute resolution, tariff schedules, and service commitments. For this reason, it is useful for telecommunications personnel working in the international arena to be aware of what exists in these documents.

5.5.1 Volume 1—The Legal Texts

Volume 1, at 550 pages long, contains the legal texts of the three main accomplishments of the Uruguay Round: (1) the Final Act, (2) The Marrakesh Agreement Establishing the WTO, and (3) the ministerial decisions and declarations. The first document, the Final Act, surprisingly is only one page long, but is the cover statement that summarizes the documents that follow and emphasizes that the Uruguay Round was the most extensive GATT round to date and made significant steps toward free trade in both goods and services. It also noted that this was the first time that trade in services was officially included in GATT negotiations and that the results of the Uruguay Round will considerably increase global trade.

The second document in the Legal Texts is the Marrakesh Agreement Establishing the World Trade Organization, and its six annexes. At 430 pages long, it is the main text of Volume 1. In addition to creating the WTO, its annexes contain four extremely important results of the Uruguay Round. *Annexes* address special situations of individual services. For example, Annex 1 contains three of the four main sections of the WTO. First, Annex 1A contains the Multilateral Agreements on Trade in Goods, which updated and modified the role of the 1947 GATT and absorbed it into the WTO. Second, Annex 1B created the new General Agreement on Trade in Services (GATS), and provided the new rules concerning the trade in services. Third, Annex 1C created the Agreement on Trade-Related Intellectual Property Rights (TRIPs), which strengthened intellectual property protections, including issues of copyright, trademarks, patents, industrial designs, and trade secrets, as well as performer's rights, geographic indications, and industrial designs. Fourth, Annex 2 contains the Understanding on Rules and Procedures Governing the Settlement of Disputes, enhancing the rules and procedures for the WTO's dispute-resolution mechanism. Fifth, Annex 3 provides the Trade Policy Review Mechanism important to updating the agreements, and sixth, Annex 4 provides the Plurilateral Trade Agreements negotiated by various countries.

The third set of Legal Text documents includes several ministerial decisions and declarations, totaling approximately 19 pages. These three sections, and their contents, are organized as seen in Table 5.2.

TABLE 5.2 LEGAL TEXTS OF WTO AGREEMENT (ESTABLISHING THE WTO AND INCLUDING GATT 1994)

I.	**Final Act**
II.	**Agreement Establishing the World Trade Organization (WTO)**
Annex 1A:	Agreements on Trade in Goods
	1. General Agreement on Tariffs and Trade 1994 (GATT 1994)
	2. Uruguay Round Protocol (GATT 1994)
Annex 1B:	General Agreement on Trade in Services (GATS) and Annexes
Annex 1C:	Agreement on Trade-Related Aspects of Intellectual Property Rights (TRIPs) (including Trade in Counterfeit Goods)
Annex 2:	Understanding on Rules and Procedures Governing the Settlement of Disputes
Annex 3:	Trade Policy Review Mechanism
Annex 4:	Plurilateral Trade Agreements
III.	**Ministerial Decisions and Declarations**
	1. Decision on Measures in Favour of Least–Developed Countries (LDCs)
	2. Declaration on the Contribution of the WTO to Achieving Greater Coherence in Global Economic Policymaking
	3. Decision on Notification Procedures
	4. Customs Valuation
	5. Technical Barriers to Trade
	6. Decision on Measures Concerning the Possible Negative Effects of the Reform Programme on Least–Developed and Net Food–Importing Developing Countries
	7. General Agreement on Trade in Services
	8. Decision on Implementation of Article XXIV:2 of the Agreement on Government Procurement
	9. Decision on the Application and Review of the Understanding on Rules and Procedures Governing the Settlement of Disputes
	10. Decision on Improvements to the GATT Dispute Settlement Rules and Procedures
	11. Agreement on Implementation of Article VI of GATT 1994
	12. Decision on Dispute Settlement Pursuant to the Agreement on Implementation of Article VI of GATT 1994 or Part V of the Agreement on Subsidies and Countervailing Measures 1994

5.5.2 WTO Established and Absorbed GATT

The second document in Volume 1 of the results of the Uruguay Round created the WTO. This was one of the most significant results of the negotiations. It was formally approved on April 15, 1994, by The Marrakesh Agreement Establishing the World Trade Organization (WTO Agreement)[59] and became effective on January 1, 1995.

The WTO is the successor to the GATT and absorbed the GATT into Annex 1A of the WTO Agreement. However, the WTO is not simply an extension of GATT. Instead, it completely replaces the GATT and has a very different character covering a much broader scope. Some of the principal differences include the following:

1. From its beginning, the GATT was a set of rules. It was a multilateral agreement that, following the failed attempt to establish an International Trade Organization in the 1940s, had no solid legal, institutional foundation. Instead, it operated as an ad hoc organization with a small secretariat in Geneva. The GATT members created the WTO to establish an international institution with a firm legal basis. The WTO is a permanent international organization, with its own secretariat, also located in Geneva.
2. Because the GATT was an agreement and not an organization, its commitments could only be applied on a *provisional basis.* After more than 40 years, the member governments chose to treat their commitments to the GATT as permanent, but the WTO, as a formal international organization, makes their WTO commitments full and permanent.
3. The GATT applied only to the global trade in goods or tangible merchandise. The WTO covers both the trade in goods and the trade in services, plus adds intellectual property protections, and provides solid legal rights for the participants.
4. While the GATT was a multilateral agreement, by the 1980s, many new, plurilateral agreements had been added. Nearly all WTO agreements are multilateral and therefore more universal. In addition, the WTO agreement calls for a single institutional framework that encompasses the GATT, including all agreements previously made under the GATT and the complete results of the Uruguay Round. This *single-undertaking approach* means that membership in the WTO requires accepting all of the results of the Uruguay Round without exception.[60]
5. The WTO's dispute settlement system is faster, more automatic, and therefore much less susceptible to blockages than the previous GATT system. The implementation of the WTO's dispute findings is also more assured.

Purpose

Given these differences, the WTO is the primary international organization overseeing the principles and rules of international trade. It is "the legal and institutional foundation of the multilateral trading system." Through collective debate, negotiation, and adjudication, it is the platform on which trade relations among countries evolve. As such, it provides the principal contractual obligations determining how governments frame and implement their domestic trade legislation and regulations.[61]

The purpose of the WTO is to help world trade flow smoothly through a system based on rules, organize trade negotiations, and settle trade disputes between governments. It oversees the implementation of the agreements based on the GATT before 1994 and the results of world trade talks since 1994.

The WTO is the first formal multilateral agreement to address the trade in services. Its initial objectives included coverage, national treatment, market access, and the concepts included in the services framework. As of 1999, it was still negotiating issues such as maritime shipping and the movement of people across borders. To contribute to the achievement of its objectives, the WTO members agreed to enter into "reciprocal and mutually advantageous arrangements directed to the substantial reduction of tariffs and other barriers to trade and to the elimination of discriminatory treatment in international trade relations."

Essential Functions

The essential functions of the WTO include (1) administering and implementing the multilateral and plurilateral trade agreements that make up the World Trade Agreement (WTA), (2) acting as a forum for multilateral trade negotiations, (3) seeking to resolve trade disputes, (4) overseeing national trade policies, and (5) cooperating with other international institutions involved in global economic policy making.

Policies and Fundamental Principles of the WTO

To accomplish these functions, the WTO has six policies on which it focuses. The six are as follows: (1) to assist developing and transitional economies; (2) to provide specialized help for export promotion through the International Trade Centre; (3) to promote regional trading arrangements; (4) to encourage cooperation in global economic policy making; (5) to publish reviews of members' trade policies; and (6) to provide routine notification when members introduce new trade measures or alter old ones. The WTO's policies translate into four fundamental principles of the WTO: trade without discrimination, predictable and growing access to markets, promoting fair competition, and encouraging development and economic reform. The application of each by the WTO is discussed in the following four sections.

Trade without Discrimination

Under the most-favored nation (MFN) clause in the GATS Article 1, WTO member countries commit to granting to the products of other members no less favorable treatment than that accorded to the products of any other country. In the WTO, MFN means that no member can discriminate among the other members. Except under special conditions, if one country grants another country a special favor, such as a lower duty rate for an imported product, that favor must also be extended to other WTO members so that they are all equally most favored. This form of nondiscrimination is one of the most important principles of the WTO trading system.

Predictable and Growing Access To Markets

While quotas generally are outlawed, tariffs or customs duties are legal in the WTO. However, tariff reductions are a key goal of the WTO. More that 120 countries reduced their tariffs in the Uruguay Round, placing the changes in some 22,500 pages of National Tariff Schedules that are an integral part of the WTO. Tariff reductions by industrial countries on industrial products generally are phased in over five years and can be quite significant. The Uruguay Round also increased the percentage of product lines bound to these reductions to nearly 100% for developed nations and countries in transition and 73% for developing countries. WTO members have also agreed to an initial set of commitments covering national regulations affecting various services. Like tariffs, these commitments are contained in binding national schedules.

Promoting Fair Competition

The WTO Agreement extends and clarifies previous GATT rules that allowed governments to impose compensating duties on two forms of unfair competition: dumping and subsidies. It also addressed intellectual property issues to protect ideas and inventions in a competitive environment.

Encouraging Development and Economic Reform

The Preamble of the Agreement Establishing the WTO states that

> members should conduct their trade and economic relations with a view to raising standards of living, ensuring full employment and a large and steadily growing volume of real income and effective demand, and expanding the production of and trade in goods and services, while allowing for the optimal use of the world's resources in accordance with the objective of sustainable development, seeking both to protect and preserve the environment and to enhance the means for doing so in a manner consistent with their respective needs and concerns at different levels of development.[62]

Furthermore, the WTO members recognized the "need for positive efforts designed to ensure that developing countries, and especially the least–developed among them, secure a share in international trade commensurate with the needs of their economic development." To accomplish this, the WTO (1) retained previous GATT provisions favoring developing countries, especially those encouraging industrial nations to enable trade with developing nations; (2) gave developing countries transition periods to adjust to the more difficult WTO provisions; and (3) gave the least-developed countries (LDCs) additional flexibility plus accelerated implementation of markets for their goods.

WTO Members

As of February 10, 1999, the WTO had 134 member countries,[63] 34 observer countries, and 7 observers to the General Council. Observer countries may attend the meetings and participate in the negotiations, but may not vote. China and Russia applied for membership, but were not members as of 2000. A list of the member countries is provided in Appendix E. Members enjoy the privileges and security the trading rules provide, but must, in return, abide by the WTO's rules and make commitments to open their markets. The commitments result from the country's membership, or *accession,* negotiations.

WTO Structure and Decision-Making Process

The WTO is run by its member governments. Continuing the GATT's tradition, the WTO members make all major decisions by consensus rather than voting. The highest authority in the WTO is the Ministerial Conference, composed of representatives of all WTO member countries. It is required to meet at least once every two years and can make decisions on all matters under any of the multilateral trade agreements.

The routine day-to-day work of the WTO is supervised by the General Council, composed of all WTO members and responsible for (1) overseeing the decisions made at the Ministerial Conferences, (2) managing the general operation of the World Trade Agreement, (3) comprising both the Dispute Settlement Body and the Trade Policy Review Body, (4) considering the full range of trade issues covered by the WTO, and (5) making regular reports to the Ministerial Conferences. The General Council also delegates some of its responsibilities to several subsidiary bodies including the Council for Trade in Goods (Goods Council), the Council for Trade in Services (Services Council), and the Council for Trade in Trade–Related Aspects of Intellectual Property Rights (TRIPs Council). In addition, three other bodies, established by the Ministerial Conference, report to the General Council: the Committee on Trade and Development, the Committee on Balance of Payments, and the Committee on Budget, Finance, and Administration.

Numerous other councils, committees, working parties, and negotiating groups address the full range of WTO issues. For example, a standing, seven-person Appellate Body provides members the possibility of appealing decisions made by the Dispute Settlement Body. The plurilateral agreements of the WTO, covering civil aircraft, government procurement, dairy products, and bovine meat, each has its own management body that reports to the General Council.

The administrative and technical support of the WTO is provided by a 500-person WTO Secretariat located in Geneva, Switzerland. A director-general heads the WTO Secretariat, with the assistance of four deputy directors-general. The WTO Secretariat (1) supports WTO delegate bodies during negotiations and the implementation of agreements, (2) provides technical support to developing countries, especially the least-developed countries, (3) provides economists and statisticians who analyze trade performance and policy, (4) provides legal staff to assist in the resolution of trade disputes involving the interpretation of WTO rules and precedents, (5) assists new members with *accession negotiations,* and (6) provides advice to governments considering membership.

The WTO budget for 1998 was around $93 million ($U.S.), with individual member contributions calculated on the basis of each country's share of the total trade conducted by all members.

5.5.3 Annex 1A—1994 GATT

Annex 1A contains the amended and updated GATT called the 1994 GATT. It integrates issues concerning the trade in goods into the WTO, including numerous previous and revised agreements on issues concerning: agriculture, textiles, antidumping practices, customs valuation, import licensing, subsidies, and countervailing measures. These revised agreements (1) provided new rules on the various issues; (2) clarified when such measures could be used, to what extent, and how; and (3) added various mechanisms to ensure transparency, such as notification procedures to ensure that members' information is continually updated.

5.5.4 Annex 1B—1994 GATS

Annex 1B contains the 1994 GATS[64] developed from the work of the GATT's Group of Negotiations on Services held throughout the eight years of the Uruguay Round of negotiations (1986–1994). It is the final version of the framework document for the worldwide trade in services, including the specific commitments of each member country, requested by the GATT members in the Uruguay Declaration at the start of the Uruguay Round. For this reason, it is critical for telecommunications companies participating in the international environment to be familiar with this document.

The 1994 GATS contains 29 articles, eight annexes, four protocols, and 130 countries' Schedules of Commitments for trade in services, as noted in Table 5.3 on the next page. The sections discussed in this chapter are italicized in the Table.

The twenty-nine articles contain the items agreed to by the member countries. They are divided into six parts to clarify their applications. The eight

TABLE 5.3 THE 1994 GENERAL AGREEMENT ON TRADE IN SERVICES (1994 GATS) APRIL 23, 1994 (PLUS UPDATES)

Part	Article	Title
Part I:	Scope and Definition	
	Article I:	Scope and Definition
Part II:	General Obligations and Disciplines (Articles 2–15)	
	Article II:	Most-Favored-Nation (MFN) Treatment
	Article III:	Transparency
	Article IV:	Increasing Participation of Developing Countries
	Article V:	Economic Integration
	Article VI:	Domestic Regulation
	Article VII:	Recognition
	Article VIII:	Monopolies and Exclusive Service Suppliers
	Article IX:	Business Practices
	Article X:	Emergency Safeguard Measures
	Article XI:	Payments and Transfers
	Article XII:	Restrictions to Safeguard the Balance of Payments
	Article XIII:	Government Procurement
	Article XIV:	General Exceptions
	Article XV:	Subsidies
Part III:	Specific Commitments (Articles 16–18)	
	Article XVI:	Market Access
	Article XVII:	National Treatment
	Article XVIII:	Additional Commitments
Part IV:	Progressive Liberalization (Articles 19–21)	
	Article XIX:	Negotiation of Specific Commitments
	Article XX:	Schedules of Specific Commitments
	Article XXI:	Modification of Schedules
Part V:	Institutional Provisions (Articles 22–26)	
	Article XXII:	Consultation
	Article XXIII:	Dispute Settlement and Enforcement
	Article XXIV:	Council for Trade in Services
	Article XXV:	Technical Cooperation
	Article XXVI:	Relationship with Other International Organization
Part VI:	Final Provisions (Articles 27–29)	
	Article XXVII:	Denial of Benefits
	Article XXVIII:	Definitions
	Article XXIX:	Annexes

TABLE 5.3 THE 1994 GENERAL AGREEMENT ON TRADE IN SERVICES (1994 GATS) APRIL 23, 1994 (PLUS UPDATES) *(continued)*

1. *Annex on Article II Exemptions*
2. Annex on Movement of Natural Person Supplying Services under the Agreement
3. Annex on Air Transport Services
4. Annex on Financial Services
5. Annex on Financial Services
6. Annex on Negotiations on Maritime Transport Services
7. *Annex on Telecommunications*
8. *Annex on Negotiations on Basic Telecommunications*

Ministerial decisions forming related instruments:

Decision on Institutional Arrangements for the General Agreement on Trade in Services

Decision on Certain Dispute Settlement Procedures for the General Agreement on Trade in Services

Decision on Trade in Services and the Environment

Decision on Negotiations on Movement of Natural Persons

Decision on Financial Services

Decision on Negotiations on Maritime Transport Services

Decision on Negotiations on Basic Telecommunications

Decision on Professional Services

Understanding on Commitments in Financial Services

Protocols:

Second GATS Protocol: (October 1995)	Revised Schedules of Commitments on Financial Services
Third GATS Protocol: (October 1995)	Schedules of Specific Commitments Relating to Movement of Natural Person
Fourth GATS Protocol: (April 1997)	*Schedules of Specific Commitments concerning Basic Telecommunications (Integrating the WTO's Agreement on Basic Telecommunications of February 15, 1997 into GATS)*
Fifth GATS Protocol: (February 1998)	Schedules of Specific Commitments and Lists of Exemptions from Article II Concerning Financial Services

annexes address special situations affecting certain services and add some obligations for those services. They are made an integral part of the GATS by the last article, Article 29 (XXIX). The five protocols were added later, in 1995, 1997, and 1998, but were made an integral part of the 1994 GATS when adopted by the member countries.

The articles, annexes, and protocols establish the overall format and basis for the members' specific commitments concerning the services. The 130 Schedules of Commitments cover more than 20,000 pages detailing each country's acceptances and exemptions to the GATS. The details of a country's Schedule of Commitments are critical to any provider seeking to enter that country's telecommunications market.

Like the fuller Uruguay Round, the 1994 GATS covers a broad range of services, not just telecommunications services. However, many of the general issues also affect telecommunications equipment and services. In addition, the GATS contains two annexes specifically addressing telecommunications issues: the Annex on Telecommunications, describing market entry, and the Annex on Negotiations on Basic Telecommunications (NBT), in which the WTO members committed to additional negotiations on basic telecommunications.[65]

Part I—Scope and Definition of GATS

Part I, Article I of the 1994 GATS,[66] adopted on April 23, 1994, defines the *trade in services* as the supply of a service:

1. from the territory of one member into the territory of any other member;
2. in the territory of one member to the service consumer of any other member;
3. by the service supplier of one member, through commercial presence in the territory of any other member; and
4. by a service supplier of one member, through presence of natural persons of a member in the territory of any other member.

These four methods of supply are more familiarly known as (1) cross-border supply; (2) consumption abroad, (3) commercial presence abroad, and (4) presence of natural persons abroad. They establish the scope of GATS, define the organization and application of numerous other sections of the GATS, and are the categories used by the GATS members to identify their Service Commitments. They are described briefly in the following four sections.

Cross-Border supply

Cross-border supply occurs when telecommunications services are provided over a network that crosses national borders.[67] It is the most common mode of telecommunications trade, but some countries do not allow it. For example, Hong Kong did not commit to giving market access or national treatment for cross–border supply, even through it made commitments for all of the other modes of receiving telephone service.[68]

Consumption Abroad

With advances in mobile communications, including cellular telephones, satellite telephones, and calling cards, consumption abroad is an increasingly important mode of providing telecommunications services. Each of these technologies allows geographic movement of customers and movement of the customer premise equipment that originates the call traffic, two key aspects of consumption abroad.[69]

Commercial Presence Abroad

Providing service through commercial presence abroad[70] involves any corporation, trust, partnership, association, branch office, representative office, or joint venture within the territory of a member for the purpose of providing a service. The concept of a telecommunications company having a commercial presence abroad is relatively new and occurs only when countries lift their foreign ownership restrictions and liberalize their rules concerning interconnection with the historically monopoly–provided networks.

Presence of Natural Persons Abroad

When a company has operations abroad, or provides personnel for technical assistance, construction, or training, the foreign company is providing service through the presence of natural persons abroad. For example, in build-operate-transfer (BOT) arrangements, the foreign investor and/or builder must place its employees in the foreign country to operate the facilities on a temporary basis until they are transferred to the local operator.[71]

Part II—General Obligations in the 1994 GATS

The General Obligations contained in the 1994 GATS are called *general* because they apply to all service sectors covered by the GATS. In contrast, the Specific Commitments apply only to those service sectors that each member includes in its schedule.

Of the 14 general obligations listed in Part II of the GATS 1994, 7 are of particular interest to telecommunications providers and users: (1) Article II: Most-Favored Nation (MFN) Treatment; (2) Article III: Transparency; (3) Article IV: Increasing Participation of Developing Countries; (4) Article VI: Domestic Regulation; (5) Article VII: Recognition; (6) Article VIII: Monopolies and Exclusive Service Suppliers; and (7) Article IX: Business Practices. The importance of each is described briefly below. Some obligations, such as safeguards (Articles X and XII), government procurement of services (Article XIII), and subsidies (Article XV), include commitments to continue negotiations on various issues affecting those obligations. For this reason, they are important to watch, but are not discussed here.

All of the General Obligations implement the underlying principles of the GATT and the principles developed in the services framework in 1986 by the

Group of Negotiations on Services. As identified in the preamble of the 1994 GATS, these principles include the following: (1) trade without discrimination, (2) predictable and growing access to markets, (3) promotion of fair competition, and (4) encouragement of development and economic reform.[72]

Article II: Most-Favored Nation Treatment

Article II of the 1994 GATS requires nondiscriminatory, Most Favored Nation treatment. It mandates that GATS members "accord immediately and unconditionally to services and service suppliers of any member treatment no less favorable than that it accords to like services and service suppliers of any other country." This requirement exists whether or not the member countries have state-owned or privatized infrastructures and regardless of the degree of openness of the markets in the other countries. For example, if the United States permitted any country to provide services in the United States, it must allow all member countries to provide service in the United States, even if some of the member countries have not opened their markets to the United States and other members. In addition, the obligation requires members to be nondiscriminatory when negotiating interconnection agreements, establishing fees, issuing licenses to operate or own telecommunications facilities in the country, and assigning radio frequencies to wireless companies. Unlike a plurilateral agreement, the MFN is not granted on a conditional basis, unless a member accepts the MFN provision in its Schedule of Commitments, as provided for in a GATS annex. All members making basic telecommunications commitments are also bound to treat all WTO members on an MFN basis, regardless of whether the country participated in the negotiations or not. Thus, all WTO members receive the benefits of the negotiations, but not all are bound by the resulting negotiations.

Article III: Transparency

Concerning transparency, Article III requires transparency of all regulations pertaining to trade in services. This means that each member must publish promptly all relevant measures pertaining to the trade in services in its country. If publication is not practical, the member must otherwise make the information available. In addition, the member must keep this information up-to-date and respond to all requests by other members for specific information about doing business in its country.

Article IV: Increasing Participation of Developing Countries

To facilitate the increasing participation of developing country members in world trade, the GATS members agree to give special consideration to developing countries, including their needs for access to: new technology, information networks, and export markets. The members commit to providing these whenever possible to help strengthen the developing countries' services capacity, efficiency, and competitiveness. To clarify the scope and intent of their commitments, members include them in their Schedules of Commitments for both market access and national rules on telecommunications.[73]

Article VI: Domestic Regulation

GATS Article VI, concerning domestic regulation, requires that where specific commitments are undertaken, each member shall ensure that all measures of general application affecting trade in services are administered in a reasonable, objective, and impartial manner.

Article VII: Recognition

In licensing and certifying service suppliers, a member may recognize the education, experience, licenses, or certifications granted in other member countries. The member must inform the Council for Trade in Services of its recognition measures.

Article VIII: Monopolies and Exclusive Service Suppliers

Article VIII requires each GATS member to ensure that any monopoly supplier of a service in its territory does not act in a manner inconsistent with the member's obligations and does not abuse its monopoly position. If another member believes this obligation has not been met, it may request consultation and dispute settlement pursuant to Part V provisions.

Article IX: Business Practices

For all providers, not just monopoly providers, Article IX requires GATS members to eliminate all anticompetitive business practices within its territory and to participate fully in addressing concerns by other members about a country's business practices.

Part III—Specific Commitments

The specific commitments of each GATS member apply only to those service sectors that the member included in its schedule. The three specific commitments include market access, national treatment, and additional commitments.

Article XVI Market Access

Article XVI requires that market access be provided through the four methods of service supply discussed above: (1) cross–border, (2) consumption abroad, (3) commercial presence abroad, and (4) presence of natural persons abroad; and when a member opens its market to access, it must treat the suppliers of any other member in a manner no less favourable than that provided for in its schedule. Concerning market access, Article XVI of the 1994 GATS does not require members to open their service markets to foreign service suppliers, but rather states that when a member does open its market, it must do so within certain parameters. Generally, opened markets are considered to be entirely open, but members can specifically exempt service sectors or elements of the market.

The concept of market access applies differently to services than to goods. For goods, the objective is to remove tariffs and other barriers to trade such as unreasonable technical standards, content requirements, and packaging specifications. For services, the objective is to allow members to provide services through the four modes of supply. Market access is a critical and ongoing issue for service providers, especially telecommunications companies. Historically, market access concessions were fairly narrow for the telecommunications sector. However, after 1997, members who signed the February 15, 1997, agreements committed to opening their telecommunications markets.[74]

Article XVII: National Treatment

Concerning national treatment, Article XVII notes that the national treatment commitments do not "require any member to compensate for any inherent competitive disadvantages which result from the foreign character of the relevant services or service suppliers."[75] This reflects the more personal and unique characteristics of services as compared to the more fungible characteristics of goods.

Article XVIII: Additional Commitments

Article XVIII allows members to negotiate additional commitments not subject to scheduling under Article XVI or XVII. These commitments must be included in the member's schedule. Some members of GATS are willing to undertake these additional commitments in order to encourage higher levels of competition in their country's services market. They predict that this progressive liberalization will expand trade in services in markets open to them.

Exceptions to the Obligations and Commitments

Similar to some exceptions contained in GATT, and consistent with the concerns raised during the trade in services negotiations, the GATS allowed the members to declare certain exceptions to the general obligations as discussed in Article XX. The specific commitment exceptions are sector specific and include security exceptions and safeguards for the balance of payment. They are to be listed by the members in their schedules.

The Final Act covered all of the negotiating areas cited in the Punta del Este Declaration except two. The first exception was the results of the market access negotiations in which individual countries made binding commitments to reduce or eliminate specific tariffs and nontariff barriers to merchandise trade. The second exception was the initial commitments made by each country on liberalization of trade in services. Both types of commitments are recorded in each country's national schedule, which, with all other schedules, form an integral part of the Final Act.

Part IV—Progressive Liberalization

Articles XIX–XXI of GATS, comprising Part IV, address the Schedules of Specific Commitments to be made by each GATS member, containing their sectoral service commitments and exemptions. These schedules are a key part of the GATS.

As of March 1999, 130 countries have filed schedules, which are summarized in Appendix F. Each GATS member must ensure that any activities or measures taken by its domestic authorities (local, regional, or national) or nongovernmental bodies are in accordance with the obligations of GATS. The only exception permitted is for those suppliers: (1) providing the service in the exercise of governmental authority, (2) when not on a commercial basis, and (3) when not supplying the service in competition with another supplier.[76] This includes any activities that affect a service supplier's ability to provide service in one of the four modes of supply described in Article I.

Article XIX discusses the negotiation of each member's Schedule of Specific Commitments, while Article XX describes the format in which the schedules should be filed. Article XXI delineates the process to modify a schedule and the parameters concerning such modifications. Each member's Schedule of Commitments becomes part of GATS, making each commitment binding. Generally, these commitments cannot be withdrawn, so any reservations or exceptions must be included in the schedule.

Part V—Institutional Provisions

The GATS retained several administrative procedures from the GATT, but changed others. The five articles in Part V provide the details of the GATS administration. The following four sections describe the GATS administrative procedures.

Articles XXII and XXIII: GATS Consultation and Dispute Settlement

Like the GATT, the GATS includes a structure for consultation and dispute settlement to address any problems or concerns that arise among members concerning the trade in services. When a member believes that the terms of the GATS have been violated, the member can invoke the GATS dispute resolution principles described in Articles XXII and XXIII.[77] The GATS Council for Trade in Services, acting as the Dispute Settlement Body, has the authority to suspend free trade principles against significant violators.[78]

Article XXIV: Council for Trade in Services

The principal managing body in GATS is the Council for Trade in Services, comprised of all members. The chairperson of the council is elected by the members.

Article XXV: Technical Cooperation

Members in need of technical cooperation may call upon the contact points referred to in paragraph 2 of Article IV. Developing countries in need of technical assistance are further provided with such assistance at a multilateral level by the Secretariat.

Article XXVI: Relationship with Other International Organizations

As directed in Article XXVI, a key obligation and focus of GATS and the work of the general council is to cooperate and coordinate their efforts with the work of the United Nations and its specialized agencies, as well as with other intergovernmental organizations concerned with services.

Part VI—Final Provisions

The last section of the GATS articles provides closing exceptions in Article XXVII (Denial of Benefits), and key definitions in Article XXVIII. Article XXIX makes the Annexes to GATS an integral part of the Agreement.

5.5.5 Annex on Article II (Most-Favored Nation) Exceptions

Since Article II of the 1994 GATS requires Most-Favored Nation Treatment, it mandates that GATS members grant the same treatment to all other members or service suppliers from any member country regardless of the degree of openness of the markets in the other countries. Thus, if the United States permitted any country to provide services in the United States, it must allow all member countries to provide service in the United States, even if some of the member countries have not opened their markets to the United States and other members. Similarly, this obligation affects interconnection agreements, fees, licenses to operate or own telecommunications facilities in the country, and assignment of radio frequencies to wireless companies.

The Annex on Article II Exceptions specifies the conditions under which a member is exempted from the most-favored nation obligations in Article II of GATS. The United States advocated such exemptions because it was concerned that GATT members with closed markets could enter other markets under most-favored nation provisions.[79] The article limits exceptions to 10 years, with a review of all exceptions by the Council for Trade in Services approximately every 5 years. The exceptions, which exclude access to specific sectors of a member's market, not the entire market, are entered in each member's Schedule of Commitments. The most-favored nation status is not granted on a conditional basis, unless a member accepts the most-favored nation provision in its Schedule of Commitments.

5.5.6 April 1995—Annex on Telecommunications

"Recognizing the specificities of the telecommunications services sector, and in particular, its dual role as a distinct sector of economic activity and as the underlying transport means for other economic activities," the members agreed to the Annex on Telecommunications.[80] It elaborates on the provisions of the GATS affecting access to and use of public telecommunications networks and services. It does not apply, however, to cable or broadcast distribution of radio or television programming.[81]

Access and Interconnection

Primarily, the GATS Annex on Telecommunications addresses how foreign telecommunications service providers can "establish, construct, acquire, lease, operate, or supply telecommunications transport networks or services" if they desire. It requires each member to "ensure that any service supplier of any other member is accorded access to and use of public telecommunications transport networks and services on reasonable and non–discriminatory terms and conditions, for the supply of a service included in its Schedule."[82]

However, the annex permits members to "take such measures as are necessary to ensure the security and confidentiality of messages" and the "technical integrity of the network," but such measures are subject to the requirements that they "are necessary" and "not used as a disguised restriction on trade in services." These measures include the following: (1) restrictions on resale or shared use of services, (2) requirements to use specified technical interfaces and protocols for interconnection with the member's networks and services, (3) requirements for interoperability and to encourage the achievement of the goals for coordinated technical standards, (4) type approval of terminals or other interface equipment, (5) restrictions on the interconnection of private leased or owned circuits, and (6) requirements for notification, registration, and licensing of services, providers, and equipment.[83]

Relation to International Organizations

Paragraph 7 of the Annex on Telecommunications also requires that GATS members recognize the importance of other international, intergovernmental, and nongovernmental organizations and agreements in ensuring the efficient operation of domestic and global telecommunications. It requires that GATS members make appropriate arrangements, where relevant, for consultation with such organizations on matters of telecommunications service exchange and interconnection.[84] In particular, the annex recognizes the importance of international standards for global compatibility and interoperability of telecommunications networks and service. It encourages the members to promote such standards through the work of relevant international bodies, including the International Telecommunications Union and the International Organization for Standardization.[85]

5.5.7 Annex on Negotiations on Basic Telecommunications (NBT)

The 1994 GATS Annex on Negotiations on Basic Telecommunications (Annex on Basic Telecommunications)[86] was announced on April 15, 1994, as an annex to the documents ending the Uruguay Round of Negotiations. It was a compromise agreement among the members, in which they agreed not to include basic telecommunications services in the 1994 GATS, but rather to include only enhanced services and to continue negotiations on basic services over the next several years. The concept and structure of the compromise was first proposed in 1991 by the Group of Negotiations on Services in its Dunkle Draft of the services framework. The Annex on Basic Telecommunications ensured the members that no commitments concerning basic telecommunications applied if they approved the 1994 GATS, unless a member specifically included such commitments in its Schedule.[87] This meant that countries were not required to take Article II most-favored nation exceptions on basic telecommunications at the time they approved the 1994 GATS.[88]

The compromise was required because, in 1994, most member countries used state-owned operators or state-sanctioned monopolies to provide basic telecommunications services over government-owned or regulated infrastructures. These basic services provided significant revenue to the member governments. Neither privatization nor liberalization of these infrastructures was widely approved in 1994. Enhanced services, on the other hand, frequently were provided by competitive providers in 1994 and could, therefore, more immediately be included in each member's schedule.

Definition of Enhanced Services

For purposes of the compromise discussions, the GATS member countries defined *enhanced services* as services in which (1) the voice or nonvoice information underwent an end-to-end restructuring or format change while being transferred from one point to another before it reached the customer; (2) value was added to the consumer's transmission of information; (3) the message transmitted was upgraded in form or content; or (4) information storage and retrieval was provided. Enhanced services included electronic mail, voicemail, on-line information, electronic data interchange, value-added facsimile services, code and protocol conversion, and data processing. The United States, in its Schedule of Specific Commitments, defined enhanced services as "services, offered over common carrier transmission facilities which employ computer processing application that: 1) act on the format, content code, protocol or similar aspects of the subscriber's transmitted information; 2) provide the subscriber additional, different, or restructured information; or 3) involve subscriber interaction with stored information."[89]

Definition of Basic Services

However, the definition of *basic telecommunications services* differed widely among the GATS members. The most common definition was "voice and nonvoice services consisting of the transmission of information between points specified by a user in which the information delivered by the telecommunications agency to the recipient is identical in form and content to the information received by the telecommunications agency from the user."[90] This definition included the resale of these services and was adopted in part by the U.S. Congress as the definition of telecommunications in the Telecommunications Act of 1996.[91]

Many of the LDCs considered basic services as including only voice telephony, while Canada included the following in its list of basic services, because all were services traditionally or routinely provided by the telephone company over the existing telecommunications infrastructure: voice telephone services, packet-switched data transmission services, circuit-switch data transmission services, telex services, telegraph services, facsimile, privately leased circuit services or mobile services.[92] Many of the other schedules, such as those of Japan, the European Community, and the United States, contained similar listings.

Decision on Negotiations on Basic Telecommunications

At Marrakesh, the ministers elaborated on the Annex for Basic Telecommunications, clarifying certain objectives, recognizing participants, and setting specific procedural due dates. Their comments were placed in their Decision on Negotiations on Basic Telecommunications,[93] announced in Marrakesh in April 1994 and attached to GATS with the other Ministerial Decisions.

In paragraph 1 of the Decision, the Ministers stated that the continued negotiations would work toward "progressive liberalization of. . .'basic telecommunications' within the framework of the GATS."[94] However, they did not impose requirements for progressive or eventual privatization.

In paragraph 2, the ministers stated that the scope of the negotiations should be comprehensive. The participating members decided early in the negotiations that they would include all basic telecommunications services including the following: (1) local, (2) long-distance, (3) international services for public and private use, (4) facilities-based and resale services, and (5) services over all networks, including satellite, cable, wireless, mobile, and cellular.

In paragraph 3, the ministers established the Negotiating Group on Basic Telecommunications to conduct the negotiations and make periodic reports to the Council for Trade in Services on the progress of the negotiations.[95]

Paragraph 4 identified 19 members, representing many of the countries with the most advanced telecommunications networks in the world, who had agreed to participate in the continued negotiations on basic telecommunications. The 19 members were Australia, Austria, Canada, Chile, Cyprus, the European Community and its 15 member states (counted as one member country), Finland, Hong Kong, Hungary, Japan, Korea, Mexico, New Zealand, Norway, the Slovak Republic, Sweden, Switzerland, Turkey, and the United States.[96] Actually 32 countries participated when the European Community countries are each counted separately. The Decision invited any other interested countries to attend, and three years later, by February 1997, over 72 members were participating.

Paragraph 5 of the Decision set the schedule for the continued negotiations on basic telecommunications to start "no later than 16 May 1994" and to conclude, with a final report, "no later than 30 April 1996." For each member country, this due date initiated serious review and revisions of the member's domestic laws on telecommunications. A string of revisions occurred worldwide including the U.S. Telecommunications Acts of 1996 and the Pakistan Telecommunications Act of 1996.

5.6 APRIL 24, 1996—INITIAL RESULTS OF THE NEGOTIATIONS ON BASIC TELECOMMUNICATIONS

By the mandated deadline of April 30, 1996, the negotiations on basic telecommunications had produced[97] a reference paper, market access commitments from members, and a plan for completion of the negotiations. These are discussed in the following three sections.

5.6.1 Negotiating Group on Basic Communications' Reference Paper to the Council for Trade in Services

The "Negotiating Group on Basic Telecommunications (NGBT) Reference Paper on Regulatory Principles," submitted to the Council for Trade in Services, provided a framework for adding basic telecommunications to the GATS.[98] It included (1) definitions and principles concerning competitive safeguards, (2) universal service assurances, (3) independence of regulators, (4) allocation and use of scarce resources, (5) procedures for interconnection negotiations, (6) licensing criteria, (7) a transparency clause, and (8) a dispute settlement mechanism. The Reference Paper was a significant multilateral agreement that addressed many of the regulatory concerns that the governments of the member countries and potential service providers had in considering competition in a market that was not privatized or fully liberalized.[99] It also documented commitments, negotiated by WTO members regarding the regulation of basic telecommunications services.

The Negotiating Group on Basic Telecommunications' Reference Paper was adopted by the Council for Trade in Services on April 24, 1996 and renamed the ministerial "Decision on Commitments in Basic Telecommunications." The NGBT also provided a draft "Fourth Protocol to the General Agreement on Trade in Services" to be used when incorporating the Reference Paper and the Agreement on Basic Telecommunications into the 1994 GATS. Thus, in approving the Reference Paper, the WTO members could accept it in whole or in part. Fifty-four of the sixty-five countries adopted the Reference Paper and its regulatory principles in full.[100] Several countries, including Brazil, committed to adopt it at a later date. Others, including India, Pakistan, Malaysia, the Philippines, and Venezuela, adopted some of the principles. Still others, including Ecuador, did not make any regulatory commitments.

The members agreeing to the Reference Paper attached it to their Schedules of Commitments as an additional commitment. In this manner, the members could incorporate the Reference Paper as a whole and still take particular exceptions to some of its provisions.

5.6.2 Market Access Offers—Country Commitments to Include Basic Telecommunications

In addition to the Reference Paper, the Negotiating Group on Basic Telecommunications obtained market access offers to include basic telecommunications from 47 countries.[101] The countries were Argentina, Australia, Austria, Belgium, Brazil, Canada, Chile, Columbia, Czech Republic, Dominican Republic, Ecuador, Finland, France, Germany, Greece, Hong Kong, Hungary, Iceland, India, Ireland, Israel, Italy, the Ivory Coast, Japan, Korea, Luxembourg, Mauritius, Mexico, Morocco, Netherlands, New Zealand, Norway, Pakistan, Peru, the Philippines, Poland, Portugal, Singapore, the Slovak Republic, Spain, Sweden, Switzerland, Thailand, Turkey, the United Kingdom, the United States, and Venezuela.

However, the quantity and quality of the offers were not sufficient to convince the United States to make a final commitment to provide unlimited market access to its basic telecommunications sector.[102] Thus, the Negotiating Group on Basic Telecommunications' deadline to conclude the negotiations and make a final report was not met. The U.S. Trade Representative, Mickey Kantor, stated that the negotiations failed because market access was not committed to by enough members to constitute a "critical mass" of the telecommunications market.[103]

Of the 47, only 11 offered to provide open market access for all domestic and international services and facilities, allow 100% foreign investment, and adopt the procompetitive regulatory principles contained in the Reference Paper. The 11 were Austria, Denmark, Finland, Germany, Iceland, Luxembourg, the Netherlands, New Zealand, Sweden, the United Kingdom, and the United States.

Several countries in Latin America and Europe had made offers, but the offers were limited to certain services, contained significant investment restrictions, or set a date of implementation significantly beyond the agreed–upon implementation date of January 1, 1998. Chile, for example, offered access only to long–distance and international services, but not to local services.[104] France limited foreign ownership to 20% in radio networks,[105] and Spain's market access commitments were not to become effective until 2003.[106]

Of the Asian countries, Japan, Korea, Hong Kong, Thailand, and the Philippines made offers, but none offered 100% market access or national treatment. India, on the other hand, claimed a most-favored nation exception for the application of variable accounting rates for terminating international traffic in its market.[107]

5.6.3 Agreement To Set Second Deadline for Continued Negotiations on Basic Telecommunications

Since the original deadline of April 30, 1996, for the completion of negotiations on basic telecommunications had not been met, but the negotiating members felt they were close to agreement, the NGBT members agreed to continue meeting. The WTO set a new, second deadline for the negotiations and gave the group a new name, the Group on Basic Telecommunications (GBT).

Over the next nine months, from May 1996 to February 1997, 53 countries participated in the negotiation process, and another 24 nations observed. The extra months provided by the second deadline increased the number of participating countries willing to discuss including basic telecommunications in their Schedule of Commitments from 19 countries in April 1994 to over 86 countries by February 1997.

5.7 DECEMBER 1996—SINGAPORE MINISTERIAL CONFERENCE

In December 1996, the WTO ministers met in Singapore to discuss outstanding issues left unresolved by the Reference Paper. These issues included:

1. potential anticompetitive distortion of trade in international services;
2. ways to ensure accurate scheduling of commitments—particularly with respect to the supply of services via satellites and the management of the radio spectrum;
3. the status of intergovernmental satellite organizations in relation to GATS provisions; and
4. the extent to which basic telecommunications commitments include transport of video and/or broadcast signals within their scope.[108]

Concerning satellite service, the group decided that no commitments would apply to satellite services unless satellite services were included in a member's Schedule of Commitments. The issue of spectrum management was probably the most controversial issue for the group. However, the group decided to use a *technology-neutral* approach and create a *scarce resource exception* to the basic telecommunications commitments as it relates to spectrum management. In addition, while accounting rate reform was not an issue on the negotiating table, it was heavily discussed among the developed and developing nations throughout the entire period of negotiations.

5.8 FEBRUARY 15, 1997—WTO AGREEMENT ON BASIC TELECOMMUNICATIONS

The extra nine months of negotiations on basic telecommunications were successful and the Group on Basic Telecommunications concluded the negotiations and filed its *Report of the Group on Basic Telecommunications*[109] by the second deadline. On February 15, 1997, 69 countries signed the WTO Agreement on Basic Telecommunications (Basic Telecom Agreement or GATS 1997). Within a few months, another 3 countries signed, bringing the total to 72 signatories; 15 European countries were represented under the one signature for the European Communities. Counting each European Community country separately, a total of 86 countries agreed to the Basic Telecom Agreement. At the time, these signatory countries generated nearly 90% of the world's telecommunications revenues.

In actuality, the WTO Agreement on Basic Telecommunications, is not a separate, stand–alone agreement, but rather is the compilation of the commitments made by the individual countries as a result of the negotiations on basic telecommunications. Each country filed an offer for an individual Schedule of Specific Commitments for Basic Telecommunications with its List of Article II (MFN) Exemptions for Basic Telecommunications, if any. See Appendix F.

On the same day, February 15, 1997, 65 of the 69 countries also signed the *WTO Reference Paper on Regulatory Principles,* to be used for consideration as additional commitments in offers on basic telecommunications. The signatories of the Reference Paper cited it in their Schedules on Basic Telecommunications and committed to regulatory reform to level the playing field for new entrants in the areas of (1) prevention of anticompetitive practices, (2) interconnection, (3) transparency, (4) licensing, (5) independence of regulators and review of decisions, and (6) dispute resolution. Agreed upon exceptions included universal services and the allocation and use of scarce resources.

5.9 APRIL 15, 1997—FOURTH PROTOCOL TO GATS ADOPTED

On April 15, 1997, the members of the WTO adopted the Fourth Protocol to the GATS,[110] which integrated the WTO Agreement on Basic Telecommunications into the GATS. It also kept the Agreement on Basic Telecommunications open for signature until November 1997 and set the date of implementation for the members' Schedules of Commitments at February 8, 1998. On the same date, the MFN suspension would end, and members could include Article II exemptions in their Schedules of Specific Commitments.

5.9.1 November 1997

The signatories had nine months, until November 1997, to demonstrate to the Council on Trade in Services that they were legislatively and technically capable of implementing their 1997 offers. At that time, they could also offer any updates or improvements to their February 1997 offers that they desired. By the end of November 1997, the signatory countries were supposed to adopt the necessary laws or regulations to implement their commitments.

For the United States, no laws needed to be changed, but the FCC needed to revise its regulations to comply with the commitments made by the United States on foreign ownership.[111] From February to November 1997, the United States worked with several developing countries to determine whether they were ready to submit offers. The United States also worked with several other participants of the negotiating group to obtain better offers from the developing countries.

After November 1997, the signatory countries had another three months, until February 1998, to implement their commitments. Thus, the provisions of the Agreement on Basic Telecommunications did not enter into force until February 5, 1998.[112] However, while the member countries did not withdraw their offers, many required another year, until February 1999, and longer to implement their 1997 GATS commitments. By mid-1999, many of the signatory countries still had not filed their final Schedules of Specific Commitments. A summary of the WTO member countries' schedules that have been filed are provided in Appendix 5-C. These new Schedules replace the members' original GATS schedules and reflect each member's commitments concerning basic telecommunications.

CONCLUSION

Since its creation in April 1994, the WTO has been the forum for successful negotiations to open worldwide markets in telecommunications services and equipment. During its first three years, the WTO dealt only with enhanced telecommunications, but in 1997, its members signed an Agreement on Basic Telecommunications committing to market access for basic telecommunications and financial services.

In 1998, telecommunications was one of the largest potential markets in the world, second only to the financial services market.[113] As an industry in its own right and an essential infrastructure for nearly all other economic and social activities, the market will continue to expand. As telecommunications moves from local telecommunications systems and services to global systems and services the WTO and its agreements will be the principal set of rules guiding the new world of telecommunications services and equipment. It will be a major part of the telecommunications world in the Internet Age.

ENDNOTES

[1]The WTO was created on April 15, 1995 by representatives (GATT ministers) from most of the countries of the world at the end of the Uruguay Round of international trade negotiations. It was announced in the The Marrakesh Agreement Establishing the World Trade Organization, a part of the Final Act Embodying the Results of the Uruguay Round of Multilateral Trade Negotiations," Apr. 15, 1994, The Results of the Uruguay Round of Multilateral Trade Negotiations: The Legal Texts 2 (GATT Secretariat 1994), 33 International Legal Materials (I.L.M.) 1125 (1994). [hereinafter WTO agreement]. It went into effect on January 1, 1995.

[2]General Agreement on Tariffs and Trade (GATT), International Trade 90–93, Vol. 1, at 1 (1992).

[3]*Telecommunications, Electronics Sectors to Thrive Under GATT, Commerce Says,* Satellite Week, June 27, 1994.

[4]The only concern raised as a potential disadvantage of GATT for the United States was the possibility of reduced government revenue from lowered tariffs agreed to in the GATT. The Commerce Committee estimated that the U.S. Treasury Department could lose $30 billion between 1995 and 2004. However, the measure passed because Congress believed that this loss would be more than offset by revenue generated from the increased sales and U.S. presence in the expanding worldwide services market.

[5]Ben Petrazzini, Global Telecom Talks: A Trillion Dollar Deal 11 at (1996).

[6]Paul B. Stephan, III, Don Wallace, Jr. and Julie A. Roin, International Business and Economics Law and Policy, The Michie Company, Contemporary Legal Education Series, Charlottesville, VA, 1993, 643.

[7]*www.WTO.org,* and Larry D. Sanders, et al., "The GATT Uruguay Round and the World Trade Organization: Opportunities and Impacts for U.S. Agriculture," SRDC No. 198–7, at. 2.

[8]*www.WTO.org* /FAQ(Frequently-asked questions)What is the WTO?.

[9]Note, *The United States Participation in the General Agreement on Tariffs and Trade,* 61 Colum L. Rev. 505 (1961); John H. Jackson, *The General Agreement on Tariffs and Trade in United States Domestic Law,* 66 Mich. L. Rev. 250 (1967); Robert E. Hudec, *The Legal Status of GATT in the Domestic Law of the United States, in the European Community and GATT* 187 (Meinhard Hilf, Francis G. Jacobs & Ernst–Ulrich Petersmann eds. 1986); Alan O. Sykes, *Protectionism as a "Safeguard": A Positive Analysis of the GATT "Escape Clause" with Normative Speculations,* 58 U. Chi. L. Rev. 255 (1991); *The Status of the General Agreement on Tariffs and Trade in United States Domestic Law,* 26 Stan.J. Int'l. 479 (1990); Ronald A. Brand, *GATT and United States Trade Law: The Incomplete Implementation of Comparative Advantage Theory,* 2 J.Leg. Econ. 95 (1992).

[10]Werner Antweiler, Jr.; 21 1996 Feb. *www.pacific.commerce.ubc.ca/trade/GATT-rounds.htm.*

[11]Jimmie V. Reyna, The GATT Uruguay Round: A Negotiating History (1986–1992) 2343 (Terence P. Stewart ed., 1993).

[12]*Id.* at 2343-45 citing the Trade Act of 1974, codified at 199 U.S.C. § 2411 (1990); the Trade and Tariff Act of 1984, codified at 19 U.S.C. § 2102 (1990); and Omnibus Trade and Competitiveness Act of 1988, codified at 19 U.S.C. § 2901 (1990).

[13]*The Report of the Consultative Group of Eighteen to the Council of Representatives,* GATT Doc. No. L/5210, reprinted in GATT B.I.S.D. (28th Supp.) at 71, 74 (1980–81).

[14]Reyna *supra* note 11, at 2345.

[15]*Id.*

[16]U.S. Proposal to Add Services to GATT and the GATT Ministerial Declaration of 1982 Adopted on November 29, 1982, GATT Doc. No. L/5424, *reprinted* in GATT B.I.S.D. (29th Supp.) at 9, 21 (1982), [hereinafter Ministerial Declaration of 1982].

[17]Taunya L. McLarty, *Liberalized Telecommunications Trade in the WTO: Implications for Universal Service Policy,* Fed. Com., Vol. 51, No. 1, Dec. 1998, 1–59.

[18]Reyna, *supra* note 11, at 2347.

[19]*Id.* at 2347–48.

[20]Ministerial Declaration on the Uruguay Round, GATT Doc. No. MIN.DEC, (Sept. 20, 1986).

[21]McLarty, *supra* note 17, and Reyna *supra* note 11, at 2359.

[22]Ministerial Declaration on the Uruguay Round, *supra* note 20, at 10.

[23]*Id. See also* Reyna, *supra* note 11, at 2359 and McLarty, *supra* note 17, at 13.

[24]Ministerial Declaration on the Uruguay Round, *supra* note 20, at 10.

[25]*General Agreement on Trade in Services* (GATS), Dec. 15, 1993, 33 International Legal Materials (I.L.M.) 44, art. III, para. 1, Apr. 15, 1994, WTO Agreement, Annex 1B, The Results of The Uruguay Round of Multilateral Trade Negotiations: The Legal Texts 325 (GATT Secretariat 1994), 33 I.L.M. 44 (1994) [hereinafter GATS].

[26]GATT, *supra* note 2, art. XXXVI.

[27]Reyna *supra* note 11, at 2361.

[28]*Id.* at 2400.

[29]GATS, *supra* note 25, para. 1; and McLarty *supra* note 17, at 29.

[30]*Draft Multilateral Framework for Trade in Services,* GATT Doc. No. MTN.GNS/35 (July 23, 1990), at 1 [hereinafter Draft Framework].

[31]Draft *Final Act Embodying the Results of the Uruguay Round of Multilateral Trade Negotiations,* GATT Doc. No. NTN.TNC/W/FA (Dec. 20, 1991) [hereinafter Dunkel Draft].

[32]*Id.* at 2413.

[33]*See* United States-Canada Free Trade Agreement Implementation Act of 1988, Pub. L. No. 100-449, 102 Stat. 1851 (1988).

[34]North American Free Trade Agreement (NAFTA) Implementation Act, Pub. L. No. 103-182, 107 Stat. 2057 (1993), codified at 19 U.S.C. § 2411 *et seq.* (1988 and Supp. 1993). [hereinafter NAFTA].

[35]Shefrin, *The North American Free Trade Agreement: Telecommunications in Perspective,* Telecommunications Policy, at 14 (Jan./Feb. 1993).

[36]*Id.* at 20.

[37]NAFTA, *supra* note 34, at ch. 11.

[38]*Id.* at ch. 13 (Telecommunications), Arts. 1301-1310.

[39]Shefrin, *supra* note 35, at 15.

[40]NAFTA, *supra* note 34, at ch. 13 Art. 1302-1304.

[41]*Id.* at Art. 1310.

[42]*Id.* at ch. 18.

[43]*Id.* at ch. 13, Art. 1304.

[44]*Id.*

[45]*Id.* at 1304-1307.

[46]*Id.* at 1304-1308.

[47]*Id.* at 1301-1302.

[48]*Id.*

[49]Bureau of National Affairs, Inc., *FTC Chairman Explores Harmonization of North American Statutes Under NAFTA,* Daily Report for Executives, at A17 (Jan. 26, 1995).

[50]NAFTA, *supra* note 34, at ch. 17.

[51]19 U.S.C. § 2411 *et seq.* (1988 and Supp. 1993).

[52]19 U.S.C. § 2418 (1988 and Supp. 1993).

[53]Brotman, *Markets See Opportunity in NAFTA,* National Law Journal, at 29 (March 7, 1994).

[54]Text of speech by Vice President Al Gore prepared for delivery on March 21, 1994 at the International Telecommunications Union (ITU) Meeting in Buenos Aires, Argentina, Daily Report for Executives, at 122 (March 22, 1994).

[55]*Benefits of NII Will Expand Overseas with Development of GII, Brown Says,* Common Carrier Week (Aug. 1, 1994).

[56]Testimony of Larry Irving, Assistant Secretary for Communications and Information, U.S. Department of Commerce, House Energy, Telecommunications and Finance, Global Telecommunications (July 28, 1994).

[57]*Final Act Embodying the Results of the Uruguay Round of Multilateral Trade Negotiations,* GATT Doc. MTN/FA (Dec. 15, 1993), reprinted in 33 International Legal Materials (I.L.M.) 1 (1994) [hereinafter Final Act].

[58]WTO Agreement, *supra* note 1.

[59]*Id.*

[60]*www.WTO.org.*

[61]*Id.*

[62]General Agreement on Trade in Services, Dec. 15, 1993, 33 (I.L.M.) 44, preamble, at 48, (Apr. 15, 1994), WTO Agreement, *supra* note 1, at Annex 1B, The Legal Texts 325 (GATT Secretariat 1994), 33 I.L.M. 44 (1994) [hereinafter 1994 GATS].

[63]*www.WTO.org* /About the WTO.

[64]1994 GATS, *supra* note 62.

[65]GATS, Annex on Negotiations on Basic Telecommunications (NBT), Apr. 15, 1995, 1994 GATS, *supra* note 62, at Annex 1B, Results of the Uruguay Round of Multilateral Trade Negotiations: The Legal Texts 364 (GATT Secretariat 1994), 33 I.L.M. 44, 77 and Part IVG (1994) [hereinafter GATS, Annex on Negotiations on Basic Telecommunications].

[66]1994 GATS, *supra* note 62, art. 1, para. 2.

[67]*World Telecommunications Development Report 1994,* International Telecommunications Union (ITU) at 26, (1994) [hereinafter ITU World Report].

[68]McLarty, *supra* note 17, citing, WTO Secretariat, *Hong Kong, Schedule of Specific Commitments,* GATS/SC/39/, 94-1037, at 11 (Apr. 15, 1994).

[69]ITU World Report *supra* note 67, at 26.

[70](1994) GATS, *supra* note 62, art. 1, para. 2(c), and art. XXVIII(d) and (l).

[71]ITU World Report *supra* note 67, at 26.

[72]1994 GATS, *supra* note 62.

[73]*Id.* at art. IV, para. 1(a)-(c).

[74]McLarty, *supra* note 17, at 28–29.

[75]1994 GATS, *supra* note 62, at para. 1, n. 10.

[76]*Id.* at art.1, para. 3(b).

[77]Understanding on Rules and Procedures Governing the Settlement of Disputes, Final Act, *supra* note 57, at Part II, Annex 2.

[78]Bliss, *GATT Dispute Settlement Reform in the Uruguay Round: Problems and Prospects,* 23 Stan.J.Int.L. 31 (1987).

[79]General Accounting Office, "*The General Agreement on Tariffs and Trade—Uruguay Round Final Act Should Produce Overall U.S. Economic Goals,*" GAO Reports, at 29 (July 24, 1994) [hereinafter GAO Report].

[80]GATS, Annex on Telecommunications, Apr. 15, 1995 [hereinafter GAO Report]. WTO Agreement *supra* note 1, at Annex 1B, The Legal Texts 359 (GATT Secretariat 1994), 33 I.L.M. 44, 73 (1994) [hereinafter GATS, Annex on Telecommunications].

[81]*Id.* at para. 2(b).

[82]*Id.* at para. 5(a).

[83]*Id.* at para. 5.

[84]*Id.* at para. 7.

[85]*Id.*

[86]GATS, Annex on Negotiations on Basic Telecommunications *supra* note 65, at 77.

[87]*Id. See also* McLarty, *supra* note 17, at 17.

[88]*Id.,* Annex on Negotiations on Basic Telecommunications, at para. 1.

[89]WTO Secretariat, *The United States of America, Schedule of Specific Commitments,* GATS/SC/90, 994-1088, II.C (Apr. 15, 1994).

[90]Jonathan David Aronson & Peter F. Cowhey, *When Countries Talk,* International Trade in Telecommunications Services 86 (1988).

[91]47 U.S.C. 153(43).

[92]WTO Secretariat, *Canada, Schedule of Specific Commitments,* GATS/SC/16/Suppl.3 (Apr. 11, 1997). *www.wto.org/wto/ddf/ep/public.html.*

[93]Decision on Negotiations on Basic Telecommunications, Ministerial Decisions and Declarations, The Results of the Uruguay Round of Multilateral

Trade Negotiations: The Legal Texts 439, 461 (GATT Secretariat 1994), 33 I.L.M. 136, 144 (1994) [hereinafter Decision on Negotiations on Basic Telecommunications].

[94] *Id.* at para. 1.

[95] *Id.* at para. 4. *See also,* GAO Report, *supra* note 79.

[96] Decision on Negotiations on Basic Telecommunications, *supra* note 93, at para. 4.

[97] *Id.*

[98] WTO, *Negotiating Group on Basic Telecommunications (NGBT), Reference Paper,* April 24, 1996, 36 I.L.M. 367 [hereinafter Reference Paper].

[99] McLarty, *supra* note 17.

[100] Shefrin, *supra* at 88-89.

[101] *Id.*

[102] *Id.*

[103] Petrazzini, *supra* note 5, at 13.

[104] WTO, Negotiating Group on Basic Telecommunications, *Communication from Chile, Draft Offer on Basic Telecommunications,* S/NGBT/W/12/Add. 16 (May 6, 1996).

[105] WTO, Negotiating Group on Basic Telecommunications, *Communication from the European Communities and Their Member States, Draft Offer on Basic Telecommunications,* S/NGBT/W/12/Add. 10 (Oct. 16, 1995).

[106] *Id.*

[107] WTO, Negotiating Group on Basic Telecommunications, *Communication from India, List of Article II Exemptions,* S/NGBT/W/19 (Apr. 26, 1996).

[108] WTO, *Report of the Group on Basic Telecommunications,* S/GBT/4 (Feb. 15, 1997) *www.wto.org/wto/ddf/ep/public.html.*

[109] *Id.*

[110] The WTO Basic Telecom Agreement was entered into force on February 5, 1998 by the WTO's Fourth Protocol to the General Agreement on Trade in Services. *WTO's Fourth Protocol to the General Agreement on Trade in Services* (WTO 1997), 36 I.L.M. 354, 366 (1997) [hereinafter Fourth Protocol to GATS].

[111] See Rules and Policies on Foreign Participation in the U.S. Telecommunications Market, *Report and Order and Order on Reconsideration,* 12 F.C.C.R. 23,891, 10 Comm. Reg. (P & F) 750 (1997) (hereinafter Foreign Participation Report and Order and Order on Reconsideration). For a brief summary of the changes required in the FCC's regulations, see Glenn S. Richards and David S. Konczal, *A New World Order Comes to Telecommunications, 15* Cable TV & New Media L. & FIN., Dec. 1997, at 1.

[112] Fourth Protocol to GATS, *supra* note 110.

[113] McLarty, *supra* note 17.

CHAPTER 6 Participating in Global Telecommunications Trade: U.S. Import and Export Laws

LAPTOP COMPUTERS, WIRELESS PHONES, AND PAGERS ARE SUCH COMMON tools in business today that many international business travelers routinely carry the equipment with them as they travel to foreign countries. What most international travelers do not realize, however, is that by simply carrying these items from country to country, they are exporting and reexporting products controlled by U.S. trade law.[1] This is true whether or not a foreign national has access to the equipment.

Second, while most people think of exports as only the sale of equipment to foreign buyers, in actuality, any movement of hardware, software, or data across international borders is an export. Thus, overseas data links, file transfers from U.S. Internet servers, and the receipt of facsimiles while abroad are considered exports.

Third, exports occur whenever technology is released to a foreign person.[2] Many years ago, Congress realized that the technology used to create a product is as important to U.S. interests as the product itself. Thus the export of technology is as closely controlled as the export of products and services.[3] The word *released* is also defined very broadly in the law and includes persons performing technical services, providing technical assistance, or otherwise applying technical knowledge overseas. Many telecommunications companies routinely send technicians to install or service systems worldwide, unaware that they are exporting U.S. technical expertise.

Fourth, even activities performed in the United States, without ever crossing a border or conducting a sale, can be deemed to be exports. For example, providing goods to or sharing technological knowledge with a foreign national in the United States can be an export. As such, exports can occur in the course of technical meetings, the sharing of product literature with potential customers, telephone conversations, fax transmissions, data transfers, emails, Web site data postings, publications, hosting of facility visits, or hiring of foreign nationals.[4]

In most cases, these exports occur routinely, without customs declarations, because the U.S. government has granted exceptions for them. However, the exceptions do not cover all circumstances. For example, the exceptions may cover the equipment, such as the computer, but not the data on the computers' hard drives or floppy disks or the software residing on the computer. This is especially true if the data contains highly technical information or encryption software. Generally, export controls on items not covered by the exceptions are more stringent than controls on items covered by the exceptions. In addition, all exceptions apply only to certain areas of the world. The most restricted areas are countries such as Cuba, North Korea, Iraq, Iran, Sudan, and Libya.[5] Exports to Russia and China are somewhat restricted, while exports to Europe and Latin America are the least restricted. Exports to a participant in a boycott against a United States–friendly nation or associated with a bribe are considered illegal.

Thus, awareness of U.S. trade law governing both imports and exports is critical for both telecommunications providers and the average user of telecommunications products and services, whether they export or not. Even inadvertent, harmless mistakes can cause companies to be fined and travelers to be detained or placed on a "watch list" that may result in routine delays each time the traveler goes through customs. In some cases, the mistakes can result in the confiscation of equipment, including personal computers and wireless phones.

In addition, the signing of the World Trade Organization (WTO) Agreement on February 15, 1997, opened the international market to U.S. telecommunications companies as their largest market for the next several decades. Thus, trade law affects every operation of telecommunications companies even if their international activities are small or infrequent. Since the WTO is a two-way door and also opened the U.S. market to foreign companies, U.S. trade law also impacts foreign companies as both competitors and international partners of U.S. companies.

Finally, U.S. trade law affects even nontraveling users, nonexporting companies, and nontelecommunications industries because it impacts the type of telecommunications products and services available to users and thus their choices of design, quantity, quality, and price of the products and services.

The purpose of this chapter, therefore, is to provide (1) a basic description of the key U.S. trade laws governing the import and export of telecommunications goods, services, and technical knowledge; (2) the government agencies and procedures involved in trade law; and (3) key items for all companies, but especially telecommunications companies, to include in their basic practices to ensure compliance with the various trade laws. To do this, Section 6.1 describes the major U.S. trade laws that have opened markets for U.S. companies, established fair trade protections, and managed import tariffs. However, the trade laws that create the greatest confusion and severest penalties for U.S. citizens and telecommunications companies are the U.S. export laws. For this reason, the remainder of this chapter is devoted to a description of the export process.

Section 6.2 describes the two primary U.S. export laws, while Section 6.3 identifies the agencies with which telecommunications companies and individuals must work to understand and comply with the laws. The agencies review *what* is exported, *to whom* and *how*. Therefore, Section 6.4 discusses the product categorization and licensing requirements controlling what telecommunications products and services can be exported. Section 6.5 describes the customer-screening requirements affecting to whom the products may be sold; and Section 6.6 discusses the anti-boycott and anti-bribery laws dictating how items can be exported. Section 6.7 outlines the reporting requirements for companies, and Section 6.8 describes the criminal and civil penalties for noncompliance with these laws. The best ways to ensure compliance with the U.S. trade laws are to write solid trade agreements and to have compliance programs within each company. Therefore, Section 6.9 suggests important items to include in an international trade contract, while Section 6.10 provides some general considerations for an exporter's compliance program.

6.1 U.S. TRADE LAWS

Like most nations, the United States has long encouraged active trade of its goods and services with other countries. The drafters of the U.S. Constitution recognized that this concept was so important to the health of the United States that they placed it in the Constitution.[6] Congress reiterated this throughout U.S. history, stating, for example, in the U.S. Export Administration Act[7] the following:

(1) The ability of United States citizens to engage in international commerce is a fundamental concern of United States policy.

(2) Exports contribute significantly to the economic well-being of the United States and the stability of the world economy by increasing employment and production in the United States, and by earning foreign exchange, thereby contributing favorably to the trade balance.

(3) The restriction of exports from the United States can have serious adverse effects on the balance of payments and on domestic employment, particularly when restrictions applied by the United States are more extensive than those imposed by other countries.

(4) It is important for the national interest of the United States that both the private sector and the Federal Government place a high priority on exports, consistent with the economic, security, and foreign policy objectives of the United States.

(5) Unreasonable restrictions on access to world supplies can cause worldwide political and economic instability, interfere with free international trade, and retard the growth and development of nations.[8]

As such, the United States's trade laws focus on (1) promoting foreign trade, (2) challenging unfair practices in import trade,[9] (3) blocking the import of counterfeit goods in order to protect American trademarked goods,[10] and (4) implementing tariffs on foreign products offered to the U.S. market at an unfairly low price. These unfairly low-priced products come from two sources: *dumped goods,* sold by foreign companies or individuals at a price below the cost to produce them or below the price they sell for in their own country, and goods whose production or sale was *subsidized* by their government. Both procedures continually plague the telecommunications industry. To neutralize the negative impact of both dumped and government-subsidized goods on the price of U.S. products and services, the U.S. government imposes antidumping duties, and countervailing duties, respectively, on the products to bring their retail prices up to a fairer level in the market.

Most of the U.S. trade laws appear as chapters within Title 19 of the U.S. Code entitled Customs Duties.[11] These chapters provide a good view of the evolution of U.S. trade law. Later trade laws amend and update certain aspects of the earlier laws, but many aspects of the earlier laws still apply today. Therefore, it is important that telecommunications managers and attorneys are familiar with all of the trade laws, not just the most recent. For this reason, a list of the chapters is provided in Appendix G. Also, many of the laws address specific trade issues or regions of the world, which must be noted. Of the Title 19 chapters, seven directly impact the expanding trade of telecommunications goods and services. Those seven include: Chapter 4, The Tariff Act of 1930; Chapter 12, The Trade Act of 1974; Chapter 13, The Trade Agreements Act of 1979; Chapter 17, Negotiation and Implementation of Trade Agreements; Chapter 19, The Telecommunications Trade Act; Chapter 21, The North American Free Trade Agreement; and Chapter 22, The Uruguay Round Trade Agreements implementation. The significance of each law to the import and export of telecommunications goods and services is described briefly below.

6.1.1 Chapter 4—Tariff Act of 1930

The Tariff Act of 1930,[12] also known as the Smoot-Hawley Tariff Act, was passed in response to the perception that imports from other countries were undercutting domestic sales of U.S. products. To counter the economic depression of the 1930s in the United States, Congress raised U.S. tariffs on foreign products to the highest levels in U.S. history. When other countries responded with their own elevated tariffs, the United States was unable to sell its products abroad. Instead of easing the economic depression, the high tariffs significantly deepened it. Therefore, Congress reversed this policy and lowered the tariffs. That lesson still guides U.S. trade policy today, including the trade of telecommunications equipment services and was clearly evident in the focus of the first six rounds of the General Agreement on Tariffs and Trade (GATT), held between 1947 and 1967. In each case, tariff reduction was the main agenda item of the rounds, directly impacting the price of telecommunications products, services, and technologies.

6.1.2 Chapter 12–Trade Act of 1974

To implement the policy of low tariffs between the 1930s and the early 1970s, Congress gave the U.S. president the power to unilaterally negotiate international trade agreements and reduce tariffs through executive agreements, both within preset congressional limits. This streamlined trade negotiations, agreements, and implementation, allowing the United States to move more quickly within congressional guidelines.

During the early 1970s, however, the Vietnam War produced high inflation in the United States and resulted in increased prices for U.S. goods. These high prices made the products and services too expensive for other countries to purchase. As with high tariffs, the high-priced goods resulted in a deepening trade balance against the United States. The Nixon administration responded with various plans and programs to reduce the inflation, but the oil crisis in 1973 overwhelmed these efforts and deepened the trade deficit to a chronic level. The Watergate crisis also divided and distracted the U.S. government, worsening the situation.

These events shaped the Trade Act of 1974[13], the "first comprehensive restructuring of U.S. trade law since 1934."[14] Among the many changes was a reduction of the president's power to negotiate unilaterally international agreements. Instead, the Act created a fast-track process under which the president was required to give Congress 90 days notice before signing a trade agreement and submit the proposed agreement for legislative review and approval. For its part, Congress promised to act within the 90 days.

The Trade Act of 1974 also made several changes that directly impact the telecommunications industry. First, it created the office of the U.S. Trade Representative[15] to work with Congress on trade matters[16] and established safeguards in Title II and protections against unfair trade practices in Title III. Significantly, Section 201 of the Trade Act of 1974 established relief of injury from import competition including adjustment assistance for workers, firms, and communities,[17] and Section 301 enforced U.S. rights under its trade agreements.[18] When the U.S. government uses this law to protect the rights of U.S. companies, including telecommunications companies, the efforts are known as *Section 201* or *Section 301 Actions.*

6.1.3 Chapter 13–Trade Agreements Act of 1979

As the 1974 Act was being discussed, and in the midst of domestic upheaval and the increasingly protectionist attitudes of industries impacted by strong international competition, such as the automobile, steel, and consumer electronics (including telephones and computers), the United States entered into the seventh round of the GATT, the Tokyo Round. The round began in 1973 with a strong agenda to reduce tariffs, technical trade barriers, antidumping and countervailing duties, and customs valuations. By 1976, however, the Tokyo Round seemed stalled. The change of administrations from President Ford to President Carter renewed energy in the round and commitment toward its competition. President Carter appointed Robert Strauss as the U.S. Trade Representative to represent the United States in the Tokyo Round, and by the spring of 1979, the participants had reached 11 separate agreements on nontariff measures, plus some tariff reductions.

The U.S. Congress codified the Tokyo Round agreements in the Trade Agreements Act of 1979,[19] using the fast-track legislative process established five years earlier in the Trade Act of 1974. The freer trade provisions energized the U.S. telecommunications industry, expanding opportunities for U.S. telecommunications products and services worldwide.

6.1.4 Chapter 17—Negotiation and Implementation of Trade Agreements (The Omnibus Trade and Competitiveness Act of 1988)[20]

During President Reagan's first term in office, the United States's trade deficit continued to grow. In response, the Reagan administration initiated several international economic policies codified in the Trade and Tariff Act of 1984. This Act, however, was generally criticized as inadequate to control the trade deficit.[21] Thus two years later, in 1986, when the GATT participants launched the Uruguay Round with an ambitious agenda to increase world trade, both the president and Congress were eager to participate, especially since many U.S. companies, including telecommunications, computer, and software companies, faced increased worldwide competition. To protect these companies, products, and services, the president and Congress passed the U.S. Omnibus Trade and Competitiveness Act of 1988[22], which, among many issues affecting international trade, had three important aspects affecting telecommunications. First, it gave special attention to updating Section 301 of the 1974 Trade Act. The new language is very broad and states that the president of the United States or the U.S. Trade Representative *must* take action whenever "the rights of the United States under any trade agreement are being denied" or when "an act, policy, or practice of a foreign country. . . violates or is inconsistent with, the provisions of or otherwise denies benefits to the United States under, any trade agreement, or . . . is unjustifiable and burdens or restricts United States commerce."[23] Second, when a telecommunications or other U.S. company has a major trade or intellectual property dispute with another country, the 1988 Omnibus Act created "Super 301" and "Special 301" procedures through which the United States may address the offending nation to protect U.S. products and services. Third, the Act amended provisions in Title VII of the Tariff Act of 1930,[24] regulating countervailing and antidumping duties, to broaden their scope and make it easier for U.S. telecommunications companies to obtain protection from unfair foreign competition or discrimination.

6.1.5 Chapter 19—Telecommunications Trade Act of 1988

The Telecommunications Trade Act of 1988,[25,26] passed the same year as the Omnibus Act, is a short act, containing only 10 sections that focus on creating opportunities for telecommunications in the world market. It codified the concepts leading to the WTO and the addition of services to the General Agreement on Tariffs and Trade (GATT). The Act requires that the U.S. Trade Representative review each year the operation and effectiveness of every "trade agreement regarding telecommunications products or services that is in force with respect to the United States."[27]

It also declares any acts, policies, or practices of a country that deny "to telecommunications products and services of U.S. firms mutually advantageous market opportunities" in the foreign country will be challenged by the U.S. government.[28] This section was used, for example, in 1993 to challenge Korea Telecom. In response, Korea made public all procurement regulations and assured the United States that no other closed-market regulations would be implemented.[29] Subsequently, the United States signed telecommunications trade agreements with Korea and other countries to more solidly delineate agreement on these issues.

6.1.6 Chapter 21—North American Free Trade Agreement Implementation Act of 1993

The details of the North American Free Trade Agreement[30] (NAFTA) were discussed in Chapter 5, but the North American agreement also opened free trade in telecommunications goods, services, and capital. Perhaps most importantly, NAFTA established a model for the WTO.

6.1.7 Chapter 22—Uruguay Round Trade Agreements

As also discussed in Section 5.5, the elements of the Uruguay Round Trade Agreements[31] of the GATT agreed to by the United States were codified in Chapter 22 of Title 19. It created the WTO, added services to GATT, and opened the largest telecommunications market in history.

6.2 U.S. EXPORT LAWS

To encourage participation in the increasing world market, rather than to restrict it, the United States has fewer export laws than import laws. However, the government also recognizes that certain exports may threaten the national security of the United States and undermine certain aspects of the economy. Congress's concern with telecommunications, computers, other electronics products, and some software, such as encryption software, is that while these products and technologies have civil uses, they may also be used in military applications or at least contribute to the military potential of other countries. These are known as *dual-use items.* Congress's concern is increased when countries considered to be enemies of the United States and/or its allies acquire such goods and technologies.[32]

On the other hand, Congress also recognizes that the availability of telecommunications, computer, and software products in the United States and abroad makes control of these products more difficult and that restrictions negatively affect U.S. exports and the robustness of the domestic economy. To control these potentially harmful effects on world trade, certain export restrictions exist. The United States has two main laws that govern the export of U.S. products, services, and technologies: the Export Administration Act[33] and the Arms Export Control Act.[34] Both impact telecommunications trade.

6.2.1 Export Administration Act of 1979 (EAA)

The Export Administration Act (EAA)[35] enacted in 1979, has three main U.S. objectives: (1) "to restrict the export of goods and technology which would make a significant contribution to the military potential of any other country and prove detrimental to the interests of the United States,"[36] (2) to restrict exports "where necessary to further significantly the foreign policy of the United States or to fulfill its declared international obligations,[37] and (3) to restrict the export of goods where "necessary to protect the domestic economy from the excessive drain of scarce materials and to reduce the serious inflationary impact of foreign demand."[38] Most of the Export Administration Act (EAA) focuses on the first two objectives and thus directly affects telecommunications issues. The third objective, addressing the short supply areas, is a smaller part of the Act, but given the world wide demand for telecommunications facilities, it is still a rich area for telecommunications products and services.[39] The Export Administration Regulations implements the EAA.[40]

6.2.2 Arms Export Control Act (AECA)

The Arms Export Control Act (AECA)[41] was enacted to control the export of "defense articles and . . . services"[42] such as weapons, weapon components, and technical data regarding weapons[43] in order to further world peace, security, and the foreign policy of the United States. The Arms Export Control Act (AECA) is completely separate from the Export Administration Act, but works with it to monitor and control exports consistent with each of their objectives. Since telecommunications and computer equipment are often interpreted as potentially contributing to weapons, and missile guidance systems and/or directly impacting the development of weapons, this Act directly impacts telecommunications. The Arms Export Control Act (AECA) is implemented by the International Traffic in Arms Regulations (ITAR).[44]

6.3 IMPLEMENTING AGENCIES

In implementing these two laws, one criticism of U.S. trade policy is that no single department or agency in the government has primary responsibility.[45] Instead, within Congress, several committees draft the laws that codify U.S. trade policy. For example, the Agricultural, Armed Services, Banking, Commerce, Foreign Relations, House International Relations, House Ways and Means, and Senate Finance Committees all address international trade issues. They coordinate their activities only through normal congressional committee communications, but that frequently occurs late in the legislative process. Within the executive branch, Congress has named the Departments of Agriculture, Commerce, Defense, Interior, Justice, Labor, State, and Treasury as the agencies responsible for developing and implementing regulations and advising the President and Congress on trade matters.[46]

Any telecommunications attorney or corporate officer who has worked in this area can attest to the confusion, overlapping areas of responsibilities, and numerous coordination and separation of power issues that this "diffusion of authority" creates for companies seeking compliance information.[47] However, proponents of the diffused system argue that it keeps individual interest groups from dominating U.S. trade policy. Most frustrating for telecommunications providers and users is that, while the various departments, agencies, and committees try to work together in a cooperative manner, they frequently issue conflicting directives. In addition, approval from one department does not relieve the exporter from the statutory requirements imposed by other departments.

Either way, telecommunications providers and users must consult this variety of agencies regarding trade. However, in overall trade policy, three key agencies have particularly influential roles: (1) the U.S. Trade Representative, (2) the International Trade Administration, and (3) the International Trade Commission. In export law, the key agencies are the (1) Department of Commerce, (2) Department of Defense and (3) Department of State. Awareness of the agencies and their primary roles is essential to telecommunications companies in licensing and complying with the details of the law. The role of each of these six agencies is described in the following six sections.

6.3.1 U.S. Trade Representative

The U.S Trade Representative is so named because she represents the United States in international trade negotiations. As such, the representative works closely with Congress, the president, and all of the agencies and committees involved in U.S. international trade issues. The U.S. Trade Representative accredits members of Congress participating in trade negotiations[48] and nominates members of advisory committees to assist Congress in its evaluation of trade agreements.[49]

In the Trade Act of 1974, Congress established the office of the U.S. Trade Representative[50] to track U.S. trade opportunities and to report to Congress annually on issues such as (1) anticipated trade levels for the next year,[51] (2) foreign barriers to market access by U.S. firms,[52] and (3) inadequate foreign protection of U.S. intellectual property rights.[53] The representative monitors imports to determine if any have harmed U.S. companies or violated the United States's antidumping and countervailing rules.[54] If any imports have, the U.S. trade representative drafts the U.S. response to the offending countries.[55] She also imposes sanctions on countries that unfairly interfere with U.S. exports.[56] Her work has been especially valuable to the telecommunications industry in opening and strengthening markets for products and services.

6.3.2 International Trade Commission (ITC)

The International Trade Commission (ITC), also created by Congress in the Trade Act of 1974,[57] was directed to evaluate whether imports posed any injury or threat of injury to U.S. interests, the injury or threat was sufficient to justify remedial action, and the impact of some of the import restrictions imposed by U.S. trade law were appropriate and effective. The ITC publishes reports on the results each year, which are particularly important to tracking global telecommunications issues and markets.

6.3.3 Coordinating Committee for Multilateral Export Control (COCOM)

Since international trade involves cooperation with other countries, the U.S. participates in many international trade groups such as GATT, the WTO, the International Monetary Fund (IMF), the World Bank, the United Nations, and the International Telecommunications Union (ITU). The primary international group addressing exports, however, is the Coordinating Committee for Multilateral Export Control (COCOM), located in Paris, France.

The COCOM is composed of delegations from 16 countries: Australia, Belgium, Canada, Denmark, France, Germany, Greece, Italy, Japan, Luxembourg, the Netherlands, Norway, Portugal, Spain, Turkey, and the United States of America. All are members of NATO, except Australia and Japan. The committee was established in 1949, during the early stages of the Cold War, to control products that were scarce or strategic or restricted exports to the Soviet Union, the People's Republic of China, and their allies.[58]

The U.S. delegation to the COCOM is led by the State Department, Office of COCOM Affairs with support from members of the Department of Commerce Office of Technology and Policy Analysis (OTPA) and the Department of Defense, Defense Technology Security Administration (DTSA).

6.3.4 The Defense Technology Security Administration and the Office of Defense Trade Control

In addition to assisting with the COCOM, the Defense Department's Defense Technology Security Administration (DTSA) works with the State Department's Office of Defense Trade Control (DTC) to track the export of weapons and defense materials. Both agencies review all applications filed under the Arms Export Control Act (AECA).

6.3.5 Department of Commerce

Of all of the agencies involved in implementing trade law, Congress has assigned the bulk of the responsibility to the U.S. Department of Commerce. Within the Department of Commerce, many offices work with trade issues, but three offices have specific functions that directly affect telecommunications companies & users: (1) the Office of Technology and Policy Analysis (OTPA), (2) the International Trade Administration (ITA), and (3) the Bureau of Export Administration (BXA).

Office of Technology and Policy Analysis (OTPA)

In addition to the office supporting the U.S. delegation to COCOM, the Department of Commerce's Office of Technology and Policy Analysis (OTPA's) engineers review the exporter's suggested product classifications submitted from the Commodity Control List. They compare it to the other pertinent lists and make a final determination on a product's classification. Given the technological convergence of (1) telecommunications and computers; (2) voice, data, and video systems; and (3) advanced electronics systems and weaponry, this assistance in classification makes billions of trade dollars of difference each year plus impacts access to markets for U.S. telecommunications and computer companies.

International Trade Administration (ITA)

The International Trade Administration, in the Department of Commerce, implements U.S. import law. Along with the U.S. Trade Representative, it is responsible for tracking the impact of countervailing and antidumping duties and determining the fair market value of sales. Based on the International Trade Administration's findings, duties are imposed on imported goods. This office is particularly useful to the telecommunications industry in tracking competitive prices, practices, and market opportunities for U.S. companies.

Bureau of Export Administration (BXA)

The Bureau of Export Administration (BXA), among its other duties, (1) develops policies to implement the Export Administration Act, (2) recommends changes to the Export Administration Regulations, and (3) represents the United States at international export meetings such as were held by the COCOM. To accomplish these tasks, the Bureau of Export Administration (BXA) is subdivided into three offices: the Office of Export Enforcement, the Office of Export Licensing, and the Office of Foreign Availability.

The Office of Export Enforcement manages the actual enforcement of the Export Administration Act and the Export Administration Regulations. As such, it monitors compliance with the EAA by exporters and tracks special products and dual-use items, including telecommunications.

The Office of Export Licensing licenses all exports, including telecommunications, computers, other advanced electronic products, and software. It processes the license applications, determines the appropriate type of license required by the statute, and manages any exceptions that remove the need for licenses.

The Office of Foreign Availability was created because Congress determined, in the Export Administration Act, that export controls should not be placed on goods that are readily available from other sources such as telecommunications systems. The OFA tracks the flow and availability of goods on the world market and recommends items from which controls should be removed.

Through these six agencies, the three main agencies from the U.S. Department of Commerce and the three subagencies from the Bureau of Export Administration (BXA), the degree of control imposed by U.S. export law on U.S. telecommunications products, services, and technology is determined by three things: *what* product, service or technology is to be exported, *to whom* and where the export is to go, and *how* the export is to be conducted.[59] These three levels are described in the next three sections.

6.4 WHAT IS EXPORTED?

In determining the three levels, both the Export Administration Act and the Arms Export Control Act require that all products, services, and technologies be categorized and placed on lists for export control, the lists referred to include (1) the U.S. Department of Commerce's Commodity Control List, (2) the COCOM International Industrial Core List, and the COCOM Munitions List, both maintained by the Department of State, and (3) the Department of Defense's Munitions List required by the Arms Export Control Act. Particularly confusing to telecommunications providers and users is that the three lists are maintained by three different departments of the government, the Departments of Commerce, Defense, and State, respectively, and they are not the same or even closely coordinated. Therefore, it is necessary for exporters, including telecommunications companies, to check all three of the lists to determine if their products fall into any category on one of them. If so, permission must be obtained from that department, regardless of any approvals granted by the other two. Descriptions of the four lists are provided below.

6.4.1 Department of Commerce's Commodity Control List (CCL)

The U.S. Commerce Department's Commodity Control List (CCL) organizes all exported products, services, and technologies into commodity groups identified by a numbering system called the export control classification number (ECCN) system. The ECCN system has existed for years, but was updated after World War II to reflect technological developments. At that time new commodity groups were created, all products and technologies were reorganized into these new groups, and the ECCNs changed from four digits to five digits. These changes allowed: (1) exporters to classify products, services, and technologies more efficiently than before; (2) interested persons to find information concerning products more easily; and (3) researchers to track the number of items exported in each category. This has proven useful to the tracking and pricing of telecommunications products and services. Telecommunications products, computers, and software use several numbers.

The process of categorizing each exported product by assigning an ECCN generally occurs through a two-step process. First, the creators and manufacturers of the product to be exported select the export control classification number (ECCN) from Commerce's Commodity Control List (CCL) that they believe best describes the export and its components. This can be difficult because many of the categories overlap.

The manufacturers then submit their selected ECCN to the Department of Commerce's Office of Technology and Policy Analysis (OTPA) in a Commodity Classification Request. OTPA engineers review the manufacturer's suggested product classification, check the other lists, and make a final determination on the classification. Each component and technology used in developing or producing the product must also be included in the assigned product classification. The product's final assigned ECCN determines which licenses are required to export the product and to which countries the product may be shipped.

6.4.2 COCOM's International Industrial and Munitions Lists

COCOM maintains two products lists: an International Industrial List, called the COCOM Core List (CCL), and the COCOM Munitions List (CML). In 1991, the Department of Commerce began incorporating COCOM's Core List into Commerce's Commodity Control List as an additional component. However, COCOM continues to provide the international perspective by maintaining both of its lists.[60] COCOM's two lists are particularly important to the communications industry because so many telecommunications products, services, and software are considered dual-use items.

6.4.3 Department of Defense's Munitions List

Beyond COCOM's Munitions List, the Arms Export Control Act (AECA) requires that the U.S. Department of Defense also maintain its own "Munitions List." However, the AECA mandates that the Defense Department's list be coordinated with the COCOM munitions list.[61] The International Traffic in Arms Regulations (ITAR) provide the definitions and interpretations to be used by U.S. companies in classifying items for these lists. Since telecommunications products, services and software packages, such as encryption schemes, can be used in defense systems, the Munitions List is of special importance to the telecommunications industry.[62]

6.4.4 Classification of Re-exports

Just as with the initial product, service, or technology, the Export Administration Regulations require that re-exported items or technologies also be categorized. The Code of Federal Regulations defines *re-exports* as "the shipment, or transmission of items from one foreign country to another foreign country or the release of technology or software to a foreign national."[63]

Thus, the export controls affecting the original product, service, or technology apply to all re-exported items. If the original item required only a general license, a re-export requires no further approval. These are known as *permissive re-exports*. However, if an item required prior approval before being exported to a particular country, the exporter must (1) notify the Office of Export Administration (OEA) of the re-export, (2) obtain authorization to re-export on the original license application form, and (3) inform the consignee on the Destination Control Statement that the consignee must also secure authorization from the United States prior to re-export. Fortunately, if an exporter discovers a violation from a consignee, and, in good faith did not know of the re-export beforehand, the exporter is not liable. Instead the liability falls to the foreign party, which usually is then placed on the Table of Denial Orders, discussed in Section 6.5.

Finally, if an item has been incorporated into a foreign-made product, even if the original item has lost its identity and regardless of whether it is only a minor part or amount of the total new product, approval from the Office of Export Administration is needed for re-export. With electronic chips and telecommunications components built in one country and installed in products in another, this area of the law directly impacts the telecommunications, computer and software industries.

6.4.5 Licensing Requirements

For all exports, one of two types of licenses issued by the Bureau of Export Administration (BXA) is required: a *general license* or a *validated license*. Which type of license is needed is determined by the ECCN of a product, service, or technology. An exception to either of these may be granted in certain cases.

General Licenses

A general license is the most common type of license granted for U.S. exports, authorizing items to be exported under normal conditions. Most telecommunications products fall under this license category. No application is required by the government to obtain a general license, and hence most companies are unaware of the more than 20 different types of general licenses that address different types of *goods*[64] and two different types of general licenses that license *technologies*.[65]

In most cases concerning laptops and wireless phones, Temporary License Exceptions are issued as a general license. This is especially true when the computer is owned by the traveler's employer and is a "tool of the trade necessary for the person to perform his or her obligations while traveling abroad." The Bureau of Export Administration (BXA) defines *tools of trade* as the "usual and reasonable kinds and quantities of commodities and software" intended for use by individuals or exporting companies in furthering enterprises abroad. Most travelers take advantage of these permissive regulations from the Bureau of Export Administration without ever knowing that they exist.

Validated Licenses

If an export does not qualify for a general license under the Export Administration Regulations, then a validated license is required. A validated license is required only in a limited number of export situations, and thus only five main types of validated licenses exist: (1) individual, (2) distribution, (3) project, (4) chemical, and (5) service supply. *Individual validated licenses* allow "the export of technical data or a specified quantity of commodities during a specified period to a designated consignee."[66] *Distribution licenses* are similar but allow "the export of *certain commodities* to approved consignees . . ." for the period of one year.[67] *Project licenses,* on the other hand, allow all goods relating to a specific activity to be exported for approximately one year.[68] Special *chemical licenses* allow "the shipment by approved exporters of certain chemicals and chemical and biological equipment to approved consignees . . . ."[69] *Service supply licenses* allow the export of needed spare or replacement parts for equipment that were previously made or exported.[70]

Validated licenses require that a written application with supporting documentation be filed with the Department of Commerce's Office of Export Licensing (OEL).[71] Validated licenses are not automatically granted to the exporter and can be denied in whole or in part to remain consistent with the purpose of the export regulations.

Facilitating the Licensing Process

Knowing which license is needed and completing the licensing process can be confusing and time consuming for exporters, including telecommunications companies and individual users. To make it easier, the Department of Commerce has implemented several important procedures over the years, including an Exporter's Bill of Rights.[72] The Bill states that those who export have the right to expect (1) an accurate and consistent licensing analyses, (2) a prompt decision regarding licensing determination, (3) full access to licensing and regulatory information, and (4) responsive and courteous service.

Second, the Bureau of Export Administration (BXA) significantly reduced the amount of time required for exporters to receive licensing decisions from months to an average of one to two weeks. To do this, it installed an optical character recognition system (OCRS) that reads the submitted forms and thus increases the speed in which the forms are processed.

Third, the BXA installed three electronic systems to help exporters: (1) the Export Licensing Automated Information Network (ELAIN), to allow electronic submissions of applications; (2) the System for Tracking Export License Applications (STELA), in which digitized voice-response units provide exporters current information on the status of their applications; and (3) the Export Licensing Voice Information System (ELVIS), which answers many commonly asked questions by a recorded message. ELVIS addresses all aspects of licensing information and, if needed, allows the exporter to speak with an export counselor. All three systems can be reached by touch-tone telephone by calling the Bureau of Export Administration (BXA) or via the Internet.

Fourth, the BXA provides the *OEL Insider,* a quarterly publication, seminars, and specialized publications to help the public understand how to comply with the regulations.

Finally, in the event that an application is denied, and the exporter disagrees with the final determination, the commerce BXA provides for an administrative appeal procedure and limited judicial review.[73]

6.5 TO WHOM IS THE PRODUCT EXPORTED?

U.S. export law also dictates *to whom* exports can be sent.[74] Two statutes, the Trading with the Enemy Act (TWEA)[75] and the International Emergency Economic Powers Act (IEEPA),[76] require that the Departments of Commerce, State, and Treasury maintain several additional lists, beyond the commodities and munitions lists, to control to whom the exports are sent. If a customer appears on one of the lists, the U.S. exporter must refuse to sell to or discuss products with the denied party. The lists include (1) the Department of Commerce's Table of Denial Orders (2) the Department of Treasury's Office of Foreign Assets Control List of Specially Designated Nationals (SDN), and (3) the Department of State's Country Lists, noting against which countries embargoes are currently in effect.

Therefore, exporters, such as telecommunications companies, must continually screen their customers and foreign contacts based on these lists, to ensure that they do not violate these laws. In addition, U.S. exporters, including all telecommunications companies, may not repair United States–originated items and technology owned or held by a denied party.[77]

To ensure compliance with the Trading with the Enemy Act and the International Emergency Economic Powers Act, U.S. exporters must establish careful procedures within their companies to check these lists on a regular basis. For telecommunications companies, this is especially important because they provide such broad-based services to so many different customers. Each company's sales department must obtain information from the customer and retain it a customer database. When the product or service has been readied, a final check must be made by the shipping or service personnel to ensure that the addressee was not recently placed on a restricted list.

The best way for telecommunications managers to ensure these tasks are done is to incorporate the screening process into a routine clerical task that is checked at each step in the export process. It is critical that each employee involved in the process understands the importance of the reporting requirements and is aware of the required compliance procedures.

6.5.1 Trading with the Enemy Act (TWEA) and the International Emergency Economic Powers Act (IEEPA)[78,79]

The Trading with the Enemy Act (TWEA) was enacted at the beginning of World War I and has been amended several times since then.[80] It prohibits trade with an enemy of the United States. *Trade* is defined as "trade or attempt to trade, either directly or indirectly, with knowledge or reasonable cause to believe that the trading partner is an enemy." *Enemy* is defined as the enemy, ally of the enemy, or a party conducting or taking part in such trade for the benefit of the enemy."[81]

The TWEA required the wartime declaration of an enemy, but in 1976, more than 60 years after the TWEA was enacted, the International Emergency Economic Powers Act (IEEPA) changed this. It gave the U.S. president authority to impose export regulations against a foreign country without having to declare a national emergency.

The Department of Treasury Office of Foreign Assets Control (OFAC) administers both the TWEA and the IEEPA. In general, the OFAC regulations control trade and financial transactions with targeted countries by imposing embargoes against specific countries and restricting exports to other countries. The OFAC also, at times, freezes the assets of a country or its nationals, prohibiting their removal from the United States. In recent decades, countries such as Cuba, North Korea, Libya, South Africa, Iraq, Iran, Cambodia, and Vietnam have all been affected by the restrictions imposed by both Acts and the OFAC. While this is changing, it is important for telecommunications providers to understand that not all restrictions change at the same time, yet companies are responsible for complying with each. For example, even when the U.S. president lifts the sanctions against a country, the OFAC regulations may still apply. Technically, an exporter may still be liable if products, services, or technologies are exported to the restricted country. Therefore, exporters and travelers must track the OFAC regulations, rather than news announcements, to update the lists described below.

6.5.2 Country Lists

The United States restricts trade with countries listed on the Commodity Control List maintained by the U.S. Department of Commerce. Each country is ranked according to an analysis of the political and military risks the country poses to the United States. The Departments of Commerce, State, and Defense conduct various aspects of these analyses, creating a list that is temporary and changes periodically with U.S. foreign policy. The country lists are preliminary screenings of countries, not individuals, to whom the exporter is exporting.

6.5.3 Table of Denial Orders

The Department of Commerce also maintains a Table of Denial Orders listing both individuals and companies that are barred from trade with the United States because of past infractions of trade law.[82] While most Denial Orders apply for several years, they do vary in their scope and duration. The Commerce Department prints the full Table of Denial Orders semiannually, usually each March and October, but also prints updates each day in the *Federal Register.*[83] Exporters are deemed to be on notice of all denial orders upon publication of the *Federal Register.* Accordingly, most exporting companies receive the *Federal Register* each day, extract any names that have been added to the table, and update all intracompany lists.

6.5.4 Specially Designated Nationals List

A similar list is published by the Department of Treasury's Office of Foreign Assets Control (OFAC) which identifies the individuals or companies that OFAC consider to be Specially Designated Nationals (SDN) of countries under foreign policy-related embargoes. Transfer of products, services or technologies to these entities is prohibited without authorization from the OFAC. Like the Table of Denial Orders, the SDN list is updated through the *Federal Register.*

6.6 HOW ARE EXPORTS CONDUCTED?[84]

In addition to the laws controlling what products can be exported and to whom, three important U.S. laws govern *how,* or on what terms, U.S. companies can export to another country. The three laws are (1) the Anti-Boycott Amendments to the Export Administration Act (EAA),[85] (2) the Ribicoff Amendments to the Tax Reform Act, and (3) the Anti-Bribery Foreign Corrupt Practices Act (FCPA). Each directly impacts the trade of telecommunications products, services, and technologies.

6.6.1 Anti-Boycott Amendments

The anti-boycott rules in international trade are amendments to the Export Administration Act (EAA)[86] that address the opposite of a United States–sanctioned boycott. Rather than prohibiting trade with a country, the Anti-Boycott Amendments make it illegal for U.S. companies to participate in a boycott against a country with whom the United States has friendly relations. They also prohibit trade between U.S. companies and foreign entities participating in a non–United States-sanctioned boycott. Thus, the law affects each company involved in the manufacture, sale, purchasing, installation, and operation of telecommunications equipment and service.

The Anti-Boycott Amendments have an interesting history and make more sense when one understands the origin of the law. In the 1930's and 1940's, the Arab League initiated a boycott against trade with the territory that was to become Israel. In the boycott, the Arab nations and Arab companies refused to do business with the nation of Israel and Israeli companies. The United States had very little response to this boycott because it involved the only Arab League and Israel. The United States, therefore, had no jurisdiction to act against the boycott. At the time, governments around the world generally acknowledged that the Arab restriction against trade with Israel and Israeli companies was a legitimate sovereign issue between the Arab nations and Israel.

However, by the late 1950's and early 1960's, the Arab nations began refusing to do business with third-country companies that traded with Israel. The Arab Boycott Committee, representing the Arab nations, compiled a blacklist of companies, including U.S. companies, that traded with Israel. This second phase of the boycott began interfering with the business transactions of U.S. companies, and as such, expanded enforcement of the boycott from solely the Arab League to foreign companies.[87]

In response, in 1965, the U.S. Congress amended the United States's Export Administration Act (EAA) by adding a declaration that it was the official policy of the United States to oppose boycotts imposed against countries friendly to the United States. The amendments, however, did not actually prohibit U.S. firms from participating in the Arab boycott of Israel.[88] Instead, they simply discouraged U.S. exporters from taking any action in support of the boycott.[89]

Eight years later, during the oil crisis of 1973, the price of oil increased more than 500%. This resulted in a serious trade deficit for the United States and required increased trade with the Arab countries. However, to trade with the Arab League, U.S. companies could not trade with Israel. This third phase of the boycott clearly impacted both U.S. trade and foreign policy. In addition, the imposed Arab prejudices against the "race, religion, and national origin" of Israelis violated the spirit and intent of U.S. civil rights, equal opportunity, and antitrust laws.

The U.S. government acted to stop the practice by requiring companies, including telecommunications, computer, and software companies, to report anyone attempting to recruit them to participate in a boycott, telling them that a sale was conditioned on the seller joining the boycott, or asking them to provide information regarding nonparticipating companies.[90] The request, whether oral or written, was required to be reported to the Department of Commerce by the end of the month following the calendar quarter in which the request was received.[91] The report required documentation of the request, disclosure of what action was taken, and a description of the commodities or technical data involved. In addition, in the landmark case of *U.S. v. Bechtel,*[92] the court considered the legality of U.S. companies' participating in the Arab boycott. The case was settled out of court, but it served as notice to U.S. companies that their participation in nonsanctioned boycotts could be illegal.[93]

Over the next four years, however, until 1977, the Arab boycott affected increasing numbers of U.S. businesses with greater impact. Congress, therefore, added stronger anti-boycott provisions to the Export Administration Act[94] prohibiting a U.S. person or business from "willfully taking, or agreeing to take, actions to comply with any boycott imposed by a foreign country against a country that is friendly to the U.S."[95]

Under the anti-boycott amendments today, U.S. companies and citizens, including telecommunications companies and users, are prohibited from the following: (1) refusing to do business with anyone pursuant to a request from or agreement with a boycotting country;[96] (2) refusing to employ or otherwise discriminate against any U.S. person on the basis of sex, race, religion, or national origin;[97] (3) furnishing information regarding the race, religion, sex, or national origin of customers;[98] (4) providing information about someone's business relationships or membership in other organizations;[99] or (5) drafting letters of credit that request any of the above information.[100] U.S. courts have upheld these regulations in various challenges.[101] If found guilty, U.S. companies can have their export licenses revoked and other civil and criminal penalties applied. This is particularly important given the additional information, beyond sales information, that the telecommunications companies have access to and can provide.

6.6.2 Ribicoff Amendments to the Tax Reform Act

One of the sanctions against U.S. companies participating in a boycott against a United States–friendly country includes the Ribicoff Amendments to the Tax Reform Act of 1976.[102] Normally, U.S. taxpayers who also pay taxes in another country are permitted to take that amount as a credit on their U.S. taxes. In the Ribicoff Amendments to the Tax Reform Act, however, Congress denied these tax benefits to companies that "participate in or cooperate with an [non-sanctioned] international boycott."[103]

Section 999 of the Internal Revenue Service regulations[104] defines that a person is participating in or cooperating with an international boycott if that person agrees

> *(A) As a condition of doing business directly or indirectly within a country or with the government, a company, or a national of the country:*
>
> *(i) to refrain from doing business with or in a country which is the object of the boycott or with the government, companies, or nationals of that country;*
>
> *(ii) to refrain from doing business with any U.S. person engaged in trade in a country which is the object of the international boycott or with the governments, companies, or nationals of that country;*
>
> *(iii) to refrain from doing business with any company whose ownership or management is made up, all or in part, of individuals of a particular nationality, race, or religion, or remove (or refrain from selecting) corporate directors who are individuals of a particular nationality, race, or religion; or*

(iv) to refrain from employing individuals of a particular nationality, race, or religion; or

(B) As a condition of the sale of a product to the government, a company, or national of a country, to refrain from shipping or insuring that product on a carrier owned, leased, or operated by a person who does not participate in or cooperate with an international boycott. . . .

These amendments apply to all U.S. companies, including telecommunications companies, and to any foreign subsidiary if the U.S. company owns more than 10% of the subsidiary's stock. In this era of rapidly merging U.S. and international telecommunications companies, this point is especially important.

6.6.3 Foreign Corrupt Practices Act (FCPA)

The Foreign Corrupt Practices Act (FCPA)[105] is also a series of amendments to a previously existing law, the Securities Act of 1934. The FCPA is a relatively short act that prohibits U.S. companies and individuals from paying bribes and using other corrupt business practices. The unique aspect of the FCPA is that it affects the actions of individuals as well as the operations of companies, whether either is in the United States or not. This is of particular importance to telecommunications companies because in many countries, bribes are part of the culture and thus a required part of the sales and installation of telecommunications systems.

The need for the FCPA was raised in 1976 after a report, submitted by the Securities Exchange Commission (SEC) to the Senate Banking, Housing, and Urban Affairs Committee, revealed that some U.S. companies involved in international trade had adopted various questionable and illegal business practices. These practices included paying bribes to foreign government officials,[106] accumulating off-the-book slush funds by not recording transactions, and falsifying records to disguise various transactions, such as recording payments to one person when the money was actually given to another person.

The report stated that "legislation addressing these issues would be desirable to demonstrate a clear Congressional policy with respect to this thorny and controversial problem."[107] To assist Congress, the SEC proposed amendments to the Securities Act of 1934.[108] The proposed amendments were adopted by Congress as the Foreign Corrupt Practices Act and signed into law by President Carter on December 20, 1977.

The FCPA, enforced by the Justice Department, governs all U.S. companies and individuals subject to the jurisdiction of U.S. law.[109] It uses the term *domestic concern* to define who is within its grasp. Domestic concern means any individual who is a citizen, national, or resident of the U.S. or any corporation, partnership, association, which has its principle place of business in the U.S. or is organized under the laws of the United States. Superimposed on this is the requirement that interstate commerce be involved in the violation. The term *interstate commerce* has been interpreted to include the use of the telephone, mails, or other system of communication, including the Internet.

The FCPA has two main portions: anti-bribery prohibitions[110] and accounting mandates.[111] It prohibits U.S. companies, whether they are registered with the SEC and publicly traded or not, from paying bribes to a foreign official or political party for the purpose of obtaining business.[112] Numerous interpretations have evolved over the years concerning what constituted a violation, but the FLPA requires that the payment must have been made to a foreign official who had "discretionary authority." "Grease payments," or payments made to a person who does not have discretionary authority, in order to simply expedite routine matters, are not covered by the statute.

In addition, in the FCPA, Congress mandated that *all* companies involved in international trade keep books, records, and accounts in sufficient detail to accurately and fairly reflect the transactions and dispositions of the company's assets. To do this, each company is required to develop and maintain a system of internal accounting to ensure that (1) all transactions are executed in accordance with management's authorization; (2) the transactions are recorded to permit preparation of financial statements in accordance with the generally accepted accounting principles (GAPP); (3) access to assets is permitted only with management's authorization; (4) recorded assets are routinely compared with existing assets; and (5) appropriate action is taken when inconsistencies are discovered.[113] These mandated accounting systems are also to be inspected by outside auditors on a regular basis to prevent off-the-book slush funds, rather than merely to find the funds after they exist. The systems also affect all amounts of assets, not just material amounts.

Several criticisms of the FCPA were raised almost as soon as it was enacted in 1977. First, many people argued that the FCPA requirements were already being met, especially by publicly traded corporations. Second, some foreign trade experts criticized the FCPA as a major disincentive to exports. Third, many companies and individuals were confused about how to comply with the new law.[114] Three years later, in 1980, the complaints were officially raised in two General Accounting Office Reports and a third to the President's Export Council of 1980.

As a result of this criticism, Congress amended the FCPA in 1988.[115] The amendments, enacted as part of the Omnibus Trade and Competitiveness Act of 1988 (OTCA),[116] accomplished several important changes to help the FCPA function properly. First, the 1988 Amendments to the FCPA clarified ambiguities such as to what degree the FCPA's accounting requirements applied to foreign subsidiaries and what responsibility the parent corporation had for a subsidiary. The Amendments stated that if a U.S. company owned 50% or less of a foreign firm and the parent reasonably and in good faith required the subsidiary to make and keep accurate books within a system of internal accounting controls, then the U.S. firm had no liability for the foreign subsidiary's practices.

Second, the 1988 Amendments lowered the standard of *knowledge* that a payment was an illegal bribe. The 1977 standard had been "knowing or having reason to know" that a payment would be used by a third party for a purpose unlawful under the Act. After 1988, the standard was just *knowing*.[117] Since 1988, a person is deemed to *know* that the thing of value will be given to a foreign official if the person "is aware of a high probability . . . that the funds will be so used."[118]

Third, the 1988 amendments expanded the "grease payments exception." Originally, *grease payments* were allowed only if the official's duties were ministerial or clerical. After 1988, the FCPA allows payments to *any* foreign official for facilitating or expediting any routine nondiscretionary action. The Amendments also require that the payment be used to retain some business or something be received in return.

Fourth, the 1988 Amendments clarified that, when enforcing the FCPA, the SEC would not impose criminal penalties for insignificant, technical, or inadvertent violations.

Fifth, while certain aspects of the FCPA were relaxed by the 1988 Amendments, the penalties for violation of the antibribery provisions of the FCPA were made significantly more severe by the 1988 Amendments. Congress raised the maximum corporate fine from $1 million to $2 million and the maximum individual fine from $10,000 to $100,000. The maximum prison term remains five years for any officers or directors who willfully violate these provisions.

Sixth, Congress strengthened the FCPA's regulations regarding the liability of individuals in international trade when it repealed the Eckhardt Amendment, which had required that a company actually be convicted of a FCPA violation before any of its directors or officers could be prosecuted. Since 1988, a company's employees or agents may be prosecuted regardless of whether the company has been convicted or prosecuted.

As such, the FCPA impacts all departments and levels within a company and, since communications are so broadly considered exports, it is important that all telecommunications personnel be aware of the law, as it changes, in order to comply with it.

6.7 RECORD RETENTION REQUIREMENTS

Since each of the various trade laws requires companies to keep specific records of their international transactions and contacts, it is important for telecommunications companies to review what records they keep and how they are organized and to establish updated internal procedures to ensure compliance. The three main categories of transactions subject to record-keeping requirements are (1) exports or reexports of U.S. products, services, or technologies; (2) transactions that involve restrictive trade practices or boycott requests; or (3) any other transactions subject to U.S. trade law, regardless of whether an export or re-export is made, or proposed to be made, by any person with or without authorization by a validated license, general license, or any other export authorization.[119] For example, records on the following items must be retained:

1. All written matter pertaining to trade including memoranda, correspondence, notes
2. Invitations to bid
3. Export control documents
4. Restrictive trade practices or boycott documents and reports
5. Financial records
6. Contacts names, addresses, and phone numbers

Generally, records must be retained for at least two years after the "termination of the export,"[120] but anti-boycott materials must be retained for three years after the termination of the export. *Termination of the export* has been interpreted as the latest time (1) the good was exported from the United States, (2) any known re-export or diversion of the goods took place, or (3) any other termination of the transaction occurred, whether formally or by other means. However, most companies retain records longer than the required time, because the statute of limitations for criminal actions brought under the Export Administration Act (EAA) is five years. In addition, the time limits may be extended if the Department of Commerce or any other governmental agency so requests. In such case, the records may be destroyed only with written authorization from the agency that originally requested the materials.

6.8 PENALTIES FOR VIOLATIONS OF EXPORT LAWS

As noted in the material presented above, the United States's trade laws are detailed, confusing, and at times inconsistent, but compliance with them is required of all U.S. companies involved in the international trade of U.S. products, services, and technologies. If compliance is not achieved, the laws carry severe criminal and civil penalties, including the loss of all export rights and confiscation of equipment. The penalties vary with the law.

6.8.1 Criminal Penalties

The Export Administration Act (EAA), for example, defines violations as "violations, attempts, or conspiracies to violate the EAA, or any regulation, order, or license under the Act." The EAA divides criminal violations of export laws into two groups: *knowing* violations and *willful* violations. Knowing violations include the mere possession of goods or technology with the intent to export in violation of U.S. export control laws, or knowing or having reason to believe that the goods or technology would be so exported. For each *knowing* violation,[121] the EAA may assess: (1) a fine of up to $50,000 or five times the value of the export involved, whichever is more; (2) imprisonment for up to five years; or (3) both.

Willful violations[122] include each occurrence, with knowledge, that the intended destination of an export is restricted for national security reasons or foreign policy purposes, or the export will be used for the benefit of a restricted country. The penalties for *willful* violations are harsher than for *knowing* violations. For individuals, (1) fines up to $250,000 can be assessed, (2) the person can be imprisoned for up to 10 years, or (3) both.[123] For companies, fines of $1,000,000 or up to five times the value of the exports involved, whichever is greater, may be assessed.[124]

The same fines apply to an individual or company holding a validated license "for the export of any goods or technology to a controlled country, and who, with the knowledge" of its unauthorized use for "military or intelligence gathering purposes," fails to report the facts to the Secretary of Defense. However, the maximum prison term in these cases is five years.[125] For the Arms Export Control Act (AECA), willful violations carry criminal penalties of (1) up to $1,000,000, (2) imprisonment for up to 10 years, or (3) both.[126] For the Trading With the Enemy Act (TWEA),[127] violations carry penalties of up to $50,000 in fines, imprisonment up to 10 years, or both.[128]

In addition to the fines and prison terms, *knowing* and *willful* violators of national security controls under all of the laws face forfeiture of the violator's interest in the "goods or tangible items that were the subject of the violation,"[129] and the proceeds of the violation itself.[130] Violations of short supply controls or the anti-boycott regulations, however, are not considered violations of national security controls and therefore are not subject to these forfeitures.

6.8.2 Civil Penalties

In addition to criminal penalties, civil penalties can be assessed for violations. The Export Administration Regulations (EAR) state that a respondent found to have violated the Export Administration Act (EAA) before December 29, 1981, is subject to civil sanctions up to $10,000. For violations after December 29, 1981, especially involving national security controls, the fines can be 10 times greater, or up to $100,000 per violation.[131]

In addition to a fine, four additional sanctions may be levied against the respondent. First, illegal exports may be seized by authorities, together with any vessel or aircraft used in the export or the attempt to export.[132] Second, "[a]ny outstanding validated export license affecting any transaction in which the respondent may have any interest . . . may be suspended or revoked."[133] Third, the respondent may be denied the privilege of participating, either "directly or indirectly, in any manner or capacity, in any transaction involving commodities or technical data exported" from the United States.[134] This is known as a "general denial of export privileges." Fourth, the offending party and its affiliates may be given a Denial Order,[135] which prohibits them from exporting goods or technical data from the United States. This penalty in effect places the offender on a "blacklist" because it means that other exporters and importers must avoid all transactions with any person or company given a Denial Order.

6.9 DRAFTING INTERNATIONAL TRADE CONTRACTS

One way for companies to avoid trade law compliance problems is to develop solid, well-written contracts before the trade occurs. This increases the importance of the negotiation process in the new competitive telecommunications environment. When drafting a contract for trade, several critical items must be included. These items are discussed in the following eight sections.[136]

6.9.1 Government Regulations and Licensing Requirements

All import and export requirements imposed by all relevant governments should be specifically delineated in the contract. This is especially true if the trade agreement is contingent on receipt of a license or export permit within a specific contract period. If exceptions apply, it is important to record the exceptions in the body of the document to ensure that all details in the exceptions are accurately observed.

6.9.2 Use of *Incoterms*[137]

Second, the International Chamber of Commerce publishes a manual of common trade and shipping terms called *Incoterms*. The manual provides standard provisions for (1) delivery and storage details, (2) allocation of transportation costs, (3) apportionment of the risk of loss, (4) allocation of liability and insurance costs, and (5) delineation of when the title passes. International contracts should use these *Incoterms* to reduce the likelihood of mistakes and misunderstandings.

6.9.3 Choice of Law and Jurisdiction—The United Nations Convention on Contracts for the International Sales of Goods (CISG)[138]

Third, each international contract must also state whose national law will govern the contract. Usually, each party wants its own national law to apply, but this is negotiable. If the parties include a U.S. company or the foreign subsidiary of a U.S. company, the contract is automatically subject to U.S. law. However, since other countries' laws may also apply, it is clearer to specifically state in the contract that U.S. domestic law, and no other international or national law, applies.

Fourth, if a contract does not expressly designate which country's law applies, the United Nations' Convention on Contracts for the International Sales of Goods (CISG) will often govern the international contract. Enacted on January 1, 1988, the CISG codifies private international law that has evolved from both common and civil law jurisdictions. The primary purpose of the CISG is to allow the foreign parties to negotiate a transaction under the CISG rather than under the laws of any one particular country, delineate the obligations and rights of each party to the contract, and resolve problems arising in international trade contracts.[139] However, the CISG will not apply, depending on the signatories and certain types of transactions. First, for the CISG to apply, all signatories to the contract must be from CISG-ratifying countries. If one party to the contract is from a non–CISG-ratifying country, the CISG generally cannot be used. Currently, many countries have agreed to the use of the CISG, and more are expected to follow. An ongoing, current list of ratifying countries is maintained by both the State Department and the UN. Second, the CISG does not apply to consumer transactions,[140] securities, goods sold by auction, or contracts concerning electricity or ships.[141]

If the CISG is not used, then the Uniform Commercial Code (U.C.C.) generally applies, unless the parties agree otherwise. The CISG and the U.C.C. are similar in their purpose, scope and content, but several important differences exist that may influence the preference of the contract negotiators. For example, the U.C.C. requires a contract be in writing and signed if the value is over $500.[142] The CISG does not require such formality and allows a contract to be proved by any means, including witnesses.[143]

The U.C.C. also includes the "mailbox rule" to determine the timing of offers and acceptances, the revocability of an offer, the battle of the forms, the requirement of consideration in contract formation, and the right to recover for defective goods. The CISG does not include these items.

If the CISG is not used, every attempt should be made by U.S. telecommunications companies to have U.S. law govern in their contracts. This will help ensure use of the U.C.C., a significant reduction in the expenses of hiring local counsel in a foreign country where the exporter may be unfamiliar with the laws, and determination of jurisdiction and venue in the case of a dispute. In each case, the parties are able to retain flexibility in developing international contracts which best serve their needs.

6.9.4 Payment

Each contract should also state the manner in which payments will be made. While most individuals prefer that they be paid before they ship the goods, this is rarely acceptable to the buyer. Instead, most payments in international trade occur through *letters of credit* issued by the buyer's bank. The advantages of letters of credit are that they ensure payment to the seller and safe delivery of the goods to the buyer before the buyer's bank actually pays the seller.

To initiate a letter of credit, the buyer places the payment with his bank. The bank then places the funds in escrow until the goods are delivered or issues credit to the buyer if the bank assesses that the buyer is a good credit risk. Once delivery is made, the seller presents documents, such as the bill of lading, invoices, or carrier's papers, to the buyer's bank as evidence of the delivery of the goods. The buyer's bank then pays the seller, usually more promptly than any other method. The details of letter of credit transactions are provided in the *Uniform Customs and Practices for Documentary Credits.*[144]

If the buyer does not agree to payment via a letter of credit, the seller may secure the transaction with a *sight draft* or *time draft,* drawn from the buyer's bank and enforceable against the buyer independent of the original sales contract. This "documentary credit" requires the buyer's bank to underwrite the risk of default by the buyer.

A third way to provide payment protection is through the use of the UN's CISG. The CISG will enforce the contract and allow cancellation of the contract only when a "fundamental breach" occurs.[145] However, nonpayment generally is not considered a fundamental breach. Therefore, to protect the exporter, the contract should explicitly state that nonpayment will be considered a fundamental breach.

6.9.5 Currency To Be Used

Fifth, an international contract must also specify the currency in which payment for the goods or services will be made. In deciding this, companies and individuals must consider the risks of changes in the foreign exchange rates between the countries and restrictions placed on converting currencies. If the currency in one party's country is not stable, most contracts will opt for payment in U.S. dollars. Other alternatives include payment in a stable, third country's currency; agreement on a specified exchange rate in the contract, although this provides less security against fluctuations; or payment in the foreign currency but with the price stated in U.S. dollars. This requires the foreign participants to pay the stated amount regardless of fluctuation in the exchange rate.

6.9.6 Warranties

Sixth, an international contract further needs to identify and describe any warranties associated with the exchange of goods or services. If no warranties exist, this should be stated in the contract in order to avoid confusion. A warranty clause should state what is warranted, how long the warranty will be in effect, what type of notice is needed to file a claim, and an identification of agreed upon remedies to satisfy the warranty.

6.9.7 Unforeseeable *Force majeure* Events

Seventh, in any trade transaction, certain events, such as war, riots, embargoes, acts of nature, and changes of government, create risks for the parties that are beyond the control of either party. These are known as *force majeure* events. An international contract should state who will be responsible for which part of (1) the loss or delay from these events, (2) compensation after the event, (3) recovery from the event, and (4) the future of the contract if such events were to occur. If the contract does not include these, the national law governing the contract often fills in the void.

In the United States, the U.C.C. delineates who will be responsible for *force majeure* events,[146] but the UN's CISG does not include the term *force majeure,*[147] increasing the need for the parties to include it in the contract.

6.9.8 Dispute Resolution

Finally, the trade contract should delineate what process will be used to resolve any disputes concerning the contract should they arise. The options include mediation, arbitration, and litigation. Of these, mediation and arbitration are the most widely used because they are quicker and less expensive than litigation. Mediation leaves the decision with the parties and thus requires cooperation between them. Arbitration places the decision with an arbitrator. Some of the options for arbitrators include the International Chamber of Commerce (I.C.C.), the United Nations Commission on International Trade Law (UNCITRAL), and the International Commercial Arbitration Rules of the American Arbitration Association. A contract's arbitration clause should preselect an arbitrator and the forum in which the arbitration will occur. This is especially important in telecommunications contracts because of the varied products and services, the short timeframes, and the importance to ongoing communications. Each helps telecommunication companies to move forward to resolve issues.

6.10 ESTABLISHING AN EXPORT COMPLIANCE PROGRAM

To comply with the trade laws, U.S. exporters and individuals must also establish some means of staying current with the changes in the trade laws. For companies, this requires a compliance program to ensure that all of the company's procedures and activities comply with all applicable laws and regulations pertaining to the export and re-export of goods, services and technologies. The program must ensure the following: (1) proper licenses are obtained for all exports, (2) problems are properly handled as they arise, (3) the company's export procedures are reviewed on a regular basis, and (4) proper reports are filed with the appropriate trade offices on a timely basis.[148]

To accomplish this, several basic items should be considered. First, management must mandate that compliance with the export laws and regulations is essential to the firm's success and communicate this to all levels of the firm.

Second, the compliance program must establish a process for exporting and a management chain of command to ensure that all trade regulations are met. The focus of this group should be to work with all departments in the company to integrate compliance into their routine work procedures. Experience has shown that the marketing, finance, legal, and distribution departments are among those critical to the process. The marketing department initially contacts customers and thus is where most violations occur. Employees must be aware of the export regulations before they contact potential clients. The finance department tracks information about customers for billing and therefore is key in assessing if any customer screening problems exist. The legal department must incorporate the trade laws and regulations into the company's trade contracts and license agreements. Finally, the distribution department is the last opportunity for a company to control the export. Preparation of the final shipment or provision of service must be the last check before the product leaves the company's possession or the service is provided.

Third, the company must establish a method and a specific team to keep the company's compliance information up-to-date. Specific items to track include (1) all changes in the export laws, regulations and lists; (2) a list of strategic products, services, or repairs provided by the company; (3) the Table of Denial Orders; (4) a list of proscribed countries; and (5) a list of *red flags* that alert the employee to suspicious circumstances. The list of red flags should include the following: (1) receipt of orders from companies not known in the trade, especially if the company has no obvious use for the items; (2) order amounts, packaging or delivery routes that are unusual in the industry; (3) reluctance to provide end-user information; (4) desire to pay cash for the order; and (5) a purchasing agent's refusal to accept standard installation or service contracts. These are crucial because they are events that tend to trigger audits or investigations of a company's trade practices.

Fourth, the process must emphasize to employees that each employee is responsible for compliance and that the laws contains penalties for individuals violating the law. Ongoing education and training should be provided to all officers and employees, especially new ones, to review the relevant laws and regulations, their updates, and how they relate to the company's activities. This training should contain information on the firm's procedures, destinations of the products, use of the *Federal Register,* Table of Denial Orders, and an organizational chart delineating who is responsible for carrying out the different parts of the compliance program. In addition, legal counsel should be made available to assist employees in understanding the laws. All employees, as a condition of employment, should be asked to sign an agreement stating that they have attended export compliance training and will conform, to the best of their ability, to the company's exporting requirements.

Compliance under the regulations works only if the company is properly organized and actions are taken quickly. Spot checks and audits must be performed to ensure that the actions of the company follow the compliance program as intended. If there is any discrepancy, immediate remedial action must be taken.

CONCLUSION

The movement of equipment, data, software, services, and technologies across borders are considered exports under U.S. trade law. In addition, the sharing of technical information on U.S. soil and the transfer of phone calls, emails, and faxes between countries are considered exports. Thus, U.S. trade law affects every telecommunications user, provider of telecommunications products and services, and the business operations of most U.S. companies and their partners abroad. It is critical, therefore, for U.S. companies and individuals to be aware of these laws and to take steps to comply with them. Fortunately, in most cases, exports are exempted from licensing, or provided with temporary licenses and thus occur without incident. However, the exceptions do not cover many items and contain serious pitfalls and penalties.

However, the information and procedures needed to ensure compliance are confusing and are often hard to find from the multiple government agencies involved. Compliance is also difficult because trade law is constantly changing, requiring ongoing awareness. For these reasons, travelers and telecommunications companies must educate themselves about these laws and integrate compliance into their daily processes and procedures.

ENDNOTES

[1]"Congressional Findings," *supra* note 1, U.S. Export Administration Act, Pub.L.96-72, Sept. 29, 1972, 93 Stat. 503; Pub.L. 99-64, Title I, § 102, July 12, 1985, 99 Stat. 120; codified at 50 U.S.C. § 2401 (1988).

[2]15 C.F.R § 779.1(b) (1991).

[3]15 C.F.R § 770.3(a) (1991).

[4]15 C.F.R § 779.1(a) (1991).

[5]15 C.F.R. 740.9(a)(2)(i) (1991).

[6]Constitution of the United States, Section 8(3).

[7]"Congressional Findings," U.S. Export Administration Act, Pub.L.96-72, Sept. 29, 1972, 93 Stat. 503; Pub.L. 99-64, Title I, § 102, July 12, 1985, 99 Stat. 120; codified at 50 U.S.C. § 2401 (1988).

[8]*Id.*

[9]50 U.S.C. § 337 (1988).

[10]50 U.S.C. § 526 (1988).

[11]19 U.S.C. § 1 *et seq.* (1988 and Supp. 1993).

[12]Tariff Act, 19 U.S.C. § 1202-1677 (1988 and Supp 1993).

[13]Trade Act, 19 U.S.C. § 2101-2495 (1988 and Supp. 1993).

[14]Stephan, Wallace, Roin, at 644.

[15]Trade Acts § 141, codified at 19 U.S.C. § 2171, Supp. P. 192 (1988 and Supp 1993).

[16]19 U.S.C. §§ 2193, 2211, 2213, 2241, and 2242 (1988 and Supp 1993).

[17]Section 201, codified at 19 U.S.C. § 2251, Supp. P. 214).

[18]Section 301, codified at 19 U.S.C. § 2411, Supp. p. 230, as noted in 19 U.S.C. § 3106(c)(2).

[19]19 U.S.C. § 2501-2582 (1988 and Supp 1993).

[20]19 U.S.C. § 2901-2906, Supp. pp. 276-291 (1988 and Supp. 1993).

[21]The Matsushita Electric Industrial Co. v. Zenith Radio Corp., 475 U.S. 574 (1986).

[22]Pub. L. No. 93-618, 88 Stat. 1978, 2041, amended by Pub.L. No. 98-573, sec. 304, 98 Stat. 2948, 3002, and Pub. L. No. 100-418, sec. 1301, 102 Stat. 1107, codified at 19 U.S.C. § 2411 (1994).

[23]Pub. L. No. 100-418, sec. 1301, 102 Stat. 1107, codified at 19 U.S.C. § 2411 (1994).

[24]19 U.S.C. §§ 1671 *et seq.* (1988 and Supp. 1993).

[25]19 U.S.C. § 3101-3111. (1988 and Supp. 1993).

[26]Pub. L. No. 100-418, 102 Stat. 1217, §§. 1371 *et seq.*, codified at 19 U.S.C. §§ 3101 *et seq.* (1992).

[27]Pub. L. No. 100-418, 102 Stat. 1217, § 1377, codified at 19 U.S.C. § 3106.

[28]19 U.S.C. § 1377 and "Section 301 Actions."

[29]*Korea Publishes Phone Regulations Following U.S. Retaliation Threat,* 19 Int. Trade Reporter, at 574 (April 7, 1993).

[30]19 U.S.C. § 3301-3473 (1988 and Supp. 1993).

[31]19 U.S.C. § 3501-3624 (1988 and Supp. 1993).

[32]Export Administration Act (EAA), 50 U.S.C. §§ 2401 *et seq.* (1988).

[33]50 U.S.C. §§ 2401 *et seq.* (1988).

[34]22 U.S.C. §§ 2777-79; C.F.R. § 120.1-3 (1995).

[35]50 U.S.C. §§ 2401 *et seq.* (1988).

[36]50 U.S.C. app. § 2402(2)(A) (1988).

[37]*Id.* at 2(B).

[38]*Id.* at 3(C).

[39]15 C.F.R. §§ 730-799 (1991).

[40]15 C.F.R. Parts 768-799 (1991).

[41]22 U.S.C. §§ 2777-79; C.F.R. § 120.1-.3 (1995).

[42]22 U.S.C. § 2778(a)(1) (1995).

[43]22 U.S.C. §§ 2777-79; 22 C.F.R. § 120.1 -.3 (1995).

[44]22 C.F.R. § 120 (1995).

[45]Stephan, Wallace, Roin, *supra* note 14, at 651.

[46]*See* Section 132 of the Trade Act; codified at 19 U.S.C. § 2152, Supp. p. 185.

[47]Stephan, Wallace, Roin, *supra* note 14, at 651.

[48]Trade Act of 1974 § 161, codified at 19 U.S.C. § 2211, Supp. p 205 (1988 and Supp. 1993).

[49]Trade Act of 1974 § 135, codified at 19 U.S.C. § 2155, Supp. p 186 (1988 and Supp. 1993).

[50]Trade Act of 1974 § 141, codified at 19 U.S.C. § 2171, Supp. p 192 (1988 and Supp. 1993).

[51]Trade Act of 1974 § 163, codified at 19 U.S.C. § 2213, Supp. p 207 (1988 and Supp. 1993).

[52]Trade Act of 1974 § 181, codified at 19 U.S.C. § 2241, Supp. p 210 (1988 and Supp. 1993).

[53]Trade Act of 1974 § 182, codified at 19 U.S.C. § 2242, Supp. p 212 (1988 and Supp. 1993).

[54]Trade Act of 1930 §§ 701(c), 782(c), codified at 19 U.S.C. §§ 1671(c), 1677k(c), Supp. pp. 152, 177.

[55]Trade Act of 1974 § 202, codified at 19 U.S.C. § 2252, Supp. p 214.

[56]Trade Act of 1974 § 301, codified at 19 U.S.C. § 2411, Supp. p 230.

[57]Trade Act of 1974 § 201*(a),* codified at 19 U.S.C. § 2251(a), Supp. p 214.

[58]For a history of COCOM, *see* ROOT & LEIBMAN, U.S. EXPORT CONTROLS 10/1-10/10 (3d ed. 1991).

[59]Matthew H. Wenig, *Exporting U.S. Products, Services and Technologies: An Overview of the Regulations and Considerations Regarding Compliance Programs,* Den. J. of Int'l. Law and Policy, Vol. 23, No. 3, (Summer 1995), pp. 569-598.

[60]56 *Fed. Reg.* 42,824 (1991), codified as the "Commodity Control List and Related Matters" at 15 C.F.R. § 799 (1995).

[61]22 C.F.R. § 121.1 (1995).

[62]*Id.*

[63]15 C.F.R. § 734.2(b) (1991).

[64]15 C.F.R. § 771 (1991).

[65]15 C.F.R. § 779 (1991).

[66]15 C.F.R § 772.2 (b)(1) (1991).

[67]15 C.F.R § 772.2 (b)(3) (1991).

[68]15 C.F.R § 772.2 (b)(2) (1991).

[69]15 C.F.R. § 772.2 (b)(4) (1991).

[70]15 C.F.R. § 772.2 (b)(6) (1991).

[71]15 C.F.R § 772.1 (1991).

[72]*See, generally,* Ian S. Baird, *The U.S. Department of Commerce Export Licensing System: Making Life Easier,* 606 PLI/Comm 81 (1992).

[73]15 C.F.R. § 789 (pertaining to administrative review). 50 U.S.C. app. § 2409(j)(3) (pertaining to limited judicial review).

[74]Wenig, *supra* note 59, at 576.

[75]50 U.S.C. App. § 1 *et seq*. (1988).

[76]50 U.S.C. §§ 1701-1706 (1988).

[77]15 C.F.R. § 781.12 (1991).

[78]50 U.S.C. App. § 1 *et seq*. (1988).

[79]50 U.S.C. §§ 1701-1706 (1988).

[80]*See, e.g.,* Historical Notes following 50 U.S.C. app. § 9. *See generally,* Swan, *A Road Map to Understanding Export Controls: National Security in a Changing Global Environment,* 30 AM. BUS. L. J. 607 (1993).

[81]50 U.S.C. App. § 3(a) (1988).

[82]15 C.F.R. § 787.12(c) (1991).

[83]15 C.F.R. § 788 (1991).

[84]Wenig, *supra* note 59, at 580.

[85]50 U.S.C. App § 2402(5) (1988).

[86]50 U.S.C. App § 2402(5) (1988).

[87]Pamela P. Bread & Pleasant S. Bronax, III, *The Anti-Boycott Provisions of the Export Administration Act,* in THE COMMERCE DEPARTMENT SPEAKS 1990: THE LEGAL ASPECTS OF INTERNATIONAL TRADE (PLI Corporate Law and Practice Course Handbook Series 1990), at 781.

[88]30 *Fed Reg*. 12,121 (1965).

[89]Pub. L. No. 89-63, 79 Stat. 209 (1965).

[90]50 U.S.C. app. § 2407(b); 15 C.F.R. § 769.6 (1988).

[91]15 C.F.R. § 769.6(b) (1991).

[92]*U.S. v. Bechtel,* 648 F.2d 660 (9th Cir. (Cal.) 1981).

[93]*See, U.S., Bechtel Reached Accord on Boycott Case,* WALL ST. J. Jan. 11, 1977 at 14.

[94]50 U.S.C. app. §§ 2401-2420. The EAA sets forth the Anti-Boycott provisions at 50 U.S.C. App. § 2407.

[95]*See* 50 U.S.C. app. § 2407(a) for prohibitions on actions in support of the boycott. Restrictive Trade Practices on Boycotts, 15 C.F.R. Pt. 769 (1995).

[96]15 C.F.R. § 769.2(a) (1991).

[97]15 C.F.R. § 769.2(b) (1991).

[98]15 C.F.R. § 769.2(c) (1991).

[99]15 C.F.R. § 769.2(d-e) (1991).

[100]15 C.F.R. § 769.2(f) (1991).

[101]*See, e.g.,* Briggs & Stratton Corp. v. Baldrige, 728 F.2d 915 (7th Cir. 1984); Trane Co. v. Baldrige, 552 F. Supp. 1378 (W.D. Wis. 1983).

[102]IRC § 999, the Revenue Act of 1971, § 501, 85 Stat. 497, codified at 26 U.S.C. § 861 (1988).

[103]*Id.*

[104]IRC § 999. IRS Temp. Reg. § 7-999-1 and Proposed Reg. § 1-999-1.

[105]Pub. L. No. 95-213 95th Cong., 91 Stat. 1494, codified at 15 U.S.C. § 78 [hereinafter referred to as FCPA].

[106]Report of the Securities and Exchange Commission on Questionable and Illegal Corporate Payments and Practices, S. Comm. on Banking, Housing and Urban Affairs, Committee print, 94th Cong., 2d Sess. (May 1976), at 2.

[107]*Id.* at 57.

[108]*Id.* at 58.

[109]15 U.S.C. § 78dd-1 (1990).

[110]15 U.S.C. § 78m (1990).

[111]15 U.S.C. § 78q(b) (1990).

[112]15 U.S.C. § 78dd-1 (for issuers of registered securities) and 15 U.S.C. § 78dd-2 (for domestic concerns that are not issuers of registered securities).

[113]15 U.S.C. § 78m, note (1990).

[114]*See* Julia Christine Bliss & Gregory J. Spak, *The Foreign Corrupt Practices Act of 1988: Clarifica-*

tion or Evisceration, 20 L. & pol'y int'l bus. 441, 442 at n.3 (1989).

[115]*See* Conf. Report, Omnibus Trade and Competitiveness Act of 1988, H.R. Conf. Rep. No. 576, 100th Cong. 2d Sess. 916-917 (1988) [hereinafter OTCA].

[116]The FCPA amendments appear at Title V, subtitle A, pt. i §5001-5003 of the OTCA.

[117]Pub. L. No. 100-481, § 5003, 102 Stat. 1416, codified at 15 U.S.C. § 78dd-1(a)(3).

[118]*Id.* at 102 Stat 1418 codified at 15 U.S.C. § 78dd-2(h)(3)(A).

[119]15 C.F.R. § 787.13 (1991).

[120]15 C.F.R. § 787.13 (1991).

[121]50 U.S.C. app. § 2410(a), 15 C.F.R. Part 787.1(a)(1)(i) (1988).

[122]50 U.S.C. app. § 2410(b)(1)(A), 15 C.F.R. § 787.1(a)(1)(ii) (1988).

[123]50 U.S.C. app. § 2410(b)(1)(B), 15 C.F.R. § 787.1(a)(1)(ii) (1988).

[124]15 C.F.R. § 787.1(a)(1)(ii) (1991).

[125]50 U.S.C. app. § 2410(b)(2), 15 C.F.R. § 787.1(a)(1)(ii)(B) (1988).

[126]22 U.S.C. § 2778(e). Some civil and administrative sanctions exist in the ACEA.

[127]31 C.F.R. Parts 500-520.

[128]No civil penalties apply to TWEA violations.

[129]50 U.S.C. app.§ 2410(g)(1)(A) (1988).

[130]50 U.S.C. app. § 2410(g)(1)(C) (1988).

[131]50 U.S.C. app. § 2410(c)(1), 15 C.F.R. §§ 787.1(b)(3) and 788.3(a)(4) (1988).

[132]22 U.S.C. § 2401(g) (1994).

[133]15 C.F.R. Part 788.3(a)(1) (1991).

[134]15 C.F.R. Parts 788.3(a)(2) (1991).

[135]15 U.S.C.A. § 788.19 (1991).

[136]Wenig, *supra* note 59, at 591.

[137]International Chamber of Commerce, Incoterms (3rd ed. 1990).

[138]*United Nations Convention on Contracts for the International Sales of Goods, April 11, 1980,* 52 Fed. Reg. 6262 (1987), 15 U.S.C. app. at 29 (1994) *reprinted* in I.L.M. 668 (1980), also called the Vienna Convention [hereinafter CISG].

[139]*Id.* at Article 4.

[140]*Id.* at Article 2(a).

[141]*Id.* at Article 2(b-f).

[142]U.C.C. § 2-201(1).

[143]CISG, *supra* note 138, at Article 11.

[144]International Chamber of Commerce, *Uniform Customs and Practice for Documentary Credits,* I.C.C. publication #400.

[145]CISG, *supra* note 138, at Articles 59, 61-65.

[146]U.C.C. § 2-615.

[147]CISG, *supra* note 138, at Article 79.

[148]Wenig, *supra* note 59, at 595.

CHAPTER 7 Licensing To Protect Telecommunications Intellectual Property

IN ADDITION TO IMPORT/EXPORT ISSUES, A SECOND MAJOR ISSUE FOR companies participating in the expanding international telecommunications market is protection of their intellectual property. Typically, companies seek patents, copyrights, trade secrets, and trademarks to protect their inventions, new services, and unique contributions to the capabilities and efficiencies of the telecommunications industry. However, these protections are sometimes not easy to obtain and they generally do not follow the products beyond national borders.

Since the details of intellectual property law vary in each country or region, companies who sell internationally or participate in several markets must file separate patents, copyrights, and trademarks wherever their equipment, software, or services are offered. Understandably, this is a difficult, expensive, and time-consuming process that is prone to errors. While various international intellectual property agreements[1] have been signed to ease this process, national laws still control. These agreements include: the 1883 Paris Convention for the Protection of Industrial Property; the 1970 Patent Cooperation Treaty; the Universal Copyright Convention, revised in 1971; the 1971 Berne Convention for the Protection of Literary and Artistic Works,[2] the Trade-Related Aspects of Intellectual Property Rights (TRIPs) in the General Agreement on Tariffs and Trade (GATT); and the World Intellectual Property Organization (WIPO).

Where intellectual property protections are not obtained, however, a standard practice in high-tech industries, such as telecommunications, is for companies to acquire a piece of a competitor's equipment or a copy of a competitor's software and to take it apart to discover its underlying design.[3] This process, known as "reverse engineering," is legal in the United States, the European Union, and most other areas of the world.

In the United States, the Supreme Court has defined *reverse engineering* as starting with a "known product and working backward to divine the process which aided in its development or manufacture."[4] In the European Union, the analogous concept of "*decompilation*" is defined by the European Software Copyright Directive as "the reproduction of the code and translation of its form . . . to obtain the information necessary to achieve the interoperability of an independently created computer program with other programs."[5] In many countries, even if intellectual property protection is granted, the protection may not restrict legal reverse engineering. Thus, markets established by distributors and value-added resellers (VARs) can be undermined when ex-employees or other nationals reverse engineer an imported product and develop a competitive "local" product that appeals to the nationalistic concerns of the recipient country and may be less expensive.

Hardware and software developers that have invested a tremendous amount of time, energy, and money into developing their products view reverse engineering and other forms of copying as "legal theft" of their work efforts. Their concerns about reverse engineering and other challenges to their intellectual property rights increase as more hardware, software, and services are traded among nations.

To protect themselves, developers continuously seek ways to legally prohibit or limit the copying of their products, protect their intellectual property claims, and place themselves in the best possible legal position to argue a claim if their product is copied and a similar product created. For these reasons, the licensing of products and services has risen to new prominence in the telecommunications community. Developers desire licenses because licenses create an immediate written contract with the user concerning use of the product or service. In doing so, licenses accomplish the three goals listed above and allow developers to collect royalties on their inventions. Frequently, licenses also provide sufficient information about the product to remove the incentive to reverse engineer, copy or undersell the product, and thus often work where other intellectual property protections may not. Finally, licenses facilitate progress in misuse cases because *breach of contract* is generally easier for the developer to prove than *intellectual property infringement*.

Nonetheless, since intellectual property law varies in each country or region, intellectual property owners must still have separate, specific licenses for each market. This opens the opportunity for confusion and error, mainly from not tracking which license should be used in each market. Thus, three important questions arise with licenses: (1) What should be included to create a solid license?; (2) Can a license be drafted that would work in several markets? and (3) What language is required to protect telecommunications products in the two most immediate markets, the United States and Europe?

To address these questions, the purpose of this chapter is to review the role of patents, trademarks, trade secrets, and copyrights in protecting intellectual property rights and the most effective use of licenses. To do this, Sections 7.1 through 7.4 briefly review U.S. patent, trademark, trade secret and copyright law, respectively. Section 7.5 discusses why reverse engineering is legal and how licenses can best protect international telecommunications providers from it. Section 7.6 compares U.S. intellectual property law with that of the European Union, while Section 7.7 outlines what should be included in a license to provide the best possible protection in these two markets. Section 7.8 provides a model license with specific language incorporating these suggestions and is written in such a manner that it could be adjusted for use in Asia, Latin America, Africa, the Middle East, Pacific Rim, or other areas of the world.

7.1 U.S. PATENT LAW

Patents offer an inventor the greatest level of protection from any form of copying, including reverse engineering, because patents grant the holder the exclusive right to make, use, and sell the invention for a certain number of years—essentially a monopoly on the product during that time. Alexander Graham Bell's patent of the telephone launched Bell's success in the industry and was so critical in the development of telecommunications that it was the topic of the first court case in the telephone industry to be heard by the Supreme Court. Patents also played a critical role in development of telegraph, radio, television, satellite, microwave, and numerous other aspects of telecommunications technology.

Despite of the number of patents issued in the telecommunications industry, however, obtaining a patent is not easy. To patent an invention, the invention must be "a new and useful process, machine, manufacture, or composition of matter, or any new and useful improvement thereof, . . . subject to the conditions and requirements of Section 101 of the U.S. Patent Act."[6] To prove this, the developer must prove that the invention is novel, nonobvious, and useful. An invention must pass all three tests to be patentable.

In the early days of telecommunications equipment, proving these elements was fairly straightforward. However, as the telecommunications industry has become more computerized, the proof has become more difficult.

From their beginning, computers clearly were useful and thus easily passed the "useful" test. However, to be "novel," *all* elements of an item must *not* have been previously patented, described in a printed publication, known by others or used by others before it was invented by the patent applicant anywhere in the United States or the world.[7] Since computers are essentially electronic machines that do mathematical operations, nearly all of the electronic and mathematical elements that cause computers to work have been "known or used by others in this country long before the invention"[8] of computers. They were well described in printed publications prior to the development of any specific software program. For this reason, most early computers and software could not pass the 'novel' test and thus were found to be unpatentable.

For similar reasons, computers also failed the "nonobvious" test. To be nonobvious, the operation of an invention, product, or process must not be readily deducible to persons skilled in the craft or significantly reflected in the "prior art" of the industry. Again, the concepts underlying computers and their related software were well known and thus early computers could not pass the "nonobvious" test for patentability.

For these reasons, most computers and software programs have not been patented.[9] An equally important reason, however, is that many developers decided that the cost and time required to obtain a patent often were unrealistic for the dynamic, rapidly changing nature of the telecommunications industry. They felt that the patent process took so long that items risked becoming obsolete before a patent was granted. Instead, many favor hardware and software licenses, although they do not view the situation as "either/or." Where an item can be patented, the significant protection provided by the Patent Act makes it clearly the intellectual property protection of choice, here in the United States and in Europe.[10]

7.2 U.S. TRADEMARK LAW

A *trademark* is a distinctive name, mark, motto, or logo, which a manufacturer affixes to the goods produced, so that they may be identified in the market, and their origin vouched for.[11] Exclusive rights to use a trademark in the U.S. are granted by the U.S. government for a specific number of years, often 10 years, with the possibility of additional renewal periods.[12]

Traditionally, the trademarks used in telecommunications were generally well known. And the reputations of the known trademarks helped to sell products and services. Occasionally competitors *rebranded* cheaper imitations in order to make a profit. Therefore, as companies merge, divest, and enter new markets, trademark law is an increasingly important area of the law.

7.3 U.S. TRADE SECRET LAW

Trademark law should not be confused with trade secret law. When a product or software program is not "novel" or "nonobvious" enough to qualify for a patent, the Uniform Trade Secrets Act (UTSA) may provide a second level of intellectual property protection.[13] The Uniform Trade Secrets Act defines a *trade secret* as "any information that derives economic value," such as a competitive advantage, "from not being generally known by others" and is "the subject of efforts that are reasonable under the circumstances to maintain its secrecy."[14]

"Reasonable steps" may include: (1) contractually imposing an "explicit duty of confidentiality" on any person with whom the secret is shared through the use of *nondisclosure* and *noncompete* agreements with business investors, product partners, and employees or (2) by an "implied duty of confidentiality" if the relationship between the parties warrants. Of particular importance to software developers is that if a party in such a relationship uses the secret to his or her own advantage or reveals it to another, the trade secret owner will be able to secure judicial relief, usually through tort law.

Trade secret protection is widely used, but has some pitfalls that telecommunications developers should consider. First, unlike a patent, which protects an invention for a specific number of years, a trade secret is valid only so long as the secret remains a secret.[15]

Second, state and federal trade secret laws protect only material that "is confidential and does not conflict with the *Policy of Free Copyability* that applies to material in the public domain." Thus, trade secret law protects only information not in the public domain. Since trade secrets are "secrets," they are considered confidential, not "public," and not in the public domain. However, the downside of this definition is that during the time the trade secret is still a secret, the public at large remains free to discover the secret by independent creation or through reverse engineering of the resulting product in the public domain.[16] Conversely, trade secret protection does not cover any unpatented information that is in the public domain.

Third, trade secrets can be stolen in two ways: by "friends" of the owner and by "strangers." *Friends* are those who learn of the secrets legitimately, with the consent of the owner, but then who use the secret to compete with the owner. *Strangers* are those who learn of the secret in an illegal or improper manner that successfully overwhelms the owner's reasonable efforts to keep the matter confidential.[17] To be protected under trade secret law, the owner of an alleged trade secret must also take "reasonable steps" to preserve the secrecy of the item, but absolute secrecy is not required.[18] For theft of trade secret by strangers, no federal criminal law exists at this time, but approximately 20 states have statutes prohibiting such activity. To protect secrets filed with the federal government, an exception to the Freedom of Information Act (FOIA) permits the government to refuse to reveal the trade secrets. If a government agency inadvertently releases the trade secret, the owner of the secret may file suit under the Administrative Procedure Act (APA) to recover damages from the disclosure.[19]

Fourth, trade secret law does not cover the innocent use of stolen secrets. This occurs when a party uses the secret without knowing that the secret has been stolen or misappropriated. In those cases, the "innocent user" is not liable to the owner until the innocent user becomes aware that the material is a trade secret.[20] This exception is certain to become more important as information of all types is released on the Internet and through electronic bulletin boards.

Fifth, a trade secret also appears to be contrary to the policy of making information widely available to the public, because it encourages people to keep their inventions secret. Nonetheless, the Supreme Court decided in *Aronson v. Quick Point Pencil Co.*[21] that trade secret law is not inconsistent with or preempted by public policy or patent law because, by definition, trade secrets are not in the public domain.

Sixth, state law governs trade secrets, and the states are free to regulate them "in any manner not inconsistent with federal law."[22] Thus, trade secret laws vary throughout the states.

Despite these apparent shortcomings, trade secret protection is widely considered today as "probably the single most important protection against reverse engineering for computer software."[23] In an unpublished case known as *Stac Electronics v. Microsoft Corp.*,[24] trade secret law proved to be the key element in one of the largest settlements ever awarded in a software protection case.[25] The jury concluded that software reverse engineering constitutes "willful and malicious misappropriation" of the original ideas and design in the software.[26] The verdict placed this issue on the agenda of companies throughout the computer industry.[27] Attorneys reviewing the case concluded that

> As long as the owner of software . . . is diligent in the prosecution of its rights, trade secret protection may be available unless or until such information in fact becomes generally known. . . . At a minimum, trade secret law may afford software developers the valuable head start against competitors that the trade secret laws were designed to protect.[28]

7.4 U.S. COPYRIGHT LAW

Copyrights provide protection for products or works that are original, but not necessarily "novel" or "useful." The Copyright Act of 1976[29] provides certain "exclusive rights" to authors for their unique expression of otherwise well-known ideas. Thus, while the electronics and mathematics behind computers and software programs are well known, the unique manner in which they are used in telecommunications equipment fits reasonably well under this definition.

Initially, the Copyright Act granted protection to the authors or creators of seven explicit categories of published works: (1) literary works; (2) musical works, including any accompanying words; (3) dramatic works, including any accompanying music; (4) pantomimes and choreographic works; (5) pictorial, graphic, and sculptural works; (6) motion pictures and other audiovisual works; and (7) sound recordings.[30] Therefore, before 1980, copyright law did not specifically cover modern technology such as computer software. However, early programs were granted copyrights under the category of literary works,[31] extending to programs the same rights granted to books and other writings.[32] Courts also found that some software generated displays that could be protected as audiovisual works, another explicit category in the Copyright Act.[33]

In 1980, however, Congress amended the Copyright Act of 1976 to explicitly include computer programs in the definitions section of the copyright law.[34] At that time, copyright law officially became the preeminent intellectual property protection for software and some resulting telecommunications services.[35] This was important because computers were shrinking to personal computers and laptops. Software similarly moved from an integrated part of the equipment to special programs sold separately on diskettes.[36] Since software was no longer acquired in face-to-face transactions, "shrinkwrap" license agreements began evolving as the common form of copyrighted hardware and software protection.[37]

By the time of the 1980 Amendment, Congress had also abolished several previous requirements. First, for example, it no longer required that a copyright be *published* before it was effective. Instead, the 1980 amendment attached copyright protection to an item or work the instant the work was "fixed in a tangible medium."[38] Section 102 of the Copyright Act states that "Copyright protection subsists . . . in original works of authorship fixed in any tangible medium of expression, now known or later developed, from which they can be perceived, reproduced, or otherwise communicated, either directly or with the aid of a machine or device."

Second, Congress clarified that the *originality* requirement is met if the work required some degree of effort on the part of the author, but no judgment about the artistic merit of the work would be required. In all copyrighted works, Congress determined that the specific *expression* of the idea is protected, not the *idea* itself.

Third, the updated publication requirement was emphasized where it affects copyright holders' rights set by the *notice, registration,* and *deposit* requirements of copyrights.[39] That is, Congress stated that "*notice*" of the copyright, usually a "c" within a circle, must appear on all copies of the work once it is published, or the copyright protection may be lost. In addition, Congress determined that registration of a copyright within the U.S. Copyright Office is optional, but is required before an author can sue for copyright infringement. While a copyright owner may register after learning of infringement and then file suit, certain remedies will be limited, so it is always in an authors' best interests to register their copyright. This is easy to do since, unlike a patent, which costs thousands of dollars to obtain, registration of a copyright costs less than $50.00. Once a work is published, the author must deposit two copies in the Library of Congress within three months after publication.

With a copyright today, a hardware or software developer acquires at least five *exclusive* rights related to the copyrighted program. The developer may do or authorize any of the following[40]

1. Make copies of the work
2. Prepare derivative works based upon the copyrighted work
3. Distribute copies of the copyrighted work to the public by sale or other transfer of ownership, or by rental, lease, or lending

4. Publicly perform literary, musical, dramatic, and choreographic works, pantomimes, and pictorial, graphic, or sculptural works, including the individual images of a motion picture or other audiovisual work
5. Display the copyrighted work publicly

These rights exist for the *lifetime* of the creator, plus 50 years after death. During that time, the developer may transfer or license the copyright just like any other personal property. Licensing of a copyright is the most common form of allowing others to use the work once they have paid a specific amount of money called a *royalty*. To protect the hardware and software developers, transfers and license agreements automatically end at an agreed upon date.

7.4.1 Exceptions to the Rights of Copyright Owners

At least four exceptions to these rights exist, however, impacting copyright holders. These exceptions include (1) certain uses of a work by educational or charitable institutions or libraries, (2) limited uses by a software owner,[41] (3) public access to the information, and (4) fair use of the product.[42] Of these, the last three exceptions play the most crucial role in assessing the level of protection provided to telecommunications products and services and are discussed in the following three sections.

The Limited Uses by a Software Owner Exception

First, it is not a copyright infringement for the owner of a copy of a computer program to make one archival copy of the program, or to make adaptations, enhancements, or modifications to a particular copy of the program for use by the owner.[43] The copier, however, may not transfer the copy or changes to someone else without the permission of the copyright holder.[44]

Public Access Exception

In a series of copyright cases[45] spanning the more than 100 years between 1879 and 1991, courts followed the federal public policy that favored free access by the public to the underlying ideas and functions of a work rather than tighter protection of creative expression. This concept is the major rationale for the approval of reverse engineering.

In 1992, two important reverse engineering cases from the Ninth Circuit Court, *Atari Games Corp. v. Nintendo of Am., Inc.*[46] and *Sega Enter. Ltd. v. Accolade, Inc.*,[47] changed this. In the two cases, the court was asked to review the four exceptions to copyright. In both cases, the defendant companies reversed engineered the plaintiffs' software to develop noninfringing, compatible end-product computer game cartridges that worked on the hardware game consoles of the plaintiff companies.

In *Atari*, the Federal Circuit Court decided that the "fair use" exclusion to the exclusivity of the Copyright Act permits an individual in rightful possession of a copy of a work "to undertake necessary efforts to understand the work's ideas, processes and methods of operation."[48] Second, the court also stated that an author cannot restrict access and "acquire patent-like protection by putting an idea, process, or method of operation in an unintelligible format and asserting copyright infringement against those who try to understand that idea, process, or method of operation."[49] Third, the court further cited that to protect "processes or methods of operation, a creator must look to patent laws."[50] Fourth, the court thus held that interim copies to be used for reverse engineering are a fair use exception to copyright infringement.[51]

In *Sega,* the court followed *Atari* and concluded, "Where disassembly is the *only* way to gain access to the ideas and functional elements embodied in a copyrighted computer program and where there is a legitimate reason for seeking such access, reverse engineering is a fair use of the copyrighted work as a matter of law."[52]

The Fair Use Exception

For certain socially beneficial uses, such as criticism, comment, news reporting, teaching (including multiple copies for classroom use), scholarship, or research,[53] the copying of copyrighted works is considered legal so long as the use does not deprive the copyright owner of appropriate rights and economic rewards. These are known as "fair use" of the copyrighted works and are a major exception to an author's exclusive rights to reproduce a copyrighted work or to create a derivative work.[54]

Hardware and software users, in particular, are affected by this exception because they (1) must copy the object code of a copyrighted program into their computer's memory to run the program;[55] (2) copy the program in order to access its underlying ideas, as permitted by public policy; and (3) may create an interim copy and modify parts of the original copyrighted program to create compatible programs.[56]

In copyright infringement cases, the courts generally have ignored these "interim copies" as a necessary part of the computer process and instead focused on: (1) a comparison of the end products, as a whole, to determine if the resulting end product is so similar to the original software program that it "infringes" on the original copyright, and (2) on the use of the end product to determine if it qualifies as a fair use under the fair use exception.[57] The key result of these cases for hardware/software developers is that the developers should not focus on restricting reverse engineering of their products, but rather developers should seek to understand the criteria the courts use in evaluating these two focus areas so that they may comply with them in their licenses and use them to place themselves in the best possible legal position to win if their software is infringed.

Comparison of End Products—A Two-Part Test

Since unauthorized copying is often difficult to detect or to prove, and reverse-engineered software programs can result in unique programs that are independent expressions of an idea, but still perform the same tasks or functions as the original program, the courts apply a two-part test to evaluate the end products of reverse-engineered programs.

Substantial similarity

First, the courts consider the "substantial similarity" of the products. This standard is not clearly defined and often depends on the opinion of the court. However, illegal copying is obvious when (1) the entire structure of the original work is duplicated in some detail, (2) specific portions are copied verbatim, (3) both works contain common errors, or (4) both works contain the original programmer's "marks"—nonfunctional identifiers incorporated into the code by programmers specifically to discern later copying.

In all copyright cases, the resulting end product must be sufficiently different from the original product that it is clear only the unprotected underlying ideas were used, and not the original programmer's specific expression. Where the expression of the same idea is sufficiently different in the two works, the court will find no infringement.[58] Where the resulting end product, however, is too similar to the original copyrighted work, it will be considered an infringement.

Access

Second, since it is also possible for two programmers to develop very similar software programs, even though they worked independently of one another and had no information from reverse engineering, the courts also required that a second criteria of "access" to the original copyrighted work be proven before copying infringement is found. Substantially similar programs can exist even if no access is found. For example, in many telecommunications development projects, "clean rooms"[59] are used to keep product development isolated from all other influences or possible leaks. Clean rooms describe the procedure of developing a product by a person or group of persons who have no access to the original, protected program, but rather write their own code using only functional specifications, or other unprotected elements, given to them by others from the original program. So long as they can prove that their work was not contaminated by access to any of the protected elements in the original work, no infringement will be found. Nonetheless, if all or parts of the works are identical or the degree of similarity is overwhelming, access will be presumed. Intent is not required for copyright infringement.[60]

It should be noted that this law affects even user manuals. In a case known as *Williams v. Arndt*,[61] the defendant (Arndt) wrote a program for commodities market that implemented a method described in the plaintiff's (Williams) copyrighted manual.[62] The court held that Arndt's resulting software was a mere translation of the English in the manual into computer language and thus violated Williams' copyright "as a derivative work." In its decision, the court noted that "a 'derivative work' is a work based upon one or more preexisting works, such as a translation. . . ."[63] This decision is important for telecommunications hardware and software developers because the manual that accompanies a vendor's hardware or software often describes many details of the product. It can, therefore, be the first source of information for those evaluating possible copyright infringement.

Fair Use Evaluation

The concept of "fair use" is defined as "the privilege for others to use the copyrighted material in a reasonable manner without the owner's consent," for certain public benefit uses. To determine if a copy qualifies as fair use exception to copyright, courts use the following four-factor test:[64]

1. First, was the defendant's "intended use" of a commercial nature or for nonprofit educational purposes?
2. Second, what was the nature of the copyrighted work?
3. Third, how much of the original work was copied and how substantial a portion of the whole was it?
4. Fourth, what effect did the defendant's use of the copy have on the potential market for or value of the copyrighted work?

All four factors, not just a few, must always be considered to determine whether a particular use constitutes a fair use.[65] If the analysis of these four factors, however, determines that the use of the copy is fair, then the fair use exception applies, and the copy does not infringe the copyright.[66] If not, the copy is determined to be illegal and the copyright infringed upon.

7.4.2 Remedies for Infringement

If infringement is found, possible remedies include the following: (1) injunctive relief, where the copyright owner may stop the infringing use or sales;[67] (2) monetary relief, where a copyright owner may recover damages plus the defendant's profits or statutory damages from between $250 and $10,000 as directed by the court;[68] (3) impoundment, seizure and destruction of all unauthorized copies;[69] and (4) criminal sanctions against the infringer, including imprisonment, if the motive for the copying was commercial advantage or financial gain.[70] To assess copyright protections, the courts consider (1) access to the original product, (2) intended use of the copy, (3) use of the underlying ideas of the original product,[71] (4) how "substantially similar" the resulting end product is to the original product,[72] (5) whether interim copies are kept by the "infringer" only or distributed or sold,[73] and (6) the impact of the resulting software on the potential market of the original product. These are the key elements that hardware and software developers, interested in protecting their work product and positioning themselves for the strongest possible claim in infringement cases, must include in their licensing agreements. The most practical way to protect their designs is through licenses. However, these must be carefully written and avoid certain pitfalls as discussed in the next section.

7.5 THE LEGALITY OF REVERSE ENGINEERING AND RESTRICTIVE LICENSE CLAUSES

In deciding that reverse engineering and certain copyright exceptions are legal, the United States Supreme Court cited the U.S. Constitution's directive to balance the public's need to access and use new ideas and the designers' and developers' need to protect their creative efforts.[74] The Court stated that "the competitive reality of reverse engineering and these exceptions may act as a spur to the inventor, creating an incentive to develop inventions that meet the rigorous requirements of patentability."[75]

7.5.1 "Only Means" of Access

Reverse engineering is thus permitted in the United States where it provides access to the underlying ideas in unpatented, or otherwise unprotected, items in the public domain.[76] The Supreme Court stated that copying or reverse engineering of a copyrighted item, such as a computer program, is a fair use of the work if the person seeking the understanding has a legitimate reason for doing so and the copying or reverse engineering provides the *only means of access* to the unprotected elements of the work.[77] In the European Union (EU), this "only means of access" language is echoed in the decompilation concept of the European Software Directive, since the Directive only allows "indispensable" decompilation.[78]

What is not clear from the U.S. court decisions and the European Software Directive, however, is what happens if reverse engineering/decompilation is *not* the only means of access to the unprotected information and, therefore, not indispensable. What are the rules, for example, when a developer of an original product voluntarily provides relevant portions or the entire source code, through a license, to parties desiring it? This question has yet to be answered by the courts.

7.5.2 Restrictive Clauses

A second question is Can reverse engineering be prohibited or limited by a restrictive clause in a license, sales contract, or shrinkwrap agreement? In the

United States, the courts have ruled that clauses prohibiting reverse engineering were illegal because they exceeded the authority granted to the equipment and software developers by the copyright or other applicable law.[79] However, none of the defendants in these cases provided the source code to their clients, and all exceeded other aspects of the applicable laws. Additionally, under trade secret law, software developers are required to take "reasonable steps," including contracts and licenses, to protect their secret. In the European Union, the European Court of Justice has not interpreted the scope of the Directive's applicability or the legal meaning of indispensable. The question, therefore, becomes What contract restrictions can legally be included in a software license?

7.5.3 Noncompete Clauses

First, a contract cannot contain *noncompete clauses.* In a case known as *Lasercomb Am., Inc. v. Reynolds,*[80] the plaintiff (Lasercomb) produced software for designing dies to make boxes.[81] The company then distributed the software under a license that contained a *noncompete clause*[82] that restricted the licensees from developing their own die-making software or assisting others in developing such software for a period of 99 years. The defendant (Reynolds) did not sign the license agreement but obtained a copy of the software. Reynolds reverse engineered Lasercomb's software to remove certain safeguards and then sold infringing copies of the software.[83] Lasercomb sued for copyright infringement and Reynolds pled copyright misuse as a defense.[84] After several appeals, the court refused to enforce Lasercomb's copyright,[85] viewing the 99 year noncompetition clause as an "anticompetitive restraint"[86] that sought to control competition beyond the level granted by copyright law. The court was unswayed by the fact that Reynolds was not harmed by the clause.

7.5.4 Overreaching Clauses

In another case, *PRC Realty Systems Inc. v. National Ass'n of Realtors,*[87] an unpublished opinion, a company called PRC Realty Systems, Inc., developed and licensed software that allowed realtors access to real estate multiple listing information. PRC's license included a clause requiring each licensee to exert its best efforts to promote the multilisting publishing business.[88] The National Association of Realtors licensed the PRC software and then independently developed a desktop publishing system that allowed licensees of PRC's software to print and publish in-house multiple listings on a laser printer.[89] PRC sued for breach of contract and copyright infringement.[90] The district court decided that realtors had copied PRC's work, but the Fourth Circuit Court reversed, emphasizing the public policy concerns in *Lasercomb*. The Fourth Circuit Court stated in its "best efforts" clause, PRC "used its copyright as a hammer to crush all future development of an independent idea by the defendant, National Association of Realtors, or any other licensee."[91] It thus refused to enforce the contract, but did uphold one count for breach of contract based on the fact that the National Association of Realtors, made nonexclusive license arrangements with parties other than PRC.[92]

7.5.5 Appropriate License Clauses

Courts have upheld some license clauses, however, where the license clause is not overreaching. For example, in one case, *Telerate Sys., Inc. v. Caro*[93], for example, the plaintiff, Telerate, developed a financial database and marketed it in two ways: as a higher-priced, standard personal computer-based software and a lease-based special terminal through which the user could access the database at a lower fee. One user, Caro, developed an alternative access software package that ran on any PC and allowed Telerate customers to purchase Telerate's database-access rights at the lower-leased terminal rate, disconnect the terminal, and still access the database.[94] Additionally, Caro's software provided several desirable access improvements,[95] but also caused some performance problems for a number of other Telerate customers.[96] The court found that Caro's reverse engineering of Telerate's software to develop the package violated various contract provisions[97] and granted Telerate's motion for a preliminary injunction finding that Caro's software infringed Telerate's database copyrights when it copied data from the database.[98]

If a court finds one or more terms of the contract impermissible, normally the court will eliminate only the offending portions of the contract and enforce the rest.[99] However, the court may also void the entire contract under the "doctrine of unclean hands."[100] In this doctrine, the court will consider any misconduct or fault of the plaintiff when deciding if a winning plaintiff is entitled to any remedy. Normally the misconduct of the plaintiff must also directly harm the defendant.[101] However, in cases of fraud, misrepresentation, or other conduct that is illegal, unfair, or against public policy,[102] no injury to the defendant is required. In the case known as *Atari,* the doctrine of unclean hands was a significant factor in the outcome of the case, which permitted the reverse engineering of Atari's product because Atari was found to have had unclean hands after it made false statements to the Patent, Trademark, and Copyright Office.[103]

Thus, copyright or patent holders must write licenses and contracts that do not impose restrictions beyond the protection provided by copyright and patent law or in conflict with antitrust laws. If they do, they may be considered to have unclean hands and lose all protection.[104] This can include any attempt to withhold ideas from the public, because doing so imposes patentlike protection on software without meeting the rigorous requirements of patentability. It is therefore unfair and inconsistent with the purpose and policy of copyright law. A copyright may also imply market power and thus open antitrust issues.[105] Similarly, the practice of "tying," or the requirement that a purchaser of one patented or copyrighted product also purchase a nonprotected product or service, would be considered to cause unclean hands since the practice violates antitrust law.[106]

However, a license that does not impose restrictions beyond the protection provided by copyright and patent law is generally considered to be legal. Telecommunications hardware and software developers, therefore, should use these protections to preserve important legal coverage of their products.

7.6 THE EUROPEAN SOFTWARE DIRECTIVE

The objectives, rights, and restrictions of the European Union regarding hardware patents and software copyrights parallel most of those in the United States. For example, in the preamble to its original proposal for a European Directive on the Legal Protection of Computer Programs, submitted to the European Council on January 5, 1989, the European Commission noted that "the size and growth of the computer industry is such that its importance in the economy of the European Community cannot be overemphasized" and that "unless a legal environment is created which affords a degree of protection against the unauthorized reproduction of computer programs . . . comparable to that given to works such as books and films, research and investment in that vital industry will be stifled."[107] The objectives of the commission's proposal for the Software Directive, therefore, were

to promote the free circulation of computer software within the Community and allow the industry to take advantage of the single market by harmonizing the national laws of the Member States relating to the use and reproduction of computer software, and to prevent the unlawful copying of computer software, or "computer piracy," within the Community by ensuring an adequate level of protection for those who create computer software.[108]

7.6.1 Similarities of European Software Copyright Law to U.S. Copyright Law

In its final version of the Software Directive,[109] adopted by the European Council (EC) on May 14, 1991, the EC paralleled many definitions used by the United States. For example, the EC defined the terms as listed here

1. *Interconnection:* A logical and, where appropriate, physical interconnection required to permit all elements of software and hardware to work

with other software and hardware to enable full use of computers' intended function.[110]

2. *Interoperability:* This functional interconnection and interaction is generally known as 'interoperability'; whereas such interoperability can be defined as the ability to exchange information and mutually to use the information which has been exchanged.[111]
3. *No protection of underlying ideas:* In accordance with the principle of copyright, and to the extent that logic, algorithms and programming languages comprise ideas and principles, those ideas and principles are not protected under the European Directive.[112]
4. *Exclusive rights:* Copyright law provides authors the exclusive right to do or to authorize the making of copies or derivative works, and/or the distribution of their copyrighted work[113] for the life of the author plus 50 years after his death or the death of the last surviving author in cases of multiple authors.[114]
5. *Interim copies:* In the absence of specific contractual provisions, lawful acquirers of computer programs may make interim copies of the program where necessary: to run the program in accordance with its intended purpose, including error correction,[115] to make a back-up copy,[116] to "observe, study or test the functioning of the program in order to determine the ideas and principles which underlie any element of the program,"[117] "provided that these acts do not infringe the copyright in the program."[118]
6. *Reverse engineering:* Decompilation or reverse engineering of software code and the translation of its form are permitted by authorized persons, with certain restrictions.[119]

7.6.2 Differences Between European Software Copyright Law and U.S. Copyright Law

The main differences between the European and U.S. copyright laws include

1. *Access:* The European Union limits the need for access to the underlying ideas as those needed for "interoperability." The U.S. limits for access are not as well defined.[120]
2. *Fair use exceptions:* The European Union's Directive limits the "fair practice exceptions"[121] to those that "make it possible to connect all components of a computer system, including those of different manufacturers, so that they can work together."[122] It further states "this exception may not be used in a way which prejudices the legitimate interests of the rightholder or the normal exploitation of the program."[123]
3. *Decompilation or reverse engineering:* The Directive further limits decompilation or reverse engineering of software code and the translation of its

form to those circumstances where reverse engineering is indispensable to obtain the information necessary to achieve the interoperability of an independently created computer program with other programs.[124]

4. *Authorized users:* An authorized user may "observe, study or test the functioning of the program in order to determine the ideas and principles which underlie any element of the program" only if done so while performing any of the acts of loading, displaying, running, transmitting or storing the program which the user is entitled to do.[125] This effectively limits access to the underlying ideas only through the "black box" of normal use of the program—not the more in-depth dumping, flow charting, and rigorous analysis allowed in the United States.
5. *Disclosure:* This process also requires adequate disclosure by the user.[126]
6. *Dominant reseller:* The Directive also recognizes "dominant resellers" in the provisions of articles 85 and 86 of the European Treaty.[127]
7. *Infringement:* Infringement of the exclusive rights of the author is defined as the "unauthorized reproduction, translation, adaptation or transformation of the form of the code *in which a copy of a computer program has been made available.*"[128] (Emphasis added).

The publication of this proposal resulted in one of the largest controversies ever experienced by the European Union over a Directive. The concern focused mainly on the conditions under which a computer program could be reverse engineered and copied for profit. The Europeans recognized that some copying is necessary to run a program, but they also sought to protect the creative efforts of the programmers to encourage further development.

7.6.3 Importance of the Similarities and Differences between European Software Law and U.S. Law to Software Developers

Since hardware and software developers have legal protections in the European Union that are similar to U.S. protections, the same or similar license contracts can often be used for customers in both the United States and Europe if carefully drafted. The European Directive appears to provide greater protection for software developers than does the United States because the Directive limits a user's access to the underlying ideas to the "black box" approach of reverse engineering. That is, reverse engineering using decompilation for interoperability is permitted only if indispensable to obtaining the underlying ideas. Therefore, it can be argued that if a software developer makes the source code available to the user, infringement may be found if the user reverse engineers the code, since such action would not be indispensable. This is significant for companies with unique software products, which, for marketing purposes, may be better off releasing the source code, encouraging compatibility, and licensing in this manner. For example, a product with major yearly revisions would be a good candidate.

The Directive also makes clear that patents, trade secrets, trademarks, and other intellectual property and contract laws are available to the software developer. The full protection of patents issued in the United States and Europe for software, however, is still unknown, and both constituencies are somewhat guarded about issuing software patents because of public policy.

The only major limitations to these European protections are that the Directive recognizes the "dominant reseller" provisions of Articles 85 and 86 of the European Treaty, and these protections have not been widely tested in the courts. Thus, while the statutes provide critical protection and guidance to software developers in licensing their software, the developers should be aware that this is still a new and unsettled area of the law.

7.7 "MUST INCLUDE" CONCEPTS FOR SOFTWARE LICENSES

Hardware and software developers seeking to protect their products from being copied and reverse engineered and to strengthen their legal position in a product infringement case should therefore implement the following action items.

1. Hardware and software developers cannot completely prohibit reverse engineering of their programs, since public policy requires access to all unprotected information in the public domain. Hardware and software developers can, however, license their works. A license proactively provides access to the information that both limits certain activities and contractually obligates the licensee to honor the information accessed as a trade secret.
2. Case law indicates that contract or license clauses, which limit reverse engineering, are legal so long as the language of the clause does not exceed the authority granted under copyright and other laws. Hardware and software developers, therefore, should include a clause prohibiting reverse engineering in their license. The plaintiffs in *Sega* did not do so and appear to have lost certain claims from this omission. In Europe caution should be observed in including a reverse engineering prohibition without a thorough analysis of Articles 85 and 86 of the European Directive.
3. In the license, require users to state their intent on reverse engineering, and to provide "appropriate notice." In *Sega,* the court required that the defendant, Accolade, have a "legitimate interest" in or "reason for" gaining access to the reverse engineered information.
    a. If the user's interest is to develop a compatible product that benefits the developer by providing added market recognition and product sales, reverse engineering should be encouraged and/or appropriate source code provided.
    b. If the user's interest is to develop a competing product, that is legal and cannot be restricted. In *Sega,* the court stated that "public pol-

icy plus the authorization of section 117(1) [is] not limited [only] to [the] use intended by [the] copyright owner."[129] The information gained from this statement in *Sega,* however, provides the software developer with important "notice" and establishes the "access" element for a possible later infringement case. The plaintiff, then, need only show that the reverse-engineered product is "substantially similar" to the original product and that the resulting product negatively impacts the potential market of the original product.

c. If the user states one's intent but carries out another, the software developer has stronger breach of contract, misappropriation, trade secret theft, and/or "unclean hands" arguments.

4. Write the license to convey that everything needed for the user's legitimate, intended purpose is provided. In *Sega,* the court stated that reverse engineering of a copyrighted software program was "fair use" because it was the *only* means of gaining access to the unprotected aspects of the program. The European Software Directive allows decompilation only when indispensable to achieving interoperability. However, if the developer of the original software voluntarily provides the object and source code to the party desiring it, this leaves no opportunity for the user to claim later that reverse engineering was necessary or indispensable to gain access, or that the copyright owner was using the code or the contract to "hide information as in a patent." In drafting the license, software developers should be certain to avoid words that indicate ambiguity, such as "sample programs" or "all needed to do a specific task." The wording of the clause must satisfy U.S. public policy that all unprotected ideas in public domain are available. To convey compliance with public policy, developers should offer to provide any additional information that the users discover as lacking for their intended purpose. This offer, of course, must be accompanied by a *caveat* that allows the software developers to use commercial discretion in providing the additional information, so as not to overwhelm their staffs. The users may still access the information through reverse engineering, but the time and expense required to do so may discourage some users.
5. Clearly define "interoperability" in the definitions section of the contract. Doing so will avoid misunderstandings and establish "mutual intent" between two skilled persons in the field.
6. Note in the license that, while verbatim interim copies of copyrighted material are permitted as a "fair use exception," they are not to be sold or distributed.
7. Take all necessary and "reasonable" steps to protect your trade secret rights, including provisions in the license to protect trade secret rights. These may include restricted use and/or confidentiality agreements. If the user then reveals the secret, the software developer can make a breach of contract and/or misappropriation claim. *Sega* did not do this, which contributed to its loss.

8. Be certain to maintain "clean hands."
9. Track the market for the product involved and maintain an accurate, detailed, historical record. This will facilitate clear documentation of the impact of competitive products and the value of trade secrets.
10. In negotiating and signing the contract, use persons with the skill to reverse engineer and to understand the contractual restrictions. This will strengthen contractual evidence if problems arise later.

7.8 MODEL LICENSE

To demonstrate how the concepts presented above might be incorporated into a portion of the software license, the following sample software license is provided.

Company A
License Agreement for Product X

Agreement made between Company A, Inc., a Delaware corporation with offices at 1234 Main Street, Anytown, Delaware 12345 (hereinafter "Company A") and [Licensee's name] located at [address] (hereinafter "Developer").

1. LICENSE GRANT. Company A hereby grants to Developer, and Developer hereby accepts, a nonexclusive, non-transferable worldwide license ("this license") to use Company A's XYZ Package as follows:
   A. Definitions of Terms Used in this License:
      1. "Developer XYZ": Software written by a Developer that works in conjunction with products owned by Company A.
      2. "Company A's XYZ Package": A set of tools, instructions, and sample programs which enables the user to write Developer's XYZ and which includes all source code, header files, object code, written descriptions, and instructions needed by Developer to produce Developer's XYZ. If Developer requires something that is not in Company A's XYZ Package to accomplish their intended use, Company A will work with Developer to provide the information needed.
      3. "Designated Product": Company A's XYZ Product developed by Company A, Inc.
      4. "Related Trademarks": Company A's Publishing System(TM), Product B(TM), Company A's Product C(TM), and XYZ(TM).
      5. "Related U.S. Registered Trademarks": Company A's (TM), Company A's "Product B" (TM), and Company A's Product C (TM). These trademarks are registered in the U.S. Patent and Trademark Office.
   B. Limitations of License:
      1. Developer must describe the purpose and function of the Designated Product and provide a description of how the Company A's XYZ Package is to be used to produce Developer's XYZ for the Designated Product.

2. Company A affirms, by its authorized signature below, that Company A's XYZ Package provides Developer with all of the information and code needed by Developer to produce the needed Developer's XYZ for the Designated Product. If Developer requires anything additional, Company A will work with Developer to obtain the information needed. Therefore, Developer may not modify, translate, reverse engineer, disassemble, or decompile the object code or tools provided in the Company A's XYZ Package.
3. Developer may use and/or modify the source code contained in Company A's XYZ Package only to create Developer's XYZ for the Designated Product. This license does not apply to use of Company A's source code, object code or header files for any other purpose. If Developer desires to create Developer's XYZ for any other Company A product, Company A and Developer must execute a separate license agreement for that product.
4. Developer may not share, transfer, release, distribute or otherwise disclose the information provided in the Company A's XYZ Package to anyone except other registered developers who are included on the most recent list of registered developers approved by Company A in writing.
5. Developer may distribute Developer's XYZ for the Designated Product which have been developed using the Company A's XYZ Package if:
   (a) the Developer's XYZ are only distributed in object code form (not source code);
   (b) the Developer is in complete compliance with the terms of this License including the restrictions set forth in paragraph 11.
6. Developer may use Company A's trademark "XYZ" in the name of Developer's product and in any documentation, packaging, advertisements or promotional activities (including, but not limited to trade show activities for Developer's XYZ. THIS LICENSE GRANT DOES NOT INCLUDE THE USE OF "COMPANY A'S XYZ" OR ANY OTHER OF COMPANY A'S TRADEMARKS.
7. Import and export filters:
   (a) Developer may include import and export filters for the Designated Product in Developer's XYZ that allow the transfer of text between the Designated Product and other programs.
   (b) With the prior written consent of Company A, Inc., Developer may include export filters for the Designated Product in Developer's XYZ which allow the transfer of page geometry from the Designated Product to other programs. Consent will be granted only when Developer also includes an equivalent import filter that allows the transfer of the equivalent geometry to the Designated Product from the other program.
   (c) Developer shall not use the Company A's XYZ Package to develop any product which converts files of any Company A product for use with programs which are competitive with any Company A product, including, but not limited to, any other page layout or typesetting programs or publication systems which would be sold in the same end-user market or markets as any Company A Product, without the prior written consent of Company A, Inc.

CONCLUSION

It is evident from the U.S. Constitution, subsequent U.S. statutes and case law, and the European Software Directive, that the United States and Europe are seeking to balance access to unprotected ideas for the public and sufficient protection to hardware and software developers to encourage them to continue producing innovative products. Reverse engineering and other forms of copying are permitted under certain circumstances in the United States, within the European Union, and around the world, but developers still maintain a number of well-supported options to achieve their three goals of legally prohibiting or limiting the reverse engineering of their product, protecting their intellectual property claims, and placing themselves in the best possible legal position to argue a claim if their product is reverse engineered and an infringing software product is created.

The initial step in claiming these options is the protection of the product under a license that should be able to withstand a court test, based at least on current U.S. and European law. Concepts for such a license were provided. The defendant in a U.S. infringement case challenging this approach would likely argue that the limiting clauses in the license are illegal.[130] The plaintiff, on the other hand, would likely argue "fair use" parameters, contract law, and the intellectual property defenses of misappropriation, trade secret, and clean hands.[131] If a company writes the license to win even just one of these claims, that is enough to prevail for infringement protection. This is also known as the "one strike and you're out!" aspect of intellectual property law. The final determination on the validity of these concepts, both in the United States and the European Union, however, will be determined as international telecommunications trade and such intellectual property challenges increase.

ENDNOTES

[1]See specific provisions in the Paris Convention for the Protection of Industrial Property (1883, as amended until 1979), the Patent Cooperation Treaty of June 19, 1970, the Universal Copyright Convention (as revised at Paris on 24 July 1971), and the Berne Convention for the Protection of Literary and Artistic Works (Paris Act, 24 July 1971). See also the Trade-Related Aspects of Intellectual Property Rights (TRIPs) in the General Agreement on Tariffs and Trade as implemented by Congress, December 1, 1994.

[2]Paris Act, 24 July 1971.

[3]Programmers do this by disassembling the program. That is called "dumping," or copying the software code into computer memory and analyzing it with flow charts and line-by-line code comparisons. *See generally,* E.F. Johnson Co. v. Uniden Corp. of Am., Inc, 623 F. Supp. 1485, 1501 n. 17 (D. Minn. 1985).

[4]Kewanee Oil Co. v. Bicron Corp., 416 U.S. 470, 476 (1974).

[5]The *European Software Directive* at art. 6, as implemented by each country. The European Software Directive, as of September 24, 1993, compiled by Clifford Chance (an international law firm), 200 Aldersgate St., London EC1A 4JJ, ENGLAND. Phone: (44 71) 600-1000.

[6]35 U.S.C. §§ 1-376, 101 (1984).

[7]*Id.* §§ 102 and 103.

[8]*Id.* § 102(a).

[9] Peter B. Maggs, John T. Soma, and James A. Sprowl, Computer Law, West Publishing Co., St. Paul, MN, 1992, pp. 185-86.

[10]Ronald L. Johnston, and Allen R. Grogan; *Trade Secret Protection for Mass Distributed Software,* The Computer Lawyer, 1995, p 5.

[11]Trade-Mark Cases, 100 U.S. 82, 87, 25 L.Ed. 550.

[12]15 U.S.C. §§ 1058, 1059.

[13]The Uniform Trade Secrets Act with 1985 Amendments, §§ 1-12.

[14]*Id.* at §§ 1(4)(i) and (ii).

[15]Kewanee Oil Co., 416 U.S. at 489-90.

[16]*Id.* at 490.

[17]Uniform Trade Secrets Act with 1985 Amendments, §1(2).

[18]*Id.* at §1(4)(ii).

[19]*Id.*

[20]*See, e.g.,* Computer Print Systems, Inc. v. Lewis, 422 A.2d 148, 155-56 (Pa. Sup. 1980) (innocent recipient became aware, in course of dealing with owner, that plaintiff's former officer had not been authorized to make the duplicate software program); Components for Research, Inc. v. Isolation Products, Inc., 241 Cal. App. 2d 726 (1966) (defendant innocent directors put on notice, thus entitling relief, including injunction, as against them and defendant corporation on whose board they sat).

[21]Aronson v. Quick Point Pencil Co., 440 U.S. 257, 99 S. Ct. 1096 (1979).

[22]*Id.*

[23]Johnston, *supra* note 10, at 7 citing Computer Print Systems, Inc. v. Lewis, 422 A.2d 148 (Pa. Sup. 1980) and Components for Research, Inc. v. Isolation Products, Inc., 241 Cal. App. 2d 726 (1966).

[24]Stac Electronics v. Microsoft Corporation, 1994 U.S. App., Lexis 18042 (1994).

[25]Johnston, *supra* note 10, at 2; noting that pursuant to the settlement agreement between the parties, all orders, verdicts and judgments in the case have been vacated.

[26]"If willful and malicious misappropriation exists, the court may award exemplary damages in an amount not exceeding twice any award made under subsection (a)." Uniform Trade Secrets Act with 1985 Amendments, § 3(b).

[27]Johnston, *supra* note 10, at 1.

[28]*Id.* at 3-4.

[29]17 U.S.C. §§ 101-810 (1988 & Supp. IV 1992).

[30]*Id.*

[31]*Id.* at § 102.

[32]John T. Soma, Gus Winfield, Letty Friesen, *Software Interoperability and Reverse Engineering,* 20 Rutgers Computer & Tech. L.J. 189, 203 (1994).

[33]*See, e.g.,* Williams Elecs., Inc. v. Artic Int'l., Inc., 685 F.2d 870, 874 (3d Cir. 1982).

[34]17 U.S.C. § 101 (1988 & Supp. IV 1992).

[35]In Apple Computer, Inc. v. Formula Int'l. Inc., 725 F.2d 521, 525 (9th Cir. 1984).

[36]Johnston, *supra* note 10, at 5.

[37]*Id.*

[38]*Id.* at § 102. Section 102 of the Copyright Act of 1976.

[39]*Id.* at Sections 401-412.

[40]17 U.S.C. § 106 (1988 & Supp. IV 1992).

[41]*Id.* at § 117(2).

[42]*Id.* at § 106.

[43]*Id.*

[44]*See* Foresight Resources Corp. v. Pfortmiller, 719 F. Supp. 1006, 1010 (D. Kan. 1989) (defendant enjoined from selling enhancements of plaintiff's products to other entities) and Sega, 977 F.2d at 1520 ("Section 117 does not purport to protect a user who disassembles object code, converts it from assembly into source code, and makes printouts and photocopies of the refined source code version.").

[45]Baker v. Selden, 101 U.S. 99 (1879); Mazer v. Stein, 347 U.S. 201 (1954); Sony Corp. v. Universal City

Studies, Inc., 464 U.S. 417 (1984); and Feist Publications, Inc. v. Rural Tel. Serv. Co., 111 S. Ct. 1282 (1991).

[46]975 F.2d 832 (Fed. Cir. 1992).

[47]785 F. Supp. 1392 (N.D. Cal. 1992), *aff'd in part, rev'd in part,* 977 F.2d 1510 (9th Cir. 1992).

[48]In Atari, 975 F.2d at 842, the court stated that the purpose and policy of the copyright law is to encourage "authors to share their creative works with society."

[49]*Id.*, citing Bonito Boats, 489 U.S. 141 (1989). *infra* note 74.

[50]*Id.*

[51]Atari, 975 F.2d at 843.

[52]Sega, 785 F. Supp. at 1527-28.

[53]17 U.S.C. § 107 (1988 & Supp. IV 1992).

[54]*Id.* at § 107.

[55]One unique aspect of computer software programs is that they are not readable by humans while they are in their "object code" format of 1s and 0s. To be read by humans, the software must be in "source code" or "word" format. This requires copying the targeted software and translating it from object code to source code. While this (act of) copying violates the copyright prohibition against copying, the courts have held that this form of interim copying is a "fair use" exception to copyright infringement. Atari, Sega, Galoob, Foresight, and NEC effectively overturn the interim copy analysis in Hubco and Walt Disney Prods. v. Filmation Assocs. (1986) (which held that the Copyright Act prohibits the creation of interim copies). From the interim copy, new insights are incorporated and new emphasis added, resulting in a fair use derivative product (Maggs, *supra* note 9, p. 2).

[56]*Id.* at §§ 106 and 117.

[57]In E.F. Johnson Co. v. Uniden Corp. of America, 623 F. Supp. 1485 (D. Minn. 1985); Midway Mfg. Co. v. Artic International, Inc., 547 F. Supp. 999 (N.D. Ill. 1982); Williams v. Arndt, 626 F. Supp. 571 (D. Mass. 1985); Hubco Data Prods. Corp. v. Management Assistance Inc., 219 U.S.P.Q. (BNA) 450 (D. Idaho 1983); and Telerate Sys., Inc. v. Caro, 689 F. Supp. 221 (S.D.N.Y. 1988) the court found infringing end products and held against the defendant even though reverse engineering had occurred.

[58]NEC Corp. v. Intel Corp., 10 U.S.P.Q.2d (BNA) 1177, 1184 (N.D. Cal. 1989).

[59]"Milton R. Wessel, *Introductory Comment on the Arizona State University Last Frontier Conference on Copyright Protection of Computer Software,* 30 Jurimetrics J. 1, 23 (1989).

[60]Illustrations of this test can be seen in E.F. Johnson, 623 F. Supp. 1485; Midway Mfg. Co., 547 F. Supp. at 1014; and Hubco, 219 U.S.P.Q. (BNA) at 452. But, also see Foresight Resources Corp. v. Pfortmiller, 719 F. Supp. 1006 (D.Kan 1989), for contrast to Hubco.

[61]Williams, 626 F. Supp. 571, 579 (D. Mass. 1985).

[62]Soma, *supra* note 32, at 207, citing *Williams,* 626 F. Supp. at 574.

[63]17 U.S.C. § 101 (1988).

[64]17 U.S.C. § 107 (1988 & Supp. IV 1992). *See also,* Rosemont Enterprises, Inc. v. Random House, Inc., D.C.N.Y., 256 F. Supp. 55, 65, 66.

[65]Sega, 977 F.2d at 1521-27. Software developers should note the Sega analysis of these four factors. Further, the market impact analysis of Harper & Row, 471 U.S. at 562; Sony, 464 U.S. at 451; and Lewis Galoob Toys, Inc. v. Nintendo of Am., Inc, 964 F.2d 965, 969 (9th Cir. 1992), *cert. denied,* 113 S.Ct. 1582 (1993) should be explored.

[66]*Id.* at § 501(a). Infringement is defined as a violation of any of the exclusive rights of the copyright owner.

[67]17 U.S.C.A. § 502(a) (1984).

[68]*Id.* § 504, 505.

[69]*Id.* at § 503.

[70]17 U.S.C.A. § 506 (1984).

[71]The Copyright Act does not extend to the ideas underlying a work or to the functional or factual aspects of the work. 17 U.S.C. § 102(b). "In determining whether a product feature is functional, a court may consider a number of factors, including—but not limited to—the availability of alternative designs: and whether a particular design results from comparatively simple or cheap method of manufacture." [Sega 1977 F.2d at 1531

citing Clamp Mfg. Co. v. Enco Mfg. Co., Inc., 870 F.2d 512, 516 (9th Cir.), cert. denied, 493 U.S. 872107 l. Ed. 2d 155, 110 S. Ct. 202 (1989)].

[72]This standard was set in See v. Durang, a nonsoftware, basic copyright case where the plaintiff author claimed that the defendant's play was based on a draft script written by the plaintiff. Rather than looking at the interim copies or process used to write the play, the court compared the two plays (end products) as a whole and found no infringement of expression.

Where the interim copies do not meet the two-prong standard and are thus held to be technically illegal, they are subject to standard royalty payments or statutory damages. (Soma *supra* note 32, p. 212), but the noninfringing end products do not need to be licensed from the plaintiff. This occurs when a program is loaded into a computer's memory thus making a copy. *See also* Bly v. Banbury Books, Inc., where the court assessed minimum statutory damages of $250.00 and partial attorney's fees for making such a copy.

[73]Bly v. Banbury Books, Inc., 638 F. Supp. 983 (E.D. Pa. 1986) (holding that the defendant used the interim copy for commercial gain.)

[74]Bonito Boats, Inc. v. Thunder Craft Boats, Inc., 489 U.S. 141 (1989).

[75]Kewanee, 416 U.S. at 489-490.

[76]Bonito Boats, 489 U.S. at 157 citing their decisions in Sears and Compco.

[77]Sega Enterprises v. Accolade, Inc., 977 F.2d 1510, 1518 (9th Cir. CA 1992).

[78]European Software Directive, *supra* note 4, at art. 6.

[79]Soma, *supra* note 32, at 223.

[80]Lasercomb Am., Inc. v. Reynolds, 911 F.2d 970 (4th Cir. 1990).

[81]*Id.* at 971.

[82]*Id.* at 972.

[83]*Id.* at 971.

[84]*Id.*

[85]Lasercomb, 911 F.2d at 979.

[86]*Id.* at 978.

[87]PRC Realty Systems Inc. v. National Ass'n of Realtors (766 F. Supp. 453, 456 (E.D. Va. 1991), aff'd, No. 91-1125, 91-1143, 1992 WL 183682 (4th Cir. 1992) (per curiam).

[88]*Id.*

[89]*Id.*

[90]*Id.* at 458.

[91]PRC Realty Systems Inc. v. National Ass'n of Realtors, No. 91-1125, 91-1143, 1992 WL 183682 (4th Cir. 1992) (per curiam) at 12.

[92]*Id.*

[93]689 F. Supp. 221 (S.D.N.Y. 1988).

[94]*Id.* at 224.

[95]*Id.*

[96]*Id.* at 225.

[97]*Id.* at 226.

[98]*Id.* at 240.

[99]Maggs, *supra* note 9, at 224 citing Somerset Importers, Ltd. v. Continental Vintners, 790 F.2d 775, 781-82 (9th Cir. 1986); Quiller v. Barclays American/Credit, Inc., 764 F.2d 1400, 1403 (11th Cir. 1985) (dissenting opinion), cert. denied, 476 U.S. 1124 (1986).

[100]Atari, 975 F.2d at 846.

[101]Atari, 975 F.2d at 846-47.

[102]*See, e.g.,* Campbell Soup Co. v. Wentz, 172 F.2d 80, 83 (3d Cir. 1948).

[103]*Id.*

[104]*See* Jere M. Webb & Lawrence A. Locke, *Intellectual Property Misuse: Developments in the Misuse Doctrine,* 4 Harv. J.L. & Tech. 257, 257-58 (1991).

[105]*See* William M. Landes and Richard A. Posner, *Market Power in Antitrust Cases,* 94 Harv. L. Rev. 937 (1981).

[106]In 1988, Congress passed the Patent Misuse Reform Act, which requires that before misuse will be found for tying, a patent owner must have market power in the relevant market for the patented product on which a license or sale is conditioned. If so, a "rule of reason" analysis must be presented for tying activity. Copyright misuse analysis follows that of patent misuse.

[107]Preamble: European Directive on the Legal Protection of Computer Programs, submitted by the

European Commission to the European Council on January 5, 1989.

[108]*Id.*

[109]European Software Directive, *supra* note 4.

[110]*Id.* at Recital 10.

[111]*Id.* at Recital 12.

[112]*Id.* at Recital 14.

[113]*Id.* at art. 4.

[114]*Id.* at art. 8 and Recital 25.

[115]*Id.* at art. 5.1 and Recitals 17 and 18.

[116]*Id.* at art. 5.2.

[117]*Id.* at. art. 5.3.

[118]*Id.* at Recital 19.

[119]*Id.* at art. 6.1(a) and Recitals 21 and 22.

[120]See Sega and Atari, *supra* notes 48 and 52.

[121]*Id.* at Recital 22.

[122]*Id.* at Recital 24.

[123]*Id.* at Recital 24.

[124]*Id.* at art. 6(1) and (2) and Recitals 21 and 22.

[125]*Id.* at art. 5.3.

[126]*Id.* at art. 5.2.

[127]*Id.* at Recital 27.

[128]*Id.* at Recital 20.

[129]Sega, 977 F.2d at 1520 n6.

[130]See the Fourth Circuit cases.

[131]See the Ninth Circuit cases.

CHAPTER 8 Privacy

Scenario 1: *A man stole Carroll Oberst's Social Security number and other personal data off the Internet, opened bank and credit card accounts in Oberst's name, and ran up $35,000 in bills. With the easy availability of Social Security numbers and maiden names, the likelihood of this happening to more people is increasing. What laws exist to protect individuals from this?*

Scenario 2: *For years, Bronti Kelly, 34, could not figure out why no one would hire him. An erroneous report, mislabeling him a shoplifter and uncorrected by the May Co., had continually been made available on the Internet to potential employers. Because of its failure to correct the misinformation, the May Co. was ordered in January 1998 to pay more than $73,000 to Kelly, who ended up homeless when he could not get a job. By law, what notices must be provided to individuals regarding information distributed about them? What opportunity must they be given to correct misinformation?*

Scenario 3: *Veteran sailor, Timothy R. McVeigh, faced expulsion for homosexuality from the U.S. Navy in early 1998, based on evidence the Navy gathered from America Online (AOL), the United States's largest computer online service. McVeigh, since reinstated, had put his sexual preferences in an online profile he thought was confidential. When system managers at AOL released his name at the Navy's request, McVeigh sued, claiming that his privacy was breached. A judge agreed and ruled that the Navy violated the 1986 Electronic Communications Privacy Act by obtaining the confidential information about McVeigh from AOL without a warrant. AOL paid a significant fine for the violation.*

8.1 INTRODUCTION

Although modern telecommunications technologies, such as computers, the Internet, and wireless communications provide tremendous convenience and tools for productivity, they also raise numerous concerns and legal issues. These concerns and issues generate new responsibilities for system managers, new challenges for law enforcement, and new questions for individuals. One of the most critical of these legal issues is privacy.

Nearly every one of us willingly provides very detailed private information about ourselves on loan applications; on medical, dental, and insurance forms; and to friends, clergy, and counselors. In doing so, however, we expect control over (1) what information is collected, (2) how it is collected, (3) how it is used, (4) for what purpose it is used, (5) to whom the information is revealed or shared, (6) what information is retained in files, (7) how long the information is kept, (8) the security of the information while it is being stored to protect against the theft of the information in the system, (9) continued accuracy of the information in the file, and (10) our right to review the information and to correct errors in it. In short, we are willing to share private information under certain circumstances, but we do not want that information to be widely publicized without our permission. We expect our communications, whether by phone, email, the Internet, or other means, to be as secure as messages in a sealed envelope.

However, with computerized databases, technology provides for the first time in history the capability to collect, organize, and store massive amounts of personal information in databases over which individuals have little or no control. Modern telecommunications networks allow that information to be accessed, analyzed, and transported across the country or around the world at speeds never before possible. Information that people once believed was private and confidential is now being compiled and used without their knowledge or control. These issues affect both customers and providers of the systems.

The term *privacy,* used in telecommunications law today, encompasses these expectations by individuals to control the confidentiality, accuracy, and application of their personal information. Currently, however, U.S. law allows the extensive collection and use of such information without the permission of the individual. This is because U.S. law tends to view personal data as a resource to help companies identify potential customers and to serve existing customers. Other countries, such as the members states of the European Union, however, view privacy as more of a human rights issue. U.S. privacy law operates mainly on privacy statements offered by companies to their customers and industry codes of conduct, rather than a concerted, directed body of law. This is unacceptable to other countries, and as the Internet and electronic commerce expand, other countries, including the European Union, have placed embargoes on personal data flows to the United States until the United States increases its protections for personal data.

The purpose of this chapter is to provide an overview of (1) the current status of U.S. privacy law, (2) major areas not addressed by U.S. law, (3) the laws' application to modern telecommunications technologies, (4) a brief comparison of U.S. law to privacy law in the European Union, and (5) the direction that U.S. privacy law is heading.

To do this, Section 8.2 discusses the evolution of the concept of privacy in U.S. law, highlighting the conceptual and philosophical foundations behind the major federal and state privacy laws, including tort law and the evolving constitutional recognition of privacy. Sections 8.3 through 8.8 describe the major U.S. privacy laws existing at this time. Two key points in these sections are that the existing U.S. privacy laws cover very few (six) items and that they address mainly government action, not private action. As discussed in Section 8.3, U.S. privacy laws address the privacy of communications from wiretapping or interception while "in transit" or "in storage." As discussed in Section 8.4, U.S. privacy laws address the privacy of personal information in government databases, but not in most private or corporate databases. The privacy of personal financial information held by banks, credit-reporting companies, and other financial institutions insured by the Federal Deposit Insurance Corporation (FDIC) is discussed in Section 8.5. The protection of information in private databases from search and seizure by government officials, but not by private inquirers is outlined in Section 8.6. Section 8.7 discusses "protected" computers, while Section 8.8 discusses protections against harassing communications and Section 8.9 surveys the privacy of information about children. Section 8.10 provides a list of the privacy laws in the various states, while Section 8.11 explores major privacy concerns not covered by current U.S. law. Section 8.12 discusses solutions to this areas, while Section 8.13 applies the privacy laws to modern telecommunications technologies such as cellular and cordless phones, pagers, and email and voicemail systems. It includes the new responsibilities and limitations placed on the systems' operators and managers by the privacy laws. Section 8.14 then compares the U.S. and international privacy laws, focusing mainly on the European Union's approach and citing proposals to resolve the resulting issues.

8.2 THE EVOLUTION OF A LEGAL RIGHT TO PRIVACY

Privacy law in the United States has evolved only in the past 100 years, since the late 1800s, brought about by the development of technology. It has occurred in five main phases: (1) no specific law; (2) tort law—establishing damages for the "invasion of one's personal privacy"; (3) Supreme Court decisions recognizing a constitutional right of privacy; (4) recognition by the Supreme Court of other balancing considerations including consent, the public's right to know certain information, and law enforcement's need to counter the use of technology in committing crimes; and (5) the recognition of the public's "expectation of privacy" with technology. The next five sections address each of these phases.

8.2.1 Privacy in the U.S. Constitution

The concept of individual or personal privacy, as a legal "right," is a relatively new concept in U.S. law. For most of human history, the right of individuals to be private about personal matters was not a significant legal issue.[1] Residents of villages and small towns lived with the fact that everyone knew their business, including their past, their families, their current activities, and their expected potential for the future. Individuals lived under old, often undeserved animosities or reputations, for both themselves and their families. The only advantage was that no system existed to collect or track significant amounts of personal information, so individuals generally could leave town and establish new identities elsewhere. Persons who migrated to a new village, country, or frontier region, such as the American West, the Yukon, the outback of Australia, or other sparsely populated areas, experienced relative degrees of privacy. As a result, the United States did not inherit a concept of a "legal right of privacy for individuals" from its roots in Europe and thus made no specific mention of privacy in the U.S. Constitution or early common law.

8.2.2 Invasion of Privacy—Tort Law

As telephones evolved in the 1880s and newspapers increased their circulation, personal privacy became an increasingly important issue. Early telephones were considered by many to be "an invention of the devil" because they interrupted an individual's privacy. For the first time in history, people could call others at any time. As Ambrose Bierce stated in his book *The Devil's Dictionary, "a ringing phone interrupts both domestic peace and thoughtful reflection* [and] *abrogates some of the advantages of making a disagreeable person keep his distance."*[2]

The first generally recognized legal discussion of personal privacy in the United States appeared in 1890, in a *Harvard Law Review* article written by two Harvard law professors, Samuel D. Warren and Louis D. Brandeis. Brandeis later became a U.S. Supreme Court justice, serving from 1916 to 1938. The article was written in reaction to the Boston tabloids' coverage of Professor Warren's wife's social activities.[3] In the article, Warren and Brandeis advocated that U.S. law should recognize a civil tort action for the "invasion of one's privacy."[4] The two law professors cited then-existing common law cases in which "relief had been afforded on the basis of defamation, invasion of some property right, or breach of confidence or an implied contract." They concluded that the cases were based on a broader principle of a "right to personal privacy" that was entitled to separate recognition in U.S. law.[5]

Their idea was so compelling that over the next 90 years, by the end of the 1970s, 49 of the 50 states passed laws recognizing the tort of Invasion of Privacy.[6] Only Rhode Island did not. Four specific tort actions define the more general tort of Invasion of Privacy. The four are unreasonable *intrusion* into another person's private life, unreasonable publicity or public *disclosure* of embarrassing facts about another person's private life, publicity that unreasonably places another person in a *false light* before the public, and *appropriation* of another person's name or likeness, without the person's permission, for commercial purposes.[7] Each is described briefly in the following four sections.

Unreasonable Intrusion

Intrusion is defined as the intentional and unreasonable invasion or intrusion upon the seclusion of another.[8] Seclusion can be either physical seclusion or the disclosure of an individual's private affairs in a manner that would be "highly offensive to a reasonable person." However, if the disclosed information is public, or if the conduct is embarrassing, but not unreasonable, no liability is found. Like more recent privacy laws, the tort of unreasonable intrusion limits unauthorized access to communications and information. For years, telephone wiretaps have been suspect under this tort.[9] In the Internet or on-line context, hacking, viruses, and junk email are all considered forms of intrusion.

Unreasonable Public Disclosure of Private Facts

Public disclosure is the intentional and unreasonable publication or disclosure of private information in a manner that makes the information public knowledge. The expectation of privacy required for this tort is lower than the level required for intrusion. However, the level of publicity required is greater than the level required for the tort of defamation, because it requires disclosure to more than one other person. For example, the names of rape victims are not made public to avoid such public disclosure and the names of possible suspects in a crime generally are withheld until an arrest is made. Unwilling participants in shock radio, where the interviewer intentionally asked, suggested, or disclosed private information in order to provide provocative programming, have also considered this tort. It is also used to protect information that a person has a disease, such as AIDS.[10]

False Light

False light is publicity that unreasonably places another person in a false light or gives false perceptions of the person in the eyes of the public, even if the publicity is not defamatory. It is not necessary that the information revealed be private or that the false information embarrasses or injures the reputation of the person. It must be publicity that would offend a reasonable person. To prevail in a "false light" case, the plaintiff must prove that the defendant intentionally published or disseminated a false statement with reckless disregard for the truth to a reasonable number of third persons. This is a higher standard of proof than the common law standard of "knowing" behavior.

A case that demonstrates "false light" is *Douglass v. Hustler Magazine, Inc.,*[11] in which *Hustler Magazine* published photographs of Douglass without her permission. The court found that the implication that Douglass was associated with the magazine could be highly offensive to a reasonable person. "False light" is of special importance in the Internet age, where private or false information about a person can be easily disseminated with the touch of a button.

Appropriation of a Person's Name or Likeness

Appropriation of another person's name or likeness for personal or commercial gain is similar to other property cases. It recognizes that each of us has certain unique identifying information, including our name and image, that cannot be used by another person for personal gain without our permission.[12]

A 1902 example of "appropriation" occurred when a New York flour company used a woman's picture to advertise their flour. She had not given the company permission for its use and sued under this tort. The lower court, however, denied her any recovery, citing the flour company's right to free speech. However, public outcry disapproved of the decision and, in response, the New York legislature passed a law making such unapproved use of a person's name or likeness a misdemeanor and a tort.

In a more modern case, however, a "human cannonball" protested the full public broadcast of his act because such broadcast threatened the economic value of his performance. The court agreed, stating that no First Amendment right exists to do so. In general, however, public persons have less privacy because of implied consent by becoming a public figure.[13]

The best-known examples today usually involve celebrities, whose names and pictures have commercial value. If a product or Web site were to use the celebrity's name or image, even just as a reference rather than an endorsement, the celebrity is entitled to damages for interference with his or her right of publicity.[14]

These four tort concepts have been widely applied to the developing technologies of telephone, radio, and newspaper and established privacy law for the first 70 years, from the late 1800s to the 1970s. However, as new technologies were developed, other, more specific privacy laws have been enacted.

8.2.3 Supreme Court Interpretations of Constitutional Amendments Recognizing a Right of Privacy

During the 50 years from the 1920s through the 1970s, the United States Supreme Court heard a series of cases in which it considered personal privacy issues.[15] As the court stated in *Roe v. Wade*[16]

The Constitution does not explicitly mention any right of privacy. In a line of decisions, however, . . . the Court has recognized that a right of personal privacy, or a guarantee of certain areas or zones of privacy, does exist under the Constitution. . . . These decisions make it clear that only personal rights that can be deemed "fundamental" or "implicit in the concept of ordered liberty,"[17] are included in this guarantee of personal privacy. The court also makes it clear that the right has some extension to activities relating to marriage, *Loving v. Virginia,*[18] procreation, *Skinner v. Oklahoma,*[19] contraception [*Griswold v. Connecticut,*[20] *Eisenstadt v. Baird,*[21] and *Carey v. Population Services Int'l.*]; family relationships, *Prince v. Massachusetts* [and *Moore v. City of East Cleveland, Ohio*];[22] and child rearing and education, *Pierce v. Society of Sisters,*[23] and *Meyer v. Nebraska.*[24]

Twenty years later, a California court echoed the interpretation of these amendments in a 1994 case, *Hill v. National Collegiate Athletic Ass'n.,*[25] stating that modern privacy law evolved from both common law and constitutional law. In it, the court noted that privacy is "a federal constitutional right, derived from various provisions of the Bill of Rights." Hence, these Supreme Court decisions created the underlying constitutional interpretation of privacy law of the United States.

Zones of Privacy—*Griswold v. Connecticut*[26]

One specific concept that was noted in the later decisions is that of *zones of privacy*. This concept was first raised in 1965 in the case of *Griswold v. Connecticut*. In *Griswold,* the U.S. Supreme Court reviewed a Connecticut statute that prohibited the use of contraceptive devices by married couples and the dissemination of medical advice about contraceptives by the couples' physicians. The Supreme Court invalidated the Connecticut law, citing the Due Process Clause of the Fourteenth Amendment and stating the following:

> *Marriage is a relationship lying within the 'zone of privacy' created by several fundamental constitutional guarantees. And, it concerns a law which, in forbidding the use of contraceptives, rather than regulating their manufacture or sale, seeks to achieve its goals by means having a maximum destructive impact upon that relationship . . . Would we allow the police to search the sacred precincts of marital bedrooms for telltale signs of the use of contraceptives? The very idea is repulsive to the notions of privacy surrounding the marriage relationship. . . . We deal with a right of privacy older than the Bill of Rights—older than our political parties, older than our school system. . . .*
>
> *. . . In reaching the conclusion that the right of marital privacy is protected, as being within the protected penumbra of specific guarantees of the Bill of Rights, the Court refers to the Ninth Amendment. I add these words to emphasize the relevance of that Amendment to the Court's holding.*
>
> *. . . This Court, in a series of decisions, has held that the Fourteenth Amendment absorbs and applies to the States those specifics of the first eight*

> *amendments which express fundamental personal rights. The language and history of the Ninth Amendment reveal that the Framers of the Constitution believed that there are additional fundamental rights, protected from governmental infringement, which exist alongside those fundamental rights specifically mentioned in the first eight constitutional amendments.*

In this landmark decision the Supreme Court stated that "zones of privacy" exist into which the government may not easily enter.[27]

Inviolability of Privacy

In *NAACP v. Alabama*,[28] the Supreme Court recalled that in *American Communications Assn. v. Douds*,[29] it recognized the "vital relationship between freedom to associate and privacy in one's associations." In *NAACP*, the Supreme Court ruled against compelled disclosure of membership in an organization engaged in advocacy of particular beliefs stating that "inviolability of privacy in group association may in many circumstances be indispensable to preservation of freedom of association, particularly where a group espouses dissident beliefs."

8.2.4 Balance of Personal Privacy with Other Considerations

In discussing a "right to privacy," the Supreme Court also recognized other balancing considerations including consent, the right to collect certain information for valid purposes, the public's "right to know" certain information (which has led to the Freedom of Information Act), and law enforcement's need to know certain information to protect the public against such things as terrorists, drug dealers, and child pornographers. This has led to wiretap and similar laws. The requirement to balance these four "needs" against privacy concerns complicates privacy law and raises many of the current privacy issues today. These four needs are discussed in the following four sections.

Consent

In any privacy case, the issue of consent, both express and implied, must be considered. With express consent, one provides verbal or written permission to access or use the personal information in question. Implied consent, on the other hand, is given by one's actions or inactions.

All of us explicitly provide detailed private information about ourselves on applications and to friends. However, we may not realize that we also implicitly share information in the use of credit cards, the Internet, and public phones.

In general, courts have ruled that many public activities suggest implied consent because the public knowledge of the action gives it less protection than a private action. However, each case depends on the details of the situation and must be decided on a case-by-case basis. In today's telecommunications environment, many companies include statements requesting written permission to use certain customer information in their sales and use agreements. However, customers often do not notice these or sign the request thinking it is a service agreement.

Right to Collect Information to Provide Service

The courts have also pointed out that all companies have a legitimate need to collect, retain, and possibly share personal information about their customers in order to bill them for the service they provide and to offer items appropriate for each customer. Noting this, Congress has avoided writing broadly worded privacy laws.

Public Right to Know

Third, the courts have pointed out that the public has a right to know certain information, especially about the activities of its government agencies and some public persons. This has led to "sunshine" laws, concerning open records, files and meetings and the Freedom of Information Act (FOIA). FOIA is discussed in more detail in Section 8.4.

8.2.5 Supreme Court Interpretation of Fourth Amendment Regarding Wiretapping

As telephones, computers, encryption, and other new technologies have evolved, criminals have used them to commit crimes. To counter this use, law enforcement officials have adopted techniques such as wiretapping and restrictions on encryption. However, these affect personal privacy and raise questions about what information should be collected and how. In addressing these questions, the courts have had to balance law enforcement's reasonable needs against the rightful privacy needs of individuals. In doing so, the courts have considered numerous arguments, including whether such activity is protected by Fourth Amendment restrictions against government searches and seizures.

The Fourth Amendment of the United States Constitution states that

> *The right of the people to be secure in their persons, houses, papers, and effects, against unreasonable searches and seizures, shall not be violated, but no Warrant shall issue, but upon probable cause, supported by Oath or affirmation, and particularly describing the place to be searched, and the persons or things to be seized.*

Thus, the Fourth Amendment requires court-issued warrants for all searches and seizures, based on probable cause and a specifically identified purpose and place for each search. Furthermore, since the courts issue the warrants, all searches and seizures are reviewed and enforced by the courts. This has proven fortunate for telecommunications issues because the courts usually can adapt to the requirements of new technology faster than Congress or state legislatures can legislate. As a result, most privacy law in the United States established in the 90 years between 1880 and 1970 was based on common law decided by the courts rather than on statutory law enacted by Congress. Following are descriptions of some of the key cases in the evolution of wiretap versus privacy law in the U.S.

Olmstead v. United States (1928)

In 1928, the U.S. Supreme Court addressed the legality of the new practice of "wiretapping" by law enforcement officials in a case known as *Olmstead v. United States.*[30] Specifically, the case addressed whether the Fourth Amendment of the U.S. Constitution protected individual privacy from wiretaps.

Surprising to many, the Supreme Court ruled that the Fourth Amendment did not protect the interception of messages on telephone wires, but only protected the physical entry of a home and the seizure of tangible items. Thus, in *Olmstead,* the Supreme Court determined that wiretapping or "the interception of messages passing over telephone wires" did not constitute a "search" or "seizure" subject to the Fourth Amendment and, as such, required no court-issued warrants for wiretaps.

Supreme Court Justice Brandeis dissented, pointing out that constitutional "clauses guaranteeing to the individual protection against specific abuses of power must have a . . . capacity of adaptation to a changing world," in which "subtler and more far-reaching means of invading privacy have become available."

The Communications Act of 1934

Responding to the *Olmstead* decision, Congress, in 1934, sought to provide greater privacy for communications transmitted over telephone and telegraph lines by including a provision in Section 605 of the Communications Act of 1934, providing that "no person not being authorized by the sender shall intercept any communications and divulge or publish [its] existence, contents, substance, purport, effect, or meaning . . ." The term *communication* was defined to include telephone, telegraph, and radiotelegraph communications. Justice Brandeis agreed with this direction, but considered Section 605 too narrow because "the progress of science in furnishing the Government with means of espionage will not stop with wiretapping."[31]

Goldman v. United States (1942)

In 1942, eight years after Congress added Section 605 to the Communications Act of 1934 restricting interception of a communication during transmission, the Supreme Court heard the case of *Goldman v. United States,*[32] in which it considered the question of whether electronic eavesdropping of a voice conversation in a phone booth was a violation of the Fourth Amendment or not.

The Supreme Court ruled that such electronic eavesdropping was not a violation of the Fourth Amendment. This decision continued for the next 25 years until the "expectation of privacy" cases in 1967. Thus, from 1876–1967 wiretapping by the federal government was legal, even without a warrant.

Expectation of Privacy (1967)

In 1967, 40 years after *Olmstead* and 25 years after *Goldman,* the Supreme Court reconsidered wiretapping in two cases in which it reversed itself by deciding that the Fourth Amendment does apply to "people, not places" where a "reasonable expectation of privacy" exists. The two cases were *Katz v. United States*[33] and *Berger v. New York.*[34]

In *Katz v. United States,*[35] the Supreme Court again reviewed whether the Fourth Amendment applied when federal agents obtained a recording of Katz's end of a telephone conversation by attaching an electronic listening device to the outside of a public telephone booth. The Court decided that the Fourth Amendment does cover the interception of telephone conversations by the government when a "reasonable expectation of privacy" exists.[36] The Court distinguished *Katz* from *Goldman* by ruling that technology and the use of phones had advanced to the point where Katz had "justifiably relied on an expectation of privacy" in the phone booth, and in such cases, the Fourth Amendment protected "people, not just places, such as homes."

In the second case, *Berger v. New York,*[37] heard the same year, the Court expanded coverage of the Fourth Amendment by deciding that it protected all electronic eavesdropping of "aural" conversations. It defined *aural communications* as "any wire or oral communications that can be overheard or understood by the human ear."

In these two cases the Supreme Court established three important legal principles. First, the Court decided that a communication is private when at least one party reasonably expects it to be private. Second, the Court prohibited the interception by government and electronic eavesdropping of aural communications carried by common carriers. Third, the Court created a flexible standard adaptable to changing technologies. With these three principles, the Supreme Court established a legal foundation in electronic privacy that was codified one year later in Title III of the Omnibus Crime Control and Safe Streets Act of 1968 and expanded nine years later in the Electronic Communications Privacy Act of 1986.

8.3 FEDERAL PRIVACY LAWS PROTECTING COMMUNICATIONS IN TRANSIT AND IN STORAGE

Acknowledging the Supreme Court decisions in *Katz* and *Berger,* Congress passed two important federal laws protecting communications from eavesdroppers while the communications are in transit from the sender to the receiver and/or in storage. These two are the Omnibus Crime Control and Safe Streets Act of 1968 and the Electronic Communications Privacy Act (ECPA) of 1986. These two laws are balanced by the Communications Assistance for Law Enforcement Act (CALEA) of 1994, Congress' acknowledgment of law enforcement's need to counter the use of technology in committing crimes. The three laws are discussed in the following three sections.

8.3.1 Omnibus Crime Control and Safe Streets Act of 1968,[38] Including the Wiretap Act of 1968[39]

The Omnibus Crime Control and Safe Streets Act (Crime Control Act) of 1968 was signed into law on June 19, 1968 in response to the Supreme Court decisions in *Katz v. United States*[40] and *Berger v. New York*.[41] Title III of the "Crime Control Act" is known as the Wiretap Act of 1968.[42] The Wiretap Act, codified at 18 U.S.C. §§ 2510-21, prohibits unauthorized wiretaps or "the actual or attempted willful interception of aural communications while being transmitted by common carriers."[43] Title III specifies the "circumstances and conditions under which wiretaps or the interception of wire and oral communications may be authorized," stating that to protect the privacy rights of individuals, all wiretaps require a court order and careful supervision by a federal court.

8.3.2 Electronic Communications Privacy Act (ECPA) of 1986[44]

The Electronic Communications Privacy Act (ECPA) of 1986[45] updated the Wiretap Act of 1968 and thus is also codified at18 U.S.C. §§ 2510-21 and 2701-10. Specifically, it expanded the coverage of the Wiretap Act by adding protection of electronic communications "in storage" to the privacy protection of the Wiretap Act. Thus, the ECPA has two main parts: Title 1,[46] addressing the unauthorized interception of electronic communications while "in transit," and Title 2,[47] addressing the unauthorized acquisition of electronic communications while "in storage." Other smaller sections of the ECPA also exist addressing (1) satellite communications,[48] (2) federal surveillance,[49] (3) mobile tracking devices,[50] and (4) pen registers.[51] Pen registers, also called "dialed number recorders," record dialed telephone numbers as ink dashes on a paper tape. They also include telephone numbers dialed by touch-tone telephones, although these numbers are actually recorded by touch-tone decoders.

In addition, the ECPA expanded the coverage of the Wiretap Act from public common carriers to all private networks, thereby including all intracompany communications in the protections of wire communications provided by the ECPA.[52] Interceptions of communications made outside the United States, however, are not within the scope of the ECPA, although U.S. interstate communications "affecting interstate or foreign commerce" are included.[53]

With these updates and expansions of the Wiretap Act, the ECPA became the predominant federal law protecting electronic communications from unauthorized interception, use, and disclosure. As such, it (1) regulates the activities of all transport or conduit communications carriers regarding privacy, (2) outlines the duties, rights, and privileges of the custodians of information, and (3) allows government officials to access, use, and disclose electronic communications only with a valid court order.[54] Although the name of the ECPA mentions only "electronic communications," the ECPA covers both wire communications and electronic communications.

Wire (Voice) Communications

Wire communications, as defined in the Electronic Communications Privacy Act (ECPA)[55] are limited to voice or oral communications over wire, cable, and other like connections. This definition of wire communications excludes three items. First, it excludes voice communications over wireless systems, such as cellular telephones or the radio portion of cordless telephones,[56] because the conversation over wireless transmission is too easy to intercept.[57] Second, it excludes communications transmitted using modulation techniques withheld from the general public[58] to protect the privacy of communications such as "spread spectrum" radio communications, scrambled or encrypted communications. Third, the definition excludes wire used only in terminal equipment at either end of the conversation as, for example, to amplify one end or because the recorder contains wires. The definition of wire communications does not exclude, however, voice conversations converted from analog to digital form for transmission. Such technical conversion does not take the matter outside the scope of Title I. Similarly, the definition includes "voicemail" because it is electronic storage of aural communications that is simply incidental to its transmission.

Electronic (Data) Communications

Electronic communications are defined in the ECPA as data communications carried over both wire and wireless systems. Electronic communications differ from wire communications in that they are communications that are not transmitted by sound waves and cannot be characterized as containing a human voice. Instead, they include telegraph, telex communications, electronic mail, nonvoice digitized transmissions, and the portion of video teleconferences that do not involve the hearing of voice or oral sounds.

Electronic communications include

> *any transfer of signs, signals, writing, images, sounds, data, or intelligence of any nature transmitted in whole or in part by wire, radio, electromagnetic, photo electronic, or photo-optical system that affects interstate or foreign commerce, but does not include: (A) any wire or oral communications, (B) any*

communication made through a tone-only paging device, or (C) any communication from a tracking device as defined in section 3117 of this title.[59]

Mixed Communications

Although the ECPA does not define "mixed communications," the term is frequently used to describe communications that are partly electronic and partly wire communications. For example, a telephone call containing some voice communications and some data communications over the same circuit would be considered an electronic communication for the data portion and a wire communication for the voice portion.[60]

Similarly, closed-circuit television systems frequently are *mixed communications,* since the television picture, transmitted over either wire or wireless systems without the voice or "oral" portion, would be considered an "electronic communication." This is often the case with video surveillance cameras. Since such transmissions have no oral portion, they are not considered wire communications, even though they might be carried over wire. Also, without an oral portion, viewers cannot be guilty of intercepting or wiretapping the communications. Hence, surveillance cameras are not covered by the ECPA. As such, providing closed-circuit television pictures of a meeting is not "illegal interception" under the ECPA because no wire communications were intercepted. On the other hand, if voice were transmitted with the picture, the ECPA would apply to the audio portion as an interception of an oral communication.[61] Thus, video cameras typically do not qualify as surveillance devices under the law.[62]

Paging systems also vary. The ECPA acknowledges three types of paging: tone only, voice, and display pagers. *Tone only* pagers are not covered by the ECPA because they provide no oral or stored communication to be intercepted and users have no reasonable expectation of privacy for the tone. *Voice pagers,* on the other hand, are addressed in Title 1 as a continuation of the original wire communications,[63] and *display pagers* are covered as electronic communications since they have no oral portion. Acquiring the telephone number from a digital pager also does not constitute interception under the ECPA because no separate device is used and because transmission over this system is considered to have ceased and the number stored.[64]

Title I—Protects Voice and Data Communications in Transit

Title I of the ECPA protects both "wire" (voice) and "electronic" (data) communications streams while in transit. To do this, it prohibits intentional[65] (1) interception of a protected communication, (2) use of an interception device related to oral communications, (3) disclosure of the contents of a transmission to others by service providers providing electronic communication service to the public, (4) use of the contents of an intercepted communications, (5) distribution of a communication to anyone other than the addressee or intended recipient by public electronic means without permission, and (5) manufacture, assembly, distribution, possession, or sale of devices designed to be "primarily useful for the purpose of the surreptitious interception of wire, oral, or electronic communications."[66]

However, Title I allows service providers and system operators to intercept, disclose, and use communications when necessary to (1) record or note the fact that a wire or electronic communication was initiated or completed; (2) create log files noting the sender, addressees, time, and destination for billing purposes;[67] (3) prevent fraudulent, unlawful, or abusive use of such service; (4) protect the rights and property of the system or provider with some limitations; and (5) assist law enforcement authorities and other authorized groups under the ECPA to intercept communications or engage in electronic surveillance. It thus permits law enforcement or other persons "acting under color of law" or engaged in authorized foreign intelligence surveillance activities to intercept communications in certain cases.

The privilege to "record" or note that a communication was initiated or completed, however, does not include the authority to publicly disclose the contents or information within the communication. Disclosure of such information is permitted only with the *consent of either the addressee or originator.* Consent, however, cannot be assumed from (1) employees who were not fully informed that they may be monitored and the manner in which the monitoring may occur, (2) anyone who originated the communication by using a pen register or trap device,[68] (3) forwarders of communications, (4) communications inadvertently obtained by a service provider, or (5) communications that appear to pertain to the commission of a crime.

Parties to a communication may tape-record that communication only with the consent of the other parties, unless the purpose of the interception is to commit a crime or tort. Congress intended that the omission overrule *Boddie v. American*[69] concerning a tape-recorded conversation. In *Boddie,* a party to the communications had permission but was still liable because the court found that his purpose in making a recording of the conversation was to embarrass the subject—the tort of defamation, considered an "injurious act."

Anyone may intercept or access (1) public communications; (2) wire or electronic communications to the extent necessary to identify the source of harmful interference to "any lawfully operating station or consumer electronic equipment;" (3) communications over a conventional telephone extension, not a tap to the extension; and (4) radio communications on the same frequency unless the intercepted communications are scrambled or encrypted.

Title II—Protects Voice and Data Communications in Storage[70]

Title II protects wire and electronic communications while in storage. The definition of *electronic storage* includes computer random access memory, magnetic tapes, disks, magnetic and optical media,[71] video conferencing,[72] and videotape rental records.[73]

Like Title I, Title II prohibits unlawful or unauthorized access to a wire or electronic communication while it is being stored in the facility or to the facility through which electronic communications service is provided and disclosure of a communications contents. [74] *Unlawful access* involves both a "means element" and a "result element." Both are described more fully in the statute.

Update—Electronic Communications Privacy Act (ECPA) of 1996

In 1996, the Electronic Communications Privacy Act (ECPA) was updated to clarify federal privacy protections in light of the dramatic changes in new computer and telecommunications technologies, including the Internet. It added a new category of "electronic mail" to the class of stored "electronic communications," enhanced protections for unauthorized access to computer databases,[75] and ECPA protection to communications carried by any provider of wire or electronic communications service, not just common carriers.[76] This is an important change considering the many new carriers in the new competitive environment.[77]

Operators of electronic communications services offered to the public were again barred from disclosing the content of a message in storage, except that electronic communications in criminal activities may be disclosed to law enforcement officers if inadvertently intercepted by a service provider.[78] However, the ECPA of 1996 rolled back privacy rights in that it narrowed the definition of the *contents* of a communication. Today, the definition includes only "information concerning the substance, purport, or meaning of the communication."[79] Thus, it does not include or protect from interception or disclosure, information about the identity of the parties or the existence of the communication.[80] This exception has been used a number of times in recent years to track threatening email messages.[81]

8.3.3 Communications Assistance for Law Enforcement Act of 1994[82]—The Digital Telephony Act

The last section of the ECPA, Title 1 provides for enforcement of the Communications Assistance for Law Enforcement Act (CALEA) of 1994,[83] also called the Digital Telephony Act. In the CALEA, enacted on October 25, 1994, Congress recognized that the rapid advances experienced in telecommunications technology affected law enforcement's ability to conduct court approved wiretaps and other forms of electronic surveillance. In the CALEA, Congress required all telecommunications providers and carriers to ensure that law enforcement personnel are able to continue legal interceptions of communications and electronic surveillance.[84]

To do this, the CALEA requires all telecommunications carriers first, to ensure that their equipment, facilities, and services are capable of meeting the law enforcement assistance criteria specified in Section 103 of CALEA; and second, to disclose detailed subscriber account information, including the communicators' identities and addresses.[85] However, it allows three significant exceptions to the Section 103 requirements: (1) "information services," (2) private networks, and (3) interconnection services and facilities.[86]

Congress authorized $500 million to implement the Act for the three-year period between 1995 and 1998, but in 1998 it did not fund CALEA.[87] Thus, in most cases, the CALEA considers carriers to be in compliance if the attorney general is not in a position to pay for the carriers' compliance costs.[88]

8.4 FEDERAL PRIVACY LAWS PROTECTING PERSONAL INFORMATION IN GOVERNMENT DATABASES

In the 1960s and 1970s, with the development of computers and modern telecommunications networks, privacy concerns took a quantum leap. The ability of computers to collect large amounts of information, aggregate the information, and find patterns in that information is known as the *Hegelian dialectic*.[89] As the quantity of information processed becomes great enough, the quality and usefulness of the information changes, creating a huge threat to privacy. Congress responded to these increased privacy concerns with a flurry of new laws affecting the collection, retention, and destruction of personal information in government databases. These laws include the Federal Records Act of 1950, Title 2 of the Electronic Communications Privacy Act (ECPA) of 1986, and the Privacy Act of 1974. Balanced against these laws, however, is the Freedom of Information Act (FOIA). It is important to note, however, that these new laws affect only information in government files and not information in private files. Each is described in the following four sections.

8.4.1 Federal Records Act of 1950[90]

How information is stored is critical to the security of the information, including its protection against theft or inappropriate use by others while it is in the custody of another. Interestingly, the first statute to address the protection of information in files or storage was the Federal Records Act of 1950, enacted years before the appearance of computers.

Toward the end of World War II, President Truman recognized the need to establish a government-wide file control program and issued Executive Order No. 9784 on September 25, 1946, requiring the "head of each agency to establish and maintain an active and continuing program for the successful management and disposition of its records." In 1948, the Hoover Commission Task

Force on Records Management studied the problems of managing files for the entire government and made the following three recommendations. First, a central staff and service agency should be created in the government with responsibility for leadership in the field of records management. Second, a new federal law concerning record management should be enacted to provide for more successful preservation, management, and disposal of government records. Third, an adequate records management program should be established in each department and agency.[91]

Congress approved the task force's recommendations in 1949 in the Federal Property and Administrative Services Act (FPASA) of 1949,"[92] focusing on two items: (1) the prompt and orderly disposal of records of temporary usefulness and (2) the transfer of retained records to less-costly space and storage equipment.

The following year, in 1950, a new Title V, called the Federal Records Act (FRA) of 1950[93] was added to the Federal Property and Administration Services (FPASA), and enacted to: (1) standardize the creation, management, and disposal of federal agency records; (2) ensure the accurate and complete documentation of the policies and transactions of the federal government concerning stored personal information; (3) provide judicious preservation of information in storage; and (4) ensure careful disposal of records when no longer needed.[94]

In the Federal Records Act, a *record* is defined as any document that, first, was "made or received by an agency of the U.S. Government under Federal law or in connection with the transaction of public business," and second, was "preserved or is appropriate for preservation by the agency . . . as evidence of the organization, functions, policies, decisions, procedures, operations, or other activities of the Government or because of the informational value of the data in them."[95] Thus, a federal agency may not dispose of a record without the approval of the archivist of the United States.[96]

Many cases tested the Federal Records Act, but one in 1993, *Armstrong v. Executive Office of the President, Office of Administration,* specifically addressed the Federal Records Act's impact on modern electronic documents such as emails and voicemails.

Armstrong v. Executive Office of the President, Office of Administration (1993)[97]

In *Armstrong v. Executive Office of the President, Office of Administration,*[98] information held by the Executive Office of the President and the National Security Council were sought. The President of the United States claimed "executive privilege,"[99] but lost when a federal court of appeals confirmed that most records in U.S. government email systems are subject to the Federal Records Act preservation requirements and agencies must retain and manage electronic documents.

As a result of this case and others supporting it, public officials and employees should understand the impact of the Federal Records Act and the many similar state laws on their use of email, voicemail, faxes, and the Internet at work. As public employees, most of their work is public record and subject to disclosure. They do not have the same privacy rights and protections as other groups of employees.

8.4.2 The Privacy Act of 1974[100]

The Privacy Act of 1974[101] specifically addresses the government's protection of information held about individuals in government databases. In the Privacy Act, Congress mandated that each government agency maintaining a system to collect or retain personal data about individuals may not conceal the existence of the system, but instead *must* publish a notice of its existence including a description of the following types of information:[102]

The name and location of the system

Categories of individuals on whom records are maintained in the system

Types/categories of records maintained in the system

Each routine use of the records contained in the system, including the categories of users and the purpose of such use

The policies and practices of the agency regarding storage, irretrievability, access controls, retention, and disposal of the records

The title and business address of the agency official who is responsible for the system of records

The agency procedures whereby an individual can be notified at his request if the system of records contains a record pertaining to him

The agency procedures whereby individuals can be notified at their request how they can gain access to any record pertaining to them contained in the system of records, and how they can contest its content

The categories of sources or records in the system

In addition, the Privacy Act of 1974 mandates that the government agencies may not disclose their records to other individuals or agencies without a court order except when it involves civil or criminal law enforcement activity, circumstances affecting the health or safety of an individual, or consumer reporting agencies acting in accordance with applicable law. As an additional privacy protection, the law also requires that the disclosing agency keep records of its disclosures and make those records available to the individual named in the record, who may review the records and request correction of any incorrect information.[103]

8.4.3 The Computer Matching and Privacy Protection Act

In 1988 and 1989, Congress amended the Privacy Act of 1974 by adding the Computer Matching and Privacy Protection Act in which it prohibited federal agencies from comparing their databases with the databases of other federal agencies or with private databases. Congress did so to avoid the power of computer matching of information, especially about individuals.

Supreme Court Review of the Privacy Act of 1974

In the first three years following the enactment of the Privacy Act, from 1974 through 1977, the U.S. Supreme Court considered three cases related to the Privacy Act of 1974. The three cases were *California Bankers Association v. Schultz*[104] in 1974, *Whalen v. Roe*[105] in 1977, and *Nixon v. Administrator of General Services*,[106] also in 1977.

In each case, the Court considered the law concerning (1) what information about individuals should exist in government files; (2) under what circumstances personal information about individuals existing in government files and databases could be disclosed to others; (3) what legitimate governmental interests affect disclosure of the information to others; and (4) whether an individual's right to privacy includes that individual's ability to control how, to whom, and to what extent personal information is disclosed by the government to others. In each case, the Court decided that the U.S. government's interest, or need for the information, outweighed the individual's asserted right of privacy and ordered the release of the information. However, each decision also affirmed that such information is definitely protected by the Fourth Amendment. Each case is summarized below.

California Bankers Assn. v. Schultz (1974)[107]

In 1974, in *California Bankers Assn. v. Schultz*, the Supreme Court considered the issue of privacy rights when personal information about individuals existing in government files and databases was disclosed. While the release of information in this case was upheld, the case clarified the limits and standards that guide the distribution or sharing of information held by the government.

Whalen v. Roe, 429 U.S. 589, 599 (1977)

In *Whalen v. Roe*, the Supreme Court considered whether a state's centralized computer files were legal when they contained the names and addresses of all persons who had obtained, pursuant to a doctor's prescription, certain drugs for which there was both a lawful and an unlawful market. The court decided that the files were legal, but required tight controls on their use.

Nixon v. Administrator of General Services, 433 U.S. 425, 457 (1977)

In *Nixon v. Administrator of General Services*, former President Nixon, both as an individual and under Executive Privilege, challenged the public release of his presidential papers and tape recordings. The Supreme Court upheld their release.

8.4.4 Freedom of Information Act (FOIA) of 1976[108]

Two years after enacting the Privacy Act of 1974, Congress added a new section to the same code, entitled the Freedom of Information Act (FOIA) of 1976. Thus, FOIA is found at 5 U.S.C. § 552b, the section immediately following the Privacy Act of 1974. In passing the FOIA and placing it immediately after the Privacy Act, Congress sought to more accurately delineate the balance between the need for privacy and the need for open information from government agencies.

Like the Privacy Act of 1974, the Freedom of Information Act (FOIA) of 1976 required all federal agencies to reveal the existence of their records, procedures, and statements of policy to any member of the public that requests such information. However, the FOIA went one step beyond the Privacy Act in that it mandated that each agency publish, in the Federal Register, a description of the place and method by which the public may obtain certain government-held information for review. Parties who are refused the right to inspect the requested information may sue the government agency. If the suit is successful, the party may receive court costs and attorney's fees in addition to the information originally sought from the agency. Nearly every state followed Congress's lead by enacting similar laws, called state "open records" or "public records" laws.

Exemptions to the Freedom of Information Act

To balance the need for privacy, Congress also included several important exemptions to the information available through the Freedom of Information Act. The exemptions address situations when the release of information would (1) constitute a "clearly" unwarranted invasion of personal "privacy"; (2) jeopardize national defense matters; (3) impinge upon internal personnel rules of a government agency; (4) release confidential financial information, trade secrets, personnel information, medical files, geological information, and interoffice or intraoffice agency memoranda; (5) reveal investigatory records that can be obtained only by a valid subpoena; and/or (6) fall within the parameters of Executive Privilege.[109]

State and Federal Court Review of the Freedom of Information Act

While the definitions of these exemptions are vague, over the years the courts have clarified them in various cases. The following five cases demonstrate the courts' comments and parameters. In the first two cases, the courts determined that the exemptions were valid, while in the last three, the courts determined that the exemptions were not valid. However, in each case, the courts recognized the importance of exemptions for "work in progress" or the "deliberative process" and offered certain protections for privacy in email, correspondence, and telephone conversations.

California Case: *Times Mirror Co. v. Superior Court* (1991)[110]

In 1991 a California newspaper known as the *Times Mirror Co.,* requested a copy of the California governor's schedules and appointment calendars under the California State Public Records Act. When the governor's office refused to release the documents, the *Times Mirror* sued. In its decision, the California Superior Court stated:

> Disclosing the identity of persons with whom the Governor has met and consulted is the functional equivalent of revealing the substance or direction of the Governor's judgment and mental processes; such information would indicate which interests or individuals he deemed to be of significance with respect to critical issues of the moment. The intrusion into the deliberative process is patent. . . .
>
> If the law required disclosure of a private meeting between the governor and a politically unpopular or controversial group, that meeting might never occur. Compelled disclosure could thus devalue or eliminate altogether a particular viewpoint from the governor's consideration.[111]

California (Appellate) Case: *Rogers v. Superior Court* (1993)[112]

Two years later, in 1993, the California Appellate Court upheld a second similar California Superior Court decision in which it affirmed the right of a city council to refuse a newspaper's request for the telephone numbers of persons with whom city council members had spoken.

The Appellate Court stated that the "deliberative process privilege" exists to "protect materials reflecting deliberative or policymaking processes, and not 'purely factual, investigative matters.'"[113] The court added that "routine public disclosure of such records would interfere with the flow of information to the government official and intrude on the deliberative process."[114]

Thus, the court determined that in such situations, government officials can shield the names and contact information of persons with whom they communicate and the content of certain conversations, emails, and correspondence.

Arizona Case: *Star Publishing Co. v. Pima County Attorney's Office* (1993)[115]

The same year, 1993, an Arizona court cited decisions of the federal and California courts when an Arizona newspaper, *Star Publishing,* requested computer backup tapes of the Pima County Assessor's office, including email communications. The county refused disclosure, arguing that the material was either protected by the "deliberative process privilege" or immune from disclosure to protect public employee privacy rights.

In its decision, the Arizona court recognized and supported the deliberative process, but also decided that, in this case, no proof had been offered to substantiate these claims. Thus, based on the argument of deliberative process, the court ruled that the records had been wrongly withheld. Concerning the second argument, that employee privacy rights would be compromised by disclosure, the Court stated that it "doubt[ed] that public employees have any legitimate expectation of privacy in personal documents that they have chosen to lodge in public computer files." The Arizona court then ordered the release of the records to the newspaper.[116]

Florida Case: *In re Amendments to Rule of Judicial Administration* (1995)[117]

Similarly, two years later in 1995, the Supreme Court of Florida noted in *In re Amendments to Rule of Judicial Administration,* that under its judicial rules, emails made or received in connection with the official business of the judicial branch fall under the definition of a "judicial record," which must be directed and channeled so that they can be recorded as public records.

The only exemptions to public disclosure of email recognized by the court included proposed drafts of opinions and orders, memoranda regarding pending cases, proposed jury instructions, and certain other email sent and received between judicial employees within a particular court's jurisdiction. Thus, the court excluded most of the working communications between the judges and staff from designation as public records.

Computer Professionals for Social Responsibility v. U.S. Secret Service (1996)[118]

The following year, in 1996, a U.S. Appellate Court ruled, in part, that the U.S. Secret Service had properly invoked exemptions to the Freedom of Information Act (FOIA) in response to a request for records related to a case involving the Secret Service's breakup of a meeting of young computer hackers at a Virginia shopping mall. The court held that retention of the information was important to law enforcement and that the release of the investigatory records was not in the public's interest.

8.5 FEDERAL PRIVACY LAWS PROTECTING PERSONAL BANK AND FINANCIAL INFORMATION

One of the most sensitive privacy areas for most individuals is information concerning their bank and other financial information. For this reason, the U.S. Congress has enacted several privacy laws in this area. These include: The Bank Records and Foreign Transactions Act (BRAFTA) of 1970, incorporating The Bank Secrecy Act of 1970, The Truth in Lending Act of 1970, and The Fair Credit Reporting Act (FCRA) of 1970; and The Right to Financial Privacy

Act (RFPA) of 1988. The provisions of each law are described in the following five sections.

8.5.1 The Bank Records and Foreign Transactions Act (BRAFTA) of 1970

Bank failures and scandals in the late 1960s and early 1970s emphasized the need for Congress to reform and restructure the U.S. banking and financial system and to enhance its record-keeping and reporting procedures. These scandals included the Texas "Rent-A-Bank" scheme, massive insider dealings involving Bert Lance while he served as a bank officer in Georgia before becoming director of the U.S. Office of Management and Budget, and the use of financial institutions by drug dealers, the underworld, white collar criminals, and tax evaders to launder illegal money. To fight these crimes, law enforcement agencies required detailed personal information about individuals from their financial institutions. The agencies most concerned about the information included (1) the Justice Department, (2) the Treasury Department, (3) Internal Revenue Service, (4) Securities and Exchange Commission, (5) Defense Department, and (6) the Agency for International Development.[119]

In 1970, Congress passed the Bank Records and Foreign Transactions Act (BRAFTA) of 1970 amending the Federal Deposit Insurance Act (FDIA). In BRAFTA, Congress *required* that all banks insured by the Federal Deposit Insurance Corporation (FDIC) maintain certain records and report certain transactions to the Department of the Treasury. However, to protect the privacy of the banks' customers, Congress also required that this information be released only when probable cause of criminal activity, based on a reasonable belief, existed, and a valid warrant was issued by an appropriate court using due process. It prohibited banks from releasing the financial records of any individual to law enforcement without a warrant. Thus, Congress required that bank records are not automatically or routinely made available to law enforcement.

The BRAFTA addresses two major areas of concern to law enforcement: financial record-keeping by domestic banks and financial institutions and the use of foreign banks and financial facilities located in countries with different secrecy laws. To accomplish it purposes, the BRAFTA contains several sections, known as "titles," as follows.

Titles I-IV—The Bank Secrecy Act of 1970[120]

Title I of the Bank Records and Foreign Transactions Act (BRAFTA), named the Bank Secrecy Act of 1970,[121] sounds as if it provides secrecy for bank customers. However, it actually mandates that the Secretary of the Treasury require all banks and other financial institutions insured by the Federal Deposit Insurance Corporation (FDIC) to "maintain appropriate types of records that may have a high degree of usefulness in criminal, tax or regulatory investigations or proceedings." These records include the names of depositors and photocopies of all transactions.

In the Bank Secrecy Act of 1970, Congress required that law enforcement agencies have a court order before they access the financial records of any U.S. resident. However, fearing the loss of financial records important to law enforcement agencies, Congress also required banks routinely to make and retain copies of checks and other financial transfer instruments.[122]

Title II provides for the records and reports of domestic currency transactions, exports and imports of monetary instruments, and records and reports of foreign transactions by residents or citizens of the United States or persons doing business abroad.

Title III amends Section 7(a) of the Securities and Exchange Act (SEA) of 1934, to make it unlawful for persons to obtain or retain credit in violation of SEA rules or regulations.

Title IV established effective dates for BRAFTA and the Bank Secrecy Act.

Title V—Truth in Lending Act (TILA) of 1970[123]

Title V of the BRAFTA, known as an update to the Truth in Lending Act (TILA) of 1970,[124] amended the then existing Truth in Lending Act to add protection concerning the distribution of credit cards.

Titles VI-VII—The Fair Credit Reporting Act (FCRA) of 1970[125]

Title VI of the BRAFTA, known as "the Fair Credit Reporting Act (FCRA) of 1970,"[126] addressed privacy issues in managing consumer credit histories. In the FCRA, Congress sought to protect consumers with three important provisions. First, it specified what information could be collected, maintained, and reported by credit bureaus. Second, Congress restricted the use of the information only for issuing credit, insurance, employment, obtaining government benefits, or other legitimate business needs involving a business transaction. Third, the FCRA required that credit bureaus implement and maintain procedures to avoid reporting obsolete or inaccurate information about consumers.

Specifically, Section 603 of the Fair Credit Reporting Act (FCRA) addressed obsolete information, including how quickly information must be entered into the reporting system and for how long the information could be retained. Section 609 addressed "disclosure of credit information to consumers," while Section 615 placed requirements on the users of credit reports. Section 612 addressed civil penalties for willful noncompliance with FCRA; and Section 618 confirmed the jurisdiction of the courts. Section 620 addressed unauthorized disclosures of information by credit company officers or employees.

Title VII of BRAFTA addressed additional provisions relating to credit-reporting agencies including consumers' rights to know the contents of their credit reports, dispute any inaccuracies, and call for reinvestigation by credit-reporting agencies. In many states this information is available to the customer *only* if the customer pays a fee each time the customer asks to review his or her credit file. Two states, however, Colorado and New Jersey, *require* that credit agencies allow people to see their credit histories for free.

8.5.2 Right to Financial Privacy Act (RFPA) of 1988[127]

In 1988, Congress sought to fortify the financial privacy of individuals and to increase "safe banking" by preventing banking abuses. To do this, it enacted the Right to Financial Privacy Act of 1988 in which it strengthened the supervisory authority of federal agencies that regulate banks and other depository institutions and provided better coordination among the financial supervisory agencies. The purpose of the Right to Financial Privacy Act (RFPA) was to protect the customers of financial institutions from unwarranted intrusion into their bank records, while at the same time permitting legitimate law enforcement review of bank activity. Therefore, in the RFPA, Congress once again sought to strike a balance between customers' right to privacy and the need of law enforcement agencies to obtain financial records pursuant to legitimate investigations.

The RFPA accomplished seven main items. First, it expanded the list of financial institutions and agencies covered by the Right to Financial Privacy Act (RFPA). For example, the Securities and Exchange Commission (SEC) was placed on the list for the first time. Second, it limited the federal government's access to an individual's financial records only to situations where the financial institution has information relevant to a violation of any statute or regulation;[128] the perfection of a security interest such as proving a claim in bankruptcy, collecting a debt, or processing an application for a government loan, or loan guarantee;[129] an administrative or judicial subpoena, summons, search warrant, or formal written request in accordance with requirements in the RFPA;[130] or coverage of the individuals' accounts by the Federal Deposit Insurance Agencies (FDIA) when a financial institution fails. Third, the Right to Financial Privacy Act (RFPA) required that financial institutions receive an appropriate court-issued warrant or legal authorization before disclosing any information about a customer's financial records to the government.[131] Fourth, the RFPA required that individuals must be given notice of, and a chance to challenge, any federal government agency request for their bank records, except individuals whose financial records have been subpoenaed by a grand jury in connection with a crime or false statements. Fifth, the RFPA acknowledged the issue of consent and noted that customers may always authorize their financial institutions to disclose certain information to the government. However, the RFPA stated that such authorization may not extend for more than three months and may be revoked at any time before the records are disclosed.[132] Sixth, the government agencies requesting the financial information from an individual's financial records must reimburse the cost of providing the records to the complying financial institution. Seventh, Section 951 of the RFPA authorizes civil penalties of up to $1 million per day for conduct that violates certain portions of 18 U.S.C. § 951. If a continuing violation occurs, the fine can rise to $5 million or the pecuniary loss to a financial institution.

8.6 FEDERAL PRIVACY LAWS PROTECTING PRIVATE DATABASES FROM SEARCHES AND SEIZURES BY GOVERNMENT OFFICIALS

In addition to addressing personal information retained in government databases and financial institutions, the U.S. Congress in 1980 looked at government access to information in private databases. As a result, Congress passed the Privacy Protection Act (PPA).[133]

8.6.1 The Privacy Protection Act (PPA) of 1980[134]

The PPA establishes safeguards for materials held by "a person reasonably believed to have a purpose" such as to publish a newspaper, book, broadcast, or similar public communication. Such materials are not subject to search or seizure by the government unless there is probable cause to believe that the materials are related to a crime or immediate seizure is necessary to prevent death or serious bodily injury.

The PPA contains two parts or subchapters: First Amendment Privacy Protection and Attorney General Guidelines. Subchapter 1, First Amendment Privacy Protection, restricts searches and seizures by government officers and employees in connection with the investigation or prosecution of criminal offenses. In particular, it addresses work product materials, other documents, and objections to court-ordered subpoenas and affidavits. It provides for civil actions by aggrieved persons, including good faith defenses, admissibility of evidence, remedies, damages including costs and attorneys' fees, and jurisdiction.

Subchapter II, Attorney General Guidelines, provides guidelines for federal officers and employees concerning the procedures to be used in obtaining documentary evidence, protection of certain privacy interests, the use of search warrants, reports to Congress, and disciplinary actions for violations.

8.6.2 Court Cases Reviewing the Privacy Protection Act

Numerous court cases have challenged and/or reviewed the Privacy Protection Act (PPA) of 1980. Two of the more recent cases raise key issues concerning its application to modern technological developments.

Steve Jackson Games v. United States Secret Service

Perhaps the leading case related to the Privacy Protection Act of 1980 was first decided in 1993 and upheld on appeal in 1994. The case, *Steve Jackson Games v. United States Secret Serv.*, 816 F.Supp. 432 (W.D. Tex. 1993), 36 F.3d 457 (5th Cir. 1994), involved a raid by U.S. Secret Service agents on the offices of a small, Texas publisher of electronic games known as Steve Jackson Games. During its raid, the Secret Service agents removed the company's computers, including one running the company's bulletin board system (BBS).

When the company informed the Secret Service of this, it refused to return the company's work product. A Texas court held that the Secret Service had violated the Privacy Protection Act and ordered the immediate return of the computer. On appeal, the U.S. Fifth Circuit Court upheld the Texas court's decision.

State ex rel. Macy v. One Pioneed CD-ROM Changer

However, the opposite decision was reached the same year, 1994, by a court in Oklahoma in the case of *State ex rel. Macy v. One Pioneed CD-ROM Changer*, 891 P.2d 600, 606-07 (Okla. Ct. App. 1994). The Oklahoma Court held that the PPA had not been violated where seized equipment contained 500 megabytes of nonobscene material to be pressed into a compact disc to be published. Courts continue to review this issue as new technologies evolve.

8.7 FEDERAL PRIVACY LAWS PROHIBITING ILLEGAL ACCESS TO PROTECTED COMPUTERS

With the evolution of technology, and the increasing prevalence of computers in businesses, homes, government agencies, and financial services across the United States, Congress acknowledged in the early 1980s the emergence of a new type of criminal—one who used computers to steal, defraud, and abuse the property of others. The proliferation of computer data gave these criminals easy access to property and information that, in many cases, was unprotected against crime.

For several years, during the 1980s, Congress investigated the problems of both computer fraud and abuse and documented three key findings. First, Congress found that more than 50% of all survey respondents had been victims of some form of computer crime, resulting in hundreds of millions of dollars of loss each year.[135] Second, Congress found that computer crime posed other threats beyond financial. For example, in 1983, Congress found that a group of adolescent computer "hackers," known as the "414 Gang," broke into the computer system at Memorial Sloan-Kettering Cancer Center in New York.

In doing so, the adolescents accessed the radiation treatment records of 6,000 past and present cancer patients and had the opportunity to alter the radiation treatment levels that each patient received, creating a potentially life-threatening situation.[136] Third, Congress found that "pirate bulletin boards" existed for the sole purpose of providing the passwords and other information necessary to break into computers such as those operated by the U.S. Department of Defense and the Republican National Committee. The Committee recognized that while no apparent financial loss had occurred, multiple sites were "trafficking in other people's computer passwords."

A 1984 Report by the American Bar Association's (ABA's) Task Force on Computer Crime, chaired by Joseph Tompkins, Jr., substantiated these findings and stated that the ability of computer crime to harm people makes it one of the worst white-collar offenses in the U.S.[137] In light of these findings, Congress also determined that U.S. criminal laws at the time were not sufficient to address the problems of computer crime and set out to draft federal laws to cope more effectively with these new abuses.[138]

8.7.1 Computer Fraud and Abuse Act of 1988[139]

As a result of its findings concerning computer fraud and abuse, Congress passed an initial federal computer crime statute[140] in 1984 that made it a felony to access *classified* information in a computer without authorization and a misdemeanor to access or "trespass" into *a government computer* or *financial records or credit histories in financial institutions* without authorization. However, the initial statute had two major shortcomings. First, it was too weak, and second, during the next three years, 1985 to 1988, the Department of Justice encountered numerous jurisdictional problems in the area of computer crime because, at the time, many states had no computer crime legislation. To bridge this gap, the Department of Justice (DOJ) encouraged Congress to expand its initial federal computer crime statute into sweeping federal legislation. However, after careful consideration and review of the state action provisions in the U.S. Constitution, Congress decided that federal computer crime law should be limited to jurisdiction only over computer crimes in which there is a "compelling federal interest." Congress defined *compelling federal interest* as "cases where computers of the federal government or certain financial institutions are involved, or where the crime itself is interstate in nature."[141] Over the next several years most of the states established state computer crime laws to address the gaps left by this jurisdictional decision.

In 1988, Congress codified this definition in the Computer Fraud and Abuse Act (CFAA) of 1988 by updating its federal computer crime statute once again. In the CFAA, Congress also raised the standard required to prosecute unauthorized access of a computer from "knowingly" to "intentionally" in order to exclude mistaken, inadvertent, or careless access. This brought the standard into line with the "knowingly and with intent to defraud" standard used in 18 U.S.C. § 1029 to prosecute credit card fraud.

Specifically, the Computer Fraud and Abuse Act (CFAA) prohibited the following seven acts, or attempts to act:[142]

1. Obtaining, or seeking to obtain, national security information with the intent to use it to injure the United States or to unfairly benefit any foreign nation
2. Intentionally accessing, resulting in the obtaining of information contained in the records of a financial institution, credit card issuer, or consumer-reporting agency
3. Intentionally accessing a government computer affecting the government's operation of such computer
4. Knowingly accessing, without authorization, a federal interest computer resulting in the obtaining of anything of value beyond the mere use of the computer with intent to defraud
5. Intentionally altering, damaging, or destroying certain computerized information belonging to another
6. Intentionally accessing a federal interest computer and preventing authorized use of any information or computer services when the loss amounts to more than $1,000 in a one-year period, or involves medical treatment (However, the concept of "loss" was not limited to actual monetary losses. For example, investors may lose on a stock if the stock projections have been altered to make them appear more desirable. This section also includes other access, such as hackers of medical information.)
7. Trafficking in passwords

In reviewing these seven actions, the Department of Justice expressed concern that the term "obtains information" implied more than unauthorized access and could be interpreted to require the actual movement or acquisition of the data or a copy of it. Congress, however, stated that its intent in the use of the word "obtaining" data was to include the mere observation of the data, except for federal employees and other authorized workers. Congress also tried to clarify in its wording of the law that it wanted the Computer Fraud and Abuse Act (CFAA) to focus on "outsiders" or those lacking authorization to access any federal interest computer. It did not, however, want the law not to be used against "whistleblowers".

As part of clarifying "unauthorized access," Congress included the introduction of viruses and "worms" into the Internet or computer systems.[143] In clarifying the term *loss,* Congress defined it as "reduced performance caused by the worm resulting in more than $1,000." In addition, the government need not prove intent to cause that loss.[144]

Exceptions to the Computer Fraud and Abuse Act (CFAA) include (1) access for authorized law enforcement with appropriate court orders, (2) access to complete authorized repairs to a computer system[s], and (3) "time bombs" or automatic termination devices built into a program that automatically terminates the service if a user fails to pay his bill for the service on time. Congress did not intend nonpayment of a bill to become a criminal activity.

In addition, Congress recognized that while laws can be a deterrent, the most effective way to control the incidence of computer crime is to educate private industry, computer users, and the general public to be aware of the ethical and legal questions involved in computer crimes and not to view them as harmless pranks.[145] As such, Congress determined that comprehensive education programs for both computer users and the general public should be undertaken.

One of the first cases to review the Computer Fraud and Abuse Act of 1988 (CFAA) was first decided in 1989 and affirmed on appeal in 1991. The case, *U.S. v. Morris,* 928 Fed 2d 504 (1991), involved defendant Robert T. Morris who introduced a computer program into the Internet that later became known as a "worm" or "virus." The goal of the program, Mr. Morris stated, was to demonstrate the inadequacies of current security measures on computer networks by exploiting the security defects that Morris had discovered. Nonetheless, the district court convicted Mr. Morris of violating the Computer Fraud and Abuse and the Court of Appeals affirmed the conviction.

During this case, one weakness of the Computer Fraud and Abuse Act (CFAA) became apparent—the overly broad definition of "computer." The CFAA defines a *computer* as "an electronic, magnetic, optical, electrochemical, or other high-speed data-processing device performing logical, arithmetic, or storage functions, and includes any data-storage facility or communications facility directly related to or operating in conjunction with such device but such term does not include an automated typewriter or typesetter, or portable hand held calculator, or other similar device."[146] This definition is so broad that it includes microwave ovens and advanced telephone systems, items clearly not intended to be included in a "computer crime" law. Congress attempted to correct this definition in the National Information Infrastructure Protection Act of 1996.

8.7.2 National Information Infrastructure Protection Act (NIIPA) of 1996[147]

Eight years after the Computer Fraud and Abuse Act was enacted, it was updated in the National Information Infrastructure Protection Act (NIIPA) of 1996. The NIIPA was enacted on October 11, 1996, and codified at 18 U.S.C. §§ 1030 *et. seq*. The main difference between the two acts is the definition of computers covered by the Act as defined in Section 1030(a)(2).

While the Computer Fraud and Abuse Act (CFAA) protected mainly government computers and financial databases, the National Infrastructure Protection Act (NIIPA) extended its protection to any protected computers if the conduct involved interstate or foreign communications. It defined a *protected computer* as "a computer used in interstate or foreign commerce or communication. . . ."[148] Thus statutory construction of the National Infrastructure Protection Act (NIIPA) suggests that section 1030(a)(2)(C) would prohibit unauthorized access to *any computer* where the act of unauthorized access involved an interstate or foreign communication and the access in question was made to a computer that was itself used in interstate or foreign commerce or communication.

The NIIPA remains the primary federal law concerning Internet viruses, "worms," and "denial of service" attacks. Depending on the computers affected and processes used, certain intentional actions harming nongovernment computers may or may not be covered.

8.8 FEDERAL PRIVACY LAWS PROHIBITING UNWANTED AND HARASSING COMMUNICATIONS

As annoying, and often frightening, as unauthorized access or theft of data can be, other types of interruptions to an individual's privacy by telecommunications technology can also be problematic. These can include (1) annoying calls, such as those from jokesters; (2) harassing, obscene, or threatening calls; (3) scams over telecommunications devices; and (4) telephone solicitations and untimely polls. The Telephone Harassment Act of 1968 and the Telephone Consumer Protection Act of 1991 were passed to address these problems. Each is discussed below.

8.8.1 The Telephone Harassment Act of 1968

The Telephone Harassment Act of 1968 imposes criminal liability on the interstate use of the telephone to make "obscene, lewd, lascivious, filthy, or indecent" phone calls or to make anonymous calls "with intent to annoy, threaten, or harass," or "repeatedly or continuously to ring, with [the] intent to harass." Nearly every state has also enacted similar laws forbidding the use of the telephone to harass by making obscene, anonymous, repeated, or nonconsensual calls. However, some states prohibited only calls that are "inconvenient," as for example, late hour calls or calls that significantly kept the called party from using his or her phone.

While well intentioned, these federal and state "telephone harassment statutes" have two major shortcomings. First, they are so vague and overbroad that they frequently fail to meet the constitutionally required levels of clarity and fairness. Thus, the laws affect only the most unconscionable actions of invasions of privacy and allow many guilty callers to go free. Second, these laws did not anticipate the use of the Internet or problems such as "page jackers" or scammers who use confusing names or insert "meta tags" into sites to trick users into visiting Web sites filled with pornography. These sites then prevent users from logging off of or backing out of a site. One example of a confusing name is *www.whitehouse.com*. Internet users trying to reach the office of the President of the United States must use *www.whitehouse.gov*. The confusing ".com" address will surprise Internet users by producing a pornographic site and creating concerns for parents and teachers asking children to research American government. This practice falls outside of the laws protecting children from pornography because is not a "site directed to children."

8.8.2 The Telephone Consumer Protection Act (TCPA) of 1991

Technological advances in the 1980s made it cost effective for telemarketers to significantly increase their use of automated telephone-dialing equipment to make unsolicited marketing calls to homes and businesses. This equipment, including automatic dialer recorded message players (ADRMPs) and automatic dialing and announcing devices (ADADs), randomly or sequentially dials telephone numbers and delivers an artificial or prerecorded voice message or fax to the called party.

In the 1980s, a typical automatic dialer recorded message player (ADRMP) could dial as many as 1,000 telephone numbers each day, and hundreds of thousands were in use, owned by over 180,000 solicitor companies.[149] Each day, millions of American telephone numbers were called, including hospital, fire and police emergency numbers, and unlisted phone numbers paid for by persons wanting to avoid unwanted calls. In addition, the automated calls frequently would not disconnect the number for a long time after the called party hung up, and thus would not release the line for other calls, including emergency calls. Since phone numbers were often dialed in sequence, the automated calls frequently tied up numerous lines of businesses, affecting their ability to conduct business. In addition, the unsolicited, automated calls to cell phone numbers cost each owner of a cell phone the cost of an incoming call, while calls to pagers required the cost of a return call. Calls to fax machines unnecessarily used fax paper and blocked other, desired calls. Often, the automated calls would fill the entire tape of answering machines, preventing other callers from leaving messages and, in some cases, the called parties paid for each message based on the length of the message. This became very expensive for such consumers. To add insult to injury, the automated calls did not identify who was calling. All of these costs, concerns, and inconveniences generated increasing consumer complaints against telemarketers. Many consumers considered the calls a nuisance and an invasion of their privacy. They felt that telemarketers were "intruders" in their homes.

In response, Congress proposed the Telephone Consumer Protection Act (TCPA) of 1991 to prohibit certain practices involving the use of telephone equipment for advertising and solicitation purposes. The purposes of the Act, as stated in its legislative history, were to protect the privacy interests of residential telephone subscribers by placing restrictions on unsolicited, automated telephone calls and to facilitate interstate commerce by restricting certain uses of automatic dialers and facsimile machines. During the discussion about the proposed Act, the Direct Marketing Association and other groups argued that some consumers appreciate the information provided in the telemarketing calls and that the proposed restrictions on marketing calls would restrict various parties' First Amendment free speech rights.

In considering these arguments, Congress cited two Supreme Court decisions. First, in *FCC v. Pacifica Foundation,*[150] the Supreme Court upheld a Federal Communications Commission (FCC) ruling that prohibited the daytime broadcast of indecent language. In its decision, the Court stated that "in the privacy of the home . . . the individual's right to be left alone plainly outweighs the First Amendment right of an intruder." Second, in *Kovacs v. Cooper*[151] the Supreme Court upheld a local ordinance banning sound trucks to protect the privacy and quiet of residents.

In passing the Telephone Consumer Protection Act (TCPA) of 1991, Congress prohibited automatically dialed or prerecorded telephone calls to emergency numbers and required all autodialed calls to identify the initiator of the call, give the telephone number of the business placing the call, and disconnect the line within five seconds of receiving notice that the called party had hung up the telephone.

Restrictions on Telemarketing to Cell Phones, Pagers, and Fax Machines

In addition, the Telephone Consumer Protection Act (TCPA) prohibited all computerized calls to any number for which the called party is charged for the unsolicited call, such as to cell phones and pagers. It also prohibited the use of "any telephone facsimile machine, computer, or other device to send an unsolicited advertisement to a telephone facsimile machine"[152] unless the receiver invites or gives permission to receive such advertisements. If sent, the TCPA requires that the senders of these "junk faxes" place identification about themselves on at least the first page of each transmission and provide a telephone number for the receiving party to call to have his or her fax number removed from the solicitation list.

Restrictions on Telemarketing by Email—Junk Email or "Spam"

Similar to junk faxes and junk mail delivered by the postal service, unsolicited marketing information delivered to an email address, known as "junk email," is not yet covered by the Telephone Consumer Protection Act (TCPA). Although occasionally useful and interesting, junk email is more often an annoying nuisance in that it requires time to read, sort, and discard. In some cases, it also delays or interferes with delivery of or user action on more important messages.

Unlike postal junk mail, however, junk email takes up space in the user's system memory and thus causes additional costs to users paying measured system time.

In addition, the ease of sending mass mailings to hundreds of thousands of email recipients with one keystroke causes what is known as "junk email bombings" or "spamming." This heavy use of the system can clog or gridlock the system, seriously threatening the overall availability and usefulness of the email system and interfering with the user's access to, use of, and cost of their email account.

At present, few laws restrict email spamming, but recent court cases have reviewed the issue. The following provide examples of such cases.

Court Cases Reviewing Spamming

Cyber Promotions Inc., a junk email company, sent thousands of unsolicited email advertisements to the customers of various Internet service providers (ISPs), including CompuServe, America OnLine (AOL), and Apex Global. [153] However, in doing so, Cyber Promotions caused the messages to appear as if they originated from CompuServe, AOL, and Apex Global. Besides being confusing to the recipients, this caused a large number of undeliverable messages to be "returned" to the three ISPs for storage.

CompuServe sued Cyber Promotions for this practice and won. In the case of *CompuServe, Inc. v. Cyber Promotions, Inc.,*[154] the court granted an injunction to CompuServe based on unfair competition and conversion claims, holding that the excessive storage required for the undeliverable messages consumed the capacity of several of CompuServe's computers, thereby representing an actionable trespass to chattels for which the First Amendment provided no defense.

AOL, on the other hand, handled the situation somewhat differently. Instead of receiving the messages, AOL blocked the incoming messages from Cyber. In response, Cyber Promotions sued, and in *Cyber Promotions, Inc. v. America Online Inc.,*[155] a Pennsylvania court found that AOL's blocking of messages could not deny Cyber's First Amendment rights because AOL was not a state actor, and AOL had not violated anti-trust laws because its email service was not an "essential facility" under anti-trust law.

Apex Global Information Services, Inc. cut off Cyber's existing service without notice. Cyber then sued Apex for breach of contract and won a preliminary injunction. In *Cyber Promotions, Inc. v. Apex Global Information Services, Inc.,*[156] the court ordered Apex to restore service for a six-week period while Cyber made other arrangements for Internet access. After these cases, most ISPs refused to provide service to Cyber as a customer. When Cyber was unable to find other outlets, it became its own specialized ISP, organized solely to send unsolicited mass mailings on behalf of its advertising customers.

However, in the meantime, one case concerning the Telephone Consumer Privacy Act was initiated by Robert Arkow, a Compuserv subscriber who received advertisements from Compuserv via email. Mr. Arkow filed suit against CompuServe and CompuServe Visa claiming that they violated the Telephone

Consumer Privacy Act by sending the advertisements to him via email. The case settled for an undisclosed amount of money.

Watching the far-reaching nature and potential abuse of online resources caused by junk email, several states are considering legislation that would make the sending of unsolicited ads directly to email accounts a misdemeanor. In civil law, the increasing incidence of junk email is being considered under the tort of "intrusion". The details of this proposed legislation are still being discussed, but should appear in legislation throughout the early 2000s.

Exceptions to the Telephone Consumer Protection Act

The Telephone Consumer Protection Act (TCPA) includes four notable exceptions. First, the TCPA states that a "public school or other governmental entity may leave a prerecorded message . . . for any emergency purpose." Second, it allows fax messages to emergency numbers when the situation requires a signature or other valid use of a fax message. Third, the TCPA states that persons may give their oral or written consent to being called back by a computer, such as when stores call to say that an ordered item is ready for pickup. In nearly every case, these exempted calls are from companies with whom the consumer has previously done business. Fourth, surveys and research projects may use automated dialers and prerecorded messages, unless they include, or are combined with, a sales pitch. Many small businesses and newspapers depend on these local solicitations, but are also guided and constrained by local good business practices and standards.

8.9 FEDERAL PRIVACY LAWS PROTECTING INFORMATION ABOUT CHILDREN

Another area of concern to parents, privacy advocates, and other adults is the privacy of information about children. The Children's Online Privacy Protection Act (COPPA) of 1998, the Child Online Protection Act (COPA) of 1998, the Child Pornography Prevention Act (CPPA) of 1996, the Protection of Children from Sexual Predators Act (PCSPA) of 1998, and the proposed Children's Internet Protection Act (CIPA) of 1999 all provide examples of the current efforts to protect children in the modern flow of potentially harmful communications. These are described in the following five sections.

8.9.1 Children's Online Privacy Protection Act (COPPA) of 1998

The Children's Online Privacy Protection Act (COPPA) was enacted on October 21, 1998 as part of the Omnibus Consolidated and Emergency Supplemental Appropriations Act. Codified at 15 U.S.C. §§ 6501 *et seq,* its purpose is to stop unfair and deceptive practices used in the collection and use of personal information from and about children on the Internet. COPPA defines a *child* as any individual under the age of 13.

Specifically, COPPA requires an operator of a Web site or online service *directed to children,* or any operator that *collects or has actual knowledge* that it is collecting personal information from a child, to (1) provide notice on the Web site of what information the operator is collecting from children, how such information is used, and the operator's disclosure practices for such information; (2) obtain parental consent for the collection, use, or disclosure of personal information from children; and (3) provide to parents upon request, a description of the information collected from the child and the opportunity to refuse to permit the operator's further use or maintenance of personal information from that child. COPPA also prohibits such Web sites and online services from setting a condition on a child's participation in a game, the offering of a prize, or another activity requiring that the child disclose more personal information than is reasonably necessary to participate in such activity and requires the operator to establish reasonable procedures to protect the confidentiality, security, and integrity of personal information collected from children.[157]

COPPA also states that *parental consent is not required* when (1) online contact information collected from a child is used only to respond directly on a one-time basis to a specific request from the child and is not used to recontact the child nor is it maintained in retrievable form by the operator; (2) the information is collected for the sole purpose of obtaining parental consent; (3) the information is used only to protect the child's safety; or (4) the information is used to protect the security or integrity of the Web site, to take precautions against liability, respond to the judicial process, or to assist law enforcement in a matter related to public safety.[158] The details of these requirements are delineated at 16 CFR Part 312.

COPPA encourages operators to satisfy the above requirements by setting self-regulatory guidelines issued by representatives of the marketing or online industries.[159] Enforcement for violations of the requirements are treated as violations of a rule defining an unfair or deceptive act or practice prescribed by the Federal Trade Commissions Act at 15 U.S.C. § 57a(a)(1)(B).

8.9.2 Child Online Protection Act (COPA) of 1998

The Child Online Protection Act (COPA)[160] was also enacted on October 21, 1998, but codified at 47 U.S.C. § 231 as an update to the Communications Act of 1934. Specifically, COPA prohibits the online communication, for commercial purposes, of material that is both available to and harmful to minors. Unlike COPPA, which defines a *child* as any individual under the age of 13, COPA defines a *minor* as any person under the age of 17.

To implement COPA and determine details such as the definition of *harmful,* the following have been helpful: (1) House Commerce Committee Report on the Child Online Protection Act—*H. Rept. 105-775;* (2) Views of Justice Department on the Child Online Protection Act; and (3) The *Federal Trade Commission,* Notice of Proposed Rulemaking, 4/20/99.

8.9.3 Child Pornography Prevention Act (CPPA) of 1996

The Child Pornography Prevention Act (CPPA) of 1996,[161] enacted on September 30, 1996, and codified at 18 U.S.C. §§ 2252A, *et. seq.,* amended the criminal code (Title 18) to define child pornography to mean any visual depiction, including a photograph, film, videotape, *or computer image, produced by any means including electronically by computer,* of sexually explicit conduct if (1) its production involved the use of a minor engaging in such conduct; (2) it appears to depict a minor engaging in such conduct; (3) it has been created, adapted, or modified to appear that an identifiable minor is engaging in such conduct; or (4) it is promoted or advertised as depicting a minor engaging in such conduct. . . . [It] revises the definition of 'visual depiction' to include *data stored on computer disk or by electronic means* which is capable of conversion into a visual image. . . ."

8.9.4 Protection of Children from Sexual Predators Act (PCSPA) of 1998

The Protection of Children from Sexual Predators Act (PCSPA) of 1998,[162] enacted on October 30, 1998, and codified at 42 U.S.C. § 13032, amended the Victims of Child Abuse Act of 1990, codified at 42 U.S.C. §§ 13001 *et seq*. The updated law requires providers of electronic communications services to report any known use of their systems to engage in prohibited acts of child pornography. As part of this, PCSPA defines the term *electronic communication service* from Section 2510 of Title 18 and the term *remote computing service* from Section 2711 of Title 18.

8.9.5 Children's Internet Protection Act (CIPA) of 1999

The Children's Internet Protection Act (CIPA), introduced in the U.S. Senate in both 1998, as Senate Bill 1545, and 1999, as Senate Bill 97, would "require schools and libraries receiving universal service assistance to install systems or implement policies for blocking or filtering Internet access to matter inappropriate for minors, to require a study of available Internet blocking or filtering software, and for other purposes.

The challenges to this Act include the constitutionality of filters under the First Amendment and the quality of the filters or ability not to block desired content. Activity in this area includes the following:

1. Prohibitions of Certain Abortion-Related Communications via the Internet (18 U.S.C. § 1462)
2. Communications via the Internet (Letter from Attorney General Reno, 2/9/96)
3. Multimedia Law, Federal Republic of Germany
4. Public Library Internet Use Policy, Loudoun County, Virginia
5. Censorship in a Box: Why Blocking Software is Wrong for Public Libraries, ACLU
6. Freedom of Expression on the Internet, Human Rights Watch (cross-reference to URL)

Work continues to resolve these issues.

8.10 STATE PRIVACY LAWS

As noted in the previous section, federal privacy law addresses mainly government and interstate activity. State privacy laws, however, provide an equally important level of protection. Appendix H provides examples of current and proposed state privacy laws. It is meant only to provide a guide to the activity in each state and is not intended to be a complete or up-to-date listing of state privacy laws.

State privacy law is a rapidly evolving with great diversity from state to state. In addition, unlike the U.S. Constitution, many state constitutions include privacy concepts. Many explicitly recognize privacy concepts for public employees. Only California recognizes privacy for employees of private companies.

8.11 COLLECTION, USE, AND DISSEMINATION OF PERSONAL INFORMATION WITHOUT PERMISSION

As can be seen by the descriptions of the statutes in this chapter, privacy law in the United States today addresses mainly financial data, personal information held in government databases, illegal access to "protected" computers, harassment, and children. The statutes generally do not address information collected, used, or disseminated by private parties without the permission of the individual affected.

For example, for the convenience of its customers, Giant Food Inc., a grocery store chain in the Midwest, located pharmacies owned by CVS, a drug store chain, in its Giant Food stores. CVS sold its pharmacy customers' medical information to a marketing company that then sent coupons to the customers for drugs related to their disorders. The customers were outraged about the lack of privacy concerning their medical information and the company stopped the practice. What laws, if any, address this practice?

The problem becomes even more critical with the Internet. Most Web users believe that their activities on the Internet are anonymous and private, but this is rarely true. An individual's identity and activity on the web can be tracked by various Web sites, search engines, online services, and software packages through the use of "cookies," browsers, ISPs or other related sources of information. While many of these services are essential to productive use of the network, they also reveal private information that individuals do not wish to share. Few laws address or limit these practices. The details of areas not covered by the law are described in the following.

8.11.1 Cookies[163]

When you click into a Web site, that site frequently installs through your browser a small piece of software on your hard drive to assist the site's manager in providing you with access to the information on the site, and to track your activity and requirements at the site. This small piece of software is known as a *cookie,* shortened from its original name of "magic cookie," named after the tokens with magical powers in fantasy or role-playing computer games.[164] This "crumb" of software comes in two varieties: session and persistent.[165] A *session* cookie is deleted when the browsing session ends, while a *persistent* cookie remains in your hard drive after the session ends. Either variety of cookie provides significant information about the visitor to the Web site manager and others able to access the information. The use of this information can be beneficial, harmful, or potentially invasive. I. Herzoff, a commentator in the online field, calls these three types of use the "good, bad, and ugly"[166] uses of cookies.

A *good, or beneficial, cookie* allows the Web site operator to tailor the site to provide greater convenience and mobility for visitors. First, the cookie software may provide the Web site operator with information about the visitor's hardware and software required to enable the Web page to appear on the visitor's screen.[167] Second, by gathering "clickstream data," as it is called, cookies permit online shoppers to use "cyber-shopping carts," collecting items as they go and allowing the shopper to pay for the items at the end of the session. This avoids the inconvenience of having to pay for each item as the shopper selects it and having to request refunds when the shopper returns an item to the shelf in favor of another item. Cookies store the information about the shopper's selections until the shopper is ready to "check out." Frequent shoppers may reorder products without having to reenter key information and may collect "bonus points" for further discounts. Third, cookies can help the site manager improve the Web site by tracking what is of interest to visitors. Fourth, cookies can store logon information, such as the visitor's name, address, credit card, and password, to enable instant access to a Web site so that these need not be reentered every time the visitor visits the site. Fifth, cookies can customize Web pages by storing the last date and time a subscriber visited the site and updating the site since the last visit. This is especially useful in help centers or for situations such as online magazines, also known as "e-zines."

The *bad, or harmful, cookie* enables hackers and others to read the information stored in the cookie, including the visitors' name, address, and credit card number, and steal that information. Normally, as simple text files, cookies do not contain executable code that could be used to transmit a computer virus, read information residing on a user's hard drive, or read a cookie from another Web site. However, a flaw in version 3.01 of Netscape Navigator, for example, apparently gave Web site operators access to the hard drives of their Web site users who also used Windows.[168]

A *potentially invasive cookie* collects data about the visitor's movements in the Web site, and creates personal preference profiles or dossiers about the interests of each individual to use for marketing purposes—or to sell to other marketers. Historically, companies such as Focalink, NetGravity, Inc., and DoubleClick, Inc. have used software to assign each Web site visitor an identification number. The software then permits the private companies to track how many times the visitor clicked on each Web site, how long the visitor stayed in each site, and which items in the site interested the visitor the most. The company then uses the information to place targeted ads on the Web pages the visitor views, or sells the information to marketing companies, such as AdServer, which sends targeted ads to the visitor.

Several downsides to the use of cookies thus exist. First, beyond the obvious potential for invasion of one's privacy, Kate Komando, an Internet commentator, points out that the customization may become too narrow. She draws an analogy to a computer tracking the movements of a shopper in a supermarket. Based on the shopper's selections and interest in specific products on the first visit, the next time the shopper visits the supermarket, it would have only one aisle, containing the items selected or of interest the first time. This could be efficient, but it is also limiting and unnerving. Second, with concerns about "Big Brother," it could also severely chill people's willingness to use the Internet for information searches, especially for political information and democratic involvement. Third, in general, these cookies collect more personal information than most people are willing to share voluntarily and share it with more people than the individuals may want. For example, a number of software packages, online services, and search engines exist to use the information in cookies to provide information about other persons on the Web. These can help find a long-lost friend or a deadbeat parent, but more insidiously, they can also assist in spying on your neighbor or enable "cyber-stalkers." Some of these software and services include the following:

1. *Altavista, Yahoo, Excite, and Lycos* all provide facilities to search for individuals by name. Information provided by these search engines may include an individual's name, email address, street address, and phone number.

2. *American Information Network (AIN)* is a "for-payment service" that has access to many databases. It can provide private information not available from the free services.
3. *DejaNews* catalogs and indexes the information from more than 15,000 Usenet groups. The members of these groups were required to provide profile data before gaining membership to the groups. DejaNews makes these Usenet profiles available to searching parties in a manner often unknown to the members. For example, for a story on search software, the *New York Times* requested information on Tim May, a privacy and cryptography advocate. The search yielded Mr. Mays' phone number, email address, and 527 messages he had posted on various topics over the 18 months prior to the search.
4. *Four-One-One (Four 11)* is a search engine that maintains its own database of email addresses. Currently the database contains only addresses of individuals who have agreed to be included on the list.
5. *The Stalker's Home Page* is a Web site that contains links to other databases containing advice and information devoted to showing people how to collect information on other people through online resources.
6. *WebCrawler* is a search engine that has a "voyeur" feature that allows people to see the results of other Web users' searches for information. It does this by providing a sampling of the key words used by other users on the Web so that search results based on these keywords can be viewed. WebCrawler management states that while "the *Search Voyeur* continuously displays actual searches that people are doing on WebCrawler, [the site] receives over 5 million queries a day, making it impossible for anyone, WebCrawler staff included, to make the association between a particular search and the person who initiated it." However, given the creativity and resources of hackers and others, this assurance is not sufficient to appease many people.

Not all Web sites deposit cookies. The main exceptions are Web sites run by government agencies. Under both the Privacy Act of 1974 and the Electronic Communications Privacy Act (ECPA), these sites may not release personal information to third parties without the written permission of the person whose information will be released. All Web sites that do deposit cookies also limit the number of cookies on your hard drive by deleting old cookies beyond a preset expiration date.[169]

8.11.2 Social Security Numbers—Identity Theft

In the movie *The Net*, the character portrayed by Sandra Bullock had her identity stolen and would not have been able to reclaim it if she had been an average person, someone unfamiliar with computer systems. While such theft seems unlikely, a recent report by the U.S. General Accounting Office (GAO) stated that the number of consumer inquiries to the Fraud Victimization Department of just one company, Trans Union, a credit-reporting company, increased from 35,235 in 1992 to 522,922 in 1997. The number of inquiries from other companies was not provided.

In response, in 1998, Congress passed the Identity Theft and Assumption Deterrence Act (Identity Theft Act) to combat the increasing problem. Under this law, a conviction for identity theft carries a maximum 15-year jail sentence, a fine, and forfeiture of any personal property used to commit the crime. It also requires that the Federal Trade Commission (FTC) assist people who have had their identities stolen. Senator Jon Kyl, R-AZ, who sponsored the bill, said that it seems to be working. In 1999, 1,350 people were charged under the law and 644 people were sentenced. Of the 644, 407 entered guilty pleas.[170]

While a step forward, the Identity Theft Act (1) does not require full forfeiture of the property gained from the crime, (2) often treats incidents as misdemeanors rather than felonies, (3) is disorganized among agencies capable of pursuing convictions, and (4) does not address the general practice of making information available. Typically, identity theft occurs through credit card loss or confusion about similar sounding names, but the easy availability of Social Security numbers and maiden names on the Internet increases the likelihood that identity theft will happen to more people. For nominal fees, usually under $200, personal information such as Social Security numbers can be found on the Internet and used to create a new identity for opening an account, leaving the fraud victim with the bills.[171]

Thus, among the online services that make personal information available, those that include access to Social Security numbers are perhaps the most frightening. One such victim was Carroll Oberst of Redding, California. A man stole Oberst's Social Security number and other personal data off the Internet, opened bank and credit card accounts in Oberst's name, and ran up $35,000 in bills. The court gave the thief the maximum sentence of six years, but it has taken Oberst years to restore her credit.

U.S. laws specifically restrict the government's use of Social Security numbers in order to protect users confidentiality, but no such legal restrictions exist to limit the private sector's use of the information. As David Sobel, legal counsel at the Electronic Privacy Information Center (EPIC), an Internet civil rights organization, stated, "Only public opinion can curb the private sector." Mr. Sobel cited a 1992 case in Virginia, *Greidinger v. Davis,* in which Marc Alan Greidinger was required by the state to provide his Social Security number to register to vote, but challenged the order on grounds that it posed an unconstitutional burden. The court agreed, stating that "Armed with one's Social Security number, an unscrupulous individual could obtain a person's welfare benefits or Social Security benefits, order new checks at a new address on that person's checking account, obtain credit cards, or even obtain that person's paycheck."

Services such as West Publishing Information America, CBD Info Tek, and the International Research Bureau[172] all make Social Security information readily available. The International Research Bureau, for example, provides Social Security numbers to "people who qualify" for $13 through the regular mail. This service provides "whatever is available" within one day to whomever sends in a number to be checked. The first mass marketer of such [personal] information, however, was Lexis-Nexis, the well-established research firm used by corporations and individuals worldwide. In 1996, the company developed a new product called the P-TRAK Person Locator File, that put "300 million names right at your fingertips." The company billed P-TRAK as "a quick, convenient search [that] provides up to three addresses, as well as aliases, maiden names, and Social Security numbers" of the persons sought.[173]

Although the P-TRAK service was available "only to Lexis-Nexis," over 740,000 paid subscribers joined Lexis-Nexis, many for little as $125 per month. Privacy advocates and industry experts feared the potential of the P-TRAK service, its ease of use, and widespread access and actively worked to close the service. Lexis-Nexis defended its P-TRAK service, saying that the company could not be responsible for what is done with the information it provides. "Our company's policy has been and continues to be, that this product is to be used in a legal manner, and that's one thing that we try to stress with our customers," said Judith Schultz, Public Relations Manager for Lexis-Nexis. "If something did happen, we wouldn't deal with it because we are a third party."

Furthermore, Lexis-Nexis pointed out that the service could be useful in finding exspouses accused of kidnapping children or being behind with child-support, people who have skipped on bad debts, or criminals hiding in the population. Sobel and others, however, countered that federal law enforcement authorities already have such information through the files of the Social Security Administration, the IRS, and other databases. They do no need a third-party intermediary, such as Lexis-Nexis to provide such information.

In June 1996, Lexis-Nexis agreed to alter the P-TRAK service so that a person could not simply enter a person's name and obtain that person's Social Security number. However, by September 1996, it was still making personal information easily available. The company did provide an 800 number for people seeking to have their information removed from the service by filling out a form available at its Web site. However, the 800 number was continually jammed with calls. Persons desiring to have their names removed from the P-TRAK file could fax their name and address to 513-865-1930, although Lexis-Nexis is under no legal obligation to fulfill such requests.

In response, victims such as Oberst have formed privacy victims' groups to fight the cheap, ready access to the range of personal details that are available through computer networks. The groups did not form to "get even," but rather to help others avoid their experience by urging state and federal lawmakers to pass more stringent laws to restrict the distribution of personal information and thus to protect electronic resources from abuse by unscrupulous individuals and businesses.[174] The issue is still open and evolving, well into the early 2000s.

8.11.3 Intellectual Property Theft

Similar to identity theft, the Internet has opened new opportunities for theft of ideas, music, film rights, marketing concepts, logos, and trademarks, making each more difficult to protect. In response, laws such as the Audio Recording Act of 1992 have been passed. The Audio Recording Act of 1992 seeks to prevent multiple generations of digital audio recordings by requiring all digital recording devices to adhere to the standards of the Serial Copy Management System.[175]

8.11.4 Monitoring by Browsers and ISPs

In addition to information collected from cookies, each ISP is in a unique position to know the identification of each individual user on its server, including those using "net names" or aliases, and the usage and preferences of those individuals. As each user's access to the net, it is technically possible for an ISP to monitor each mouse click and keystroke made during a session. The ISP also has stored electronic copies of the data and emails that passed through its facility and may search these records.

Most ISPs do not monitor the details of the traffic on their site, but have been requested to do so by law enforcement and federal agencies. For example, the federal government recently proposed having the ISPs monitor such traffic as offshore gambling to assist the federal government in collecting the income tax due from the winnings. In addition, some ISPs have shared user information and preferences with outside marketing companies. For example, in 1997, AOL discarded a plan to sell millions of members' telephone numbers to telemarketers wanting "to pitch everything from discounted travel to phone service."[176] Few laws address or limit these practices.

8.11.5 Purchased Information and Merged Databases

Not all sources of information come from cookies or the Internet. For example, in November 1999, DoubleClick purchased Abacus Systems and compiled personal information about the residents of millions of households who had purchased items through mail-order catalogs. It was DoubleClick's intent to merge the information held by Abacus with the information DoubleClick had collected from the Internet. In protest, a California woman sued DoubleClick, citing privacy issues. During the suit, DoubleClick dropped its plans, but it is only one example of such mergers and sharing of information.

Similarly, in the mid-1980s, the Bell companies first were prohibited from providing information services because Judge Greene was concerned that, because their control of their customers' lines of communication, they also had access to their customers lines of credit, travel plans, credit card expenditures, medical information, and general calling patterns. Judge Greene was concerned that a Bell company could, therefore, pinpoint subscribers for Bell company–generated information and other products which would open a "Big Brother" type of relationship for their customers.[177]

This prohibition was overturned in *United States v. Western Elec. Co.*,[178] but only with the continuing restriction that the Bell companies not offer "user profile" services. Judge Greene's concerns were eventually addressed by federal and state wiretapping and eavesdropping statutes, plus FCC regulations concerning a telephone company's use or release of subscription, billing, or signaling data (§ 5.5) or "customer proprietary network information" (§ 11.8).

In other cases, groups such as radio stations, grocery stores, and associations frequently sell their customer lists and purchase information to other groups. Most of the time, the individuals affected do not know the information has been collected, how it is being used, or that it is being sold. Neither can the individual generally correct any misinformation. Few laws exist to track or control such activity.

8.12 SOLUTIONS TO COLLECTION AND USE OF INFORMATION WITHOUT PERMISSION

When the collection, use, and dissemination of personal information is discerned, it usually arouses a significant amount of public concern.

> ***Scenario 4:*** *For example, imagine that you use your computer to price vacation options over the Internet, and, after browsing from one Web site to another, you find several great destinations with excellent transportation costs, hotel packages, and entertainment options. You decide to purchase one, enter your name, address, phone number, and credit card information. You log off, excited about the unique bargains you've found and pleased with the convenience, efficiency, and ease of use of the Internet.*
>
> *Several days later, you begin to receive numerous offers through both the postal service and email from travel agencies, luggage companies, and travel magazines offering you special deals on travel-related items. You realize that this probably is not a coincidence and that someone has tracked your Internet use and sold the information. In addition, when you receive your next credit card bill, you notice several charges that you do not recognize. How can you protect yourself from these downsides to the many advantages of the Internet? What are your legal rights in these areas?*

A survey by a Boston consulting group revealed that 86% of persons polled want to control the use of personal and demographic information about themselves, 81% felt that sites should have no right to resell information without consent, and 70% said that privacy was the primary reason for not registering at Web sites. The problem is that individuals rarely know that they have received cookies, or from which Web sites, or that they are on a list. Several solutions have evolved in response to this situation. These include the following: (1) proposed legislation; (2) customer caution, aliases, and other nondisclosure; (3) industry self-regulation; (4) voluntary programs; (5) software to find cookies and purge them from your hard drive; and (6) Web sites developed to provide ongoing information about privacy and options. These are discussed below in the following seven sections.

8.12.1 Proposed Legislation

In response to the public's concern about privacy with new telecommunications technologies, including the Internet, the U.S. Congress began considering over 50 privacy-protection measures in the early 2000s. As yet, however, such bills have not passed, and few laws address or control these new technologies and capabilities. In addition, at least three problems exist with these proposed laws. First, marketers who use the electronic information to custom-tailor pitches to customers resist the proposed restrictions and actively lobby against them. Second, free speech advocates have raised concerns about the potential of these laws to restrict legitimate access to information. Third, and perhaps the largest problem, is that the proposed laws target mainly financial and medical information. Other issues, such as cookies, merged databases, and programs that track credit card use or grocery store discount programs are not addressed in the proposed legislation.

Even at the state level, the large number of proposed privacy laws address more traditional privacy concerns of financial and medical information. For example, a large number of states including California, Colorado, Massachusetts, New Jersey, and Oklahoma, are considering proposed laws that would prohibit the sale of medical records to marketers and other third parties for nonessential purposes.

To fill this gap, legislators, such as Bruce Vento of Minnesota, introduced legislation such as the Consumer Internet Privacy Protection Act of 1997. Such bills would require written consent from an individual before a computer service could disclose a subscriber's personal information to a third party. It would also allow consumers to access and correct information they believe is faulty. The chances for passage of such bills, however, are uncertain.[179]

In the meantime, the Federal Trade Commission (FTC), continues to encourage the online industry to develop its own technological safeguards and monitors progress in periodic hearings begun in June 1997. In updates to documents such as its June 1998 *Privacy Online: A Report to Congress,* the FTC continues to track cookie technology. In general, industry experts agree that the direction in which privacy law evolves will determine the public's confidence in the systems, especially the Internet.[180] Therefore, the FTC "encourages" those collecting information from customers of Internet Web pages to "adopt reasonable privacy protection practices." If this proves to not be enough, Congress likely will pass new legislation.

8.12.2 Tort Action

Some attorneys suggest that cookies are challengeable under the laws of "intrusion into seclusion," "trespass," and/or "conversion." As noted earlier in this chapter, the Restatement (Second) of Torts § 652B (1976), defines *intrusion* as the "intentional intrusion, physical or otherwise, upon the solitude or seclusion of another or his/her private affairs or concerns, that would be highly offensive to a reasonable person." Certainly, cookies (1) are intentional, (2) are invasions into the visitor's personal computer, and (3) are highly offensive to a reasonable person. *Trespass* applies because cookies actually enter and attach to the visitor's computer. *Conversion* is the illegal act of using another's property without the property owner's consent, and thus, is also a form of stealing. Since cookies use the Web site visitor's computer for a purpose to which the unknowing receiver of a cookie has not consented and occupy a small amount of space from the visitor's computer for the cookie, both actions could be considered conversion. Additionally, if information is stolen or if that stolen information is shared with others, both theft and the tort of "public disclosure of private facts" likely have been committed. Even if the parties do not litigate, the acts violate most people's sense of ethics and fair play. To date, the cases making these arguments have not been decided.

8.12.3 Customer Caution

Until laws are passed and proven to be effective, however, individuals are being cautious to protect themselves. Surveys indicate that nearly 70% of users who want access to Web sites, but fear invasion of their privacy, give false information, use aliases, or use proxy servers when required to provide information about themselves. This, of course, defeats the purpose for marketers, so it is in the interest of all parties to find a solution to consumers' concerns.

8.12.4 Industry Self-Regulation

In response to public concern about privacy and the companies' concerns about the proposed legislation, companies have developed various levels of self-regulation including (1) privacy statements to customers; (2) "opt-in" voluntary participation programs; (3) "opt-out" options requiring notification by the individual; and (4) organizations such as the "e-Trust" rating system developed by the Electronic Frontier Foundation. Each is discussed below.

Privacy Statements to Customers

Today, most companies issue privacy statements to assure their users of confidentiality. The Direct Marketing Association (DMA) strongly encourages their members to disclose their cookie practices and provides a model statement for which their members may use or amend. Several examples of these agreements are presented at the DMA's Web site at *www.the-dma.org*.

However, these privacy statements are only the companies' "voluntary best-effort" promises. They actually promise very little and have no real teeth if the promise is not kept, since no damages are available to the injured parties. An example of this was a privacy study completed by the California HealthCare Foundation (CHCF). The study reviewed 21 popular health-related sites to discern how each site handled the privacy of the information it received. Of interest to the CHCF was not the promises made, but what actually happened with the information. To its dismay, the CHCF found that most of the sites did little to protect the actual privacy of the information, despite their claims. Instead advertisers were able to obtain the names, addresses, and significant private health information of most site visitors. Details of the study and its findings can be found at *www.chcf.org*. For this reason, corporations, health-care organizations, insurers and other large collections of users frequently negotiate a contract, written with specific nondisclosure language and agreed-upon damages if the information is revealed.

Opt-In Programs—Voluntary Participation

Rather than participate in the unauthorized collection of personal information, some Web sites and consumer interest–tracking companies work only with voluntarily provided information. Proponents argue that this allows the individual to be in control of what is received, maintaining the free speech rights of individuals to receive the information they desire.

In general, this voluntary method of collecting information falls into two categories known as "opt-in" and "opt-out." Opt-in programs generally require a visitor to register before participating, such as visiting the Web site. The opt-in programs generally state what the information will be used for, how long it will be retained, and how the privacy of the information will be protected while it is retained.

Opt-in is the approach preferred by privacy advocates, because personal information must be released by the individual. In addition, it ensures that the consumer is aware of the right to voluntarily provide the information and who is collecting it. However, many others voice concerns about opt-in programs because while individuals may give the information for one purpose, such as to enter a Web site, register a product for warranty purposes, or complete a license or job application, they may not be aware that the recipient of the information legally may use that information for other purposes or sell it.

Once the information is released, the individual has little control over its dissemination. Unless the recipient specifically states otherwise, frequently this information is sold to other marketers, generating sales calls.

Two branches or variations of opt-in have evolved: *clustered data programs* and *open profiling*. With clustered data, companies such as NetCount, ask visitors to provide specific information, including their name, street address, zip code, and email address, but then the companies compile the information into generalized demographics known as "cluster data," which they sell to marketers. The second variation started in June 1997 when, led by Netscape, proposed a group of over 40 companies proposing an *Open Profiling Standard (OPS)*. The Open Profiling Standard suggested that online users complete a form providing their names, postal addresses, email addresses, and demographic information such as age, marital status, income, and hobbies/interests. It is not clear to many individuals why users would want to provide such information or what privacy concerns would be guaranteed if they did, but even if the Open Profiling Standards were successful, industry analysts expect cookies would remain.[181]

Opt-out Options—Forms and Contact Numbers

On the other hand, opt-out programs offer individuals the opportunity to have their names, addresses, and phone numbers removed from databases and marketing lists. For example, the Telephone Consumer Protection Act (TCPA) of 1991 requires that unsolicited faxes and email include a phone number or return address for individuals to request removal of their names from the distribution lists. However, many of these phone numbers do not answer and mail sent to the return addresses frequently is returned. The downside to this approach is that an individual must make a request to each and every holder of his or her information. Until then, holders of the information have free use of any personal information collected. Marketers and others tend to prefer this approach because, while notice of the option must be given to users, most people are not aware of it or its importance and thus will not respond to it. This has been the case with most bank, credit card, and insurance privacy statements mailed in 2000 and 2001. Most consumers simply throw away the privacy statements, considering them to be junk mail or advertisements slipped into their monthly statements. They usually do not read through the several pages of small print to learn that they must act to protect their privacy. Many also have tight deadlines.

The option places the burden or responsibility on the individual to stop the flow of junk mail. In addition to being burdensome and time consuming, opt-out programs tend to focus on marketing lists. Individuals may not realize how much personal information other parties have about them and therefore from whom they should request that the information be removed. Thus, the manner in which the notice or option is presented to the public can have a tremendous impact on the effectiveness of this option in protecting privacy.[182]

The Center for Democracy and Technology (CDT)

The Center for Democracy and Technology (CDT) (at *www.cdt.org*), an organization established to promote democratic values and constitutional liberties in the digital age, has developed a Web site called "Opt-out Online" (*www.opt-out.cdt.org/online*) in which it provides links to the companies that allow users to opt-out of online information gathering while online. Similarly, at the "Generate Opt-out Forms" Web site (*www.opt-out.cdt.org/submit.shtml*), the CDT provides preprinted forms for mailing to those companies that do not allow consumers to opt-out while online.

Beyond these Web sites, the CDT seeks to build expertise in law, technology, and policy in order to find practical solutions to enhance free expression and privacy in global communications technologies and to build consensus among all parties interested in the future of the Internet and other new communications media.

e-Trust Rating Organization—The Electronic Frontier Foundation

The Electronic Frontier Foundation (EFF), a San Francisco–based organization that tracks online privacy issues, advocates solutions, and supports the concept of informed consent for individuals, has, with other organizations, formed the "e-Trust." The e-Trust is an organization that rates the privacy policies of various Web businesses. e-Trust has created three different logos or ratings, called "trustmarks" available to online companies for a price, to validate whether the site: (1) collects no data on users, (2) collects data only for the site's own use, or (3) shares or sells the data collected with third parties. These trustmarks are titled "No exchange," "1 to 1 exchange," and "3rd party exchange," respectively. America OnLine (AOL) and numerous other companies have assisted in testing the e-Trust system.[183]

Internet Engineering Task Force (IETF)

Similarly, the Internet Engineering Task Force (IETF) has agreed upon voluntary rules requiring that Web software include "cookie warnings" and give users control over their "cookie trails" or electronic footprints. To do this, the IETF recommends that browsers insert default settings to reject cookies if they exceed proposed standards. The proposed standards are listed at *portal.research.bell-labs.com/dmk/cookies*. Since decisions are made by consensus, no deadline exists. However, the standards are viewed very favorably in areas with strong personal privacy cultures, such as Europe.

8.12.5 Software to Detect and Delete Cookies and Spam

Several Web sites and software packages have evolved that enable users to determine if cookies exist on their hard drives and from which sites. The software then either allows users to purge the cookies from their hard drive entirely or to sort through their cookies and retain some. This option is available because many Net users believe that some cookies are useful and should be retained. A list of such Web sites and software resources is provided in Appendix I.

8.12.6 The Individual's Ability to Correct Inaccurate Information

Accepting that information may be collected, retained, and distributed without their knowledge or permission, individuals want to be able to sort through information about them, and to correct any inaccurate information that may exist. For example, Bronti Kelly of Temecula, California, won his case against the May Co. for placing erroneous information about him on the Internet, mislabeling him a shoplifter. This erroneous information continually caused potential employers to avoid hiring him and he ended up homeless when he could not get a job. Because of the May Co.'s failure to correct the misinformation, the company was ordered in January 1998 to pay more than $73,000 to Kelly.

Background-Check Companies

Currently, employers are not required to tell candidates why they were not hired. However, a bill being considered in several states, including California, would require employers to inform candidates if a background check was the reason why they were not hired. In addition, the bill would give individuals more rights to correct any information in the background report that the individual felt was flawed. With these precautions, employers could still use the background checks, but job seekers would know that such reports were being used and be able to view the information in the reports.

In the meantime, background check companies are starting to police themselves. For example, Background America, a background-check company based in Nashville, Tennessee now requires employers that use its service to clearly notify job candidates that they are conducting a background check. This would have corrected the problem for Kelly.

Credit-Reporting Companies

In other cases, such as credit reporting, the issues of incorrect information have been raised in court. The following case provides an example of this.

In the case of *Dun & Bradstreet v. Greenmoss Builders*[184] Dun & Bradstreet released a credit report to a client stating that Greenmoss had filed bankruptcy. This was not true, but based on the report, the client denied credit to Greenmoss Builders. Greenmoss sued and the issue was eventually heard by the U.S. Supreme Court which held that "credit reports are not the subject of public controversy" and thus established a basis upon which to protect the "privacy of purely private information of no public concern."

8.12.7 Web Sites Offering Information on Privacy Protection

"With advent of the Internet, you're seeing a whole new class of victims," said Dave Banisar, staff counsel at the Electronic Privacy Information Center (EPIC), a Washington-based watchdog group.[185] However, as noted above, it is up to individuals to educate and protect themselves and their personal information. This is an ongoing effort, as both the technology and sources of information change. The following list provides a sample of Web sites and other sources of information to assist individuals in this process.

1. **ACLU Freedom Network**—The ACLU provides information for citizens on how to preserve their right to privacy in situations involving electronic data and communications. Such information is available at *http://www.aclu.org/privacy/*.
2. **Better Business Bureau Online**—The Better Business Bureau (BBB) Online's Privacy Seal Program, including the bureau's dispute resolution procedures promote trust and confidence on the internet is avilable at *http://www.bbbonline.org/* or *http://www.bbbonline.org/download/DR.PDF.*
3. **Center for Democracy and Technology**—The Center for Democracy and Technology (CDT) states on its Web site at *www.cdt.org* that it believes "that maintaining privacy and freedom of association on the Internet requires the development of public policies and technology tools that give people the ability to take control of their personal information online and make informed, meaningful choices about the collection, use and disclosure of personal information." The CDT also has a specific privacy site at *http://www.cdt.org/privacy/* and a link to a second site at *www.13x.com/cgi-bin/cdt/snoop.pl* that explains cookies and provides Internet users with updated information on how to retain their privacy and browse the Web with anonymity. As a demonstration of how easily cookie technology allows a Web site operator to gain information, this site displays information about the visitor, such as the visitor's computer model, work place, and email address. Normally, cookie information does not include a visitor's e-mail address, but if a Web site visitor is using JavaScript, it can add such information.
4. **Electronic Frontier Foundations** (EFF)—"protecting rights and promoting freedom in the electronic frontier" is the Foundation's slogan. Its Web site at *www.eff.org* provides an excellent index of laws, articles, and other information on preserving the online privacy of individuals.
5. **Electronic Privacy Information Center** (EPIC)—The Center's Web site at *www.epic.org* provides a wonderful archive plus up-to-date articles and commentary on information regarding communications privacy.
6. **Online Privacy Alliance**—The Online Privacy Alliance (OPA) describes itself as "a diverse group of corporations and associations who have come together to introduce and promote business-wide actions that create an environment of trust and foster the protection of individuals' privacy

online." Its Web site at *http://www.privacyalliance.org/* provides an industry perspective on privacy issues.

7. **PrivacyPlace**—The Web site at *www.privacyplace.com* provides daily wire stories and a library of publications on privacy issues.
8. **Privacy International** (PI)—Privacy International, based in London, England, describes itself as "a human rights group formed in 1990 as a watchdog on surveillance by governments and corporations. From London and a branch office in Washington, D.C., PI conducts research throughout the world on issues such as wiretapping, national security activities, ID cards, video surveillance, data matching, police information systems, and medical privacy." Summaries of its work can be viewed at: *http://www.privacy.org/pi/*.
9. **Privacy Rights Clearinghouse** (PRC)—The Privacy Rights Clearinghouse is a nonprofit organization based in San Diego, California, that "offers consumers a unique opportunity to learn how to protect their personal privacy. Our publications provide in-depth information on a variety of informational privacy issues, as well as practical tips on safeguarding personal privacy." The clearinghouse provides a database of speeches, Congressional testimony, privacy laws and issues at its Web site at: *http://www.privacyrights.org/AR/ar.htm*.
10. **Tech Law Journal's Privacy News**—The Journal's Web site at *http://www.techlawjournal.com/privacy/Default.htm* provides "news, records, and analysis of legislation, litigation and regulation affecting the computer and Internet industry."
11. **TRUSTe**—This third-party oversight privacy program describes itself on its Web site at *http://www.TRUSTe.org/* as "an independent, nonprofit privacy initiative dedicated to building users' trust and confidence on the Internet and accelerating growth of the Internet industry. We've developed a third-party oversight "seal" program that alleviates users' concerns about online privacy, while meeting the specific business needs of each of our licensed Web sites."

8.13 APPLICATION OF PRIVACY LAW TO MODERN TELECOMMUNICATIONS TECHNOLOGIES AND SERVICES

Scenario 5: *You return to your desk after lunch and decide to check your voicemail, email, and postal mail before you pick up the project you were working on during the morning. One of your voicemail messages is from your brother giving you detailed flight information for his trip to see you. One email message is an off-color joke forwarded to you by a friend. Another email is a ques-*

tion from a friend asking a simple question in your professional area of expertise. In your stack of postal mail are several unsolicited ads or offers for merchandise. As you quickly toss or respond to each, it never crosses your mind that while the mail in the trash is destroyed, the voicemail and email are communications that likely will be retained by your employer for over one year on tape. What are your privacy rights concerning these retained messages?

8.13.1 Email and Voicemail Systems

Another area of significant concern today is the current law surrounding an individual's privacy when using new technologies such as email and voicemail. The statutes in this area typically fall into three categories: (1) laws affecting government systems, intended for official government business; (2) laws affecting public systems offering electronic communications services to the public, such as AOL, CompuServ, Microsoft Network, and ISPs; and (3) laws affecting private systems. As noted earlier in this chapter, most existing privacy laws address government systems, fewer laws address public systems, and almost no laws address private systems. All seem to turn on the "reasonable expectation of privacy." The laws in each category are discussed in the following three sections.

Government Systems

The creation, management, transfer, and disposal of information in government files, including stored electronic messages, known as electronic mail or email, and unauthorized access to computer databases are controlled by the the Federal Records Act of 1950, Title 2 of the Electronic Communications Privacy Act of 1986, the Privacy Act of 1974, and the Freedom of Information Act.[186] The sections of the Telecommunications Act of 1996 specifically addressing email and voicemail systems support the concepts in these laws.[187] However, the Supreme Court also noted in *ACLU v. Reno* that an email or other information sent over the Internet is not secure unless encrypted because "it is no less secure than talking on a cellular phone." Therefore, no "expectation of privacy" should apply to unencrypted Internet communications or transactions.

Public Systems

Customers of companies providing electronic communication services to the public, such as voicemail and email, have a strong expectation that, as with the telephone system, their communications are private and that the service provider will simply transmit and store the message until it is retrieved and deleted by the customer. As codified in the Electronic Communications Privacy Act, the operators of public systems may not disclose the contents of a message unless an authorized agency produces a court order to review the message.[188] [189]

This prohibition is so strong that courts have upheld it even in the face of otherwise unpopular or illegal actions. For example, in *United States v. Maxwell*,[190] a child pornography case involving the seizure of files held by AOL, a U.S. Air Force Court of Military Appeals ruled that under the Fourth Amendment, the defendant had a right of privacy in "any e-mail transmissions he made so long as they were stored in [an] America Online computer." The court stated: "[The defendant] clearly had an objective expectation of privacy in those messages stored in computers which he alone could retrieve through the use of his own assigned password. Similarly, he had an objective expectation of privacy with regard to messages he transmitted electronically to other subscribers of the service who also had individually assigned passwords."

However, since the case involved child pornography, the court also affirmed that there was probable cause "within the intent and meaning of the Fourth Amendment" to support issuance of a warrant authorizing the search of America Online computers. With such a warrant, law enforcement's review of the messages would be legal, just as wiretapping is legal with a warrant.

Private Systems

While the Electronic Communications Privacy Act (ECPA)[191] prohibits the operators of *public* voicemail, email, and other electronic communications systems from disclosing the contents of messages in storage, operators of *privately owned* corporate systems usually are not restricted by the same prohibitions.[192] This is important since nearly all companies today have internal email and voicemail systems and the employees of those companies often do not have "privacy rights" under the law. Even the use of these systems for communications outside of the company are included when such communications occur over systems owned and managed by an employer.

However, as the law has evolved, it has also been suggested that employees may have a right to privacy under state tort law. Of the four torts discussed in Section 8.2, the tort of "unreasonable intrusion into the seclusion of another" is the most relevant to email interception. The elements considered in a tort liability case include (1) did the employer initiate the communication or have the power to avoid it? (2) did the employee have a reasonable expectation of privacy concerning messages received over the employer's system? (3) did a legitimate business reason exist for the employer's intrusion that was sufficient to override the employee's expectation of privacy?[193] and (4) what liability, including loss of employment, exists for messages received, whether in the employee's power to stop the messages or not?

Concerning the first element, an employee may not have requested the communication, may not be able to stop it from being sent, and should not be responsible for it. However, concerning the second element, when the communication is intended by the employee, studies have shown that the average employee today expects the same level of privacy in corporate communications as private communications.

Concerning the third element, "legitimate business reasons for intrusion or review of electronic systems, desks, and offices by employers," companies cite the corporate resources used, work time spent reviewing personal communications, and the company's liability or embarrassment when the communications contain comments, off-color jokes, racial or gender slurs, or other subject matter of concern to the employer.[194] In reviewing these concerns, courts consider whether authorized personnel, such as systems operators, supervisors, and some coworkers, have legitimate reasons to access and read messages and files on the systems. For example, such surveillance is important in cases where corporate espionage or liability may be an issue. In other cases, such as routine office work, such access may be held to be less acceptable, depending on the nature of the work and the importance of the electronic system to that work. Some middle ground exists, for example, where employers have a legitimate need to (1) monitor communications over its systems for quality control, although employees must be notified that such monitoring occurs; (2) track communications to protect its strategies, new product information, and intellectual property; (3) communicate with employees via email; (4) manage third-party access to company records; (5) manage confidential information about employees such as salaries, benefits, and emergency information; and (6) investigate issues such as work-related injury, insurance, or medical claims. To decide, courts look to what practices, policies, procedures, and agreements exist in the company that may or may not create a reasonable and legally enforceable expectation of privacy among employees concerning their interoffice email and other computer files.

In most cases, to ensure that no confusion exists for employees, courts look to whether companies provided notices to employees stating that the employees should have no "expectation of privacy" in their interoffice email, voicemail, and other files, whether on paper or in electronic form. Of the companies that have provided such statements, many (1) include them in their employee manuals; (2) require a signature from each employee stating that he or she has read, understood, and agreed to the fact that they have no reasonable expectation of privacy in their communications over employer-provided equipment and facilities; (3) train their supervisors and system operators to ensure that they make no statements, agreements, or establish customs, practices, policies, or procedures that would create a reasonable or legally enforceable expectation of privacy by employees in their workplace computers, interoffice email, voicemail, desk, office files, or other employer-provided spaces and systems.

In some cases, courts have found that when employees used a lock, password, or encryption to protect certain items, that action created an "expectation of privacy" that could be violated when companies break the lock, password, or encryption. Again, however, this is a nonissue where the employer has a written policy, signed by all employees that clarifies that the employer may override locks, passwords, and encryption to gain access to systems, files, messages, or property.[195] Two examples of these rulings follow.

California Court of Appeals: Bourke v. Nissan Motor Corp.[196]

While demonstrating the email system at a Nissan Infiniti car dealership in California, a manager randomly selected an email message on the system. It revealed a message of a "personal, sexual nature and not business-related" subject matter. The incident was reported to management and the email messages of all employees were reviewed. In the process, Nissan discovered several other personal email messages, including some with sexual comments. As a result, two employees were fired for violating the company's policy that prohibited the use of email for personal communications.[197]

The two employees sued Nissan for (1) wrongful discharge in violation of public policy, (2) violation of their constitutional right to privacy, (3) invasion of privacy, and (4) violation of the criminal wiretapping and eavesdropping statutes. The trial court granted summary judgment for Nissan stating that the plaintiffs had no "objectively reasonable expectation of privacy" in their email messages because they had read and signed a statement of the company's policy restricting use of email for business purposes, had signed a waiver, and as employees familiar with the system, were aware that their email messages would be read by coworkers. The appellate court upheld the summary judgment.[198]

The plaintiffs countered that they had a reasonable expectation of privacy because they were given passwords to access the system and were told to safeguard their passwords. The court recognized that this could give rise to a subjective expectation of privacy, but the plaintiffs failed to prove an objectively reasonable expectation of privacy.[199]

Plaintiffs also argued that Nissan violated Section 631 of the California Penal Code, which prohibited, in part, the use of any telephone, wire, line cable, or instrument to read or learn the contents of any communication while it is in transit over a wire or line or cable.[200] The court rejected the argument because (1) accessing a computer network did not constitute tapping into a telephone line, (2) Nissan had a right to connect to the system as the system operator, and (3) Nissan did not access the message during transmission.[201]

Pennsylvania (1996): Smyth v. The Pillsbury Co.[202]

In another case, Smyth was an "at-will" employee of Pillsbury who received certain email messages on his home computer from his supervisor over Pillsbury's email system. On at least one occasion, Smyth sent emails to his supervisor that contained offensive comments about the company's management and holiday party. Company executives saw a printout of one message, read all of his email messages, and terminated him for "inappropriate and unprofessional comments" over the company's email system.[203]

Smyth filed a wrongful discharge action alleging that Pillsbury had violated his right of privacy under Pennsylvania's law and that the company's reading of his email was an impermissible intrusion on his seclusion. He further stated that his employer had repeatedly assured its employees that all email communications would remain confidential and privileged. The court determined that (1) any reasonable expectation of privacy was lost once plaintiff communicated to a second person over the company's email system, (2) a reasonable person would not consider the employer's interception to be a substantial and highly offensive intrusion upon seclusion, and (3) the company's interest in preventing inappropriate and unprofessional comments or illegal activity over its email system outweighed any privacy interest the employee may have had.[204]

8.13.2 Cordless/ Wireless Phones

> ***Scenario 6:*** *In the rush of your day's activities, you pick up your cellular or cordless phone to make a series of quick personal calls. First, you make an appointment with your physician for a minor problem that you've been experiencing. You briefly describe the symptoms to the nurse, providing only enough information for the nurse to determine which specialist you should see and the amount of time needed for the appointment. Then you call a friend and, in the short conversation, mention a sensitive personal issue with which you are dealing. However, in the midst of one conversation, you hear a neighbor's voice questioning who is on the line. You realize that wireless phones broadcast your conversations and can mistakenly be overheard by others. How does the law address this?*

As also noted in the discussion of the Electronic Communications Privacy Act (ECPA) in Section 8.3, conversations over cordless and cellular phones do not have the same legal protections as wireline phones. This is because the wireless portion of the communication is broadcast over a radio frequency. [205] As Newt Gingrich and Prince Charles of England found, many parties can intercept and listen to such conversations. Thus, users of cellular or cordless phones should be aware that no "reasonable expectation of privacy" exists with wireless technology. Therefore, they should select another phone to discuss sensitive business details with colleagues, medical details with their doctors and nurses, or private, personal details with family and friends.

It is for this same reason that the definition of *wire communications* in the Electronic Communications Privacy Act (ECPA) excludes the transmission of voice by microwave, auxiliary broadcast services, satellite, or other radio communications media,[206] and communications transmitted using modulation techniques, such as "spread spectrum," that have been withheld from the general public to protect the privacy of communications. Similarly, attorneys, physicians, clergy, and counselors are warned that they lose their attorney-client, physician-patient, and counselor-counselee privileges if they use cordless or wireless phones for such communications. The broadcast component of wireless phones is also what makes them vulnerable to *cloning,* or the stealing of the number from the wireless phone while it is in use, in order to make fraudulent calls using that number later. While U.S. law addresses this issue under fraud law, most companies and users look to technological solutions to address this issue.

8.13.3 Pagers

U.S. privacy law concerning paging systems varies. The Electronic Communications Privacy Act acknowledges three types of pagers: (1) tone only, (2) voice, and (3) display pagers. *Tone only pagers* may be monitored without a court order because users have no reasonable expectation of privacy for the tone portion. *Voice pagers,* on the other hand, are addressed in Title 1 of the Electronic Communications Privacy Act as a continuation of the original wire communications.[207] Similarly, *display pagers* are covered in Title II of the Act as part of "data record". On the other hand, in *United States v. Meriwether,*[208] the court ruled that "acquiring a telephone number from a digital pager does not constitute interception within Title 1 of the ECPA because no separate device was used for the interception and because transmission over this system had ceased and [the] number was stored." In general, cases involving the privacy of pager communications are made on a case-by-case fact basis.

8.14 INTERNATIONAL PRIVACY CONCERNS

Like the United States, other countries in the world are also addressing the challenge of meeting the privacy needs of their citizens in an environment of rapidly changing technologies. However, while U.S. privacy law currently tends to view personal data as a resource to help companies identify potential customers and to serve existing customers, other countries, including the members states of the European Union, view privacy as more of a human right. Thus, while U.S. law allows the extensive collection and use of such information without the knowledge or permission of the individual, other countries find this unacceptable. While U.S. privacy law currently operates mainly on industry codes of conduct expressed in privacy statements offered by companies to their customers, other countries have developed specific privacy directives. Normally, such diverse legal systems would continue to operate independently, but as the Internet and electronic commerce expand, other countries find that they cannot meet their required level of personal data protection if that information flows to less protected areas, such as the United States. Thus, some of these more privacy-oriented countries have placed embargoes on the transmission of personal data to the United States until the United States increases its protections for such data.

8.14.1 European Community's Privacy Directive

One such example is an embargo placed on data to the United States from the European Community (EC). In October 1995, the EC adopted its' Directive on the Protection of Individuals with Regard to the Processing of Personal Data and on the Free Movement of such Data[209] (EC Personal Data Directive). The Directive became effective on October 25, 1998, and defines *personal data* as "any information relating to an identified or identifiable" person, and *data processing* as the "collection, recording, organization, storage, adaptation, alteration, use, disclosure, erasure, destruction and any other means of automatically processing personal data." It includes "manual processing" only when data are stored in a filing system, and does not apply the constraints of data processing to information concerning public security or defense.

Under the European Directive, individuals in the European Union, as the subjects of the personal data collected about them, are granted the following rights, automatically and free of charge: (1) the right to withhold their consent to the processing of their data, (2) the right to access the data collected about them, (3) the right to correct inaccuracies in the data collected, and (4) the right to bring complaints and seek redress for misuse of their data. To ensure these rights, the Directive also provides for a *data controller* or a person or entity that decides how and for what purposes data are processed. Consent of the data subject, therefore, is not valid unless the data controller provides the data subject with notice of the purposes for which the data are sought, the intended uses, and the intended recipients of the data. Such informed consent, however, is not required when the data processing is required (1) by law, (2) to perform a contract with the data subject, (3) to protect the data subject's interests, (4) to perform a task in the public interest, and/or (5) to pursue either the general interest or the legitimate interest of the controller or a third party to whom the data may be disclosed.

The EC member states are required by the Directive to enact national legislation to ensure that these basic rights of individuals are reflected in their national laws and cross-border data flow, and that the transfer of sensitive personal data to less-protected countries does not occur. Since U.S. privacy law does not provide an "adequate level of protection as measured against EC standards of privacy protection," this creates a serious challenge in the world of the Internet, global trading, and worldwide application of technology. This conflict affects all aspects of the economy, but carries special challenges for equipment vendors, service providers, and system managers.[210]

To enable trade while solutions to this problem evolve, private companies have adopted independent privacy statements such as those discussed in Section 8.12.4, that offer increased protection of personal data from the EC in the United States. In addition, the U.S. government has included these concerns in its proposed privacy legislation. These efforts are referred to as the Safe Harbor Principles advanced by the International Trade Administration and best described in documents such as the (1) International Trade Administration; Safe Harbor Principles (November 1998 Draft), (2) International Trade Administration; Safe Harbor Principles (April 1999 Draft), 3) Safe Harbor Principles Frequently Asked Questions (4/19/99 Draft); (4) US/CEC Data Privacy Dialogue Draft Paper on EU Procedures (4/19/99); and (5) Safe Harbor Principles (Frequently Asked Questions (4/30/99 Draft).

The Europeans are assisting this process and have documented their positions and processes in documents such as the (1) European Parliament Working Party Paper on Processing of Personal Data on the Internet (2/23/99); (2) European Parliament Working Party Recommendation 1/99 on Processing Data on the Internet (2/23/99); and (3) EU Directive on Distance Contracts, Directive 97/7/EC (5/20/97).

8.14.2 Other Nations

In addition, in February 1998, the Organization for Economic Cooperation and Development (OECD) held a conference on Data Protection in International Networks. The results of the workshop, including an overview of various efforts to ensure privacy protection, are discussed under the reference to "OECD Workshop on Privacy Protection in a Global Networked Society" (February 1998) available at *www.oecd.org//dsti/sti/it/secur/prod/reg985final.pdf*. The United Nations has also established a human rights Web site at *www.unhchr.ch/html/menu3/b71.htm* in which it addresses privacy issues from the perspective of other countries. Thus this is a continuing area of importance in U.S. telecommunications law that should be tracked by all parties involved in system operations and use.

CONCLUSION

As one can see from the information presented in this chapter, privacy law in the United States is a very unsettled, evolving area of the law. The current spectrum of U.S. privacy law addresses mainly personal data held in government databases, personal financial data, access to government databases, harassing calls, and personal information about children. It does not address personal information held in private databases or communicated over many of the new telecommunications technologies such as the Internet and wireless communications. As with many other areas of our society, the law has not been able to keep pace with the rapid technological development, but companies and technology have provided some partial, interim solutions and others have reduced employees' expectation of privacy.

To ensure the privacy expected by individuals concerning their identities and personal information, new laws and policies must be developed, especially following those developed in other countries. As the legislatures and courts struggle with this, they must be careful not to apply old rules to new technology.

Courts recognize the absence of developed law and have provided a rich legacy of common law, seeking the best criteria on which to evaluate the cases before them. However, the best solutions to the immediate issues are technological solutions. Encryption provides the best technical solution to privacy for anyone who desires privacy. Encryption is the generic term for various mathematical algorithms that, when built into chips or provided through software, provide very powerful codes for a modest cost to provide unbreakable protection and security in modern communications. Given its importance in this area, encryption is the topic of Chapter 9.

ENDNOTES

[1]W. Prosser & R. Keeton, Prosser & Keeton on the Law of Torts § 117 at 849 (5th ed. 1984).

[2]Ambrose Bierce, *The Devil's Dictionary,* 257 (1878).

[3]A. Miller, The Assault on Privacy: Computers, Data Banks, and Dossiers 3 (1971).

[4]Warren & Brandeis, *The Right to Privacy,* 4 Harv. L. Rev. 193 (1890).

[5]Prosser & Keeton, *supra* note 1, and Restatement (Second) of Torts §§ 652B-652E, (1977).

[6] Note entitled *Tort Recovery for Invasion of Privacy,* 59 Neb. L. Rev. 808, 809-810 (1980); Restatement (2d) of Torts §652 (1977). Annotation, 49 A.L.R. 4th 430.

[7]Prosser and Keeton *supra* note 1, at 851 & n. 17 (citing jurisdictions); and Restatement (Second) of Torts, *supra* note 5.

[8]*Id.* Restatement (Second) of Torts at § 652B.

[9]*See for example,* Hamberger v. Eastman, 206 A.2d 239 (N.H. 1964).

[10]Barber v. Time, Inc., 159 S.W. 2d 291 (Mo. 1942).

[11]Douglass v. Hustler Magazine, Inc., 769 F.2d 1128 (7th cir. 1985).

[12]*Id.* at § 652 C cmt.b.

[13]Zacchini v. Scripps-Howard Broadcasting Co., 433 U.S. 562, 97 S.Ct. 2849, 53 L.Ed.2d 965 (1977).

[14]Haelan Labs v. Topps Chewing Gum, Inc., 202 F.2d 806 (2d Cir.), *cert. denied,* 346 U.S. 816 (1953).

[15]The cases included (1) Meyer v. Nebraska, 262 U.S. 390, 43 S.Ct. 625 (1923) concerning the right of parents to select the [foreign language] curriculum to be studied by their children in private schools; (2) Pierce v. Society of Sisters, 268 U.S. 510, 45 S.Ct. 571 (1925) concerning the right to educate one's children as one chooses; (3) Palko v. Connecticut, 302 U.S. 319, 58 S.Ct. 149 (1937); (4) Skinner v. Oklahoma ex rel. Williamson, 316 U.S. 535, 62 S. Ct. 1110 (1942) concerning the right of "[habitual] criminals" to not be sterilized by the state; (5) NAACP v. Alabama, 357 U.S. 449, S.Ct. 1163 (1958) concerning the right to refuse compelled disclosure of membership in organizations; (6) Griswold v. Connecticut, 381 U.S. 479 (1965) concerning the right privacy in the decision by married couples to use contraception; (7) Loving v. Virginia, 388 U.S. 1, 87 S.Ct. 1817 (1967) concerning the right to marry interracially; (8) Eisenstadt v. Baird, 405 U.S. 438 (1972) concerning

the right of unmarried persons to use contraception; (9) Roe v. Wade, 410 U.S. 113, 93 S.Ct. 705 (1973) concerning the right of a woman to have an abortion; (10) Moore v. City of East Cleveland, Ohio, 431 U.S. 494 (1977) concerning the definition of family and to make reasonable decision about the use of personal property; and (11) Carey v. Population Services Int'l. 1977 concerning the right of minors to use contraception.

[16]Roe v. Wade, 410 U.S. 113, 726 -727, 93 S.Ct. 705, 35 L.Ed.2d 147 (1973).

[17]Palko v. Connecticut, 302 U.S. 319, 325; 58 S.Ct. 149, 152; 82 L.Ed, 288 (1937).

[18]Loving v. Virginia, 388 U.S. 1, 87 S.Ct. 1817, 18 L.Ed.2d 1010 (1967).

[19]Skinner v. Oklahoma *ex. Rel.* Williamson, 316 U.S. 535, 62 S.Ct. 1110, 86 L.Ed. 1655 (1942).

[20]Griswold v. Connecticut, 381 U.S. 479, 484-86, 14 L.Ed.2d 510, 85 S.Ct. 1678 (1965).

[21]Eisenstadt v. Baird, 405 U.S. 438, 92 S.Ct. 1029, 31 L.Ed. 2d 349 (1972).

[22]Moore v. City of East Cleveland, Ohio, 431 U.S. 494 (1977).

[23]Pierce v. Society of Sisters, 268 U.S. 510, 45 S.Ct.571, 69 L.Ed. 1070 (1925).

[24]Meyer v. Nebraska, 262 U.S. 390, 43 S.Ct. 625, 67 L.Ed. 1042 (1923).

[25]Hill v. National Collegiate Athletic Ass'n., 7 Cal.4th 1,23, 865 P.2d 633 (1994).

[26]Griswold, *supra* note 20 381 U.S. 479.

[27]Buck v. Bell, 274 U.S. 200, 47 S.Ct. 584, 71 L.Ed. 1000 (1927).

[28]National Association for the Advancement of Colored People (NAACP) v. Alabama ex Rel. Patterson, 357 U.S. 449, 78 S.Ct. 1163, 2 L.Ed.2d 1488 (1958).

[29]American Communications Assn. v. Douds, 339 U.D. 382 at 402, 70 S.Ct. 674, at 686, 94 L.Ed. 925.

[30]Olmstead v. United States, 277 U.S. 438, 464-466 (1928).

[31]Goldman v. United States, 316 U.S. 129, 134-136 (1942).

[32]*Id.* at 134.

[33]Katz v. United States, 389 U.S. 347 (1967).

[34]Berger v. New York, 388 U.S. 41 (1967).

[35]Katz 389 U.S. 347 (*supra* note 33).

[36]*Id.* at 351 ("reasonable expectation of privacy").

[37]Berger, 388 U.S. 41 (*supra* note 34).

[38]Omnibus Crime and Safe Streets Act of 1968, Pub.L. No. 90-351, 82 Stat. 213 (June 19, 1968).

[39]Wiretap Act of 1968, 18 U.S.C. §§ 2510-21 (1968).

[40]Katz, 389 U.S. 347 (*supra* note 33).

[41]Berger, 388 U.S. 41 (*supra* note 34).

[42]Wiretap Act of 1968, *supra* note 39.

[43]The Electronic Communications Privacy Act (ECPA), S. Rep. No 99-541. 99th Cong. 2d Sess. (1986) at 2, reprinted in 1986 U.S.C.C.A.N. 3555 (1986) (ECPA S. Rep.) and codified at 18 U.S.C. §§ 2510-2541 (1988) [hereinafter Senate Report on ECPA] citing United States v. New York Tel. Co. 434 U.S. 159, 167, 98 S.Ct. 364 (1977).

[44]18 U.S.C. §§ 2510-22, 2701-11 (1988).

[45]Senate Report on ECPA, *supra* note 43.

[46]18 U.S.C. §§ 2510-2521 (1988).

[47]18 U.S.C. §§ 2701-2710 (1988).

[48]18 U.S.C. § 1367 (1988).

[49]18 U.S.C. § 2232 (1988).

[50]18 U.S.C. § 3117 (1988).

[51]18 U.S.C. § 3121-3126 (1988).

[52]Senate Report on ECPA, *supra* note 43, at 3.

[53]18 U.S.C. § 2510 (1) (1988).

[54]Russell S. Burnside, *The Electronic Communications Privacy Act of 1986: The Challenge of Applying Ambiguous Statutory Language to Intricate Telecommunications Technologies,* 13 Rutgers Comp. & Tech. L.J. 451 at 497-498 (1987). *See also,* 18 U.S.C. § 2510(1)&(12)(A) (1988).

[55]18 U.S.C. § 2510 (1) (1988).

[56]18 U.S.C. § 2510 (1) (1988).

[57]Burnside, *supra* note 54, at 497-498.

[58]18 U.S.C. § 2510 (16) (1988).

[59]18 U.S.C. § 2510(12) (1988).

[60]Senate Report on ECPA *supra* note 43, at 11-12, 16.

[61]*Id.* at 16.

[62]United States v. Torres, 751 F.2d 875, 880 (7th Cir. 1984) (reversing suppression of videotapes of Puerto Rican terrorist bomb factory).

[63]Report of the Committee of the Judiciary on the Electronic Communications Privacy Act, H.R. Rep. No. 647, 99th Cong., 2d Sess., 24-25 (1986) [hereinafter House Report on ECPA].

[64]United States v. Meriwether, 917 F.2d 955, 960 (6th Cir. 1990).

[65]18 U.S.C. § 2511 (1988).

[66]18 U.S.C. § 2512 (1988).

[67]18 U.S.C. § 2511(2)(h)(ii) (1988).

[68]18 U.S.C. § 2511(2)(g)(v) (1988).

[69]Boddie v. American Broadcasting Co., 731 F.2d 333 (6th Cir.1984).

[70]18 U.S.C. § 2701-2710 (1988).

[71]J. Podesta and P. Sher, Protecting Electronic Messaging: A Guide to the Electronic Communications Privacy Act of 1986, pp.viii, ix, 1 (1989).

[72]*See* United States v. Cafero, 473 F.2nd 489, 501 and n.9 (3rd Cir. 1973) (rejecting arguments that Title III failed to satisfy the requirements of *Berger* and citing cases from other circuits).

[73]18 U.S.C. 2710 (1988).

[74]18 U.S.C. § 2702(a) (1988). *See also,* Deal v. Spears, 780 F. Supp. 618 (W.D. Ark. 1991).

[75]18 U.S.C. § 2510(l) (1988).

[76]18 U.S.C. § 2510(5)(a), § 2511(2)(a); § 2518(4) (1988).

[77]18 U.S.C. § 3571(b) (1988).

[78]18 U.S.C. § 2511(3)(b)(iv) (1988).

[79]18 U.S.C. § 2510(8) (1988).

[80]18 U.S.C. § 2510(8) (1988).

[81]18 U.S.C. § 2510(8) (1988).

[82]Public Law 103-414 [105-220], 108 Stat. 4279, codified at 47 U.S.C. §§ 1001-1010, 1021 and 18 U.S.C. § 2522.

[83]*Id.*

[84]47 U.S.C. § 1002 *et seq.*

[85]18 U.S.C. § 2703 (1988).

[86]47 U.S.C. § 1002(b) (1988).

[87]47 U.S.C. § 1009.

[88]47 U.S.C. § 1008(b)(2).

[89]Peter Maggs, J. Soma, and J. Sprowl, Computer Law, 504, West Publishing Co., St. Paul, MN (1992).

[90]Federal Records Act (FRA) of 1950; Pub. L. 152, Title V, Section 501 *et seq.,* of the Federal Property and Administrative Services Act (FPASA) of 1949, codified at 44 U.S.C. §§ 2101 *et seq.*; 2501 *et seq.*; 2701 *et seq.*; 2901 *et seq.;* 3101 *et seq.*; and 3301 *et seq.* (1950).

[91]*Id.,* Legislative history at 3547-3548.

[92]Federal Property and Administrative Services Act (FPASA) of 1949, Public Law 152, sec. 104(a), codified at 44 U.S.C. §§ 300c *et seq.* (1949).

[93]Federal Records Act *supra* note 90, at 3561 (Legislative History).

[94]44 U.S.C. § 2902 (1950).

[95]44 U.S.C. §3301 (1950).

[96]44 U.S.C. § 3303a (1950). *See also,* Armstrong v. Executive Office of the President, Office of Admin, 1 F.3d 1274, 1278-79 (D.C. Cir. 1993).

[97]*Id.* Armstrong, 1 F.3d 1274, at 79

[98]*Id.*

[99]5 U.S.C. § 552(b)(5) (Executive privilege exemption to the Freedom of Information Act).

[100]Privacy Act of 1974, codified at 5 U.S.C. § 552a (1974).

[101]5 U.S.C. § 552a (1974).

[102]5 U.S.C. § 552a(e) (1974).

[103]5 U.S.C. § 552a(b) through (d) (1974).

[104]California Bankers Assn. v. Schultz, 416 U.S. 21 (1974).

[105]Whalen v. Roe, 429 U.S. 589, 599 (1977).

[106]Nixon v. Administrator of General Services, 433 U.S. 425, 457 (1977).

[107]California Bankers Assn., 416 U.S. 21 (*supra* note 104).

[108]5 U.S.C. § 552b (1974).

[109]5 U.S.C. § 552(b)(5); NLRB v. Sears, Roebuck & Co., 421 U.S. 132, 150, 44 L.Ed.2d 29, 95 S.Ct. 1504 (1975).

[110]Times Mirror Co. v. Superior Court, 53 Cal. 3d 1325 (1991).

[111]*Id.* at 1343-44.

[112]Rogers v. Superior Court, 19 Cal. App. 4th 469, 478 (1993).

[113]*Id.* at 478.

[114]*Id.* at 480.

[115]Star Publishing Co. v. Pima County Attorney's Office, 181 Ariz. 432, 891 P.2d 899 (Ct. App. 1994).

[116]*Id.* at 901.

[117]*In re* Amendments to Rule of Judicial Admin., 2.051—Pub. Access to Judicial Records, 651 So.2d 1185 (Fla. 1995).

[118]Computer Professionals for Social Responsibility v. U.S. Secret Service, 1996 U.S. App. Lexis 14 (D.C. Cir. 1996).

[119]Federal Records Act (FRA), *supra* note 90, at 9273 and 9279 (Legislative History).

[120]The Bank Secrecy Act (BSA) of 1970, codified at 2 U.S.C. § 1829b(d) (1970).

[121]2 U.S.C. § 1829b(d) (1970).

[122]California Bankers Assn., 416 U.S. 21 (*supra* note 104). *See also* United States v. Miller, 425 U.S. 435 (1976).

[123]Truth in Lending Act (TILA) of 1970, codified at 15 U.S.C. § 1601ff (1970).

[124]15 U.S.C. § 1601ff (1970).

[125]The Fair Credit Reporting Act (FCRA) of 1970, codified at 15 U.S.C. §§ 1681-1681t (1970).

[126]*Id.*

[127]Right to Financial Privacy Act (RFPA) of 1988, 12 U.S.C. § 3401 *et seq.* (1988).

[128]12 U.S.C. § 3403(c) (1988).

[129]12 U.S.C. § 3403(d) (1988).

[130]12 U.S.C. § 3405-3408 (1988).

[131]12 U.S.C. § 3403(a) (1988).

[132]12 U.S.C. § 3404 (1988).

[133]42 U.S.C. § 2000aa(a).

[134]The Privacy Protection Act (PPA) of 1980, codified at 42 U.S.C. § 2000aa *et seq.* (1980).

[135]*Report on Computer Crime;* Task Force on Computer Crime, Section of Criminal Justice, American Bar Association; June 1984 [hereinafter ABA Report].

[136]Federal Records Act (FRA), *supra* note 90, at 2, 3 and 2480 (Legislative History).

[137]ABA Report, *supra* 135, at 36-40.

[138]Computer Fraud and Abuse Act (CFAA), Pub. L. 104-294, title II, codified at 18 U.S.C. § 1030, General Statement and History of the Legislation, Senate Report No. 99-432 (1988).

[139]*Id.*

[140]A portion of Pub.L. 98-473, codified at 18 U.S.C. § 1030.

[141]18 U.S.C. § 1030(2)(g) (1988).

[142]18 U.S.C. § 1030(b) (1988).

[143]United States v. Morris, 928 F.2d 504, 509 (2d Cir. 1991).

[144]*Id.* at 506-07.

[145]ABA Report,*supra* note 135, at 11, Table 8.

[146]18 U.S.C. § 1030(e)(1) (1988).

[147]National Information Infrastructure Protection Act (NIIPA) of 1996, Pub. L. 104-294, Title II, Sec. 201, Title VI, Sec. 604(b)(36), 110 Stat. 3491, 3508, codified at 18 U.S.C. 1030 *et seq.*, amended Oct. 11, 1996.

[148]18 U.S.C. 1030 (e)(2)(B) (1996).

[149]Federal Records Act (FRA), *supra* note 90, at 1970 (Legislative History).

[150]FCC v. Pacifica Foundation, 438 U.S. 726, 748 (1978).

[151]Kovacs v. Cooper, 336 U.S. 77 (1948).

[152]47 U.S.C. § 227(b)(1)(A)(iii).

[153]Myrna L. Wigod, *Privacy in Public and Private E-mail and On-line Systems,* 19 Pace L. Rev. 95 (Fall, 1998).

[154]Compuserve Inc. v. Cyber Promotions, Inc., 962 F. Supp. 1015 (S.D. Ohio 1997).

[155]Cyber Promotions Inc. v. America Online Inc., 948 F.Supp. 456 (E.D. PA 1996).

[156]Cyber Promotions, Inc. v. Apex Global Information Services, Inc., 1997 WL 634384 (E.D. PA Sept. 30, 1997).

[157]15 U.S.C. § 6503.

[158]15 U.S.C. § 6503.

[159]15 U.S.C. § 6504.

[160]Child Online Protection Act (COPA), Pub. L. 105-277, Div. C, title XIV, § 1405, 112 Stat. 2681-739 (Oct. 21, 1998).

[161]The Child Pornography Prevention Act (CPPA) of 1996, Pub.L. 104-208, Div. A, Title I, § 101(a), [Title I, § 121], 110 Stat. 3009-26 (Sept. 30, 1996) codified at 18 U.S.C.A 2252A, *et seq.* (West 1999).

[162]The Protection of Children from Sexual Predators Act (PCSPA) of 1998, Pub.L. 105-314, Title VI, § 604(a); 112 Stat. 2983; 42 U.S.C. 13032, as amended Oct. 30, 1998 (West 1999).

[163]Sandra Davidson, University of Missouri, "Cyber-Cookies: How Much Should the Public Swallow?", as presented in Chapter 14 of *Advertising and the World Wide Web,* edited by David W. Schumann, and Ester Thorson, Lawrence Erlbaum Assoc., Publishers, Mahwah, New Jersey (1999).

[164]S. H.Wildstrom, "Privacy and the Cookie Monster," *Business Week,* Dec. 16, 1996, p. 22.

[165]P. Kyber, "Unwitting 90's Hansels and Gretels Leave Behind Computer Cookie Trail," *The Richmond Times Dispatch,* June 5, 1997, p. D-15.

[166]I. Hertzoff, "Cookies Are Not Always a Treat for Web Users," *Network World,* Nov. 25, 1997, p. 38.

[167]Z. Moukheiber, "DoubleClick is Watching You", *Forbes,* 158(11), Nov. 4, 1996, p. 342.

[168]L. Trager, "Netscape Flaw Lets Snoops See Files; Web Site Operators Can View Browsers' Hard Drives, *The San Francisco Examiner,* June 13, 1997, p. A1.

[169]B. Glass, "Help Desk: Cookies Won't Spoil Your Diet But Might Hurt Web Security," *Infoworld,* July 15,1996, p.54.

[170]Paul Shepard, "Despite Law, Identity Theft Still a Threat," The Associated Press, March 20, 2000. [*The Denver Post,* p. 7C.]

[171]David E. Kalish, Internet Faces Backlash Over Privacy Invasions," Asso. Press, *Boulder Daily Camera,* July 1997, p. 1B.

[172]Rose Aguilar, Service Still Provides Sensitive Information, *CNET News.com,* September 19, 1996.

[173]Rose Aguilar, Research Service Raises Privacy Fears, *CNET News.com,* June 10, 1996.

[174]David Kalish, *supra* note 171.

[175]17 U.S.C. § 1002. *See also,* Goldberg and Bernstein, *The Information Infrastructure,* New York Law Journal at 3, Sept. 16, 1994.

[176]Kalish, David, *supra* note 171.

[177]United States v. Western Elec. Co., 673 F. Supp. 525, 567 n.190 (D.D.C. 1987) rev'd in part, aff'd in part, 900 F.2nd 283 (D.C. Cir. 1990).

[178]United States v. Western Elec. Co., 714 F. Supp. 1, 12 n.40 (D.D.C. 1988).

[179]David Kalish, *supra* note 171.

[180]A. Miller, *The Assault on Privacy: Computers, Data Banks, and Dossiers,* 3 (1971). *See also,* D. Burnham, The Rise of the Computer State: The Threat to Our Freedoms, Our Ethics and Our Democratic Principles (1983).

[181]"Hide Your Cookies," *Canada Newswire,* June 24, 1997 and Kyber, P., "Unwitting 90's Hansels and Gretels Leave Behind Computer Cookie Trail," *The Richmond Times Dispatch,* June 5, 1997, p. D-15.

[182]This may be especially true concerning the disclosure of medical and financial information. Special concerns are raised when collected information moves beyond simple identity or address data to more sensitive areas. *See, for example,* 19 Pace L. Rev. 95, *104.

[183]L. Bransten, "Cookies Leave a Bitter Taste: Invasive Data Collection is Widespread," *Financial Times* (London Edition), Oct. 28, 1996, p. 15.

[184]Dun & Bradstreet v. Greenmoss Builders, Inc., 472 U.S. 749 (1985).

[185]David Kalish, *supra* note 171.

[186]44 U.S.C. §§ 2101 *et seq.*; 2501 *et seq.*; 2701 *et seq.*; 2901 *et seq.*; 3101 *et seq.*; 3301 *et seq.*

[187]18 U.S.C. § 2510(l) (1988).

[188]18 U.S.C. § 2702(a).

[189]18 U.S.C. § 2702(a); Deal v. Spears, 780 F. Supp. 618 (W.D. Ark. 1991), *aff'd,* 980 F.2d 1153 (8th Cir. 1992) (employer recording of personal phone calls served no legitimate purpose).

[190]United States v. Maxwell, 42 M.J. 568 (A.F.C.C.A. 1995).

[191]18 U.S.C. § 2702(a).

[192]Flanagan v. Epson America, Inc.; Slip Op. No. BC007036, Ruling on Demurrer and Motion to Strike of Defendant Epson America, Inc. (Cal. Super, Ct., Los Angeles Cty., Jan. 4, 1991);

"ECPA and Online Computer Privacy", 4 Fed. Com L.J. 17, 39 (1989) (an employee has no ECPA claim against an employer over examination of stored communications on the company's in-house computer system).

[193]Billings v. Atkinson, 489 S.W. 2d 858 (Tex. 1973).

[194]*See, for example,* Huffcut v. McDonald's Corp., No. 94-CV-6589 (W.D.N.Y. Dec. 7, 1995).

[195]*See* Mark S. Dichter, and Michael S. Burkhardt, "Electronic Interaction in the Workplace: Monitoring, Retrieving and Storing Employee Communications in the Internet Age," Morgan, Lewis and Bockius, LLP (1996) presented at The American Employment Law Council, Fourth Annual Conference—Grove Park Inn, Asheville, North Carolina, October 2-5, 1996.

[196]Bourke v. Nissan Motor Corp; Slip Op. No. B068705 (Cal.Ct. App. July 26, 1993).

[197]*Id.* at 3.

[198]*Id.* at 6-7.

[199]*Id.* at at 8.

[200]*Id.*

[201]*Id.* at 9.

[202]Smyth v. The Pillsbury Co., 914 F. Supp. 97 (E.D. Pa. 1996).

[203]*Id.* at 98.

[204]913 F.Supp. at 101.

[205]18 U.S.C. § 2510(1)&(12)(A) (1988). *See also,* Burnside, *supra* note 54, at 497-498.

[206]18 U.S.C. § 2510 (16) (1988).

[207]House Report on ECPA, *supra* note 63.

[208]Meriwether, 917 F.2d at 960 (*supra* note 64).

[209]Directive 95/46/EC of the Parliament and the Council on the Protection of Individuals with Regard to the Processing of Personal Data and on the Free Movement of Such Data, 1995 O.J.L. 281 (Oct. 24, 1995) [hereinafter The Privacy Directive].

[210]Jerry Berman, and Deirdre Mulligan, *The Internet and the Law: Privacy in the Digital Age: Work in Progress,* 23 Nova L. Rev. 549, Winter, 1999.

CHAPTER 9 Encryption

Encryption technologies are the most important technological breakthrough in the last one thousand years. No other technological discovery—from nuclear weapons (I hope) to the Internet—will have a more significant impact on social and political life. Cryptography will change everything.

Lawrence Lessig, Harvard law professor and author of *Code: and Other Laws of Cyberspace*[1]

WITH THE INCREASED COLLECTION, USE, SALE, AND EXCHANGE OF private information about individuals, usually without the individual's knowledge or permission, it is clear that technology has introduced privacy concerns that the framers of the U.S. Constitution never dreamed about. As discussed in Chapter 8, current law has not totally addressed these concerns, and very few legal restrictions exist to control them.

Technology, however, also provides a solution to many of these privacy concerns. *Encryption* is the process of writing a message in code or other secret form so that only persons with the *key* can decode or "decrypt" the message to read it. Readily available through low-cost or free computer programs, encryption prevents eavesdroppers or interceptors of a communication, including the government, from reading that communication. This allows nearly anyone to protect email, e-commerce transactions, online banking, contracts, data files, and other electronically transmitted communications or messages stored in unsecured computer databases. Furthermore, modern encryption programs allow users to control their own encryption rather than having to go through a central government agency or clearinghouse. This provides an extra level of privacy that legislation cannot offer.

As use of the Internet increases, people will demand this greater security for their communications, credit card information, and e-commerce transactions.

They will also resist having their "net-surfing" and purchasing habits recorded. Corporations will demand increased privacy for their communications to their branch offices, customers, and suppliers worldwide. Thus, modern forms of communications such as the Internet are driving the need for greater privacy. Since encryption provides a technical solution to privacy concerns, it will profoundly affect every person and corporation existing today and in the future. For this reason, Lawrence Lessig, a Harvard law professor and international authority on cyberlaw, stated in his recent book *Code: and Other Laws of Cyberspace* that encryption technologies are the most important technological breakthrough in the last 1,000 years . . . and will change everything.

However, several critical issues surround the topic of encryption in the United States today. First, while nearly everyone, including the U.S. government, acknowledges the increasing need for individual and corporate privacy in the 21st century, law enforcement and national security agencies, such as the Federal Bureau of Investigation (FBI), Central Intelligence Agency (CIA), and the National Security Administration (NSA), argue that they must be able to intercept and decode encrypted messages in order to fight terrorism, drug trafficking, organized crime, and child pornography. This requires that the government assess the true crime fighting needs of law enforcement and national security agencies and find a balance between those needs and the privacy needs of individuals.

Second, while the U.S. government may be willing to permit the use of encryption in the United States, historically it has severely restricted the export of certain encryption products and technologies, believing that such export threatens our national security. However, strong encryption software programs are readily available worldwide for free or as little as $5.00. Therefore, many persons in the United States strongly oppose the government's restrictions on the export of encryption. They point out that the export restrictions do nothing to protect the national security of the United States but, instead, harm the competitiveness of U.S. companies to sell software or to provide the seamless privacy protections required in modern communications and e-commerce transactions. Opponents also state that the restrictions are unconstitutional constraints on the right of free speech and are inconsistent with the approach used by other democratic governments and international organizations with equal national security concerns. Finally, opponents argue that the U.S. government's definition of "export" is overly broad. The definition includes sending an encrypted email to a relative, friend, colleague, or customer abroad. Even internal memos to corporate branch offices abroad are considered "export." It can also include clicking into Web sites located outside of the United States if people use encryption in their Web systems. Basically, any encrypted transaction involving a party outside the United States, is considered an "export."

In seeking a solution to this conflict, the U.S. government has proposed several alternative concepts including (1) "key escrow," both voluntary and automatic in hardware such as the Chipper Chip, (2) "key recovery," providing "back door" access to encrypted messages for the government through both hardware and software options; and (3) legislation limiting the strength of the keys used in encryption. *Key escrow,* is a general name for regulations requiring that encryption users register or "escrow" with the government or a designated third party their "keys" to unlock or decode their encrypted messages, so that the government can, with a court-ordered warrant, read any encrypted message. This is intended to provide strong encryption capabilities to U.S. citizens while still allowing the government to be able to decrypt an encrypted message if it deemed necessary to do so. *Key recovery,* on the other hand, is a general name for the hardware or software capability built into an encryption system that allows the government, "with proper judicial authorization in the form of a court order," to decrypt an encrypted message without the knowledge of the parties to the communication. Key recovery therefore provides "back-door" access for the government to encrypted communications designed into the basic operations of the system by the system developers at their own expense.

However, public opposition to these proposals has delayed their adoption. In addition, several important court cases have held that encrypting a message is *free speech* protected by the U.S. Constitution, and on January 14, 2000, the government issued new, more relaxed encryption export regulations, but did not remove all of them. These new regulations make it somewhat easier for U.S. companies and individuals to export encryption technology in common products to certain countries, but still require companies to file applications, escrow keys, and provide key-recovery capability to the government before they can export products containing or using encryption.

The purpose of this chapter is to describe briefly how encryption works, the main issues surrounding it, and the major laws governing it. To do this, Section 9.1 explains basic encryption technology, including the importance of "keys" and key strength. Section 9.2 outlines the early use of encryption in the United States and some of the initial legislation securing its use for military and law enforcement. Sections 9.3 and 9.4 describe the two most important encryption technologies available to the public today, private-key and public-key systems, respectively. One important aspect of public-key encryption technology discussed is that of "digital signatures," which are amazingly secure and thus important to e-commerce and "identity theft" issues. For this reason, recent legislation recognized digital signatures as having the same force as handwritten signatures in contracts and other commercial documents. Section 9.5 compares the private-key and public-key technologies, noting the strengths and weaknesses of each. Section 9.6 expands on the concerns of law enforcement and national security agencies regarding public use of encryption systems and the legislation initiated by law enforcement and the security agencies to control that use. It also discusses public and private opposition to these restrictions and government efforts to relax the restrictions. Section 9.7

outlines the views and approaches of other democratic countries and international organizations concerning encryption. Section 9.8 then describes the United States's reaction to these changes and three key court cases whose decisions have impacted recent legislation and caused some U.S. restrictions on encryption use and export to be relaxed. Since this is such a rapidly changing area of technology, Section 9.9 also provides a list of the current state, federal, and international laws that exist in this area, new laws that have been proposed, and several Web sites that provide additional information and updates.

9.1 ENCRYPTION TECHNOLOGY

Encryption does not prevent someone from intercepting a message, but rather it prevents that person from reading the intercepted message. In many cases, encryption also blocks information about *to whom* the message is sent or *from where* it came.[2] Encryption is based on the ancient study of cryptology and the resulting practice of cryptography. Such systems have been in use for over 4,000 years, beginning with the written hieroglyphics of ancient Egypt. Governments since the Greeks and Romans have used encryption to protect important military and diplomatic communications, and today, computer-generated mathematical algorithms make encryption available to individuals for free or at very low cost.

In each case, the text of the original message, known as the "plaintext" or "cleartext," is encoded or enciphered into "codetext" or "ciphertext," which usually has the appearance of random, unintelligible characters. This unreadable message is called a "cyptogram." The process of reading the cryptogram requires decoding or deciphering the encrypted message with a key. If the key is unknown, "cryptanalysis" is the art of breaking the code or cipher to discover the key.

Traditionally, the term *code* described any system in which the basic unit of concealment was a word, while the term *cipher* described any system in which the basic unit of concealment was a single letter within a word. However, in Morse Code, letters are the basic unit, which in turn encodes words.

Most individuals have had some experience with cryptography in puzzles or games. Many children have used decoder rings found in boxes of cereal to write notes to their friends. However, before computers, most secure encryption systems were too expensive and complicated to be used by individuals or corporations. Only governments could afford the cost and then only for diplomatic and military, not financial or administrative communications. As computers have evolved, however, this has changed, and strong encryption systems are now readily available for private, nongovernmental use. This is fortunate since the need for highly secure private messages has also increased significantly with the proliferation of computers and the Internet.

9.1.1 Forms of Encryption

In the thousands of encryption systems that have existed over the years, all use some form of "substitution," "transposition," "concealment," "device," or "mathematical" techniques to change the original message.

Substitution

Substitution encryption systems replace the letters of the alphabet with combinations of symbols or signals to encode the message. Two familiar examples of substitution encryption systems are the Morse code, in which dots and dashes substitute for the letters of each word, and the five-row–five-column code used by U.S. prisoners of war (POWs) during the Vietnam War.

In the Vietnam POW system, the English alphabet was grouped into five rows, each containing five columns. Row 1 contained *A, B, C, D,* and *E;* row 2 contained *F, G, H, I,* and *J;* row 3 skipped *K* (represented by *C* to reduce the alphabet to 25 letters), and contained *L, M, N, O, P.* Rows 4 (*Q, R, S, T,* and *U*) and 5 (*V, W, X, Y,* and *Z*) completed the alphabet in sequential fashion. To tap out a message, the prisoners would tap first the row, then the column. For example, *C* would be one tap, followed by three quick taps representing Row 1, third column or position. Similarly, *R* would be four taps, followed by two taps, representing fourth row, second column.

Like the Morse code and the Vietnam POW system, all substitution systems are understandable or interpretable only by those who know the code or have access to a codebook. Other systems may require a decoding device or symbol manipulator.

Transposition

Transposition encryption systems rearrange the position of the letters or elements in the text of the message. For example, each letter of the original message may be replaced by a letter one or more positions to the left or right. Thus, if shifted one letter to the left, the name *IBM* becomes *HAL*. Some students of the movie *2001: A Space Odyssey* suggest that this was the source of the computer's name in the movie.

Julius Caesar used a simple code in which each letter of the text was replaced by the letter three positions to the right. Thus *CAB* became *FDE*. This system was called the "Caesarian Cipher," named after the Roman general Cipher (100–44 BC) who is credited with inventing it.[3] Other transposition encryption systems also divide the letters into groups and then reverse the letters in each group. If needed, the system then uses extra, or filler letters called "nulls" to complete block groups. For example, if a five-letter group system were used, the message "HELLO THERE" would be *OLLEH EREHT* or more hidden, *OLLEHEREHT*.

Concealment

In 1600, Cardinal Richelieu invented the *grille* or a card with holes in it. He would place the card on a sheet of paper, write his message in the holes and then fill in the rest of the paper to appear as an innocent message. Many other concealment systems exist, including patterns of letters created on a page.

Device

The Spartans, centuries ago, used a *device* encryption system in which communicators would wrap a belt around a stick, write the message along the length of the stick, and then unwind the belt. To read the message, the same size stick was needed to wrap the belt around.

In the United States, Thomas Jefferson invented an early cryptographic device known as the "wheel cipher" around 1795. The wheel cipher contained a stack of wheels loaded onto a central rod or axle. On each wheel, Jefferson wrote the letters of the alphabet in random order. To encode a message, he would shuffle the wheels and then place them on the axle, in random order. The letters of the message would then be aligned along the rotational axis of the axle, one letter per wheel. Any other row of aligned letters could then be used as the ciphertext for transmission. To avoid the very small probability that two readable messages could be spelled by the order of the wheels, the sender would review all combinations on the wheels.

To decrypt the message, the recipient aligned the letters of the ciphertext along the rotational axis and looked for a set of aligned letters that spelled words. Without knowing the order of symbols on the wheels or the order of wheels on the axle, any plaintext of the appropriate length is possible, and thus the system is quite secure for one-time use. Statistical decryption is feasible if the same wheels are used in the same order many times. Thus, the strength of the system is in the random order of the wheels when they are placed on the axle.

Although Jefferson rarely used his device himself, the U.S. Navy used a similar item for nearly 200 years, until the mid-1990s. Today, device systems generally include computerized discs, spun in opposite directions.

Mathematical Algorithms or Source Codes

Modern encryption systems use computers to arrange strings of zeros and ones, also called "bits," into groups or blocks to create keys. Each key has a preset number of bits. The arrangement of the bits in each key is based on mathematical patterns provided by the encryption algorithm or source code. These algorithms then use the keys to encrypt and decrypt messages.

In their book *Privacy on the Line,* encryption experts Whitfield Diffie and Susan Landau provide an excellent description of the manner in which algorithms use keys to encrypt and decrypt messages. The example they use is this:

If two people were trying to meet on a specific day, time, and location, the information needed to be sent from one to the other would include the year, month, day, hour, minute, avenue, and street in that pre-agreed upon order. Thus, if the meeting were to take place at 3:25 PM *on December 30, 1999, on the corner of 1st Avenue and 44th Street, and a two-digit system were agreed upon to represent each piece of information, the resulting message would be:*

19 99 12 30 15 25 01 44

Every digit in this message is significant. To encrypt it for secrecy, both persons must also agree on a key that will convert the string of 14 digits into some other string of 14 digits that is understandable by both persons, but not by anyone else. For example, if the key agreed upon by the two people is

64 25 83 09 76 23 55 72

when it is added to the original message, the resulting encrypted message is:

73 14 95 39 81 48 56 16

The above "crytogram" is what is transmitted to the receiver in whatever manner is possible, including very public, nonsecure methods such as over the telephone, on a postcard, or over the radio. Diffie and Landau point out that, in some cases, the addition produced a number greater than 10. In the third place, for example, the sum is 11. In such cases, the 1 in the 10s place is thrown away, rather than "carried" into the next place as in ordinary arithmetic.

To decrypt the message when received, the intended recipient simply subtracts the agreed-upon key, without "borrowing," to learn the actual message. Since the key was chosen at random, even if the message were intercepted in route, the interceptor could see that a series of numbers or information had been transmitted, but could not determine the key or whether the actual information was numbers or letters. Therefore, the interceptor could not decrypt the message. If the key were used a second time, however, the interceptor could use information about the first message, to determine the key. Thus it is important to use a different key each time a critical message is transmitted.

This mathematical algorithm type of encryption system is readily generated by computers and yet works well for transmitting important information, such as a credit card number, even in the midst of other less-critical information, such as the items purchased on the Internet. The remaining issue is, however: How does the intended recipient learn the key? The next several sections explain this, but *public-key* and *private-key encryption systems* complete the process in very different ways.

9.1.2 Key Structure

As noted, if the receiver has the correct key, he or she can "decrypt" any message encrypted with that key. If the receiver does not have the correct key, however, that person has three options: (1) steal the key, (2) analyze the code to determine the algorithm used to create the key, or (3) use a "brute-force attack" in which all possible keys are tried, one at a time, until the correct key is found.

In either of the two last options, the more bits a key contains, the more secure it is considered to be. This is because with Option 2, longer keys require more time and effort to analyze, while with Option 3, keys require more time to break the code using a brute-force attack. Thus, the strength of an encryption scheme increases with its "key length."[4]

For example, if a key contains only two bits, four possible key combinations exist: 00, 01, 10, and 11. If a third bit is added to the key, eight possible key combinations result: 000, 001, 010, 100, 011, 101, 110 and 111. Thus, key combinations grow exponentially as each bit is added. For this reason, the following increase in possible key combinations results as increasingly more bits are added to the key.

Key Length	Possible Number of Key Combinations
2-bit key	4
3-bit key	8
8-bit key	256
56-bit key	72 quadrillion
128-bit key	4.7 billion trillion[5]

In the same fashion, the time required to break the code with a brute-force attack, or by trying all key combinations until the correct one is found, also increases exponentially. Usually, the process requires a network of computers and an average of trying about 50% of the possible keys. It is also impacted by changes in technology. For example, in 1986, it was estimated that a fast computer would take 2,283 years to try all possible keys in a 56-bit key, the then-adopted IBM standard called the U.S. Data Encryption Standard (DES). However, advances in computer power and speed have decreased the time required to find a key using a brute-force attack to under five hours.[6]

9.2 EARLY USE OF ENCRYPTION IN THE UNITED STATES

Encryption systems in the United States historically have been the domain of government and military systems. Systems such as Thomas Jefferson's wheel system, shift systems, and the Morse code served the government well, so long as the keys were changed frequently and the key or code was not intercepted while being communicated to the intended recipient of the message.

This concern became a special problem with radio. During the early 1900s, radio became an increasingly popular broadcast medium but was not secure. Thus, during World War I, communicating sensitive information by radio required encryption. The United States realized the need for broad-based encryption systems, but in a precomputer era, such systems tended to be slow and costly. Their security also depended on unusual keys with complex key handoffs, usually coordinated by a centralized authority.

9.2.1 1930s–1950s: Encryption

Encryption remained the domain of military and diplomatic use through the Depression in the 1930s, World War II in the 1940s, and the Cold War in the 1950s. Its focus was national security, followed by law enforcement and protection of technological advances such as nuclear power and airline traffic safety. During World War II, the function of intelligence, including communications intelligence, was placed under military command.[7] However, this caused its development and oversight to be quite disjointed and uncoordinated.

Central Intelligence Act of 1947

To correct this, Congress passed the Central Intelligence Act of 1947, creating the Central Intelligence Agency (CIA). The new position of *Director of Central Intelligence* was tasked with coordinating the efforts of the entire U.S. intelligence community.

Wiretapping

As discussed in Chapter 8, Section 8.2.5, wiretapping was a common practice for law enforcement who monitored some domestic and all international calls from 1876–1967. It did not require a court order until two cases, *Katz v. U.S.*[8] and *Berger v. New York,*[9] were decided in 1967. Even then, however, a court order was not required when issues of national security or wiretaps of a foreign government were involved.

Shannon's "Communications Theory of Secrecy"

In 1948, Claude Shannon, an electrical engineering professor at MIT, published an article in the *Bell System Technical Journal* entitled "The Communications Theory of Secrecy Systems."[10] Shannon was one of the first modern cryptographers to apply advanced mathematical techniques to the study of crytography. Although the use of frequency analysis for solving substitution encryption began many years earlier, Shannon's work demonstrated several important features about the statistical nature of language that made the solution to nearly all previous codes very straightforward. Perhaps the most important result of Shannon's famous paper is the development of a measure of cryptographic strength called the *unicity distance.*

The *unicity distance* is a number that indicates the amount or quantity of an encrypted message that is required in order to decrypt that message. It is a function of the length of the key used to encrypt the message and the statistical nature of the language in which the original message was written. Given enough time, it is guaranteed that any encrypted message can be broken if the unicity distance is 1. For example, any time the length of all messages sent under a particular key in English exceeds 2.6 times the length of that key, the key and the original messages eventually can be determined. Encryption systems with an infinite length random key have a unicity distance that is infinite. While unbreakable, they are also impractical. However, Shannon found that a message created with an alphabetic substitution encryption system (1) with a random key, (2) with a length greater than or equal to the length of the message, and (3) used only one-time, cannot be decrypted from the ciphertext alone.

Shannon also introduced the concept of "workload," which is the difficulty in decrypting a message given the availability of enough ciphertext to theoretically break a code. He showed that an alternative to increasing the unicity distance is the use of systems that increase the workload required to decrypt the message. Two important concepts provided by Shannon in his paper are the *diffusion* and *confusion* properties of encryption. These form the basis for many modern cryptosystems because they tend to increase the workload of cryptanalysis.

Diffusion is the dissipation of the statistical structure behind the language being transmitted. For example, making a different symbol for each English word makes statistical occurrence for many words difficult to detect in the short run, and increases the quantity of data needed to decrypt messages. This difficulty occurs because meaningful words are rarely repeated, and common words such as "the," "and," "or," are frequently repeated. The random use of obscure synonyms for frequently repeated words tends to dissipate the ability to decipher them. Strange sentence structures have a similar effect on the cryptanalysis process.

Confusion is the obscuring of the relationship between the original message, the key, and the encrypted message. For example, if any bit of the key has a 50% chance of affecting any bit in the encrypted message, statistical attacks on the key require solving a large number of simultaneous equations.

Extensions to Shannon's basic theories include the derivation of an "index of coincidence" that allows approximations of key length to be determined purely from statistical data,[11] the development of semiautomated techniques for attacking cryptosystems, and the concept of using computational complexity for assessing the quality of cryptosystems.[12]

National Security Agency

In 1952, responding to increasing Cold War concerns, President Truman created the National Security Agency (NSA). The unexpected result was that the National Security Agency took over all work in cryptography in the United States, successfully suppressing it in other agencies of government. As a result, the use of encryption in the other agencies froze or diminished, even for nuclear command and air traffic control systems. This did not change for over a decade, until ordered to change by President Kennedy in the 1960s.

Abuses of Power

During this era, the government's control of encryption was meant to benefit all Americans through increased national security. However, the control also led to serious abuses. For example, although census information was publicized as confidential, during World War II it, and other government-held information, was used to wiretap and locate Japanese-Americans to be sent to internment camps. During the McCarthy era, Civil Rights movement, and Vietnam protests, many illegal wiretaps were placed on numerous citizens' communications, not just on those of political activists. In 1968, court orders were required for wiretaps, but these were easy to obtain, and when they could not readily be obtained, the FBI and others frequently ignored this requirement and wiretapped many law-abiding citizens, political leaders, Supreme Court judges, members of the U.S. Congress, and *all* international calls. As many readers may recall, illegal wiretaps against the Democratic National Committee, ordered by the White House, ignited the Watergate scandal.

9.2.2 1960s–1970s: Computers, Digitization, Multiplexing, Packetization, and Encryption

During this era of frequent wiretapping, individuals and companies made no practical use of encryption, even though analog communications were not very secure. It wasn't until the 1960s and 1970s, with the development of computers, that several important changes occurred in communications technology, including the *digitization of communications, multiplexing, packetization,* and affordable *encryption*.

Digitization of communications improved the security of communications because, when communications are digitized, the communication is converted to 1s and 0s, divided up into small groups called "bytes," and then transmitted over whatever circuit or route is best for that transmission. When someone tries to wiretap a digitized conversation, they hear only static.

When *multiplexed* onto shared resources, such as T-1, T-3, and higher circuits, portions of each digitized communication are mixed with digitized portions of hundreds of other communications and transmitted over the same communication line. This makes it even more difficult to wiretap single digitized communication.

By the time technology evolved to *packet* communications, hundreds of thousands of communications were first digitized, divided into "packets" of information, and then multiplexed together before being transmitted over thousands of circuits. When the packets reached the receiving end, they were placed in their proper sequence and delivered to the listener. Thus, *packetization* further enhanced the security of communications by dissembling messages and permitting their transmission over different paths, further complicating wiretaps.

With computers also came affordable *encryption*, opening possibilities for its use by individuals and companies. Computers drove the cost of encryption technology to a low enough level to be affordable for private, nongovernmental use. Second, people became excited about the revolutionary power and speed that computers brought to encryption, providing the complexity required for truly secure encryption systems. Third, computers eliminated the need for centralized control of encryption, enabling "end-to-end" user control. These three benefits brought about by computers touched off tremendous private research and development of encryption systems and created an explosion in technology that continues today.

1965—Brooks Act

In 1965, recognizing the potential of computers and their increasing importance in commerce, Congress passed the Brooks Act of 1965. Sponsored by Representative Jack Brooks of Texas, the uncontested law assigned responsibility to the U.S. Department of Commerce's National Bureau of Standards (NBS) to make recommendations to the president concerning standards for both civilian and government computers under a uniform federal automatic data-processing standard (FADPS). Given this new range of responsibilities, Congress also changed the name of the National Bureau of Standards to the National Institute of Standards and Technology (NIST).

1967—International Covenant on Human Rights

However, along with the increasing use and commercial benefits of computers came increased concerns about privacy. In 1967, two years after the Brooks Act, the international community recognized the growing use and impact of computers on human rights and renewed Article 12 of the 1948 Universal Declaration of Human Rights, which stated

> *No one shall be subjected to arbitrary interference with his privacy, family, home or correspondence, nor to attacks upon his honour and reputation. Everyone has the right to the protection of the law against such interference or attacks.*[13]

1967—Publications on Encryption Ignite Public Interest

Also in 1967, David Kahn, a journalist, published a history of cryptography entitled *The Codebreakers.*[14] Years later, in 1982, James Bamford published a history of the National Security Agency (NSA) entitled *The Puzzle Palace.*[15] Both works ignited tremendous public interest in cryptography, especially among academics, as a means of providing the needed privacy and increased security required with modern communications.

As part of this new interest in encryption, the academic community developed two main approaches to computerized encryption in the early 1970s: private-key and public-key cryptosystems. They are also known as "symmetric" and "asymmetric" systems respectively, or one-key and two-key systems, not to be confused with one-bit and two-bit key systems. Details of these systems are described in the following two sections of this chapter. The U.S. government, on the other hand, was less excited about this new civilian interest in encryption and threatened the academic community with *prior restraint* in order to prevent the publication of additional information on new encryption concepts and systems.

9.3 PRIVATE-KEY CRYPTOSYSTEMS

Private-key encryption systems are the most familiar encryption systems to people because they include the "James Bond" type of cryptosystems in which both the sender and receiver know the same code so they can exchange secret messages. Other private-key systems include the Navaho "code-talkers" of World War II. The primary risk with private-key systems is when third parties learn the code and thus compromise the secrecy of the messages.

To send a message through a computerized private-key encryption system, the sender (1) writes the message as normal, (2) encrypts it using an encryption system previously agreed upon with the recipient, and (3) sends it to the recipient. To decrypt it, the recipient must have the *same* key or code. For this reason, private-key systems are also known as "single-key" or "symmetric" systems.

To respond, the recipient uses a different, randomly generated key to protect the security of the keys, communicates it to the sender, drafts the response, encrypts it, and sends the encrypted response to the sender. When the sender receives the response, he or she decrypts it with the same key.

This requirement that both the sender and recipient must have the same key leads to several limitations. First, the sender and receiver must either know each other before they can communicate or they must have some method of establishing a secure "key hand-off." Second, private keys cannot be changed without the recipient's knowledge. Third, the secrecy of the key must be preserved so that unauthorized eavesdroppers or interceptors cannot learn the key and read the message. Thus, the security of private-key systems is more dependent on the security of the key than on the complexity of the encryption system. These limitations make private-key systems less desirable in the Internet Age since (1) parties, who often do not know one another, communicate, but perhaps only once; (2) many users want to change their keys frequently; and (3) the Internet is an open medium. Thus, the communicators have little opportunity to agree upon a key and need to have a separate key for each communication.

Nevertheless, the main encryption system in the United States for 21 years, from January 1977 through December 1998, was a computerized private-key system called the U.S. Data Encryption Standard (DES). In January 1999, the DES was replaced by a more complex private-key system known as the Triple DES (3DES), and in 2001, Triple DES is being replaced by an Advanced Encryption System (AES). All three systems are described below.

9.3.1 1970 (1977): U.S. Data Encryption Standard (DES)

In 1970, as computers were becoming increasingly used, an engineering team at IBM, led by Horst Feistel, developed a private-key encryption system that became the U.S. Data Encryption Standard (DES). The DES is a "shift-register" or "block-cipher" system that uses the fact that most modern communications systems divide each 64 bits of digitized information in a message into eight blocks of eight bits each. These blocks are called "bytes" or "octets." It then scrambles the eight bits in each block using a random selection encryption key or pattern. Third, the DES performs a complex mathematical operation on the first four of the eight blocks and uses the results to alter the remaining four blocks.[16] To avoid a possible situation where a block could contain all zeros, the DES uses only seven of the eight blocks, or 56 bits of each 64 bits of data, for the encryption scheme and thus is called a 56-bit key.

Amazingly, this 56-bit key process results in 70 quadrillion possible key combinations for each message. For this reason, any message encrypted by the DES was considered, for many years, to be secure from a successful brute-force attack. As such, the DES provided a highly reliable encryption system for both businesses and individuals. However, it was still breakable by the government if a security threat warranted the effort required to do so. In 1971, one year after the DES was developed, two Israeli scientists developed a way to break the DES under certain limited situations.[17] However, the scientists did not widely distribute their information.

The security of the DES led the U.S. government, in 1974 to select it as the winner of a request for proposals (RFP) for a U.S. encryption standard issued by the National Bureau of Standards in 1973.[18] Discussion about the request for proposal began in the early 1970s, as various federal agencies increasingly used computers and digital communications. The National Security Agency and the National Bureau of Standards recognized that federal government agencies needed standardized cryptography to safeguard their records. This was further emphasized in 1974 when the Privacy Act of 1974 was enacted. However, the National Security Agency was unwilling to develop the needed encryption system for fear that such activity would reveal too much about its capabilities and encryption philosophies. It believed that wide availability of the National Security Agency's products could compromise other national security efforts.[19]

While selected in 1974, it wasn't until three years later, on January 15, 1977, that the U.S. government adopted the DES as the official Federal Information Processing Standard (FIPS) to protect unclassified computer data in the U.S.[20] To ensure that the standard kept pace with technical growth and change, the government required that the National Institute of Standards and Technology (NIST) reaffirm the DES's designation as the official U.S. encryption standard every five years. It was reaffirmed for the last time on December 30, 1993, and remained the standard through December 1998.[21] In its 21 years as the U.S. standard, from January 1977 through December 1998, DES also became the primary international encryption system used by banks and other commercial groups for electronic funds transfer.

During those 21 years, however, computers also became faster and more powerful, reducing the time required to successfully break a DES code by a brute-force attack. This was widely recognized, to the point that even the movie *Sneakers,* released in 1992 by Universal Studies, depicted the potential impact of a fictional computer capable of breaking any encrypted information in the world.

In January 1995, the National Institute of Standards and Technology announced that the 56-bit DES would no longer be sufficient to protect the security of encrypted messages after its December 1998 end date. Instead, NIST recommended that, effective January 1999, the DES be replaced by Triple DES (3DES) as the U.S. encryption standard. The Triple DES code is three times harder to break than the original DES.

As if to underscore the need for a more powerful system, on June 19, 1997, 14,000 computer users took part in a brute-force attack on the DES code. The group tried 18 quadrillion keys before it broke the code but it required only a few weeks. Exactly one year later, on June 18, 1998, a group of Silicon Valley computer workers, using faster computers, cracked the 56-bit DES code in only 56 hours. Nineteen months later, on January 19, 2000, a group of computer users cut this time in half, requiring only 22 hours and 15 minutes to break the 56-bit DES code.

9.3.2 1995 (1999): Triple DES (3DES)

On January 15, 1999, following the recommendation made four years earlier in 1995 by the National Institute of Standards and Technology, Triple DES replaced DES as the new U.S. encryption standard. Triple DES encodes a message using DES, but does it three times in sequence, each time with a different key. It is, therefore, three times harder to break than the original DES. In the four years between January 1995 and January 1999, the banking industry had already converted to the more secure Triple DES, so it had actually become the de facto standard before it was named the official new standard.

9.3.3 1997 (2001): Advanced Encryption System (AES)

However, the continually increasing speed of computers and users' ability to crack the DES code convinced the NIST that Triple DES would only be a temporary solution and on January 2, 1997, even before Triple DES was in place as the U.S. encryption standard, NIST announced plans for an advanced encryption system (AES). Later that same year, on September 12, 1997, the NIST issued a formal RFP for designs for a more secure encryption algorithm. The RFP stated that the selected proposal must be for a private-key, block-cipher system using a minimum 128-bit key, with upward capability to192-bit and 256-bit keys. It must also be publicly disclosed, available worldwide, and distributed royalty-free. AES is expected to be completed by the summer of 2001 and to become the new federal standard, replacing Triple DES, after that time.

9.4 PUBLIC-KEY CRYPTOSYSTEMS

Also during the 1970s, while private-key DES was becoming the U.S. encryption standard, a second approach to cryptography, public-key encryption, was developed by three men in California. During the academic year, 1974-75, Whitfield Diffie and Martin Hellman, from Stanford University, and Ralph Merkle, from the University of California at Berkeley, developed a two-key encryption system that they called "public-key cryptography."[22] One key is a "public-key," which the owner may print on business cards, use in emails, place in public directories, or advertise on billboards. The second key is a "private-key," known only to the key owner. Hence, public-key systems are also called "two-key" or "asymmetric" encryption systems.

There are two ways to send a public-key encrypted message. Users choose one or the other depending on the type of communication desired. First, if the message is being sent to a specific individual, and the sender wants to ensure that only the intended recipient can read the message, the sender would use method one. For example, a customer sending credit card information over the Internet to a company from whom the customer is purchasing items would use this method. Since the company's address and other information are public, they need not be encrypted. However, numerous customers may wish to send separate, individually encrypted messages to the company. For this reason, method one is commonly called "many-to-one encryption." Senders of some email messages, personal health, insurance, banking information and tax filings would also use method one. The focus is that only the sender's side of the message requires encryption. The recipient's side of the message need not be encrypted.

On the other hand, if the message is potentially sent to numerous public recipients, but the sender wants to ensure that only he or she and no one else would be the recognized sender, the sender would use method two. For example, this would be used by a person wishing to send a digital signature on a contract over the Internet to one or more recipients with whom the sender is conducting business. Since the sender's address and other information must be encrypted in order to verify for potentially numerous recipients, method two is commonly called "one-to-many encryption." Senders of digital signatures, orders, personal email messages, and order confirmations would use this method. Again, the focus is that the sender's side of the message requires encryption, but the recipient's side of the message need not be encrypted. Both methods one and two require that both parties have encryption technology and are in contrast to the one-to-one aspect of private-key cryptography and traditional person-to-person communications.

Method One—Many-to-One Encryption

To send a public-key encrypted message using method one, a user (1) creates the message in normal text, (2) encrypts it *using the intended recipient's public-key,* and (3) sends the encrypted message to the intended recipient over any network, including the public switched telephone system (PSTN) or the Internet. When the intended recipient receives the message, that recipient uses his or her corresponding private key to decrypt the message.

To respond to the sender, the recipient (1) drafts a response, (2) encrypts it using the sender's public key (received with the message as part of the sender's ID), and (3) sends the encrypted response to the sender. When the sender receives the response, he or she decrypts it with his or her own private key.

Method Two—One-to-Many Encryption

Method two requires an extra step. With method two, a user (1) creates the message in normal text, (2) *encrypts it using his or her own private key,* (3) then encrypts it a second time with the recipient's public key, and (4) sends the encrypted message to the intended recipient over any network. When the intended recipient receives the message, that recipient uses his or her own private key to decrypt the message and then uses the sender's public key to verify that the message is actually from the person claiming to be sending it.

To respond to the sender, the recipient (1) drafts a response, (2) encrypts it using the sender's public key received with the message as part of the sender's ID, and (3) sends the encrypted response to the sender. When the sender receives the response, he or she decrypts it with his or her private key.

With either method, public-key encryption messages are always sent to a user's public key and decrypted by that user's corresponding private key. Stated another way, a user's private key unlocks any messages sent to his or her public key.

9.4.1 Mathematical Link Between Key Pairs

Public-key cryptography works because each public/private key pair is unique and linked to each other through a complex mathematical process. The complexity of the mathematical process ensures that a person's private key cannot be linked to two public keys at the same time, discovered from its corresponding public key, or copied, forged, or replaced. The following sections explain how this works.

Hash—Error Checking

The mathematical process that links each public/private key pair creates a single large number called a "hash." The primary purpose of the hash is to check for errors and it can be generated in several ways, including those used in the *checksum* and *cyclic redundancy check (CRC)* error-checking algorithms.

In the checksum system, the sending computer counts or sums the number of bits in each packet as they leave the sending computer's domain. This summing of the bits creates a single number called the *check digit*. In encryption, this single number is called the hash. The check digit/hash is then attached to the "tail" or "trailer" of the packet, and is sent to the receiving computer with each packet.

As the receiving computer receives each packet, it uses the same algorithm to count the incoming bits. At the end of the packet, if the receiving computer's count of the bits in each packet agrees with the check digit/hash, the receiving computer knows that the transmission was received correctly. If not, the receiving computer asks the sending computer to resend the packet. Since the flow of the check digit/hash is only from the sending computer to the receiving computer and does not work in the reverse, it is called a one-way function.

The cyclic redundancy check (CRC) approach is different only in that the check digit/hash is generated by dividing all of the serialized bits in a packet by a predetermined binary number. The resulting single number is then sent with each packet to the receiving computer, which checks the "remainder" to confirm that the information in the communications was received correctly. It is still a one-way function, but the CRC algorithm is less secure than the checksum algorithm because several combinations of numbers can result in the same remainder and therefore produce an identical hash. This does not occur with the checksum algorithm because only one number is possible when the bits are summed. Therefore, the CRC algorithm is not used for secure communications, but does works well for link-level, bit-oriented communications.

While the primary purpose of the hash is to check for errors, it also provides several additional important benefits in encryption. Some of these are described below.

Security of Link Between Public and Private Keys

First, the one-way function of the hash process prevents a private key from being linked to two public keys at the same time. If such a link occurs, the hash will not "add up," and the receiving computer will detect the mismatch immediately upon receipt. Thus, with public-key encryption systems, it is mathematically infeasible for two private keys to be linked to the same public key. In the same manner, the hash produces a "fingerprint" that inextricably links one key to the other.

Decryption with Private Key

Second, the hash means that a message encrypted with a person's public key can only be decrypted by that key's corresponding private key. Since the two keys are mathematically linked within the encryption program, only the owner of the public key can break the code or decrypt a message sent to his or her public key. No one else, not even the person who encrypted and sent the message, can decrypt the message. Any unintended wiretapper, eavesdropper, or recipient, including governments, can see that a message exists on the network, addressed to a specific public key, but cannot read the encrypted source or content of the message.

No Discovery of Private Key through Public Key

Third, since the hashing algorithm is a one-way function, the hash cannot be analyzed or reversed to obtain the encryption key. Thus, knowing a person's public key does not help to determine that person's corresponding private key or discover the link between the two keys.

Encrypted Messages Cannot Be Duplicated, Forged, or Replaced

Fourth, if either the hash or the text of the message is changed in any way, the hash will not "add up" when it arrives at the receiving computer, and the change will be detected immediately. In addition, with the mismatch, the message cannot be decoded. Thus, it is mathematically infeasible for someone to copy, forge, or substitute a message encrypted with a public-key encryption, checksum system. Any attempt to do so will not produce an identical hash.

Message Digests

Fifth, when a message is encrypted, the calculation of the hash also automatically produces an encrypted, abbreviated version of the message, called a "message digest."[23] The message digest is unique to each message and thus provides an exclusive "message ID" for that message that is then attached to the front of each encrypted message. Typically, a message digest uses a 128-bit encryption scheme, so it is highly resistant to brute-force attacks and thus "cryptographically strong." The one-way function of the hash further ensures that it is computationally infeasible for anyone to decrypt the message digest in order to obtain the original message or to duplicate the message.

Key-IDs

Sixth, in the encryption process, the hash produces an identifier of each key used to encrypt a message. This "key ID" is an abbreviation of the full key used to encrypt the message and usually contains the least significant 64-bits of the actual key. Like the message digest, the key ID, is attached to the front of each encrypted message as part of a "key certificate," but usually is further abbreviated to half of its 64-bits, displaying only its lower 32 bits.

Key Certificates

Seventh, each key ID is part of a key certificate and is stored on a "key ring" With the message digest, a key certificate is attached to the front of each encrypted message. However, the key certificates and key rings for public-key systems differ from those of private-key encryption systems in the following ways.

"Private-key certificates" contain (1) the key owner's name or user ID, (2) a private-key ID, (3) the actual private-key material, and (4) a unique encrypted password to protect the private key in case it is stolen. An individual's private key is intended to remain private and, therefore, can only be accessed through the encrypted password.

"Public-key certificates" contain (1) the key owner's name or user ID, (2) a public-key ID, (3) the actual public-key material, and (4) a time stamp of when the public/private-key pair was generated. An individual's public key can be accessed through either the public-key ID or the user ID. Both are useful because many users have multiple keys and thus the same *user ID,* but no two keys are identical and therefore never have the same *key* ID.

Key Rings

Eighth, the principal method of storing and managing public and private keys are known as key rings. Rather than keep individual keys in separate key files, most encryption systems collect them on key rings to facilitate the automatic lookup of keys either by *key ID* or *user ID.* A key ring may contain one or more key certificates, but public- and private-key certificates are never combined on the same key ring. Instead, public-key rings contain only public-key certificates, and private-key rings contain only private-key certificates. Each user keeps his or her own pair of private-key and public-key rings. Additionally, each public key is temporarily stored in a separate file while in transit to a requester/recipient, who will then add it to his/her key ring.

Receipt Process—Use of Key IDs and Key Rings

Ninth, when an encrypted message is received, the key ID of the key used to encrypt the message appears in the prefixed key certificate. The recipient's encryption software uses the key ID to find the correct private-key required to decrypt the message.

Private key: When a private-key encrypted message is received, the recipient's software uses the private-key ID in the prefixed key certificate to retrieve the full private key from the recipient's private-key ring The full private key is then used to decrypt the message.

Public-key: When a public-key encrypted message is received, the recipient's software uses the key ID of the recipient's public key used to send the message and provided in the prefixed key certificate, to retrieve the public key's corresponding private key from the recipient's private-key ring. The recipient's private key is then used to decrypt the message.

Validation of Sender's Identity—Digital Signature

Tenth, since both the sender's user ID and the key ID of the key used to encrypt the message appear in the key certificate attached to the front of each encrypted message, when an encrypted message is received, the recipient's software compares the sender's user ID with the sender's public-key ID on the receiver's public-key ring. If the two do not match, the mismatch immediately alerts the recipient that an imposter sent the message.

In this manner, the hash "authenticates" that the message is truly from the person who says he or she is sending it. It thus (1) validates the identity of the actual sender, (2) provides a critical part of security verification used in "firewalling," (3) prevents forgery, and (4) ensures that the response reaches the correct party. This verification is called an "electronic" or "digital signature." As an additional benefit, digital signatures also confirm that no one has tampered with or read the message en route, assuring its integrity and security. [24]

9.4.2 Digital Signatures

Sending Process—Creation of a Digital Signature

Since a digital signature is created when a sender uses his or her private key to encrypt a message, it is, therefore, an integral part of the encryption process. As such, encryption provides, at no additional cost, nonforgible electronic signatures that are far more secure, reliable, accurate, and verifiable than a handwritten signature. For this reason, they prevent confusion and fraud in the Internet Age and provide the security required by persons doing business in a non–face-to-face manner. As such, digital signatures will be an increasingly important technology in the new Internet Age. They provide the means for any intended recipient to receive a highly secure, identifiable private message that cannot be altered or decrypted by others. Without this digital verification, e-commerce and advanced uses of communication could not proceed to its fullest potential.

1991/1993—U.S. Digital Signature Algorithm (DSA)

Since the National Institute of Standards and Technology (NIST) was given responsibility for computers and related systems in the Brooks Act of 1965, it reviewed encryption and digital signature issues for the United States in the 1970s and 1980s, even before people realized the importance of digital signatures to commerce and other official communications. However, it wasn't until a quarter of a century later, in August 1991, that NIST announced its selection of the United States's digital signature standard, called the digital signature algorithm (DSA). The DSA, developed by an employee of the National Security Agency (NSA) was made available free to users with no royalties charged, but it was not broadly embraced because of its close connection to NSA.

1993—Digital Signature Standard (DSS)

As a result, two years later, in June 1993, NIST announced a second, more refined standard known as the digital signature standard (DSS). NIST first issued its Notice of Approval for the DSS on May 19, 1994.[25] When it received no negative comments, it proceeded with its recommendation to the White House and U.S. Congress. Both approved the DSS as the U.S. standard in 1994 and reconfirmed it four years later, on December 9 1998.[26] It continues to be the standard used for digital used for digital signature technology in the United States today.

1999—Millennium Digital Commerce Act (MDCA)

Based on this standard, the next year, in 1999, Senator Spencer Abraham (R, Michigan) sponsored the Millennium Digital Commerce Act (MDCA) to "ensure that individuals and organizations in different states are held to their agreements and obligations even if their respective state have different rules concerning electronically signed documents." The proposed MDCA sought to (1) prohibit individual state laws from restricting the legality of digital contracts simply because they are in electronic form, (2) establish guidelines for international use of electronic signatures that would remove obstacles to electronic transactions, and (3) allow the market to determine the type of authentication technology used in international commerce.

2000—Electronic Signatures in Global and National Commerce Act (E-SIGN Commerce Act), June 30, 2000

Springboarding off the support generated by the proposed Millennium Digital Commerce Act, the U.S. Congress passed the Electronics Signatures in Global and National Commerce Act (E-SIGN Commerce Act) of 2000, also known as the Digital Signature Act (DSA), enacted on June 30, 2000. The DSA affirmed the legal validity of digital signatures for e-commerce transactions, contracts, and court documents. For this reason, it is an enormously important law in the new Internet Age. The two major reasons Congress passed the Digital Signature Act were: digital signatures are superior to handwritten signatures, in that they cannot be forged or later denied, and numerous commerce and trade groups, plus many judicial courts petitioned in strong support of them. This new law has generated numerous activities that are best tracked through sources such as the Digital Signatures: A Practitioners Checklist found at *http://www.macleoddixon.com/04public/Articles/digsig.pdf*.

9.4.3 Key Servers

As the effectiveness and popularity of public-key systems have increased, a recipient's public key must be known in order to use it. When public-key systems were first developed, if a recipient's public key was unknown, it was difficult to find. Today, however, public keys are as easy to locate as someone's telephone number. Many companies also offer public-key information with their employees phone numbers. Public directories, or "key servers," act as white pages or information databases, providing easy-to-use, readily available, up-to-date listings of public keys. Acknowledging that not everyone wants his or her phone number or public key published, key servers require registration of each public key by the key owner. In the future, key servers are likely to be an increasingly significant part of the National Information Infrastructure (NII), especially since public keys are required to verify digital signatures. Generally, key servers fall into two basic categories: certification and "web-of-trust" systems.

Certification

Certification key servers involve a central body or certification authority with whom parties register their public keys. The central body then uses its private key to "sign" or "certify" each registered party's public key for anyone seeking confirmation that a public key is truly owned by a particular party. The U.S. Post Office has proposed that it serve as such authority for the United States in the future.

Web of Trust

In contrast, a *web-of-trust* key server involves no central authority. Instead, users upload their public keys to a key server at any time. To verify that a specific individual is the true owner of a public key, the inquirer asks other people to "certify" the key ownership by uploading authentications "signed" by their private keys.

9.4.4 Session Keys

A second issue with public-key encryption systems is that the encryption and decryption processes take longer, require more resources, and thus are slower than with one-key systems, such as DES. For this reason, in the 1970s and early 1980s, one-key systems were used for longer messages and real-time applications, while two-key systems were more useful for shorter messages and stored information.

Today, with high-speed computers, this time and resource difference is less noticeable, but still a factor. To solve this speed problem while retaining the advantages of public-key systems, "session keys" are used. Each session key is unique and used only for one communication or session. Two of the most popular types of session keys are "hybrid systems" and "key exchanges." The strengths and weaknesses of both are discussed below.

Diffie-Hellman Key Exchange

When a recipient's public key is not known and not available in a key server, or when the two parties wish to establish a session key in public, rather than in private, the parties may use a "key exchange." In 1976, a year after the original public-key concept was developed, two of the developers, Diffie and Hellman, published a paper in *IEEE Transactions* entitled "New Directions in Cryptography," in which they described a partial solution to these problems existing with the use of public-key cryptosystems.[27] With the Diffie-Hellman Key Exchange, for example, the following procedure is followed.

1. Parties A and B first send each other numbers (N1 and N2) to establish a session key.
2. In the session, Parties A and B then agree on a number (b) to serve as the basis for their calculations. This number does not have to be kept secret.
3. With the calculation number, parties agree on a large prime number, which will serve as their "modulus m" or base for arithmetic operations. Normally, we operate in "base ten," while in binary we calculate in "base two."
4. To add to their "base number," the parties A and B select two secret large random numbers (A and B).
5. Party A then sends the number she calculates as [(b × A) modulus m] to Party B.
6. Party B reciprocates with [(b × B) modulus m].
7. Finally, both parties can then compute [b × AB], which becomes their secret session key for subsequent encryption of their information.

While this sounds complicated, it is easy to calculate, especially with computers, and yet very difficult to break. It works well for real-time communication, but imposes long delays as the parties exchange the six or more messages to generate a session key.

The Diffie-Hellman Key Exchange also provided an additional feature, the reality of ephemeral keys that changed the basic power relationships in cryptography because they allow two parties to create a secret key in a very public "conversation," without the use of any centralized resources. Before public-key cryptography, encryption always required centralized facilities to manufacture and distribute equipment and keys—consistent with the military's top-down structure. After public-key encryption systems became available, businesses could communicate readily with each other even if the parties did not know each other or had never done business before.

Hybrid Systems

Hybrid session key systems are similar to the Diffie-Hellman systems except that instead of going through the multiple steps to develop a session key the following occurs.

1. Party A simply generates a session key, encrypts it with Party B's public key, and sends the encrypted message to Party B.
2. Party B then decrypts the message with his or her private key, inputs the session key into his or her single-key software or telephone, and begins the conversation or data transfer in the faster temporary, single-key mode.
3. The actual process of selecting the temporary, random session key is invisible to the users because it occurs in the mathematical algorithm contained in the encryption software each uses.

This more simplified process retains the ability for two parties to communicate in a secure manner over a nonsecure communications link such as the public phone system, wireless communications, or the Internet. For this reason, public-key systems have become the preferred encryption systems for email and e-commerce communications. The most popular systems include the RSA, Pretty Good Privacy (PGP), and Hewlett-Packard's VeriSecure. Each is described below.

9.4.5 Well-Known Public-Key Cryptosystems

RSA

In 1977, one year after Diffie and Hellman's paper was published in *IEEE Transactions,* three young faculty members from MIT, Ronald Rivest, Adi Shamir, and Leonard Adelman, began a company called the RSA Data Security, Inc. in Redwood City, California. They developed a popular two-key system called the RSA system. Like the name of their company, their software product's name came from the first initial of each developer's last name.

The RSA system uses two very large prime numbers to create the public-private-key pairs. The two prime numbers are then multiplied by each other, creating a number that is over 200 digits long. Nearly 22 years later, on January 6, 1999, RSA Data Security, Inc. established an Australian subsidiary through which it distributes its encryption products without violating the then-existing U.S. 56-bit limit on exported encryption systems. That U.S. limit has recently been increased to 128-bits, but it is still less than the RSA's capability.

Pretty Good Privacy

All encryption systems developed in the 1970s and 1980s used 56-bit keys. However, in the early 1990s, Philip Zimmerman, a computer programmer in

Boulder, Colorado, designed a public-key encryption program that conformed to the DES algorithm, but used a variable key-bit length starting at 128-bit key encryption and moving up from there. Each user may select the key length he or she desires.[28] Zimmerman named his new program the Pretty Good Privacy (PGP) encryption system and shared it with friends. The program was eventually loaded onto an Internet bulletin board.

While Zimmerman never sold PGP, it is now available worldwide from the Internet and has become the de facto standard for Internet email. For this reason, it is called the "encryption program of the people." Governments worldwide, however, oppose it because no government or law enforcement agency, including the U.S. government and the U.S. National Security Agency, can break the code.[29] In part, this is because of its variable key-bit length, but mainly because all public-key technology is so hard to break. The PGP encryption system is also based on a web-of-trust validation system, so it is completely independent of centralized control.

Hewlett-Packard's VeriSecure

Hewlett-Packard also developed a public-key encryption system known as VeriSecure, which received the U.S. government's approval to export to customers in five countries. However, the Center for Democracy and Technology opposes VeriSecure because of the extent of government access built into its design.

9.5 COMPARISON OF PRIVATE-KEY AND PUBLIC-KEY ENCRYPTION TECHNOLOGIES

Both private-key and public-key encryption systems were developed in academia in the early 1970s and provide very secure modern communications over public networks. Both use computers and many of the same technical concepts and approaches including (1) a mathematical process that generates each key, (2) an error checking scheme that also generates a hash, (3) message digests, (4) key IDs, (5) key certificates, and (6) key rings. However, the manner in which the two encryption systems do this differs significantly. The similarities and differences, advantages and disadvantages, of each are compared in Table 9.1 on pages 354 and 355.

Given today's technological environment, encryption provides a very effective, efficient, and affordable means of retaining privacy in communications and protecting users against electronic theft, fraud, and forgery. As such, it meets the privacy needs of modern communicators, whether over the Internet, satellite, optic fiber, wireless, or other paths of modern communications.

9.6 LAW ENFORCEMENT AND NATIONAL SECURITY CONCERNS WITH PUBLIC USE OF ENCRYPTION

With the development of sophisticated encryption systems creating unbreakable codes, law enforcement and government officials have become increasingly concerned about not being able to decrypt intercepted communications, especially those affecting criminal activity and national security issues. To address these concerns, law enforcement pressed Congress to pass expanded laws to control both the export of strong encryption systems outside of the United States and the domestic use of encryption systems within the United States. Each is discussed below.

9.6.1 U.S. Laws Restricting the Export of Encryption Systems

Since revolutionary days and the beginning of the United States, U.S. government officials have attempted to restrict the spread of encryption techniques to parties who would threaten its the national security. To do this, as described in Chapter 6, two laws control all U.S. exports: the Arms Export Control Act (ACEA) and the Export Administration Act (EAA). The Arms Export Control Act confers authority on the U.S. State Department to regulate the export of anything it deems to be a "munition" or "weapon of war." The State Department's definition of munitions and specific regulations affecting them appear in its International Traffic in Arms Regulations (ITAR). Specific items considered to be munitions by the State Department appear on the U.S. Munitions List (USML).[30] All items on the Munitions List require State Department approval, in the form of an export license, before they can be exported. In the early 1900s, encryption systems were placed on the Munitions List and have been highly controlled by the State Department.

The Export Administration Act (EAA) confers authority to the U.S. Commerce Department's Bureau of Export Administration (BXA) to regulate the export and re-export of all nonmunitions commodities from the United States. Its definitions and regulations appear in the Export Administration Regulations (EAR).

Both sets of regulations, the ITAR and EAA, acknowledge that certain items and technologies not normally recognized as munitions, could potentially harm the United States and/or benefit its enemies. These are called "dual-use" items and include computer software and many encryption systems. Historically dual-use items have been closely scrutinized, highly restricted, and rarely granted an export license by the State Department because, like munitions, they were considered too dangerous to be readily exported. In 1996, Congress changed the law and moved responsibility for "dual-use," nonmilitary encryption systems from the State Department to the Commerce Department. The Commerce Department

TABLE 9.1

	Private-Key Systems	Public-Key Systems
Knowledge of Keys	Both sender and receiver must know the (same) key before they can exchange encrypted information.	Sender need only know the recipient's public key to communicate. If recipient's public key is unknown, sender can locate it in a key server or the two parties can establish a one-time session key. The recipient's private key is not shared, but rather retrieved by a software-based link to recipient's public key.
Knowledge of Recipient	Sender and receiver must know each other, have had previous experience with each other, or have some way to exchange key information before they can communicate.	Communicating parties need not know each other before they communicate. Need not have communicated before. Strangers can communicate with one another in private. A sender need only know the recipent's public key to send a message to that party.
Use in Internet Age	Difficult to use outside a closed circle because of requirement for communicators to know the key prior to information exchange.	Works well in Internet Age because it allows flexible, secure communications between any two or more persons (without the need for any prior key hand-off or information about one another.) This means that two total strangers, persons who have never met one another nor had any previous communication, may immediately exchange encrypted information with one another. They need not exchange a key before this occurs. For this reason, public-key technology is perfect for the Internet Age and the modern communications between very far-flung persons. Public-key encryption is the current direction of most new encryption systems in the Internet Age.
Key Exchange	Requires secure key exchange.	No need for key exchange.
Key Creation and Control	Key creation, hand-off and control are generally provided by a centralized authority.	Keys are created and controlled by the user, not by a central authority. No key hand-off is required. No centralized authority is needed or required to operate the system.

TABLE 9.1 *(continued)*

	Private-Key Systems	Public-Key Systems
Flexibility to Change Keys	A key cannot be changed until the sender informs the receiver about the change and communicates the new key to the receiver.	A person's public key can be changed at any time simply by selecting and publishing a new public key. The encryption system then provides a new corresponding private key to be used by the owner to decrypt all messages sent to the new public key. Directories will make knowledge of the change available to anyone not directly updated by the owner.
	Thus, a key cannot be easily changed.	Private keys are managed by the key owner, and never needs to be revealed to others.
Breakability of Key	Since both sender and receiver must know the key, the security of the key is vulnerable during both hand-off and routine use.	The owner of a key controls his or her private key and never needs to reveal it to others. If the owner believes the key has been compromised, the owner can change the key quickly and easily.
	Key must be closely guarded by all parties to avoid its discovery or duplication (forgery) by others.	Since the user's private key need not be shared with or communicated to anyone else, it has fewer opportunities to be compromised and therefore is generally considered to be more secure than a private key. In addition, error-checking systems will reveal if the message or code were changed in any way, alerting the key owner to a possible discovery or duplication of the key.
Forgibility	Can be copied and/or forged.	Not forgible.
Speed	Fast.	Slow.
		Two-key systems take longer and require more resources to encrypt and decrypt information than one-key systems.
		To reduce this time requirement, session keys are used, including hybrid systems and key exchanges.
Revolutionary Characteristics	Available to corporate and individual users because of increased speed and lower cost of computing power.	Control of key by users. No requirement for centralized control.
		Provides an excellent technical solution for modern communications over the open Internet.

was generally considered to be more supportive of commercial use of encryption than the State Department. Therefore, today, if the State Department decides that something is a dual-use item, it transfers jurisdiction over the export of that item to the Department of Commerce.

Either way, from 1970 to 1994, exports of encryption systems were limited to systems using 40-bit keys or less. In 1994, the number of bits permitted in an exported encryption key was raised to 56-bits, so long as the systems were used only for commercial purposes and exported only to approved countries. The export of stronger encryption systems, over 56-bits, was still highly restricted. After 1998, encryption export licenses and domestic use permits for robust encryption systems, over 64-bits, became permitted contingent upon key holders registering or "escrowing" their private keys with the government, and the inclusion of "key-recovery" systems, meeting the U.S. government's approval, being designed into each exported system.[31] The current discussion is to raise the export limit to something over 128-bit encryption, or to acknowledge that encryption systems are already widely available worldwide and to completely eliminate the United States's encryption export controls.[32] The flow of these changes, therefore, is contingent upon the countries receiving the exported products, and key-escrow and key-recovery requirements.

Export of Commercial/NonMilitary Encryption

1970s	40-bit key limitation
1991	Seven-day approval for software employing RC2 and RC4, RSA Data Security algorithms that used 40-bit keys. Other, stronger DES continued to be restricted for export.
1994	56-bit key
1998	64-bit key (with contingent requirements)
2000	128-bit key (with contingent requirements)

In exchange, the updated laws provided for an expedited review and licensing process. Before 1994, the same product had to be approved and licensed for *each* export transaction. The new rules permitted a single review and licensing process for encryption products when distributed to multiple, but approved countries, and used only in certain industries such as banking, health care, and electronic commerce. As part of this, technology is defined as "the specific information necessary for the development, production, or use of a product." Information can take the form of either "technical data" or "technical assistance."[33]

9.6.2 U.S. Laws Restricting Private, Nonmilitary Use of Encryption Systems within the United States

In the 1950s–1970s, as communications began to be digitized, multiplexed, and packetized, government agencies including the National Security Agency, FBI, and local police departments realized that these new communications technologies negatively affected their ability to track terrorists, drug traffickers, and child pornographers. When encryption was introduced in the 1970s and 1980s, law enforcement, especially the FBI, became increasingly active in communications and encryption issues. By the time the Internet was made public in the early 1990s, the law enforcement agencies argued that these problems would now increase exponentially worldwide. Law enforcement viewed the Internet as both a national security threat and an avenue for unrestricted international crime. In response, during the 1970s through the 1990s, Congress, the White House, and various government agencies were active in restricting the domestic use of digital communications techniques and encryption within the United States and encouraging international discussions to address the worldwide issues. Several of the key domestic laws and activities include (1) the Foreign Intelligence Surveillance Act of 1978; (2) the Reagan Directive of 1985; (3) the Computer Security Act of 1987; (4) the NSA/NIST Memorandum of Understanding; (5) the FBI's Advanced Telephone Unit of 1992; (6) the Escrowed Encryption Standard and Clipper Chip, introduced in 1993; and (7) the Communications Assistance for Law Enforcement Act (CALEA) of 1994. Each of these laws and activities are discussed in the following seven sections.

1978—Foreign Intelligence Surveillance Act

As discussed in Chapter 8, following the *Katz* and *Berger* decisions in 1967, all wiretap activity in the United States conducted by the FBI and other government agencies had to be reported to the federal government. In addition, the government was required to make regular reports of wiretap activity available to public. However, in 1978, the FBI and other national security agencies convinced Congress to pass the Foreign Intelligence Surveillance Act (FISA) of 1978. In the FISA, Congress (1) legalized secret searches, (2) required only that the total number of wiretaps be reported to key agencies within the government, and (3) removed the requirement to inform the public. Therefore, wiretap details have not been published since 1978.[34]

1985—Reagan Directive NSDD-145

In its effort to prevent the use of encryption by the general public, the National Security Agency (NSA) sought to restrict publication of David Kahn's and James Bamford's books, academic distribution of encryption research, and export of most computerized encryption systems. In 1985, the NSA succeeded in convincing the Reagan administration to promulgate Directive NSDD-145, giving the NSA authority to include control over technology for handling "sensitive but unclassified information." It also partnered with the FBI on encryption issues.[35]

1987—Computer Security Act

However, even with the NSDD-145 Directive, the National Security Agency's effort "backfired," and instead, in the Computer Security Act of 1987, the National

Institute of Standards and Technology's (NIST's) responsibilities for computer security standards, including civilian cryptography, were increased. Unfortunately, the Computer Security Act of 1987 was not sufficiently funded to enable NIST to do its job. [36]

1989—Memorandum of Understanding

The National Security Agency, with its much greater resources, then forced NIST into a Memorandum of Understanding that shackled NIST's attempt to develop its own cryptography standardization program.[37] This led to a period of uncoordinated encryption activity and planning in the United States.

1992—FBI's Advanced Telephone Unit

While NIST was given the responsibility of tracking encryption technology in 1987, during the 1970s and 1980s, the FBI worked closely with the NSA in its efforts to restrain public use of encryption. Through its Advanced Telephone Unit, the FBI studied the impact of modern communications technologies, such as digitization, packetization, and encryption, on law enforcement. In the 1980s, the FBI asked the Senate Judiciary Committee for a "Digital Telephony" resolution to require telephone companies to provide the plaintext of any encrypted messages that were encountered during court-authorized wiretapping. Law enforcement, argued that in an increasingly technological age they could not protect the public against crime without this type of technological assistance. In 1992, the FBI's Advanced Telephone Unit (ATU) completed a briefing for Congress in which it stated that, by 1995, only 40% of legal, Title III wiretaps in the United States would be intelligible. At the pace of technology, the Advanced Telephone Unit warned Congress that law enforcement and national security agencies may not be able to track communications at all.[38] It again pressed for a "Digital Telephony Act" that would require communications companies to build government access capability into their networks, software, equipment, and service offerings so that new developments, features, and technologies in communications could not prevent the government from this access.[39]

In particular, the request evolved into the U.S. government's policies toward key-escrow and key-recovery concepts, which it described as the perfect solutions to the public's need for encryption and law enforcement's need to access communications with court orders. The FBI and NSA, however, did not specify the particular mechanisms to be used, but after 1998, encryption export licenses and domestic use permits for robust encryption systems over 64-bits became contingent upon key holders escrowing their private keys with the government, and the U.S. government's approval of the key-recovery capabilities designed into each exported product.[40]

9.6.3 Escrowed Encryption Standard (EES) and the Clipper and Capstone Chips

Within months after the FBI's 1992 briefing, on April 16, 1993, the Clinton administration announced a plan for a new single-key encryption standard, known as the Escrowed Encryption Standard (EES), proposing that key owners escrow their keys with the U.S. government.[41] Ten months later, on February 4, 1994, the U.S. government formally adopted the EES as the recognized, but voluntary, national encryption standard to protect voice, data, and video information carried over telephone lines. However, unlike other encryption standards then to date, the EES had several unique features.

First, the EES used an algorithm called *Skipjack,* developed by the National Security Agency. It provided an 80-bit key and thus was 16 million times more difficult to break than the previous standard, the Data Encryption Standard (DES). However, the technical details of Skipjack were classified by the government, which prohibited public review of the algorithm and concerned the U.S. software industry privacy advocates and private corporations. This was significant because, since the EES was voluntary, users could choose between it and the DES for encryption.

Second, the EES contained a *Law Enforcement Access Field* (LEAF) that provided the U.S. government and law enforcement with a "back door" to decrypt any message encrypted by EES. The EES required that this back door be used only when the government or law enforcement agency had a court order, but this requirement has not worked well in the past since it was so easily ignored.

Third, rather than being implemented through software, as occurs in most encryption systems, the government required that both the Skipjack algorithm and the Law Enforcement Access Field be encoded onto electronic chips or microprocessors and physically placed in equipment such as telephones, fax machines, or computers. For this reason, the government's plan for implementation of the EES was known as the "Clipper Chip." Another chip, known as the "Capstone Chip," was designed for installation in high-speed data communications equipment and video networks. However, the Capstone Chip was never mass-produced or installed.

Fourth, each Clipper Chip was programmed with a unique "chip identification number" and a unique "key" that encrypted the communications. If reverse engineering were attempted on a Clipper Chip to determine either the identification number or key, the chip was designed to self-destruct.

Fifth, when a Clipper Chip's unique key was programmed, the key was designed to split into two unrelated halves. The purpose of splitting each key was to create additional security for key owners by escrowing each half with two separate government agencies. The argument was that if the government believed that a serious enough threat existed to use an escrowed key, it would first have to convince two separate agencies to seek court orders to release their half of the key. Both halves were needed to reconstruct the complete key.

This was intended to provide EES key owners and Clipper Chip users with a "checks and balance" system to assure them that the government would not misuse its access capability. The two escrow agencies named in the proposed EES, however, were the National Institute of Standards and Technology (NIST) and the Automated Systems Division of the Department of Treasury. The fact that both were Executive Branch agencies caused concern for many people.

Sixth, to further ensure that the two halves of the keys were unrelated and that possession of half of the key could not provide knowledge about the second half, both halves of the EES keys were encrypted separately. With this, the key escrow agents holding each half could not read the keys in their possession, or the full key when the two halves were combined.

Seventh, the EES was announced as a voluntary standard because the government expected the EES to become the de facto encryption standard in the United States once most equipment contained the Clipper Chip. The government counted on the fact that by building the Clipper Chip into all equipment, even exported devices, mass production would help drive the cost of the equipment down, making it even more attractive to buyers. To accelerate the acceptance of Clipper Chip phones, the government applied its massive purchasing power to purchase millions of phones containing the Clipper Chip. The government also advertised to Americans and foreign companies doing business in the United States the advantages of the Clipper Chip in providing secure communications in an increasingly technological age. To further encourage acceptance of Clipper Chip phones, the U.S. government announced that, while export controls on most encryption techniques would be continued, equipment and systems using the EES could be easily exported. [42]

1994—The Communications Assistance for Law Enforcement Act of 1994

The 1993 announcement of the EES/Clipper Chip was made in a proposed law called the "Digital Telephony" bill encouraged by the FBI and NSA. When enacted, the "Digital Telephony" bill became formally titled the Communications Assistance for Law Enforcement Act (CALEA) of 1994. CALEA requires telecommunications companies, at their own expense, to make their networks, software, and equipment accessible to the government.

9.6.4 Opposition to the EES, Clipper Chip, and Encryption Export Restrictions

Reactions to the EES, Clipper Chip, and CALEA were extremely unfavorable. The public, privacy advocates, the software industry, corporate users, other departments and agencies in the U.S. government, foreign governments, and even law enforcement generally opposed the key-escrow and key-recovery proposals, but for very different reasons.

Concerns of the Public

In July 1993, the National Institute of Standards and Technology (NIST) requested public comment on the proposed EES and received 300 responses, of which only 2 were favorable. Eight months later, in March 1994, *Time Magazine* and CNN polled 1,000 people and found that 80% of those polled opposed the EES, Clipper Chip, and CALEA proposals. The Computer Professionals for Social Responsibility organized an electronic petition against the EES, which was quickly signed by over 50,000 people.[43] Opponents had several concerns.

Ongoing Surveillance Capability

First, a main concern for most opponents was the ongoing surveillance capability that ESS, Clipper Chip, and CALEA would provide to the U.S. government. If fully adopted, the Clipper Chip and CALEA would allow the government to monitor *every* telephone in the United States. The key-escrow and key-recovery capabilities mandated in the EES and CALEA would permit the government to decrypt *every* encrypted message, without the individual's knowledge. Both were viewed as uncomfortably close to the "Big Brother" capability of the government in George Orwell's *1984*.[44] Even those individuals and groups sympathetic with the needs of the law enforcement and national security agencies were concerned that the "Information Highway" was becoming a "surveillance machine"[45] of the government.

Requirement for Court Orders

Second, while the government reminded all opponents that its use of the Clipper Chip and proposed key-escrow and key-recovery capabilities required court orders, its history of illegal wiretaps and privacy abuse gave opponents little confidence in this protection. In addition, even if the government's record of compliance with wiretap law had been better, many people were concerned that the ongoing *potential* for uncontrolled wiretaps and other forms of governmental abuse would lead to a permanent erosion of basic freedoms and individual privacy rights. As part of this, opponents noted that the NSA could easily breach any privacy protections incorporated in a system because it is not required to obtain a court order if the national security is threatened or for interception of communications between or within foreign countries.[46]

Key-Escrow Agents

Third, opponents questioned the wisdom in having both named escrow agents, National Institute of Standards and Technology and the Department of Treasury, be a part of the Executive Branch of the government. They felt that this would provide no checks or balances to safeguard citizens from government abuse of the ongoing monitoring capabilities, escrowed-key information, key-recovery techniques, or future developments under CALEA. Further, since the National Security Agency is also part of the Executive Branch, opponents concerns were heightened.

Constitutional Concerns

Fourth, many people and government agencies opposed the concept of ongoing government surveillance of communications, in any form, because the U.S. Constitution guarantees certain fundamental rights of individuals[47] that are threatened by Clipper Chip, key-escrow, and key-recovery requirements. Among others, the First, Fourth, and Fifth Amendments were cited.

First Amendment—Freedom of Speech and Association

Concerning the First Amendment, opponents argued that the proposed Chipper Chip, key-escrow, and key-recovery systems violate the First Amendment guarantees of freedom of speech and the right to peaceably assemble.

Freedom of Speech: Freedom of Speech guarantees individuals the right to speak freely without government interference or punishment. Enacted in 1791, the First Amendment protects both the spoken and written word. Modern communications meet the "time, place, manner" analysis of successful free speech cases.[48]

Right to Peaceably Assemble: Cryptography is also an essential and powerful tool for human rights work, used by groups such as the American Civil Liberties Union, Tibetan Government-in-Exile, and Stop Prisoner Rape. In all cases, key-escrow and key-recovery systems are inconsistent with this work and threaten the participants' right to peaceably assemble.

Fourth Amendment—Unreasonable Search and Seizures

Concerning the Fourth Amendment, opponents argued that the ESS, Clipper Chip, and CALEA violate the Fourth Amendment criteria of probable cause, particularity, and restrictions against unreasonable search and seizure.

Probable Cause: Opponents pointed out that the U.S. Constitution requires the government to establish "probable cause" with a court before conducting surveillance. Any surveillance without reasonable cause, as would be possible with the Clipper Chip, key-escrow, or key-recovery technologies, in ESS and CALEA, is not consistent with Fourth Amendment protection of probable cause.

Particularity: Opponents further argued that a Clipper Chip would be implanted, not upon probable cause, but rather "in case" of a possible future crime. This broad-based access affecting all citizens is in conflict with the Particularity Clause that limits the scope of authorized searches. On the other hand, supporters argued that the Clipper Chip is more like regulatory drug testing, which requires no warrant. Since the government's plan required a warrant, this argument was not compelling.

Unreasonable Search and Seizure: Supporters also argued that breaking a single encrypted message or wiretapping a single telephone call rarely uncovers crimes. Instead, ongoing surveillance generally must be used. Opponents countered that ongoing surveillance is an unreasonable "search and seizure," a Fourth Amendment violation that threatens individual rights.[49] Both opponents and supporters agree that several questions exist that require court determination, including: (1) Does broad-based, on-going access, such as provided by the Clipper Chip, key escrow, or key recovery constitute a search? (2) If so, is that search reasonable? (3) Since wiretap law permits the secret seizure of a conversation with a court order, but does not permit the subsequent secret seizure of a record of that conversation, how are these modern capabilities to be defined?

Right of People To Be Secure in their Persons, Houses, Papers, and Effects: In addition, supporters argued that a key is not a conversation, but the means to decrypt one. Opponents countered that the purpose of encryption is to create or increase privacy. Therefore, the owner of a private key or telephone conversation has both a subjective and objective "expectation of privacy," and any law that requires disclosure of either to the government is a violation of that expectation.

Fifth Amendment—Self-Incrimination

Finally, concerning the Fifth Amendment protection against "self-incrimination," opponents of the Clipper Chip, key-escrow, and key-recovery capabilities point out that requiring users to make their keys and communications available to the government is analogous to forcing users to disclose their secrets in advance and/or allowing a possible waiver of any future Fifth Amendment privileges.

Ineffectiveness in Preventing or Solving Crime

While both opponents and sympathetic supporters of the needs of the law enforcement and national security agencies agree that modern communications technologies make crimes easier to commit, few people accept that escrowed encryption systems or hardware such as the Clipper Chip stop crimes or are effective in discovering crimes. First, since strong encryption programs are readily available worldwide, criminals have easy access to them. For example, sophisticated encryption software can be purchased for as little as five U.S. dollars in nearly any other country, and very strong encryption programs, such as Pretty Good Privacy, are available free from the Internet. An Australian company recently announced that strong encryption capability is built into their version of Netscape Communications Corporation's browser. Thus, any criminal who wanted to encrypt a message could easily choose from a range of effective, inexpensive encryption software completely unaffected by government-imposed key-escrow or key-recovery requirements. As Jack Kempsey, with the Center for Democracy and Technology in Washington, D.C., states, "No bad guys are being prevented from getting 128-bit encryption by U.S. export controls."

Second, even if key escrow were mandated worldwide, criminals are not likely to comply. They definitely would not comply when the key-escrow program is voluntary, as was proposed in the United States.

Third, the suspects must act on the message communicated or the discovered communication becomes "harmless" free speech. Since action by the perpetrators is required, the government's access to communications obviously does not *prevent* the criminal activity uncovered.

Voluntary Aspect of the EES and Clipper Chip

For this reason, the voluntary aspect of the EES and Clipper Chip was a general concern for both supporters and opponents of the concept. The law enforcement and national security agencies challenged the voluntary aspect of both because they felt that it was ineffective and preferred that it be made a mandatory standard from the start. Opponents were concerned that the voluntary aspect would be short-lived and, instead, would become mandatory if the Clipper Chip became so widely installed in equipment that it became the de facto equipment standard.

Balance of Need for Encryption vs. Threat

Even if certain criminal activity or security threats were prevented by the government's access to modern communications, opponents argue that the time, expense, and limitations of key-escrow and key-recovery systems, such as EES and the Clipper Chip, make them inefficient tools for law enforcement. To the average citizen, the potential for abuse of these systems is more threatening than the benefit they provide. Furthermore, as modern communications affect increasingly more areas of our lives, Americans look to encrypted communications to protect them from the threats introduced by "cookies," "identity thieves," and compiled databases.

Concerns about Security of Escrowed Keys

Added to this, many people polled stated that they do not trust the government to protect the secrecy of the escrowed keys, despite the incorporated security measures. People were concerned about the potential for loss, theft, bribery, and fraud of their escrowed key information. This concern was escalated by the fact that the standard contained no sanctions against violations of the incorporated safeguards. Instead, the written materials included a disclaimer about the government's responsibility for the loss of escrowed keys and a reminder of the existing laws prohibiting suits against the government.

Banks, Financial Services, and International Businesses

Banks and other members of the financial services industry opposed the adoption of the EES for several reasons. First, those banks and financial services institutions that had a significant investment in DES were reluctant to change their systems to comply with the new standard. Second, those that had already implemented the newer, RSA encryption system were happy with it and had just incurred the cost of a shift. Third, the U.S. Council for International Business developed a list of requirements for a flexible international policy on encryption that included (1) free choice, (2) open to the public, (3) international acceptance, (4) flexibility of implementation, (5) user key management, (6) key escrow, and (7) liability.[50] The Council determined that the EES failed to meet most these requirements and thus they could not recommend it to their members.

Software Industry and Equipment Exporters

For years, U.S. software companies dominated software development in the world, providing more than 75% of all software sold globally, including encryption software.[51] However, two key trends caused a significant decline in the U.S. share of the international software market. First, the U.S. government's desire to make the EES an international standard significantly hurt sales. Foreign purchasers would not accept equipment built on the EES or containing the Clipper Chip because they did not want to buy or use a system that the U.S. government could monitor or use to decrypt any communication. Foreign customers were hesitant to even communicate with such devices. Thus, the U.S. companies and service providers felt that if such a system were adopted, they could not compete with foreign companies who did not have such embedded monitoring capabilities. Also, it was not cost effective for U.S. companies to make two versions of a product.

Second, export restrictions on encryption prevent the sale abroad of any U.S. software products that contain encryption keys of more than a set limit. While the limit has increased over time, the restrictions caused U.S. companies to lose market share and to no longer be able to compete in foreign markets. Other countries with fewer restrictions on encryption have taken the lead in the worldwide software market. Private companies outside the United States have become adept at developing sophisticated encryption programs, and international software firms have gained an advantage by being able to sell software with encryption features built into them. These packages provide the "seamless" privacy solutions that software purchasers are seeking. According to a study conducted by the Economic Strategy Institute, a nonprofit think tank in Washington, D.C., this results in the loss of billions of dollars each year in sales for the software industry.[52]

In the future, as electronic commerce expands, especially internationally, the need for secure transactions will also increase. To provide this, encryption is becoming an essential part of transactions to protect both the companies and their customers. Desiring to compete in this market, the U.S. software companies continue to strongly oppose the key-escrow and key-recovery concepts. To communicate their concerns, over 1,000 large U.S. corporations responded to a survey conducted by KPMG Peat Marwick LLP in New York in 1998. Forty-one percent of the respondees said security was the most significant barrier to their ability to conduct electronic commerce over the Internet. As such, it undermines any export market potential.[53] Additionally, opponents of the Clipper Chip and key-recovery systems were concerned that the adoption of one system would discourage the development of other, more secure encryption systems.

Opposition by Software Users

Finally, users of U.S. software products also opposed the U.S. encryption polices. First, for multinational and domestic companies, the U.S. encryption export restrictions prohibit the use of encryption over the legal key-length limit to secure messages to their various customers, international offices, personnel, vendors, and suppliers. Second, for U.S. companies selling products or services over the Internet, the U.S. export regulations prohibit the use of competitively strong encryption, even when needed to protect the customers' personal and financial information. Companies feel that they cannot attract customers or offer competitive security assurances without the use of encryption.[54] Third, for individuals, the export restrictions prevent the use of encryption when communicating internationally over the Internet. Since nearly every person who logs onto the Internet may be routed internationally or may click into a Web site without knowing that the Web site's server is located outside the United States, these restrictions affect every individual and corporation in America that uses the Internet, even persons who do not knowingly communicate internationally.

9.6.5 Efforts to Relax Export Restrictions

In November 1993, Representative Maria Cantwell (D., Wash.) and Senator Patty Murray (D., Wash.) proposed an amendment to the Arms Export Act that would remove controls on both software and hardware that incorporate encryption.[55] On July 20, 1994, Vice President Al Gore responded to Cantwell with a letter in which Gore stated that EES, consistent with the Clipper Chip proposal, would be limited to telephone systems and expressed the administration's intent to cooperate with industry and privacy advocates to meet their needs.

1995—Clipper II Raised Key-Bit Limit for Exported Systems

In 1995, the Clinton administration proposed relaxing the U.S. export restrictions by raising the limit on the length of keys in exported encryption systems from 40-bits to 64-bits, at least through 1998. This permitted the export of stronger, more difficult to break encryption systems. However, the proposal still required that the key be escrowed with the U.S. government and that the key-escrow mechanism be protected from alteration. It did not, however, require use of the Skipjack algorithm. Instead, any algorithm up to 64-bits could be used, and the system could be implemented through software, not just through hardware, such as the Clipper Chip. Since this proposal was a variation on the Clipper-chip key-escrow concept, it became known as "Clipper II."

1995—Clipper III: Trusted Third-Party Concept

Later the same year, in August 1995, the Clinton administration announced what became known as "Clipper III." It was essentially the same as Clipper II, in that it announced that companies would be permitted to export stronger encryption systems—up to 64-bits—so long as the keys were escrowed. However, the Clinton administration, through Clipper III, announced that the difference was that the keys could be escrowed with a private *trusted third party (TTP)* rather than a government agency. The rational was that since the third party was not part of the U.S. government and was "trusted" by the key owner, key escrow would be more acceptable to key owners. The third party, however, would still be under the jurisdiction of the U.S. government, and thus, if the government, with a court order, needed a key, it could still access it through the third party.[56]

9.7 INTERNATIONAL ENCRYPTION POLICIES

While other the governments of other Western democracies, mainly allies of the United States's, have the same concerns about the impact of modern technologies on law enforcement and national security, no other Western democracy restricts the domestic use and export of strong encryption as intensely as the United States. Many other governments question the value of these tight controls and believe that key-escrow and key-recovery concepts are ineffective and contrary to basic foundations of human rights and international law.[57] In addition, all realize that the United States's restrictive policies place U.S. companies at a significant disadvantage in the global market. The more relaxed export policies of most other countries have enabled their companies to take control of the global encryption market.

9.7.1 The Organization for Economic Cooperation and Development (OECD)

The Organization for Economic Cooperation and Development (OECD) is an international organization comprised of 29 industrialized member countries committed to all aspects of international economic cooperation, including investment, conduct of multinational companies, electronic commerce, privacy, and encryption. The members of the OECD are Austria, Australia, Belgium, Canada, The

Czech Republic, Denmark, Finland, France, Germany, Greece, Hungary, Iceland, Ireland, Italy, Japan, South Korea, Luxembourg, Mexico, New Zealand, the Netherlands, Norway, Poland, Portugal, Spain, Sweden, Switzerland, Turkey, the United Kingdom, and the United States. In its work, the OECD publishes voluntary guidelines on several topics such as the following.

1. Code of conduct for multinationals, which establishes voluntary guidelines for appropriate international enterprise behavior,
2. Privacy, used as the basis of privacy policy in many European and Pacific Rim nations,
3. Transborder Data Flow, 1980,
4. Information Security, 1992, and
5. Encryption, 1996–1997.

1996—OECD Meetings on Encryption

In 1996 the OECD began holding meetings on the topic of encryption. The Clinton administration first viewed the meetings as an opportunity to press for international acceptance of the its key-escrow and key-recovery policies, and, with that goal, the United States sent a delegation composed of personnel from the National Security Agency and the National Security Council and headed by the Department of Justice's Computer Crime Unit. During the OECD meetings, personnel from the Department of Justice's Computer Crime Unit joined and worked actively on the OECD committee assigned the task of writing the draft of the OECD guidelines on encryption.[58] The same year, 1996, the United States created an ambassador-level diplomatic position to promote an international encryption export regime consistent with the United States's key-escrow principles. In February 1997, the Clinton administration appointed David Aaron to this position.[59]

1997—OECD Guidelines on Encryption

The United States's active participation in the OECD meetings, however, was not successful in that the OECD guidelines did not adopt the United States's key-escrow and key-recovery policies.[60] The reaction of individual member countries varied, but most were generally unsupportive of the approach, questioning the effectiveness of the U.S. key-escrow/key-recovery concepts and the details of their proposed implementation and management. Some semisupportive member countries noted that the OECD was perhaps not the best forum for such a proposal because the main function of the OECD is commerce, not law enforcement.

In addition, the discussion reflected the fact that the OECD member countries regulated encryption in very different ways. Some, such as the United States, regulate the *export* of encryption systems and related technical data, while others regulate the *import* of encryption products and information. Still others regulate the *use* of encryption, while a fourth group favors *no regula-*

tion, and a fifth group has just begun to study the issue. Table 9.2 on pages 370 and 371 lists some of the countries in each category. More details are discussed in "A Summary of International Crypto Controls," by Bert Jaap-Koops, available at *http://cwis.kub.nl/~frw/people/koops/cls2.htm#oecd,* and "A Graphic Summary of Import, Export, and Domestic Controls," by Bert Jaap-Koops, available at *http://cwis.kub.nl/~frw/people/koops/cls-sum.htm.*

1997—OECD Revised Guidelines on Encryption

In March 1997, the OECD issued a set of revised guidelines on encryption in which the member countries stated that (1) governments have a right to act in defense of their national interest; (2) the fundamental rights of individuals to privacy should be respected in national cryptography policies and in the implementation and use of cryptographic methods, including secrecy of communications and protection of personal data; (3) users should have a right to choose any cryptographic method, subject to applicable law; (4) cryptography should be developed in response to the needs of individuals, businesses, and governments; (5) the development and provision of cryptographic methods, technical standards, and critical protocols should be determined by the market in an open and competitive international environment; and (6) market forces should serve to build trust in reliable systems.[61] With these guidelines, the OECD skirted the issues of key escrow and key recovery, focusing instead on the importance of trust in cryptographic products.[62]

1999—OECD Workshop Overview Paper, Joint OECD-Private Sector Workshop on Electronic Authentication

Two years later, the OECD held a joint workshop with the private sector on the topic of electronic authentication. In June 1999, the OECD published an "Overview Paper" reporting on the outcome of that workshop, but resulting actions have not yet been agreed upon. Updates can be found at *www.oecd.org/dsti/sti/it/secur/act/wksp-overview.pdf.*

9.7.2 1996—1998: Wassenaar Arrangement

Recognizing that it was not making progress in the OECD and accepting that the OECD may be the wrong forum, the United States shifted to another approach. The United States had already categorized encryption systems according to the strength of each system's key structure and placed export controls on any over 64-bits, stating that stronger encryption systems were a threat as a potential weapon or element within weapons. However, the U.S. government recognized that any U.S. export restrictions are basically ineffective if other countries do not have similar export restrictions. Therefore, in a continuing effort to encourage other countries to adopt similar restrictions, the United

TABLE 9.2 INTERNATIONAL VIEWS OF REGULATED ENCRYPTION

Opposes Key Escrow	Supports U.S. Position	Regulates Exports	Regulates Imports	Regulates Use	Advocates No Regulation/ Supports Open Use	Just Beginning National Study of Issues
Netherlands: Voiced strong opposition to key escrow.	United Kingdom: Britain was the most supportive of U.S. efforts. The U.K. Department of Trade and Industry sponsored research on public-key encryption programs at the Cryptologic Research Unit of the University of London. They drafted legislation to outlaw the use of nonescrowed encryption systems and were the most open to the concept of "trusted third-party" key-escrow.*	United States	India	France	Australia: Government states that Clipper Chip digital telephony and encryption are the biggest threats to modern telecommunications interception.	Developing countries: Are beginning studies and increasing attention to encryption controls.
Japan: Japan was the most skeptical, wondering how key escrow would really stop crime since criminals have very strong encryption already. The Japanese government is not yet ready to accept			South Korea	Russia	Denmark: Denmark's Information Technology panel recommended no limits on a citizen's right to use encryption.	Pakistan: Working toward greater use of encryption technology.

(continued)

TABLE 9.2 INTERNATIONAL VIEWS OF REGULATED ENCRYPTION *(continued)*

Opposes Key Escrow	Supports U.S. Position	Regulates Exports	Regulates Imports	Regulates Use	Advocates No Regulation/ Supports Open Use	Just Beginning National Study of Issues
the concept of an international encryption standard based on a trusted third-party key-escrow system.†						
					Scandanavian countries: Representatives spoke openly for strong encryption with no trap doors.	
					Germany: German government supports strong encryption for individual users.‡ Germans are selling strong encryption software since U.S. companies cannot. This is very lucrative, and the German government did not want to restrict sales. It is, however, reviewing the need for stronger regulations on cryptography.	

*Diffie and Landau, *supra* note 7 at 220.
†Stewart Baker, *Japan Enters the Crypto Wars,* Wired, Sept. 1996, at 120.
‡Alex Lash, *Germany Urges Strong Encryption,* C/NET News (July 7, 1997), available at *www.news.com/News/Item/0,4,12203.00.html.*

States pressed its allies to include export control of strong encryption products in the "Wassenaar Arrangement."[63]

The "Wassenaar Arrangement on Export Controls for Conventional Arms and Dual-Use Goods and Technologies" (Wassenaar Arrangement) signed by 33 countries in December 1998 is a document in which the signatory countries generally agreed to restrict the international transfer of weapons to specific countries that were considered a threat to worldwide security. It is an "arrangement," not an "agreement" because it is a nonbinding and discretionary document implemented only through legislative action taken in each signatory country. As stated in the Arrangement,

> *The decision to transfer or deny transfer of any item will be the sole responsibility of each participating State. All measures undertaken with respect to the arrangement will be in accordance with national legislation and policies and will be implemented on the basis of national discretion.*[64]

The 33 signatory countries included the 17 member countries of the former Coordinating Committee for Multilateral Export Controls" (COCOM): Australia, Belgium, Canada, Denmark, France, Germany, Greece, Italy, Japan, Luxemburg, The Netherlands, Norway, Portugal, Spain, Turkey, United Kingdom, and the United States. There were 12 cooperating countries: Austria, Finland, Hungary, Ireland, New Zealand, Poland, Singapore, Slovakia, South Korea, Sweden, Switzerland, and Taiwan. In addition, Russia and others joined the Arrangement. As described in Chapter 6, COCOM was the international weapons control organization that also restricted the export of cryptography as a dual-use technology that could be used both for peaceful purposes and as part of a weapon. While COCOM was formally dissolved in March 1994, its former member countries still work together on issues of common importance, such as munitions control to certain nations.

In the final document, the Wassenaar Group agreed to follow the United States's lead and control export of encryption products with keys exceeding 56 bits (64 bits for mass-market products). However, the Wassenaar Group did not embrace the U.S. position favoring key-escrow or key-recovery systems [65] and has not in the subsequent five years. Nonetheless, the United States imposed domestic regulations on the export of encryption systems using keys over 56 bit and related technical data. In fact, the U.S. restrictions are significantly more stringent than those agreed upon in the Wassenaar Arrangement.

9.7.3 United Kingdom

The strongest supporter of the United States's position on encryption was the United Kingdom. This was partly the result of independent studies completed by the United Kingdom's Department of Trade and Industry. Two such studies include papers on proposed encryption policies released in 1996 and 1997.

1996—United Kingdom's White Paper on the Use of Encryption on Public Networks

In June 1996, during the OECD negotiations and just before the Wassenaar negotiations began, the United Kingdom's Department of Trade and Industry published a White Paper entitled "On Regulatory Intent Concerning Use of Encryption on Public Networks." The paper was an informational document for the government and did not address privacy and anonymity issues on the Internet because they were not commercial issues at the time.

1997—United Kingdom's Public Consultation Paper on Licensing of Trusted Third Parties for the Provision of Encryption Services

However, nine months later, in March 1997, the same Department of Trade and Industry published a public consultation paper entitled "Licensing of Trusted Third Parties for the Provision of Encryption Services." In it, the Department of Trade and Industry prepared draft legislation for voluntary use of a key-recovery system that closely followed the legislation in the United States.[66] Civil liberties groups in the United Kingdom and Europe actively opposed the proposal and worked hard in the United Kingdom and the European Community to defeat the proposal.[67] The issue has not yet been resolved at the time of this writing.

9.7.4 1997—European Commission Policy Paper on Encryption

Seven months after the United Kingdom's proposed legislation, the European Commission officially rejected the U.S. position in a policy paper on encryption and digital signatures, released in October 1997, entitled "Towards A European Framework for Digital Signatures and Encryption." In it, the European Commission stated that, while the European nations were sympathetic about the national security and law enforcement motives underlying the U.S. encryption policy, the European governments had studied the key-escrow and key-recovery concepts and concluded that both were inefficient and ineffective in providing solutions to the problem. The paper noted that (1) such systems are easily circumvented, (2) the involvement of a third party increases the likelihood of message exposure, (3) key escrow across national borders created special issues, and (4) any system adopted must be limited to only what is "absolutely necessary."[68] For these reasons, the European governments were not willing to embrace the policies or other law enforcement approaches. Instead they emphasized privacy protection and security of online transactions.[69] The one European Community country that disagreed with the conclusion of the paper was France, which continues to adhere to a strict national policy against encryption and enforces a form of national key escrow.[70]

Groups such as the European "Senior Officials Group on Information Security" (SOG-IS) continue to study the issue. Private groups are also involved. For example, a group of European companies, working under the name "the Trusted Third-Party Program," are exploring and developing systems with third-party key-recovery capabilities.[71] They have not selected a proposed standard or a candidate to be the trusted third party; despite some public opposition, efforts are underway in Europe to develop a third-party key-escrow system.

9.7.5 United Nations Commission on International Trade Law (UNCITRAL)

In another international forum, the United Nations addressed the issue of encryption through the United Nations Commission on International Trade Law (UNCITRAL). Created by the United Nations General Assembly in 1966 to establish a global trade organization with representatives from all geographical regions and legal systems, UNCITRAL seeks to include (1) developed and developing countries, (2) socialist and capitalist economic systems, as well as both (3) common law and civil law legal systems. While other international organizations address issues in global trade, the mandate of UNCITRAL is the unifying and harmonizing of international trade law, through efforts such as the reduction of legal obstacles to international trade and promotion of orderly development of new legal concepts to assist further growth in international trade.

While UNCITRAL consists of only 36 member nations, many other nations attend most meetings as observers. In addition, most international organizations concerned with international trade attend UNCITRAL meetings, including the International Monetary Fund (IMF). Decisions are made by consensus of both members and observers, not by vote. Thus, the process is one of compromise and persuasion rather than alignments. This makes the progress of UNCITRAL very slow, but the approach is very effective when seeking to harmonize diverse groups, and once decisions are made, they are widely accepted and readily adopted by most countries.

Under this consensus approach, and as part of UNCITRAL's mission to unify international commercial law, UNCITRAL has negotiated several important international agreements. Six of the most significant include (1) the Convention on Contracts for the International Sale of Goods (CISG), also known as the United Nations Convention on International Sales; (2) the 1974 Convention on the Limitation Period in the International Sale of Goods; (3) the 1976 UNCITRAL Arbitration Rules; (4) the 1978 United Nations Convention on the Carriage of Goods by Sea, also known as "the Hamburg Rules"; (5) the 1987 Convention on International Bills of Exchange and International Promissory Notes, adopted by the U.N. General Assembly in 1988; and (6) the 1991 Model Law on Electronics Funds Transfers (EFT).

1999—UNCITRAL Report of Working Group on Electronic Commerce

Concerning e-commerce specifically, in September 1999, UNCITRAL published a report by its Working Group on Electronic Commerce.[72] The final results are not yet available, but drafts and updates may be found at *www.uncitral.org/english/sessions/unc/UNC-33/acn9-465.pdf.*

2000—UNCITRAL Drafts Uniform Rules on Electronic Signatures

The September 1999 report led to the "United Nations Commission on International Trade Law (UNCITRAL) Draft Uniform Rules on Electronic Signatures," published in February 2000.[73] Additional information may be found at *www.uncitral.org/english/sessions/wg_ec/wp-84.pdf.*

9.8 U.S. REACTION TO INTERNATIONAL VIEWS OF ENCRYPTION

During the same four years of 1996–2000, the United States responded to these international developments and experienced several of its own revisions and court cases. The key events include (1) a proposed Encrypted Communications Privacy Act of 1996, (2) a proposed Promotion of Commerce On-Line in the Digital Era (PRO-CODE) Act of 1996, (3) the 1996 National Research Council Report on Encryption, (4) a Proposed Security and Freedom through Encryption (SAFE) Act of 1997, (5) a proposed Secure Public Networks Act of 1997, (6) an amended proposal for the SAFE Act in 1997–1999, and (7) three landmark court cases, *Bernstein, Karn,* and *Junger,* which have significantly impacted encryption issues. These events are described in Section 9.8 and 9.9.

9.8.1 1996—Proposed Encrypted Communications Privacy Act of 1996 (S 1587)

As the United States was pursuing its domestic key-escrow and key-recovery plans and preparing for the 1996 OECD meetings, Senator Patrick Leahy introduced the Encrypted Communications Privacy Act of 1996 as Senate Bill S. 1587. It was a compromise bill that (1) proposed relaxed export controls, (2) affirmed the right of citizens to use any form of encryption domestically, (3) created a legal framework for escrow agents, and (4) criminalized the use of encryption in the furtherance of a crime. It was actively discussed but quickly overshadowed by another Senate proposal, the "Pro-Code" Bill.

9.8.2 1996—Promotion of Commerce On-Line in the Digital Era (PRO-CODE) Act (S. 1726)

Less than a month after Senator Leahy proposed his Encrypted Communications Privacy Act, Senator Conrad Burns (Mont.) introduced a stronger proencryption bill called the Promotion of Commerce On-Line in the Digital Era (PRO-CODE) Act. In his support for the bill, Senator Burns noted the importance of high-quality, secure electronic communications to rural areas. His proposed legislation (1) vigorously supported liberalization of U.S. domestic and export laws on encryption, (2) promoted the freedom to sell and use any type of encryption domestically, and (3) prohibited mandatory key escrow. It did not mention criminalization of use of encryption in a crime, but subsequent discussions revealed that this was simply an oversight. However, 1996 was a presidential election year and the legislation, known as Senate Bill 1726 (S. 1726) did not make it out of committee. Senator Burns reintroduced the bill again in 1997 as Senate Bill 377 (S. 377). It was also overshadowed by subsequent proposed legislation, especially the Secure Public Network Act of 1997.

9.8.3 1996—National Research Council Report

As part of the OECD and Wassanaar talks, the National Research Council (NRC) began meeting in 1994 to discuss the United States's encryption export policy and controls.[74] Two years later, on November 15, 1996, the NRC published its report, prompting an executive order titled "Administration of Export Controls on Encryption Products," Executive Order 13026, 11/15/96, and a White House memorandum on Encryption Export Policy.[75] In the executive order, President Clinton transferred the jurisdiction over nonmilitary and dual-use items from the State Department to the Commerce Department. Nonetheless, both the NRC report and the executive order were less liberal than the Senate proposals and continued to support the key-escrow and key-recovery concepts.

9.8.4 1997—Proposed Security and Freedom through Encryption (SAFE) Act (HR 695)

Also in 1997, at basically the same time as the "PRO-CODE" bill was reintroduced in the Senate, Representative Bob Goodlatte proposed the "Security and Freedom through Encryption" (SAFE) Act in the House of Representatives as House Bill 695 (HR 695). In it, Representative Goodlatte also (1) supported encryption, (2) proposed that individuals and corporations be free to sell and use any type of encryption domestically, (3) proposed Congressional approval for the export of cryptography be moved from the State Department to the Commerce Department, (4) proposed the export of strong encryption be permitted if similar products are available overseas, and (5) criminalized the use of encryption in a crime. It met significant opposition because it still required key escrow and key recovery. Nonetheless, it passed House committee level, but was stopped and significantly amended by the full House of Representatives. Discussion continued through 1999.

9.8.5 1997—Secure Public Networks Act (S 909)

At the same time, when Congress returned from its summer break in 1997, the Congressional bills supporting liberalization of encryption laws in the United States suffered several blows. First, in the Senate, Senator Burns's "PRO-CODE" bill was set aside and replaced by the Secure Public Networks Act, known as Senate Bill 909 (S. 909). Despite its name, the Secure Public Networks Act reversed the liberalization of encryption law. Introduced by Senators Bob Kerrey and future presidential candidate, John McCain, the Secure Public Networks Act tightened rather than loosened encryption export controls, and sought to create incentives for organizations to introduce key escrow.

9.8.6 1997—Amended SAFE Act

In the House, despite repeated assurances from the Clinton administration that it would not impose domestic restrictions on cryptography and approval of the SAFE Act by the House International Relations and Judiciary Committees, FBI Director Louis Freeh successfully lobbied Congress to pass new laws severely restricting the use of encryption even within the United States. This occurred when the House National Security Committee accepted an amendment to the SAFE Act introduced by Representatives Porter J. Goss and Norman D. Dicks. The amendment completely reversed the tone of the original SAFE Act by proposing key escrow and tightened controls on exports and the use of encryption. As a compromise, the amended version of the SAFE Act proposed that businesses escrow their encryption keys to a third-party escrow agency, rather than to the government. However, it retained the same access to the keys with a court order.

The revised version of the "SAFE" bill drew immediate, strong opposition from knowledgeable members of the public, privacy advocates, the U.S. software and communications hardware industries, e-commerce proponents, and multinational companies able to readily access and use strong encryption outside of the United States. Starting in September 1997, these groups loudly challenged the effectiveness of this approach to meet national security and law enforcement needs.

In support of the protestors, Senator Trent Lott (R., Miss.) publicly opposed the FBI's approach to encryption export, stating that continued restriction of domestic use of encryption technology would (1) invade the fundamental privacy of U.S. citizens, (2) be of minimal use to the FBI, (3) require nonexistent technology, (4) create new administrative burdens, and (5) seriously damage our foreign markets.[76] President Clinton asked Senator Lott to work out a compromise on encryption legislation.

Within the next five months, some progress was made. By February 1998, the Commerce Department expanded the definition of "financial institutions" to include credit card and securities firms, but not insurance companies. This name change permitted the firms to export stronger encryption. It also approved the export of Hewlett-Packard Corp.'s VeriStrong encryption technology to five European countries. The system allows key recovery, but only upon activation by the local country, pursuant to the countries' own regulations.[77] A year later, on February 25, 1999, the "SAFE" Act was reintroduced to the House of Representatives. For the 106th Congress, this resubmitted version was known as House Bill 850 (H.R. 850). However, it was still not favorably reviewed and continues as an item of discussion.

9.9 COURT CASES REVIEWING THE U.S. ENCRYPTION EXPORT RESTRICTIONS

During these same four years, 1996–2000, while the U.S. government was pressing its position on encrypted products internationally, it was under attack in the U.S. courts for that position. The debate over whether to reduce or remove the encryption export restrictions was argued in several cases as First Amendment issues. These cases included (1) the *Bernstein Decisions I, II,* and *III,* decided between 1996–1999; (2) the *Karn Decisions* (1996–1999); and (3) the *Junger Decisions* (1998–2000). These three cases had an enormous impact on the United States's encryption export restrictions and caused recent changes in U.S. policy. They are described below.

9.9.1 *Bernstein* Cases

In the late 1980s, Daniel Bernstein, while a Ph.D. candidate in the Mathematics Department at the University of California at Berkeley, developed a mathematical encryption algorithm or source code written in the high-level programming language known as C. Bernstein named his encryption algorithm *Snuffle,* and described it in a scientific paper. After graduation, Bernstein became a computer science professor at the University of Illinois.

In order to have Snuffle tested and critiqued by the international scientific and software communities, Bernstein wanted to publish both information about Snuffle and the actual Snuffle source code in scientific papers to be distributed through the normal channels of scientific exchange, including scientific conferences and the Internet.

Knowing that some encryption programs were affected by export issues, Bernstein filed a "commodity jurisdiction request" with the State Department in 1992 to determine whether the Snuffle source code or any of its related information was controlled by the International Traffic in Arms Regulations (ITAR) under the Arms Export Control Act (ACEA).[78]

In 1992 and 1993, the State Department responded to Bernstein, stating that the Snuffle source code and all information related to Snuffle, were "defense articles" subject to being placed on the U.S. Munitions List. Without a license from the State Department, Bernstein could neither post information about Snuffle on the Internet, nor disclose the information to foreign nationals in the United States—except in very limited circumstances.[79] The prohibition against disclosing the information was a particular concern for Bernstein because "disclosure" included having foreign students in his classes hearing his lectures, reading his publications, or overhearing his technical conversations with other U.S. colleagues.

Therefore, in 1995, Bernstein filed a case in the federal District Court for the Northern District of California in San Francisco, claiming (1) that his Snuffle program was speech, protected by the First Amendment, (2) that the ITAR's export restrictions on his Snuffle program and related technical information violated his constitutional "right of free speech," and (3) *inter alia,* that the ITAR's licensing program and definitions, covering encryption software and related technical information, were so overly broad and unproven that they were unconstitutional prior restraints on free speech. He claimed that as free speech, he was free to disclose his Snuffle program and teach it in his classes, without violating federal law. In three separate opinions, U.S. District Judge Marilyn Patel agreed with Bernstein. The three cases are known as *Bernstein I, II,* and *III.*

1996—*Bernstein I*

In *Bernstein I,*[80] the district court, with Judge Patel presiding, held that encryption source code, including Bernstein's Snuffle program, is speech entitled to First Amendment protection. Judge Patel granted partial summary judgment in favor of Professor Bernstein on his constitutional claim. In doing so, Judge Patel rejected the government's argument that the program was not speech but rather conduct, because its purpose was functional—to encrypt electronic communications—rather than to communicate. She ruled against the government's assertion that "language that allows one to actually do something, like play music or make lasagna, is no longer speech."

However, Judge Patel did not rule on the second constitutional question of whether the U.S. export restrictions actually violate the First Amendment. That decision was delayed until December 1996 in *Bernstein II.*

1996—*Bernstein* II

In *Bernstein II,*[81] the district court considered the second part of Bernstein's First Amendment claim and held that the ITAR restrictions requiring an export license for the Internet publication of Bernstein's encryption algorithm, were "a paradigm of standardless discretion" that imposed unconstitutional prior restraints on speech.

Perhaps sensing the mood of the court and anticipating the likely ruling on the constitutional issues in *Bernstein,* plus responding to the comments of the United States's international partners in the OECD and Wassanauer meetings, on November 19, 1996, President Clinton ordered in Executive Order 13026,[82] 61 Fed. Reg. 58,767, that jurisdiction over the export of nonmilitary and dual-use items, including encryption software and related technical information, be transferred from the State Department to the Commerce Department.

It is significant to note that this change in regulatory authority occurred under the provisions of the International Emergency Economic Powers Act (IEEPA). The IEEPA is a temporary, emergency approach because, to use it, the president of the United States must certify annually that the the United States is confronted with a "national emergency."

One month later, in December 1996, shortly after the *Bernstein II* decision, the Commerce Department adopted amendments to the Export Administration Regulations (EAR). However, the amendments were essentially identical to the previous ITAR restrictions on the dissemination of encryption software and related technical information, including Bernstein's Snuffle. The government conceded this similarity in *Bernstein III.*

The new Commerce Department's regulations permitted the export of commercial non–key-recovery encryption products: (1) only to certain approved countries, (2) so long as the encryption was a 56-bit key length or less, and (3) the manufacturer committed to develop key-recoverable items.[83] No commercial/nonmilitary encryption products longer than the 56-bit DES could be exported at that time, 1996.

1997—*Bernstein III*

Again, Bernstein sued, this time against the Commerce Department, the agency newly appointed to review and issue encryption export licenses. In *Bernstein v. U.S. Dept. of State,*[84] or *Bernstein III,* as it became known, Bernstein claimed that Commerce's new Export Administration Regulations were as much an "unconstitutional prior restraint" as the State Department's restrictions. The district court agreed with him for a third time, granting partial summary judgment for Bernstein and reiterating that encryption software published in source code over the Internet was protected speech under the First Amendment. In her decision, Judge Patel cited the Supreme Court's statement in *Reno v. ACLU*[85] that Internet communications are "subject to the same exacting level of First Amendment scrutiny as print media."[86] The court also held that the Commerce Department's Export Administration Regulations (EAR) were deficient, among other things, because they (1) did not specify time for acting on a license application, (2) failed to contain any standards for granting or denying a license, and (3) lacked any provision for judicial review.

1999—*Bernstein* Appeal

The U.S. government immediately appealed the decision in *Bernstein III* to the Ninth U.S. Circuit Court of Appeals. Arguments were heard in December 1997, and the decision was issued on May 7, 1999. The case, *Bernstein v. U.S. Dept. of Justice,*[87] produced inconclusive results. This is because, while the decision was widely applauded by those opposing the U.S. encryption restrictions, the appellate decision was based on technical defects that could be cured. The broad effect of the case's outcome, therefore, could be short-lived. [88]

Following the Ninth Circuit Court's decision in May 1999, however, the Clinton administration announced on September 16, 1999, that a substantial number of changes in U.S. encryption export regulations would take effect on December 15, 1999. The changes would permit the export of 128-bit nonmilitary encryption, so long as it is used only for "commercial purposes" and only after a review by a "trusted-third party." In this case, the White House announced that the "trusted third party" would be the Bureau of Export Administration (BXA). The purpose of the review was to establish an *ad hoc* approach to allow regulators to determine, on a case-by-case basis, what export restrictions apply, if any. In general, the regulations required only a one-time review.

Opponents argue that this mandatory one-time review creates the opportunity for the National Security Agency or the FBI to require firms to place "back-doors" or other key-recovery capabilities in their encryption systems or to reveal information about their business practices that would allow authorities to track information about them. Since the reasoning behind this policy is classified by the government, it increases the mistrust of individuals and companies required to comply.

9.9.2 *Karn* Decisions

1996—*Karn I*

In 1994, Philip Karn, a network engineer at Qualcomm, Corp., wanted to sell copies of Bruce Schneier's book, entitled *Applied Cryptography* outside the United States. To do this, Karn applied for an export license from the then export-regulating agency, the State Department. The license was granted because, as stated in an accompanying letter, the State Department did not have authority over material already in print. Instead, the First Amendment protected published materials.

The book, as printed, contained both a discussion of certain encryption techniques and an appendix containing the source code for the encryption discussed. Karn proposed that, for the convenience of the reader, he would like the exported version of the book to include a floppy disk containing a verbatim copy of the source code provided in the book. When Karn applied for an export permit to include the floppy disc, however, the application was denied. The State Department stated that the book, with the source code in print, could be sold abroad, but that the disc, with the same information, was a defense article requiring an export license.

This decision from the State Department seemed ludicrous to many and increased widespread resentment and frustration with the government's seemingly foolish, irrational, and erratically applied encryption export restrictions. In response, Karn sued the State Department in the federal district court for the District of Columbia in Washington, D.C. The case, *Karn v. U.S. Dept. of State,*[89] is known as *Karn I*. Interestingly, Karn filed in 1996, the same year that Professor Bernstein filed his case across the country in California.

Like Bernstein, Karn cited constitutional issues. However, unlike the district court judge's decision in Bernstein's case, the district court judge's decision in Karn's case rejected Karn's claims. In fact, U.S. District Judge Charles Richey, the judge in Karn's case, ruled virtually opposite to U.S. District Judge Marilyn Patel, the judge in Bernstein's case. This gave more importance to the *Bernstein* Appeal. In his decision, Judge Richey held that source code was speech for First Amendment purposes, but that the courts should not become involved in foreign policy issues when Congress and the Executive Branch are not yet convinced that the export of encryption technology no longer endangers the national security.[90]

Karn II—Appeal

Karn appealed in the U.S. Circuit Court of Appeals for the District of Columbia.[91] The appellate court remanded the case for consideration in light of the transfer of nonmilitary encryption export regulations from the State Department to the Commerce Department and the decision was made without a published opinion. No further cases were filed, but many groups launched lobbying efforts to try to change Congress's and the Executive Branch's opinions that encryption products present a danger to national security.

9.9.3 1998—*Junger* Decisions

The third case in this area was filed by Peter D. Junger, a professor at the Case Western University School of Law, who maintains sites on the Internet that include information about courses he teaches, including a course on computers and the law. On that Web site, Junger wanted to post encryption source code that he had written to demonstrate how encryption works. However, posting encryption source code on the Internet was defined as an "export" under the then-current U.S. Commerce Department's Export Administration Regulations.[92]

Seeking a waiver, Junger submitted three applications to the Commerce Department on June 12, 1997, requesting determinations of commodity classifications for five encryption software programs, his textbook, and other items. Within weeks, on July 4, 1997, the Commerce Department's Bureau of Export Administration (BXA) told Junger that four of the five software programs he had submitted were subject to BXA regulations,[93] but the first chapter of Junger's textbook, *Computers and the Law,* was an allowable unlicensed export. Therefore, the BXA stated that Junger's encryption source code contained in his *printed book* could be exported, but the export of the book and encryption source code in *electronic form* would require a license to export.

1998—*Junger I*

Citing *Karn I,* Junger sued the Department of Commerce in district court on September 2, 1997, in *Junger v. Daley,* also known as *Junger I.*[94] In the suit Junger challenged the BXA's regulations on First Amendment grounds and sought declaratory and injunctive relief that would permit him to engage in the unrestricted distribution of encryption software through his Web site. Junger claimed that encryption source code is protected speech. The district court, under Judge Gwin, granted summary judgment in favor of the defendant BXA, dismissing the complaint on the ground that software is not expression protected by the First Amendment because its software is inherently "functional" and a "device," holding that (1) encryption source code is not protected under the First Amendment, (2) the BXA's regulations are permissible content neutral regulations, and (3) the regulations are not subject to facial challenge on prior restraint grounds.

1999—*Junger* Appeal

Junger appealed the decision, citing the outcome of the *Bernstein* and *Karn* decisions. On March 1, 1999, the American Civil Liberties Union of Ohio filed a brief for Junger as plaintiff-appellant in his appeal to the United States Court of Appeal for the Sixth Circuit from Judge Gwin's judgment. The U.S. Court of Appeals for the Sixth Circuit reviewed the grant of summary judgment *de novo* and reversed the decision of the lower court, finding that computer source code is free speech protected by the First Amendment. The case was remanded to district court to amend the Commerce Department's regulations.

9.10 CHANGES IN 2000

Partially as a result of these cases, on September 16, 1999, the White House announced a new encryption policy to simplify U.S. encryption export rules. The new policy was implemented in January 2000 when the U.S. government, under the Clinton Administration, relaxed certain encryption export restrictions, but not all. First, and perhaps most notably, Congress amended the

Export Administration Regulations (EAR) to allow the export and re-export of encryption up to 128 bits. Thus, individuals, commercial firms, and other non-government end users in all approved destinations can use any retail encryption system or software. Exports to "terrorist-supporting states" such as Cuba, Iran, Iraq, Libya, North Korea, the Sudan, and Syria, their citizens, and other sanctioned entities remain prohibited or tightly restricted.

The issue continues being discussed. Commercial proponents argue for streamlined export-reporting requirements and more effective encryption. The U.S. law enforcement and security agencies are pressing the "Predator" system to monitor email and other Internet traffic; and the international community desires U.S. laws that upgrade consumer privacy protections to avoid violating the European privacy laws and reflect the Wassenaar Arrangement. Updates on these developments can be tracked on the Bureau of Export Administration, Department of Commerce Web site at *www.bxa.doc.gov/Encryption/pdfs/Crypto.pdf.*

9.10 STATE LAWS CONCERNING ENCRYPTION, KEY ESCROW, AND DIGITAL SIGNATURES

In addition, the U.S. government, 49 states, and 15 other countries have enacted or are currently considering some form of encryption, key escrow and digital signature legislation.[95] A list of some of these activities in each state are provided in Appendix J. The following Web sites provide frequently updated compendiums of state, federal, and international law on this topic.

State of Indiana's Digital Signature Laws Information page: *www.state.in.us/digitalsignatures/index.html*

Kaye Caldwell's Software Industry Issues: Digital Signatures page: *www.softwareindustry.org/issues/1digsig.html*

McBride, Baker & Coles (Chicago) E-commerce page: *www.mbc.com/ecommerce.html*

Perkins Coie's Digital Signature Laws by State: *www.perkinscoie.com/resource/ecomm/digsig/state.htm*

Digital Signature Legislation, Version 1.4, April 1997, by Simone van der Hof, Bert-Jaap Koops: *cwis.kub.nl/~frw/people/koops/digsig.htm*

CONCLUSION

While technology has created new privacy concerns in the Information and Internet Ages, technology has also provided a solution to some of those privacy issues in the form of encryption. Modern computers have made low-cost, highly effective encryption readily available to most persons who desire it. However, that reality also raises concerns for law-enforcement. As the U.S. government seeks to find a balance between the individual privacy interests of its citizens and its law enforcement needs, its efforts are further impacted by the international view of privacy and encryption, especially in Europe and various international organizations. The realities that many other countries view encryption favorably and that encryption programs are therefore readily available throughout the world affects the direction of U.S. encryption policy and the market position of U.S. encryption software companies. The courts and the states are also reviewing the issue. As a result of these various viewpoints and interests, U.S. encryption policy continues to evolve. Its importance to individual privacy, national security, and commerce in the Internet Age, however, ensures that, as predicted by Lawrence Lessig, encryption will be one of the major issues in U.S. telecommunications law in the next century.

ENDNOTES

[1]Lessig, Lawrence, Code:and Other Laws of Cyberspace, 35-36. Basic Books, New York, New York (1990).

[2]Jill M. Ryan, *Government-Controlled Encryption,* Vol. 4:3, William & Mary Bill of Rights Jour., at 1171 (1996).

[3]David Newton, *Encyclopedia of Cryptology* 42 (1997).

[4]Ryan, *supra* note 2, citing Hoffman, et al., *Cryptography Policy,* COMM. ACM, Sept. 1994, at 109.

[5]Lee Dembart, *U.S. Removes an Encryption Barrier,* International Herald Tribune, Jan. 31, 2000.

[6]Christoffersson et al., *Crypto Users' Handbook* (1988) and Bruce Schneier, *The Cambridge Algorithms Workshop,* Dr. Dobb's J., Apr. 1994, at 18 and 22.

[7]Whitfield Diffie and Susan Landau, Privacy on the Line, at 55, MIT Press, Cambridge, Massachusetts (1999).

[8]Katz v. United States, 389 U.S. 347 (1967).

[9]Berger v. New York, 388 U.S. 41 (1967).

[10]*all.net/refs/Shannon49.html.*

[11]*all.net/refs/Knight68.html.*

[12]*all.net/refs/Diffie76.html.*

[13]1948 Universal Declaration of Human Rights, Article 12. *See also,* United Nations *1985 International Covenants on Civil and Political Rights; Human Rights Commission, Selected Decisions under the Optional Protocol, Second to Sixteenth Sessions,* at 149, United Nations Office of Public Information, New York; and Academy on Human Rights, 1993 Handbook of Human Rights, at 3.

[14]David Kahn, The Codebreakers, The Story of Secret Writing, New York: The Macmillan Company (1967).

[15]James Bamford, The Puzzle Palace, Houghton Mifflin (1982).

[16] Diffie and Landau, *supra* note 7.

[17]John Markoff, *A Public Battle Over Secret Codes,* N.Y. Times, May 7, 1992, at D1.

[18]Fed. Reg. (USDoC 1973).

[19]Diffie and Landon, *supra* note 7, at 60, citing Bayh.

[20]*See* 48 Fed. Reg. 41062 (1983).

[21]*Data Encryption Standard (DES),* National Institute of Standards and Technology (NIST), U.S. Dep't. of Com., Fed. Info. Processing Standards Pub. 46-2 (1993).

[22]Diffie and Landau, *supra* note 7, at 60.

[23]Newton, *supra* note 3.

[24]See European Commission Policy Considerations for Electronic Commerce Discussion Paper: World Telecommunication Day 1999, II, Building Trust 2.1, Security of Data Transmissions. [hereinafter EC Policy 1999]. Available at *www.cenorm.be/isss/News/mtgrecord.html.*

[25]*Digital Signature Standard (FIPS 186),* Notice of Approval, National Institute of Standards & Technology (NIST), U.S. Dept. of Com. (5/10/94).

[26]*Id.*

[27]*See* Whitfield Diffie and Martin E. Hellman, *New Directions in Cryptography,* IT-22 IEEE Transactions Info. Theory 644 (1964); Ralph C. Merkle, *Secure Communication over Insecure Channels,* COMM. ACM, Apr. 1978, at 294; and Whitfield Diffie, *The First Ten Years of Public-Key Cryptography,* 76 Proc. IEEE 560 (1988) [discussing the history of public-key cryptography].

[28]Philip Zimmerman, *Pretty Good Privacy: Public-key Encryption for the Masses,* reprinted in Building in Big Brother 93 (Lance J. Hoffman, ed., 1994).

[29]*crsc.nist.gov/encryption/aes/*

[30]22 U.S.C. § 2778(a)(1) (1994).

[31]61 Fed. Reg. 68,572-87 (1996). *See also, White House to Ease Controls on Exports of Software Codes,* Wall St. J., Oct. 2, 1996, at B5.

[32]Diffie and Laudau, *supra* note 7, at 207.

[33]15 C.F.R. Part 772 (1998).

[34]Diffie and Laudau, *supra* note 7, at 229.

[35]*Id.* at 231.

[36]*Id.*

[37]*Id.* at 207 and 231.

[38]Advance Telephone Unit, Federal Bureau of Investigation, (1992), "*Telecommunications Overview*" briefing, as noted in Diffie and Laudau, *supra* note 7, at 232.

[39]Diffie and Laudau, *supra* note 7, at 232.

[40]61 Fed. Reg. 68,572-87 (1996). *See also, White House to Ease Controls on Exports of Software Codes,* Wall St. J., Oct. 2, 1996, at B5.

[41]*Escrowed Encryption Standard (EES),* 59 Fed. Reg. at 6003.

[42]Peter Cassidy, *Reluctant Hero,* WIRED, June 1996, at 114.

[43]Diffie and Laudau, *supra* note 7, at 212.

[44]Henry R. King, Note, *Big Brother, The Holding Company: A Review of Key-Escrow Encryption Technology,* 21 Rutgers Computer & Tech. L.J. 224, 249-53 (1995); Kirsten Scheurer, Note, *The Clipper Chip: Cryptography Technology and the Constitution—The Government's Answer to Encryption "Chips" Away at Constitutional Rights,* 21 Rutgers Computer & Tech. L.J. 263, 277-80 (1995).

[45]Andrew Grosso, *The National Information Infrastructure,* 41 Fed. B. News and J. 481, 481 (1994).

[46]Rochelle Garner, *Clipper's Hidden Agenda, Unix-World's Open Computing,* Aug. 1, 1994 at 51.

[47]A. Michael Froomkin, *The Metaphor Is the Key: Cryptography, the Clipper Chip, and the Constitution,* 143 U.Pa. L. Rev. 709, 823-33 (1995).

[48]*See* Note, *The Message in the Medium: The First Amendment on the Information Superhighway,* 107 Harv. L. Rev. 1062 (1994); and Jill M. Ryan, Note: *Freedom to Speak Unintelligibly: The First Amendment Implications of Government-Controlled Encryption,* 4:3 Wm. & Mary Bill of Rights Jour. 1165 (1996).

[49]*Id.* and Henry R. King, Note, *Big Brother, The Holding Company: A Review of Key-Escrow Encryption Technology,* 21 Rutgers Computer & Tech. L.J. 224, 249-53 (1995).

[50]*Business Group Gets Specific on Encryption,* Newsbytes News Network, Oct. 11, 1994, Westlaw, Allnews Database, 1994 WL 2420643.

[51]Hoffman, *supra* note 4, at 111.

[52]*See generally,* William A. Reinsch, Under Secretary for Export Administration, Department of Commerce, (Testimony Before the House Subcommittee on International Economic Policy and Trade), *Encryption: Security in a High Tech Era* (May 18, 1999).

[53]*Id.*

[54]*Id.*

[55]Office of Technology Assessment (OTA), Information Security and Privacy in Network Environments 1-2 (1994) at 64-65.

[56]Michelle Quinn, *Encryption Relaxation Gets Mixed Reaction,* S.F. Chron., Aug. 18, 1995, at E1. *See also, csrc.ncsl.nist.gov/keyescrow/criteria.*

[57] *See* Mark Rotenberg, The Privacy Law Sourcebook 1999: United States Law, International Law, and Recent Developments, Electronic Privacy Information Center 2 (1999). [*www.epic.org*].

[58] Neil Munro, *US Underscores the Importance of Encryption in International Issues,* Capital Roundup, Wash. Tech., Nov. 7, 1996 at 8.

[59] *U.S. Crypto Czar,* WIRED, Feb. 1997, at 45.

[60] Alex Lash, *OECD Dodges U.S. Crypto Policy,* C/NET News (Feb. 5, 1997). *See also* *www.news.com/News/Item/0,4,9189.00.html.*

[61] Organization for Economic Cooperation and Development (OECD), *Cryptography Policy Guidelines,* (Mar. 27, 1997).

[62] Alex Lash, *Global Crypto Rules Laissez-faire,* C/NET News (Mar. 27, 1997), *www.news.com/News/Item.*

[63] Wassenaar Arrangement [*www.wassenaar.org*]

[64] *Id.* at [*www.wassenaar.org/docs/docindex.html*]

[65] George Leopold, *Despite Deal, Encryption Stalemate Endures,* Elect. Eng. Times, Dec. 21, 1998 at 18.

[66] Courtney Macavinta, *U.K. Seeks Voluntary Key Recovery,* c/Net News (Apr. 27, 1998), available at *www.news.com/News/Item/0,4,21538.00.html.* and *www.gilc.org/crypto/uk/dtistatement-498.html.*

[67] Ashley Craddock, *Rights Groups Denounce UK Crypto Paper,* WIRED Online (May 30, 1997) *www.wired.com/news/.*

[68] European Commission, *Towards A European Framework for Digital Signatures and Encryption,* at 16-18 (Oct. 1997).

[69] Jennifer L. Schenker, *EU is Expected to Reject U.S. Proposal for Monitoring Internet Communications,* Wall St. J., Oct. 8, 1997, at B9.

[70] Jennifer L. Schenker, *French Proposal for Encryption is Worrying EC,* Wall St. J., Oct. 8, 1997, at B9.

[71] Rolf Oppliger, *Internet and Intranet Security,* Artech House, 1998 at 77.

[72] United Nations Commission on International Trade Law (UNCITRAL) Report of Working Group on Electronic Commerce (Sept. 1999). *See also:* *www.uncitral.org/english/sessions/unc/UNC-33/acn9-465.pdf.*

[73] United Nations Commission on International Trade Law (UNCITRAL) Draft Uniform Rules on Electronic Signatures (February 2000). *www.uncitral.org/english/sessions/wg_ec/wp-84.pdf.*

[74] Diffie and Landau, *supra* note 7, at 218.

[75] Encryption Export Policy, White House Memorandum, 11/15/96.

[76] Congressional Record, Oct. 21, 1997.

[77] Department of Justice, *Frequently Asked Questions on Encryption Policy,* (4/24/98).

[78] 22 USC § 2778 *et seq.* (1994).

[79] 22 C.F.R. 120.1-130.7 (1994).

[80] Bernstein v. U.S. Dept. of State, 922 F. Supp. 1426 (N.D. Cal. 1996) [Bernstein I].

[81] Bernstein v. U.S. Dept. of State, 945 F. Supp. 1279 (N.D. Cal. 1996) [Bernstein II].

[82] Executive Order 13026, 61 Fed. Reg. 58,767 (1996).

[83] 61 Fed. Reg. 68572 (Dec. 30, 1996).

[84] Bernstein v. U.S. Dept. of State, 974 F. Supp. 1288 (N.D. Cal. 1997) [hereinafter Bernstein III].

[85] Reno v. ACLU, 117 S.Ct. 2329 (1997).

[86] Bernstein v. U.S. Dept. of Commerce, 974 F. Supp. 1288, 1307 (N.D. Cal. 1997).

[87] Bernstein v. U.S. Dept. of Justice, 192 F.3d 1308 (9th Cir. 1999) [Bernstein Appeal].

[88] *See also:* DOJ Press Release on Ninth Circuit's Bernstein Decision, 5/7/99; and BXA Press Release on Ninth Circuit's Bernstein Decision, 5/7/99.

[89] Karn v. U.S. Dept. of State, 925 F.Supp. 1 (D.D.C. 1996).

[90] *Id.* Supp. 1 (D.D.C. 1996).

[91] Karn v. U.S. Dept. of State, 107 F.3d 923 (1999).

[92] Junger v. Daley, 8 F. Supp. 2d 708 (N.D. Ohio 1998) [hereinafter Junger I].

[93] Such encryption information is regulated under Classification Number 5D002 of the BXA's export regulations.

[94] Junger, 8 F. Supp 2nd 708 (*supra* note 92).

[95] Information Security Committee, Electronic Commerce Division, Digital Signature Guidelines, 1996 A.B.A. Sec. Sci & Tech., available at *www.abanet.org/scitech/ec/iscdsgfree.html.*

CHAPTER 10 Cyberlaw: Evolving Legal Issues with the Internet

THE INTERNET IS UNIQUE IN SEVERAL WAYS FROM ALL OTHER communications phenomena that have existed before. First, the Internet is global. With a click of a mouse, communications on the Internet cut instantly across national boundaries. Second, for those with access to the Internet, this instant communications means that individuals or groups of people can communicate interactively, noninteractively, or in broadcast mode worldwide for very little cost. Third, no one "owns" the Internet. It is a collection of hundreds of thousands of networks, linked together in such a manner that any person in the world with access to the Internet can contact any other person in the world with access to communications equipment, including simple phones or radios, regardless of location, type of equipment being used, or standards controlling that equipment. Fourth, the Internet is not regulated by any specific entity. Governments can regulate certain specific activities within their countries, but the fluidity of the Internet makes regulation much more complicated than with previous systems. This means that to resolve most Internet issues, international agreements are needed. Fifth, even if a communication is between two people in the same country, the actual path of the communication may travel internationally. The system picks the best path for the communication at each moment, but "local" calls may not, in fact, be local calls. Sixth, the explosive growth of the Internet exceeds the annual rate of growth of any other system ever known and thus requires faster than normal responses to issues.

These unique aspects of the Internet create a number of new legal issues that are, for the most part, still undecided. The purpose of this chapter is to identify some of these new issues and to provide the reader with the background to appreciate the ongoing discussions around these issues and their resolutions as they occur. The issues discussed in this chapter include (1) electronic commerce, (2) jurisdiction, (3) taxation, (4) trademark/domain names, (5) copyright, (6) trade secrets, (7) defamation, (8) liability, (9) obscenity and violence, and (10) fraud. They are, by no means, all of the issues surrounding this new area of the law but they do provide an overview of the types of legal issues the Internet is presenting. Sections 10.1 through 10.10 discuss each issue, respectively.

10.1 COMMERCE ON THE INTERNET

As is commonly known, "electronic commerce," "e-commerce," and "dot-com companies" are all modern terms suggesting ease of use for shoppers and, initially, instant wealth for owners. E-commerce has completely altered the manner in which many companies reach customers, the cost to do so, the information available to both buyers and sellers, the way in which customers buy and from where they buy. For these reasons, it is the most significant development in business in decades, and, even in its current deflated state, has a significant role in the world of commerce today and in the future.

10.1.1 Uniform Commercial Code (U.C.C.)

In the United States, the Uniform Commercial Code (U.C.C.) governs the sale of goods, whether conducted in traditional "bricks and mortar" shops or electronically online. Additionally, every state government has also adopted a state version of the U.C.C. However, some aspects of the U.C.C. may cause problems for e-commerce.[1] First, for example, the U.C.C. requires that most commercial transactions have written contracts, signed by all parties, that describe the details of the transaction including the product offered, the manner in which accepted, the price agreed upon, the method and timing of delivery, and the processes provided for dispute resolution. This is known as U.C.C.'s "writing and signature requirement."

In e-commerce, however, the fact that the offer, order, delivery, and payment details for the transaction may be completed electronically rather than on paper directly conflicts with the "writing and signature requirements" of the U.C.C., contract law, and many consumer protection laws.[2] While the Digital Signature Act, discussed in Chapter 9, recognizes electronic signatures as valid for commercial contracts, resolved the signature requirement, the "writing requirement" has not been resolved.

Second, under contract law, parties must agree to deal electronically before they can make an electronic contract. This affects all leases, purchases, subscriptions, memberships, disclaimers, and warranties. It is not yet clear what methods of agreement will meet this requirement. Third, on the Internet, a "sale" may be more than just a "tangible product." It may be also be a service or information. However, not all laws accommodate this change.

Fourth, payment for the delivered goods may be made through electronic payment systems, not by traditional cash or check. Some of the new technologies that enable electronic payment for goods and services over the Internet include, but are not limited to (1) credit card and debit card systems that link electronic banking and payment systems with retailers and (2) "smart cards" that store preset values on the cards. Both are known as "digital cash" or "electronic money." Each of these systems, however, raise concerns about their safety and soundness, and it is not clear what technical systems, safeguards, and laws are needed to protect consumers. For example, how can these systems both verify a user's identity and protect that identity? The marketplace and industry self-regulation alone are not considered sufficient to address fully these issues. What government action, if any, is needed to provide these protections?

Fifth, laws to protect consumers from fraudulent, misleading, and otherwise unscrupulous business practices have exploded in the past few decades in the more developed countries, but are not as common in other areas of the world. In an increasingly global economy, with commercial transactions between people who have never met face-to-face and may not even know where the other is located, consumer protection has become an important issue. In general, consumer protection laws require full disclosure of terms and fairness in the details of transactions, but different languages, cultures, and credit systems make tracking these details more complex. Most remedies in consumer protection laws are cash payments for damages, but calculating, enforcing, and collecting the damages in an international environment pose special problems that tend to undercut the effectiveness of the laws.

In the United States, groups such as (1) the U.S. Congress, with the House and Senate Commerce Committees; (2) the Federal Trade Commission (FTC);[3] (3) state governments; and (4) private industry groups;[4] are active in resolving these issues. A sampling of the activities of each is described briefly in the following three sections.

U.S. Congress: Uniform Laws Concerning E-Commerce

To more directly address the new issues raised by electronic transactions, three new uniform laws were enacted by the U.S. Congress and adopted by numerous states. They include (1) the Uniform Commercial Code for Electronic Commerce (U.C.C-EC), (2) the Uniform Computer Information Transactions Act (UCITA), and (3) the Uniform Electronic Transactions Act (UETA). The advantage of these uniform laws is that they standardize solutions to broad-reaching issues.

The Uniform Commercial Code for Electronic Commerce (UCC-EC) is important in that it proposes revisions to the U.C.C. to resolve some of the new issues arising in e-commerce. For example, for U.C.C. Articles 1, 2, 2A, and 9, the U.C.C.-EC proposes that the word "writing" be extended to include all computer displays, including both printed and visual information such as "email" text. In contrast, the Uniform Computer Information Transactions Act (UCITA) governs the supply of computer information and electronic commerce with a "transactions" focus, and the Uniform Electronic Transactions Act (UETA)[5] governs services or information not covered by the UCITA. Both of these last two laws suggest replacing requirements for a "writing" with requirements for a "record." In their proposals, *record* means information that is inscribed on a tangible medium or stored in an electronic or other medium, retrievable in perceivable form. With the passage of these acts, the Clinton administration formed an Interagency Working Group on Electronic Commerce (IWGEC) to explore the appropriate roles of government and the private sector in promoting commerce on the Internet and to draft a conceptual framework with which to guide U.S. policy in this area. Other pertinent U.S. legislation includes (1) the Federal Trade Commission (FTC) Act at 15 U.S.C. § 45(a); (2) the Computer Fraud and Abuse Prevention Act at 18 U.S.C. § 1030; (3) the Magnuson-Moss Act concerning product warranties at 15 U.S.C. § 2301(a); and (4) Restrictions on Use of Telephone Equipment, regulating telemarketing activities, online sales, and online advertisements at 47 U.S.C. § 227.

Federal Trade Commission (FTC)

In addition to its normal regulatory activities, the Federal Trade Commission (FTC) publishes numerous publications and guides to various consumer issues. The following lists some of the FTC's resources on e-commerce.

1. "Advertising and Marketing on the Internet: The Rules of the Road," FTC Business Guide, (April 1998).
2. "Interpretation of Rules and Guides for Electronic Media: Request for Comment," 63 *Fed. Reg.* 24,996 (May 6, 1998), in which the FTC sought public comment on appropriate regulation of online commerce.
3. "Consumer Protection in Cyberspace: Combating Fraud on the Internet," Prepared Statement of the Federal Trade Commission Before the Telecommunications, Trade, and Consumer Protection Subcommittee of the House Committee on Commerce," 195th Cong. D714 (June 25, 1998) (statement of Eileen Harrington, Associate Director, Division of Marketing Practices, Bureau of Consumer Protection, FTC) in which the provided written testimony to Congress is available online at *www.ftc.gov/os/1998/9806/text.623.htm.*
4. "FTC International Web Survey: Disclosure of General Business and Contract-Related Info by Online Retailers," available at *www.ftc.gov/opa/1999/9906/internationalwebsurvey.623.htm.*

The FTC continues to be a leader in e-commerce.

State Consumer-Protection Laws

While all of the U.S. states have consumer-protection laws, only a few have enacted consumer-protection laws focusing specifically on electronic commerce.

Two of the strongest are California's Business and Professions Code (Section 17538), and Illinois' Consumer Fraud and Deceptive Business Practices Act, (810 ILCS 505/1, *et seq.*) However, it is an important area that continues to evolve.

10.1.2 End-User Contracts

Another approach to resolution of e-commerce issues is the concept of end-user contracts or licenses, including shrinkwrap and "point-and-click" licenses.

Shrinkwrap Licenses

Shrinkwrap licenses are preprinted, standard-form use contracts or licenses between the software developers and/or product manufacturers, and the end users of the software or product. The contracts are contained in the packaging of software, CD, or other items claiming intellectual property ownership. They were introduced in the early 1980s with mass-marketed personal computer software, with the terms of each contract displayed through the plastic wrapping or "shrinkwrap" sealing the package. By opening the package, or "breaking the shrinkwrap," the user agreed to abide by the terms of the contract. This created an immediate, binding contract or license to use the product.

The enforceability of shrinkwrap licenses over the past 20 years has been very fact dependent. For example, in 1988, in *Vault Corp. v. Quaid Software, Ltd.*, 847 F.2d 255 (5th Cir. 1988), a conflict between a state statute validating the terms of a shrinkwrap license and federal copyright law arose. The court held that federal copyright law preempted the state-approved shrinkwrap license, but did not otherwise strike down the concept of shrinkwrap licenses. In later cases, where the terms of the contract were not visible through the packaging, courts held that the shrinkwrap licenses were unenforceable because the contract was completed when the end user opened the package, but before the end user was aware of the terms of the contract.[6] However, by the mid-1990s, as these details were corrected and the practice of using shrinkwrap license became more common, courts increasingly have upheld the enforceability of shrinkwrap licenses.[7]

Point-and-Click Licenses

The practice of using shrinkwrap licenses has been extended to the Internet, where standard "use contracts or licenses" are displayed on Web sites. Access to the content of a specific Web site will not be granted unless the user agrees to the terms of the site's access contract by clicking on the appropriate icon or box, or typing the word *agree*. For this reason, these licenses are called "point-and-click contracts," "click-wrap licenses," and "Web wrap agreements." Two cases that considered the enforceability of these online contracts include *CompuServe, Inc. v. Patterson*, 89 F.3d 1257 (6th Cir. 1996) and *Hotmail Corp. v. Van$ Money*

Pie, Inc., 47 U.S. P.Q. 2 1020 (N.D. Cal. 1998). Both courts held that the clickwrap licenses were enforceable, citing the arguments in support of shrinkwrap licenses and the fact that Web users can read the detailed terms of the contract before "manifesting assent."

Adhesion Contracts

Both shrinkwrap and clickwrap licenses are "adhesion contracts." *Adhesion* means to join, adhere to, or cleave to. In contracts, the term *adhesion* means to agree to join or enter into the terms of an existing contract, agreement, or treaty, and to be bound to the terms of the whole contract, without an opportunity to change, bargain, or negotiate specific terms. Generally, such contracts are drafted by one party and contain terms that are most favorable to that party. They are presented to the other party with only two choices: (1) accept the contract as written or (2) refuse to enter into the agreement. Such contracts are common in sales, loan, and rental agreements, such as for cars, homes, and equipment, and in service agreements such as for medical services, use of parking lots, and the purchase of concert tickets. The consumer or contractee has no real choice in the terms of the contract and cannot obtain the offered goods or services except by acquiescing to the existing contract. Cases in this area include *Cubic Corp. v. Marty*, 4 Dist., 185 C.A.3d 438, 229 Cal.Rptr. 828, 833, and *Standard Oil Co. of Calif. v. Perkins*, 347 F.2d 379, (1965). In the past, courts tended to recognize that these contracts are not the result of traditional "bargaining" and thus often relieved parties from any onerous conditions in such contracts. However, as the language of adhesion contracts has become more standardized, many have been found to be enforceable.

In the new world of the Internet, adhesion contracts have become widely accepted as a normal and acceptable way to do business, but their "take it or leave it" approach is increasingly being challenged by the unique, interactive nature of the Internet and sophisticated e-commerce participants. Frequent participants email or key-in suggested changes to the contracts in an effort to negotiate more acceptable contract terms. This creates a very different environment from other situations where adhesion contracts are used. However, one concern is that the person receiving the email or requested changes to the contact often is a Web site technician or other employee with little or no negotiating authority. Frequently the email or request for changes is forwarded to someone with negotiating authority and agency law addresses some of the remaining issues. A second option being used by Web site owners is to place "sliding scales" or other automated terms and conditions from which users can select. For example, if customers want faster package delivery, they can opt to pay more for "overnight" delivery than for "two-day" or "bulk rate." Such options are becoming known as "online negotiating." Since online contracts are an increasing part of our global economy, these issues will continue to be an evolving part of Internet and contract law.

10.1.3 International E-Commerce Efforts

Given the global nature of the Internet, all activities surrounding electronic commerce have an international impact. Therefore, the solutions to any e-commerce issues must work worldwide and be acceptable to other nations in the form of international agreements. For this reason, it is important to be aware of how other countries and international organizations are studying these issues.

1997–2000—European Union

For example, the European Union (EU) considered most of the same issues as the United States and, on May 20, 1997, released the "European Directive 97/7/EC of May 20, 1997 on the Protection of Consumers in Respect of Distance Contracts," ILR, pg. IC-587. Three years later, in May 2000, the European Union updated the document in its "Directive of the European Parliament and of the Council on Certain Legal Aspects of Information Society Services, in Particular Electronic Commerce, in the Internal Market," [EU Information Society Services Directive] (adopted May 2000) ILR, pg. IC-597. In general, both documents identified the key issues, but did not commit the European Union members to specific solutions. The discussions toward resolution continue.

G-10 Working Group on Electronic Payment Systems

The leaders of the seven nations of the world with the strongest economies, known as the "G-7," issued a G-7 Economic Communique at the Lyon Summit in which they called for a cooperative study of the implications of new, sophisticated retail electronic payment systems. Later, the group expanded to include the deputies of the world's 10 strongest economies, forming the G-10 Working Group to continue the study. With representatives from each of the countries' finance ministries, central banks, and law enforcement authorities, plus the International Institute for the Unification of Private Law (UNIDROIT in French) and the International Chamber of Commerce (ICC). The G-10 Working Group was given the task of producing a report that identified common policy objectives among the G-10 countries and analyzing the national approaches to electronic commerce tried to date. A representative from the U.S. Treasury Department chaired the Working Group.

UNCITRAL Model Law on E-Commerce

The borderless, inexpensive, immediate nature of the Internet has made products readily available to consumers worldwide in a manner that has never existed previously and spurred the growth of e-commerce internationally. To support the resulting need for valid, internationally recognized commercial contracts in electronic commerce, the United Nations Commission on International

Trade Law (UNCITRAL) developed a model law on e-commerce that defines the characteristics of a valid electronic contract for e-commerce, provides default rules and norms for the formation and performance of such contracts, provides for the acceptability of electronic signatures for legal and commercial purposes, and supports the admission of computer evidence in arbitration and litigation proceedings. Countries and international e-merchants complying with this model law are assured that their commercial contracts, formed through electronic means, will be recognized as valid worldwide and supported in dispute-resolution proceedings in most jurisdictions. This has spurred the use of e-commerce worldwide and opened markets previously wary about such transactions.

July 1, 1997—U.S. Framework for *Global* Electronic Commerce"

For the United States's part, the White House issued a document entitled "A Framework for Global Electronic Commerce"[8] on July 1, 1997. In it, the Clinton administration advocated several principles for international Internet issues including four key ones. First, in the new, international world of e-commerce, the Clinton administration advanced the concept that the private sector should lead and governments should take a "hands-off" or minimalist approach in order to avoid undue restrictions on the evolution of e-commerce. The White House strongly urged that participants in the marketplace should define and articulate most of the rules that will govern electronic commerce. Second, where governmental involvement is needed, the White House counseled that the aim of governments should be to support and enforce a predictable environment for commerce. Governments should encourage the development of simple and predictable domestic and international rules and norms that will serve as the legal foundation for commercial activities in cyberspace. Third, the Clinton administration advocated that parties should be able to do business with each other on the Internet under whatever terms and conditions they agree upon. Fourth, to encourage electronic commerce, the White House recommended that the U.S. government should support the development of both domestic and global uniform laws that recognize, facilitate, and enforce electronic transactions worldwide. In this regard, the U.S. Departments of Commerce, Treasury, and State continue to represent the United States in international electronic commerce policy and law discussions.

10.2 JURISDICTION

Jurisdiction is the legal right, authority, or power to decide a case or to settle a disputed matter.[9] When a product or service is advertised or sold, either in a store or over the Internet, if the buyer and seller are in the same state or country, the contract, sales, consumer protection, and tax laws of that state or country apply. However, if the buyer and seller are in different states or countries, the laws of all the states and countries involved in the transaction apply. In many transactions, the laws of four or more different states or countries could apply, including (1) the state/country of the seller's residence, (2) the state/country of the buyer's residence, (3) the state/country in which the sale occurred, and (4) any state/country through which the product was produced or transited. These situations are known as transactions with *multiple jurisdictions* and they frequently occur with Internet transactions.

Traditionally, sellers control which markets they enter in order to limit the number of legal systems that have jurisdiction over them and their products. While this means limiting the distribution of products to only a few specific jurisdictions in which the companies are willing to do business, it provided the companies with more control over their legal exposure.

In the Internet Age, however, unless very specific actions are taken, information on the Internet can become available to any user, anywhere in the world, even without the sender's action or knowledge. Thus, a leading concern is that the mere establishment of a Web site generates worldwide jurisdiction and places the Web site owner under the authority of foreign decision makers and exposes the site owner to the possibility of legal action in distant jurisdictions.

For example, if a person simply places an advertisement on the Internet, that person opens himself or herself to advertising, consumer-protection, and commercial laws of nearly every jurisdiction worldwide. What has not yet been determined is to what extent and under what circumstances that advertiser can be sued in a foreign court for violating a state's or country's laws, but even if persons or companies are accustomed to limiting their operations to selected markets in order to comply with the laws in those markets, they find that they can no longer control or guarantee such limitation over the Internet and thus cannot protect themselves from jurisdiction issues.

For the purchaser of a product over the Internet similar concerns arise. Any purchaser of a product becomes immediately subject to the sales tax, import/export, consumer protection, and warranty laws of the locations and jurisdictions of both the buyer and seller. However, the extent to which states, countries, and courts are willing to pursue these issues has not yet been determined.

Similarly, when a person opens a Web page and actively offers a product, service, or information on that site, that person has no control over (1) which markets he or she is entering, (2) where the buyer lives, or (3) which states or countries the electronic transaction may go through on its way to the buyer. Once the information is available on the Internet, it typically becomes available to all users, anywhere in the world. Thus, people conducting transactions on the Internet place themselves under the numerous jurisdictions that may have control over that transaction. This lack of boundaries on the Internet makes jurisdiction an increasingly important issue in the Internet age. As an example, if a person established a Web page in Massachusetts to sell personalized greeting cards or exercise equipment, that person is open to the jurisdiction of any location involved in the transaction. It also means that the Web page owner may potentially be sued anywhere in the world if the owner or transaction violates the contract, sales, liability, tax, or consumer-protection laws of any location with jurisdiction. While fewer than 100 cases have been raised in this area, and thus it is still an unsettled area of the law, it is a significant issue in Internet law to watch. (Only three Internet jurisdiction cases existed before 1996. Nine new cases were heard in 1996, growing to twenty-four cases in 1997.) U.S. court (Judge Gertner) worried that the mere establishment of a Web site would generate worldwide jurisdiction.

10.2.1 Jurisdiction in E-Commerce

A court cannot decide a case unless it has jurisdiction over the parties.[10] The parties can be individuals, corporations, partnerships, or any other legal entity, but each is generally referred to as a "person." To establish jurisdiction, a court must make two decisions. First, it must determine if it has "personal jurisdiction" over the parties. If so, the court must then determine if it has "subject matter jurisdiction." Typically, subject matter jurisdiction is clearly delineated in the law and thus has little additional impact on the Internet transactions. Personal jurisdiction, however, becomes a primary factor in Internet law because of its borderless nature. The court's jurisdiction over each "person" is known as *jurisdiction in personam.*

10.2.2 Personal Jurisdiction in the Internet Age

In the United States, a state cannot just claim jurisdiction over any party it chooses. It must first consider whether the state has certain "minimum contact" with the party. In such cases, the claimant is known as the *forum state* and in deciding what contacts are sufficient, elements such as (1) residence, (2) incorporation, or (3) property ownership in the state are considered. While physical presence of the defendant in the state is not required, lesser contact, such as business transactions by mail or wire communications can establish jurisdiction. As the U.S. Supreme Court noted,

> it is an inescapable fact of modern commercial life that a substantial amount of business is transacted solely by mail and wire communications across state lines, thus obviating the need for physical presence within a State in which business is conducted. So long as a commercial actor's efforts are "purposefully directed" toward residents of another State, we have consistently rejected the notion that an absence of physical contacts can defeat personal jurisdiction there.[11]

These criteria are known as the states' *long-arm statutes* which must also adhere to the Due Process Clause of the Fourth Amendment of the U.S. Constitution.

Second, while "substantial, continuous, and systematic" contact establishes *general jurisdiction,* these are rare in Internet cases because online transactions typically are brief and one-time occurrences. Instead, nonsystematic contact results in *specific jurisdiction,* which is more common in Internet cases. On the other hand, a mere phone call, fax, or other electronic communication to the forum state, is not, by itself, sufficient contact to establish personal jurisdiction.[12] There are three elements used in determining specific jurisdiction: (1) the claim must arise directly out of the nonresident's activities in the forum state; (2) the contacts must represent a "purposeful availment" of the privilege of conducting activities in the forum state; and (3) the exercise of jurisdiction must be "reasonable." *Purposeful availment* focuses on the "deliberateness" of the defendant's contacts. The contacts must be such that the defendant "should reasonably anticipate being hauled into court" in the forum state.[13] Since this is a subjective test, different courts have interpreted it very differently.

The traditional tests or rules of thumb to measure this, including advertising in nationally distributed publications, TV broadcasts, and/or 800 numbers, are not, by themselves, a basis for jurisdiction. However, the Fifth Circuit Court filed a minority view that advertisements in national magazines are sufficient for jurisdiction.

In the Internet Age, therefore, companies that "purposefully [direct]" marketing and conduct sales activities in other states or countries should expect to be subject to the jurisdiction in those states and countries. The questions of what rules will apply to more passive activity such as electronic bulletin board postings that are viewed by other users is less clear. Some courts have determined that since computers and communications have made nationwide and worldwide commercial transactions simpler and more feasible, "it must broaden correspondingly the permissible scope of jurisdiction exercisable by the courts."[14] Thus, if the user has not acted to "purposefully direct" an Internet communication, jurisdiction likely would not be established. Much like with phone calls, a single unsolicited remark in a phone conversation cannot establish libel.[15] "To impose traditional territorial concepts on the commercial uses of the Internet has dramatic implications, opening the Web user up to inconsistent regulations throughout fifty states, indeed, throughout the globe. It also raises the possibility of dramatically chilling what may well be the most participatory marketplace of mass speech that this . . . world . . . has yet seen."[16] This contrasts with a case in which the user was sued in a state in which he no reason to expect he might need to defend a lawsuit.

10.2.3 International Jurisdiction

While being sued and having to defend a lawsuit in another state is of concern, a greater concern for many Internet users is having to defend a lawsuit in a foreign country. The distance and difference in language, customs, and laws makes the experience very costly and stressful. While the prospect of having to defend a lawsuit in another state is a daunting and usually very expensive challenge, it pales compared with defending a lawsuit in another country, under different laws, legal philosophies, language, and at a considerable distance.

This is a real possibility because the same "minimum contacts" standard required in the U.S. is not necessarily used by other governments and legal systems. Other countries may claim jurisdiction over a U.S. information service provider under its own criteria, including whether the defendant owns assets in the foreign country. Without this, the only way to collect a disputed judgment is to ask a U.S. court to enforce it. In general, U.S. courts do not do this unless the legal standards are consistent with those of the United States.

For example, in December 1995, German authorities notified CompuServe, an online information service headquartered in Ohio, that approximately 200 of its adult-oriented newsgroups contained material that might violate German obscenity laws. CompuServe promptly closed the newsgroups to all of its 4 million customers worldwide, an action that raised protests from subscribers and free speech advocates—especially in the United States. The complainants noted that one country's laws or moral codes should not "censor cyberspace for the rest of the world."

10.3 TAXATION

Once a product sells, whether in a bricks and mortar environment or over the Internet, the issue of taxation arises. The first issue concerning taxation on the Internet, however, is should the use of the Internet and/or products or services sold over the Internet be taxed? If so, a second issue is what should be taxed? Typically, taxes are based on the value of a transaction, the use of an item, or the income of the participants. Thus, should the taxes be for access to the Internet, bits transmitted over it, or the actual sale?

A third issue is that a significant number of Internet transactions are also interstate and international. Depending on each country's individual tax policies and principles, how should international taxation be addressed? If the buyer and seller are in different states or countries, which should receive the revenue from the tax? How should this be enforced?

Fourth, if some form of taxation is appropriate, who should collect the tax, the buyer, seller, or ISP? If casinos locate servers offshore and gamblers from around the world access the cyber-casinos, should the winnings be tracked? If so, how? To which country should the gamblers pay tax on their winnings? If a seller operates solely on the Internet, how can the host government accurately track how many sales occur? What privacy issues does this tracking raise? Can companies avoid paying sales, import/export, and other taxes either by locating "off-shore," or by sending the goods/services in digital form? How can double taxation be avoided?

Fifth, if transactions on the Internet are not taxed, what impact will this easily accessed "tax-free" zone have on competitive bricks and mortar or international providers? For example, if people begin making numerous free telephone calls over the Internet, should those also be tax free? Are these "losses" of previous tax revenue to the governments or are they new transactions that would not have occurred without the Internet? If they are losses, how significant are they, and how should governments respond to this loss of revenue?

Sixth, does the uniqueness of the Internet require a new approach to tax options or can existing tax policy and administration work for e-commerce?

Numerous court cases, statutes, and studies at the local, state, federal and international levels have provided input to the discussion surrounding these issues, but final resolution has not been reached. Several of the key inputs include (1) the 1967 U.S. Supreme Court decision in *National Bella Hess, Inc. v. Dept. of Revenue,* (2) the proposed 1987 modem tax, (3) the 1992 U.S. Supreme Court decision in *Quill Corp. v. Heitkamp,* (4) international discussions beginning in 1996 and 1997, (5) the 1997 proposed Internet Tax Freedom Act, (6) the 1998 California Internet Tax Freedom Act, (7) the 1998 U.S. Internet Tax Freedom Act, and (8) the current National Tax Association Communications and Electronic Commerce Tax Project. Each of these is described in the following eight sections, respectively.

10.3.1 1967—*National Bella Hess, Inc. v. Dept. of Revenue*

First, in 1967, long before the Internet or e-commerce was an issue, the Supreme Court heard the case of *National Bella Hess, Inc. v. Dept. of Revenue,* 386 U.S. 753 (1967), in which the State of Illinois requested that a Missouri mail-order company collect Illinois state use tax for goods sold to residents of Illinois. The U.S. Supreme Court held that the interstate clause of the U.S. Constitution prohibits a state from requiring an out-of-state vendor, who solicits sales by mail-order catalogues, to collect use tax for sales made to customers in that state when the vendor lacked outlets, sales representatives, or other significant property in the tax-requesting state.[17] A key factor for the Supreme Court in its decision was that the Missouri company had no tangible property, sales personnel, telephone listings, sales outlets, or solicitors in Illinois. In making this decision, the Court decided that, without some physical presence in Illinois, the Missouri company did not have "substantial nexus" with Illinois and could not, therefore, be required to collect its use tax. This decision established a "bright line, physical presence test" for such cases. The Supreme Court did not, however, define what constitutes substantial nexus. Its decision did indicate that some level of negligible or *de minimis* physical presence does not create taxable nexus.

10.3.2 1987—Proposed Modem Tax

The first proposal to tax Internet services came in 1987 with a "modem tax" that would have required all enhanced service providers (ESPs), including

Internet service providers (ISPs), to pay interstate access charges. However, the U.S. Congress, following discussions with the FCC, did not act on the proposal.

10.3.3 1992—*Quill Corp. v. Heitkamp*

Five years later, a second court case, *Quill Corp. v. Heitkamp,* 504 U.S. 298 (1992), influenced the discussion. The Quill Corporation, a Delaware corporation, sold more than $200 million per year in office equipment throughout the United States, including approximately $1 million to about 3,000 customers in North Dakota. It was the sixth largest office supply vendor in North Dakota, even though it had no employees or significant tangible property in the state. Instead, it delivered all of its merchandise to its North Dakota customers by mail or common carrier from out-of-state locations.

North Dakota state tax law required that every "retailer maintaining a place of business" in the state collect a "use tax" on property purchased for storage, use, or consumption in the state from its customers and remit it to the state. North Dakota defined *retailer* as any vendor who has made three or more advertisements in a 12-month period.

Quill was not collecting the required use tax and the state of North Dakota sued Quill to do so. Quill took the case to the U.S. Supreme Court. First, the court held that without some physical presence in North Dakota, a Delaware company did not have "substantial nexus" with North Dakota and could not, therefore, be required to collect its use tax. The Court stated that Quill's only contacts with the State of North Dakota were by mail or common carrier, and therefore Quill lacked the substantial nexus required by the commerce clause. The Court stated that such vendors are free from state-imposed duties to collect sales and use taxes.

The Court stated that the U.S. Constitution, Article 1, § 8, cl. 3 authorized Congress to "regulate Commerce with foreign Nations and among the several states." If Congress has not regulated or acted on an item, the Court held that the states may not act on their own to interfere with interstate commerce unless it is "fairly related" to the services provided by the state. Unless Congress passes legislation to require certain taxes, states cannot themselves require businesses to collect state taxes when the only contact of the business in the state is through email or Internet contacts.

The Court then clarified that businesses that do operate within a state and have an office or employees in the state, must collect the taxes. This applies to both Internet sales and store sales. Realizing that as Internet commerce grows, the states, fearing significant loss of revenue if too many sales migrate to the Internet to avoid taxes, are likely to pressure Congress to increase their ability to require Internet businesses to collect sales and use taxes.

This decision heightened the discussions surrounding an Internet tax. It also caused the states to review their definitions of what constitutes "substantial nexus," which varies from state to state. On February 5, 1998, in the case of *The Ohio Table Pad Co.,* New York Administrative Law Judge (ALJ) Arthur Bray decided the use of independent salespeople who provide samples and on-site assistance creates nexus. In other cases, under the traditional test for physical presence, a vendor with equipment, property, and employees in a state has substantial physical presence to establish nexus in that state. In such cases, an in-state office is not necessary.[18] For example, the New York Court of Appeals held that 12 visits by a company sales representative over three years established substantial presence. A Tennessee appellate court held that the physical presence of equipment and employees on just four occasions over a three-year period satisfied the substantial nexus test. On the other hand, the Florida Supreme Court held that the three-day per year presence of an out-of-state seller at a Florida trade show did not constitute physical presence in the state.[19] The discussion continues.

10.3.4 1996 to 1997—International Taxation Discussions

In 1996, recognizing the borderless aspect of the Internet and the tax issues raised by the new world of e-commerce; the impact of e-commerce on trading partners; countries' tax revenues, policies, and administration; and the reality that no nation can act alone in this new environment, the international community began researching options and solutions. Both the Organization for Economic Cooperation and Development (OECD) and the U.S. Treasury Department issued papers to generate discussion among international businesses and governments concerning the best way to address these issues. In December 1997, President Clinton announced an agreement with the European Union not to impose new tariffs on Internet commerce in order to allow it to develop as freely and completely as possible.

Beyond treaties, most developed countries have a bifurcated or two-part international tax system that recognizes two criteria as a basis for asserting the right to tax the income of foreign persons: *source* and *residence*. *Source-based taxes* consider the source of the income, and generally, the source country, or the country from which the income was derived, has the right to tax the income of foreign residents. Source taxes generally include such items as some sales taxes and income tax based on wages paid in the country. The details of source taxes vary somewhat from country to country, but the principle is similar throughout the world. It is a long-standing approach that is accepted by nearly all multinational corporations (MNCs), the largest international taxpayers. Typically, since the income is created where the activity occurs, most source taxes require physical presence in the host country.

Residence-based taxes consider the residence of the taxpayer. They include the financial gain earned on (1) interest income, (2) the sale of securities, or (3) the sale of the personal property of residents. This income or financial gain could be generated anywhere, but attaches to the country of the resident receiving the income. Thus, no physical presence generally is required for the tax. As one would expect, most governments try to reach a reasonable balance between the two, but historically, most governments tax *residents* on a global basis and *nonresidents* on a source basis,[20] and fewer items have generated a "residence tax" than a "source tax."

However, the Internet is changing this because it permits anyone to purchase goods and services and to conduct other transactions anywhere in the world. Therefore, some commentators and tax scholars argue that most international taxation should shift from source-based to residence-based taxation since, with only a few exceptions, every individual, corporation, partnership, or other legal entity is a resident of some specific country. Residence requires no physical presence and thus, supporters of this proposal argue, very little source-based tax should exist on the Internet. This would also shift most countries' reliance from source-based taxes to residence-based tax. Others, however, are strongly opposed to this idea, especially the less-developed countries (LDCs). The LDCs typically favor source-based tax because they tend to have smaller populations and thus fewer residents, but significantly more natural resources and inexpensive labor benefited by source-based taxation.

As of 2001, the issues have not been fully resolved, but currently are addressed through "tax treaties." Tax treaties generally are bilateral agreements between two countries covering issues such as capital imports and removal of tax obstacles in cross-border trade and commerce. Tax treaties tend to have the greatest impact on nonresident source-basis taxpayers in their host country because they narrow the scope of taxability and reduce the potential for double taxation. For this reason, their application to e-commerce would require only minor changes to existing treaties. However, the discussion continues.

10.3.5 1997—Proposed Internet Tax Freedom Act (ITFA)

In March 1997, U.S. Congressman Christopher Cox (R. Calif.) and U.S. Senator Ron Wyden (D. Ore.) coauthored a bipartisan Internet Tax Freedom Act (ITFA). The Act, introduced to the U.S. House of Representatives as HR 1054 and to the U.S. Senate as S 442, sought "to establish a [six-year] moratorium on the imposition of any taxes or fees, with specified exceptions, by any state, county, or municipal taxing authority on the Internet or any other on-line activity." It contained moratorium provisions similar to those introduced the following month in California. The bills, however, were tabled for the session to allow time for additional discussion.

10.3.6 1998—California Internet Tax Freedom Act[21]

On April 17, 1997, the California General Assembly passed a joint resolution encouraging California to lead national efforts to keep the Internet a "tax free zone."[22] Several weeks later, in June 1997, California introduced a state version of the federal Internet Tax Freedom Act, entitled the California Internet Tax Freedom Act.[23] Of particular note, the California Bill banned taxation on "services offered by a telephone company, cable television company, or cellular telephone company specifically for the purpose of receiving access to the Internet or interactive computer services," but continued to tax "basic service for telephone, cable television, or cellular telephone service even if some or all of such basic service may be used to provide access to the Internet or interactive computer services."

In September 1997, California State Senator Mountjoy introduced Senate Constitutional Amendment No. 18 to amend the California Constitution to prohibit the state or any political subdivision from levying or collecting taxes or fees on Internet communications or Internet users.[24] Mountjoy's proposal permitted, however, taxation or a fee "of a general application that applies in a uniform and nondiscriminatory manner." The proposal was referred to the California's Senate Revenue and Taxation Committee on January 6, 1998.

On February 19, 1998, Senator Vasconcellos, whose district included Silicon Valley, introduced Senate Bill No. 1908 requiring the California Public Utilities Commission to adopt policies and develop incentives for utilities regulated by the California Commission to increase the speed, capacity, and bandwidth of packet-switched networks used to serve the Internet.[25] The staff of the California Commission staff, however, was not certain what the bill meant since they did not regulate packet-switched networks or the Internet services provided by utilities under their jurisdiction. The California Senate Energy, Utilities, and Communications Committee reviewed the topic in April 1998.

Nearly a year later, on January 13, 1997, this bill was referred to California's Senate Revenue and Taxation committee and was reviewed in light of Senator Cox's revisions to the federal House Bill. The next month, on February 23, 1998, California Assembly Member Bowen introduced Assembly Bill No. 2640 proposing to provide additional civil remedies to enable Internet service providers to recover for "certain harmful actions" or "unauthorized acts" based on either the actual commercial value of the loss, or if that loss is difficult to calculate, an amount determined as either $10 per megabyte copied without authorization; $10 per account per day established, used, or given away; $10 per megabyte of storage used without authorization on the plaintiff's server or equipment.[26] The bill also suggested remedies for persons found liable for unauthorized seizure or use of the names or email accounts of the ISP's subscriber in the amount of $10 per name or account, in addition to any other criminal or civil penalties. The California Assembly Judiciary committee reviewed the proposal in May 1998.

On August 24, 1998, the California Legislature enacted the California Internet Tax Freedom Act (CITFA).[27] It created a three-year moratorium on new taxes imposed by cities, counties, and special districts on (1) Internet access, (2) online computer services, (3) use of Internet access or any online computer services, (4) bits, and (5) bandwidth and created a moratorium on discriminatory taxes on online computer services and Internet access.[28]

10.3.7 1998—Federal Internet Tax Freedom Act[29]

At the federal level, on March 23, 1998, a compromise version of the Internet Tax Freedom Act (ITFA) bill was introduced to the U.S. House of Representatives by Representative Chabot. The new bill proposed (1) reducing the six-year moratorium, proposed in the original version, to three years; (2) "grandfathering" some existing state and local Internet taxes; (3) establishing a Commission on Electronic Commerce to develop a "unified" tax structure for Internet transactions. The three-year moratorium would cover new or discriminatory taxes, such as taxes on Internet access to online services, email, bits, bandwidth, or other Internet-specific taxes. However, the bill would allow states to impose sales and use taxes if they were the same as those now levied on interstate mail-order and telephone transactions.

This was accepted and, seven months later, on October 21, 1998, the U.S. Internet Tax Freedom Act (ITFA) was signed into law. Its stated purpose is "To establish a national policy against State and local government interference with interstate commerce on the Internet or interactive computer services, and to exercise congressional jurisdiction over interstate commerce by establishing a moratorium on the imposition of exactions that would interfere with the free flow of commerce via the Internet, and for other purposes."

With the Internet Tax Freedom Act (ITFA), Congress accomplished four important goals. First, it mandated that the Internet should be free of new federal taxes (§ 201), foreign tariffs, trade barriers (§ 202), and other restrictions (§ 203). Second, it created a three-year moratorium on state and federal taxes for Internet access, unless such taxes were imposed and enforced prior to October 1, 1998 [§ 101(a)(1)]. It also created a three-year moratorium on multiple discriminatory taxes on electronic commerce [§ 101(a)(2)]. These moratoria, however, contain two exceptions. They do not apply to any person or business who knowingly engages in selling or transferring material on the Web that Congress has deemed "harmful to minors" unless they provide certain procedures to restrict access by persons under age 17 [§ 101(e)(1)]. The moratoria also do not apply to Internet service providers (ISPs) that do not offer screening software to their customers designed to allow the customers to limit minors' access to "harmful material" on the Internet. Third, the ITFA established an Advisory Commission on Electronic Commerce (ACEC) to study the effects of taxation on trade and Internet commerce [§ 102], and fourth, required that the ACEC report its findings to Congress within 18 months of the enactment of the Internet Tax Freedom Act (ITFA) [§ 103]. To complete the report, the ITFA gave the e-commerce advisory commission reasonable access to materials, resources, data and other information from the Department of Justice, Department of Commerce, Department of State, Department of Treasury, and the Office of the United States Trade Representative. Surprisingly, the Internet Tax Freedom Act is a relatively short document containing only two Titles. Title I contains only four sections, entitled Moratorium, Advisory Commission on Electronic Commerce, Report, and Definitions. Title II contains only six sections: Declaration that Internet Should be Free of New Federal Taxes; National Trade Estimate; Declaration that the Internet Should be Free of Foreign Tariffs, Trade Barriers, and other Restrictions; No Expansion of Tax Authority; Preservation of Authority; and Severability.

10.3.8 National Tax Association Communications and Electronic Commerce Tax Project

While the states must comply with the Internet Tax Freedom Act (ITFA), they may still tax certain aspects of Internet commerce. The goal of the ITFA is to avoid any new taxes on Internet commerce, but nothing in the ITFA requires changes in existing fees imposed by the Federal Communications Commission (FCC) or states under the Communications Act of 1934 or the Telecommunications Act of 1996. It also ensures, to the extent possible, that the work of the Advisory Commission on Electronic Commerce (ACEC) does not undermine the efforts of the National Tax Association Communications and Electronic Commerce Tax project. Updates on the work of this group are available at *nhdd.com/nta/ntaintro.htm.*

As an example of the numerous issues that remain in this area is that of off-shore gambling. Since the location of servers makes no technological difference, many gambling facilities have located their operations off shore and thus out of countries that tend to regulate gambling activities. However, this also means that governments are losing out on the taxes traditionally gained from such winnings. For this reason, the U.S. drafted a requirement for Internet service providers (ISPs) to monitor the gambling activities of their subscribers and to report such activity to the government. The ISPs refused, citing the privacy of their subscribers. Discussions on all of the issues will resume toward the end of 2001 as the moratorium on Internet taxes ends.

10.4 TRADEMARK/DOMAIN NAMES

A fourth major issue concerning the Internet includes the protection of trademarks. A *trademark* is any word, name, slogan, design, or symbol that is used in commerce to identify a particular product and distinguish it from others. Trademarks identify specific goods with a particular company as that product's "source of origin." Thus, the purpose of a registered "trademark" is to protect a famous or recognizable name so that it cannot be copied or misused. Trademark issues arise when other companies use a trademark in "bad faith" to mislead consumers, create a likelihood of confusion about the source of the product,[30] or dilute association with the name. However, this occurs frequently because trademark rights are national in scope and in the increasingly borderless world of the Internet, companies must register throughout the world or have their unprotected trademark used by others.

In addition, *domain names* are the way companies and individuals represent themselves on the Internet. Domain names are

> the addresses for Web sites, which are computer data files that can include names, words, messages, pictures, sounds, and links to other information. Use of a domain name takes an Internet user directly to the particular Web site or server associated with that domain name. Often domain names consist of some memorable or intuitive name or phrase related to the content of the website or the identity of the name owner. In many instances, the domain name of a corporate Web site will consist of or incorporate the company's name or trademark. Such domain names can be valuable to a company because a domain name is the most direct way of locating a Web site.[31]

10.4.1 Unique Trademark Infringement Issues in the Internet Age

While the law has established procedures for resolving trademark issues, the Internet Age has presented several unique trademark infringement issues that have not been resolved. Theses include (1) cybersquatting, (2) metatags, (3) spamdexing, (4) spoofing, and (5) corporate consistency issues. These are discussed in the following list.

1. **Cybersquatting.** When another company or individual registers a domain name containing a trademarked, famous, or recognizable name or symbol that is not their own for the purpose of selling it later to the trademark's owner, this is known as *cybersquatting*. In cases such as *Lockheed Martin Corp. v. Network Solutions, Inc.*, 985 F. Supp. 949, 959 (C.D. Cal. 1997) and *Intermatic, Inc. v. Toeppen*, 947 F. Supp. 1227, 1234 (N.D. Cal. 1996), the courts noted that Congress and the states have been slow to respond to the activities of cybersquatters, but the issue will continue to increase in importance as domain names become part of corporate and individual identities.

2. **Metatags.** Metatags are small blocks of text that are attached to Web pages and serve as a code to provide information about the Web page, such as the date it was last updated. While they are not part of the visible portions of the Web pages, they are read by Internet search engines. Metatag misuse occurs when a company uses another company's trademark as a metatag in order to trick a search engine into directing traffic to a Web site by taking a "free ride" on the popularity of the trademark. Such use can be enjoined as trademark infringement.

3. **Spamdexing.** The word *spamdexing* comes from a combination of *spam* and *indexing*. In spamdexing, a company's trademark is either placed in the text of a Web page instead of in the metatag, or placed in white print text so that it is technically part of the Web page, but not visible to readers. Either

way, the object of spamdexing is to cause search engines to find the Web site based on the popularity of the trademark and thus give the Web site a higher "hit rate" than it would normally have. This places it at the top of the list in a response to an Internet search, and based on its position of likely options drives additional traffic to the site.

4. **Spoofing.** Many corporations and individuals have software that filters out email from undesirable or unwanted Web sites such as "spam" companies that send junk email, or pornographic sites. "Spoofing" is a method used by these blocked sites to make it appear as if their email is coming from an acceptable site and thus able to pass through the filters. It requires one Web site to assume the identity of another Web site. Where such companies use another company's trademark to indicate a specific source of origin, "spoofing" is considered a clear trademark infringement.

5. **Same Company Trademark Issues.** Since trademarks identify specific products with a particular company as that product's "source of origin," trademark issues arise when the true trademark owner sells different goods on its Web site than it sells in its land-based stores; sells only products in its land-based stores, but provides services with the products over the Internet; and/or has separate legal entities for its land-based and Internet operations. These may occur innocently but are an important aspect of managing corporate trademarks online.

10.4.2 Actions by Companies To Protect Their Trademarks

Traditionally, trademark issues have been addressed under the Lanham Act's remedies for trademark infringement.[32] While the Lanham Act is still used, it applies only in the United States and does not assist in detecting infringement. Thus, companies have developed "software bots" to detect infringement, and in late 1999, two additional avenues of resolution became available to domain name registrants: (1) the International Uniform Dispute Resolution Policy (UDRP) developed by the Internet Corporation for Assigned Names and Numbers (ICANN) in October 1999 and (2) the U.S. Anticybersquatting Consumer Protection Act (ACPA) enacted in November 1999.[33] They add to, but do not replace the Lanham Act remedies, especially where the domain name is not registered in "bad faith," but nonetheless creates a *likelihood of confusion*[34] or *dilution* of the name. Software "bots" and the two new laws are discussed in the following three sections.

Software Robots (Bots) and Software Agents

To counter practices such as *cybersquatting, metatags, spamdexing,* and *spoofing,* legitimate trademark and domain name owners frequently use "*software robots (Bots)*" and/or "*software agents*" to locate unauthorized third-party use of a company's trademarks, logos, and/or recognizable designs. These include small, purposeful "defects" or variations that prove infringement. Authorization for legitimate use of the trademark or domain name is granted by the owner through licensing.

October 1999—The International Uniform Dispute Resolution Policy

The Uniform Dispute Resolution Policy (UDRP) provides a procedure that can be used by any owner of a domain name registered with a registrar accredited by the Internet Corporation for Assigned Names and Numbers (ICANN),[35] or certain country code domain name registrars that have voluntarily adopted the policy. Allegedly infringing domain names can be challenged in an administrative proceeding before an ICANN-approved "dispute resolution provider." One of the most widely used is the World Intellectual Property Organization (WIPO).

To bring a claim or dispute under the Uniform Dispute Resolution Policy, certain general requirements must be met. First, the challenged domain name must have been registered by the registrant and be in the process of being used by that registrant in "bad faith." Second, the challenged domain name must include a registered trademark or service mark. Complaints concerning bad faith use of domain names not including a trademark or service mark must seek resolution under the United States's Anticybersquatting Consumer Protection Act (ACPA). Third, the registrant's domain name must be identical or confusingly similar to a trademark or service mark in which the complainant has rights. And fourth, the registrant must have no rights or legitimate interests with respect to the domain name, trademark, or service mark.

To prove the elements of *registration* and *use in bad faith*, the UDRP offers four nonexclusive factors. First, the registrant has registered the domain name primarily for the purpose of selling it to the owner of the trademark or to a competitor for valuable consideration in excess of out-of-pocket costs. Second, the registrant has registered the domain name in order to prevent the trademark owner from reflecting the mark in a corresponding domain name, provided that there is a pattern of such conduct. Third, the registration was obtained primarily to disrupt the business of a competitor. And fourth, the use of the domain name is intended to attract, for commercial gain, users to the Web site by creating a likelihood of confusion with respect to the complainant's mark.

As with any arbitration, the advantages of using the Uniform Dispute Resolution Policy are first, that the process operates on a relatively short schedule from filing to decision and thus, a decision by an arbitrator can be obtained in a few months rather than the years often required for litigation. Second, this typically makes the process less expensive than litigation. Third, the dispute resolution process provides a means of overcoming some jurisdictional problems, such as personal jurisdiction, if the disputed domain name is registered with a foreign registrar.

The disadvantages of using the Uniform dispute Resolution Policy UDRP, however, include: First, no damages are provided. UDRP proceedings do not provide a possibility for damages, and therefore, if the plaintiff believes that he or she has provable damages and the defendant or cybersquatter has the money to satisfy a judgment, they would be in a better position to litigate. Second, the UDRP covers only limited issues. Therefore, if the cybersquatter also engaged in other illegal activities such as false advertising or unfair competition, these are not addressed by the UDRP arbitrator process. The domain name owner may not want to litigate in pieces and therefore may prefer federal court.

Third, the UDRP is not appropriate for complex or difficult cases. The UDRP process is not in a position to decide the details or nuances of trademark law because of the limited procedural filings. For complicated issues involving fair use, acquiescence, or the commercial relationship between the parties, or important factual matters, the expanded process of litigation may be needed.

Fourth, no discovery is available with arbitration. Since the arbitration process does not provide a process for discovery, if a plaintiff cannot readily prove "bad faith" on the part of the defendant, the plaintiff would be better served by litigating under the Anticybersquatting Consumer Protection Act because it does provide for a discovery process.

Fifth, tradenames are not eligible for relief under the UDRP. Since the rules supporting the UDRP require a trademark or a service mark, domain names not containing a trademark or service mark are not eligible for relief under the Policy and must use the Anticybersquatting Consumer Protection Act.

Sixth, decisions from either an ICANN, Uniform Dispute Resolution Policy (UDRP) or World Intellectual Property Organization (WIPO) proceeding are not binding in subsequent court actions, but should be given "appropriate weight" by the courts.[36] Most courts find these decisions to be helpful recommendations.[37]

Seventh, since the decisions are not binding, plaintiffs may use both avenues of redress at the same or separate times. An Anticybersquatting Consumer Protection Act action, therefore, can be brought, before, during, or after the institution of ICANN proceedings.[38]

November 29, 1999—The U.S. Anticybersquatting Consumer Protection Act (ACPA)[39]

A case can be brought under the Anticybersquatting Consumer Protection Act (ACPA) at any time. No requirement exists that the complainant must first use the Uniform Dispute Resolution Policy (UDRP). A cause of action can be brought under ACPA against anyone who with bad faith intent to profit from another's trademark (including a personal name, which is protected as a mark); "registers, traffics in, or uses a domain name" that is identical or confusingly similar to that mark or, for famous marks, is either confusingly similar to or dilutive of that mark.

The ACPA lists several nonexclusive factors to be used in determining if the use of a domain name is in "bad faith." These factors include the following.

1. Does the defendant have trademark or other intellectual property rights in the domain name?
2. To what extent does the domain name consist of the legal name or name by which the defendant is known?
3. What prior use, if any, has the defendant made of the domain name in connection with the bona fide offering of goods and services?
4. Does the defendant have a bona fide noncommercial or fair use claim on the mark?
5. Did the defendant intend to divert consumers from the plaintiff's Web site in a manner that could harm the plaintiff's mark, either for commercial gain or with intent to damage the mark by creating a likelihood of confusion?
6. Did the defendant offer to transfer the domain name to the plaintiff or any third party for financial gain without having used or intended to use the domain name?
7. Did the defendant provide material and misleading false contact information when applying for the domain name registration?
8. Did the defendant register or acquire multiple domain names that are confusingly similar to other distinctive marks or dilutive of famous marks?

Remedies

If successful under the ACPA, a plaintiff can receive at least four remedies including (1) forfeiture or cancellation of the domain name, (2) transfer of the domain name to the plaintiff, (3) actual or statutory damages from $1,000 to $100,000 per domain name, and (4) attorney fees and costs which can be awarded to the plaintiff.[40] The first two remedies, forfeiture or cancellation or transfer of the domain name to the plaintiff apply even for domain names registered before the ACPA was effective. The third and fourth remedies, damages, and attorneys' fees apply only to domain names registered or used after the ACPA became effective on November 29, 1999.

Limitations in Using the Anticybersquatting Consumer Protection Act (ACPA)

However, the ACPA can be used only for domain names registered in the United States. If the challenged domain name is registered with a foreign registrar, the plaintiff must obtain personal jurisdiction over the domain name owner or use international avenues of redress such as the Uniform Dispute Resolution Policy (UDRP). *In rem* actions under ACPA are limited to forfeiture, cancellation, or transfer of the domain name. Monetary relief is not possible with *in rem* actions. *Intent to profit* from the domain name is not sufficient. *Bad faith intent to profit* must be proven.[41]

This is a new area of the law, but so important that ongoing cases are reported in the "Domain Name Law Reports" found at *http:/dnlr.com/searchindex.html*>.

10.5 COPYRIGHT

A fifth area of concern in the Internet Age concerns "copyright."Copyright is the right of a creator's control over his or her literary or artistic creations. The U.S. Constitution grants to the U.S. Congress the power and responsibility: "To promote the Progress of Science and useful Arts, by securing for limited Times to Authors and Inventors the exclusive Right to their respective Writings and Discoveries."[42] However, U.S. copyright law does not address the international issues raised by the borderless nature of the Internet, and thus, as we move into the digital age, numerous new issues arise, and Congress must continually modify and update U.S. copyright law to address these changes.[43] These updates include (1) the Copyright Act of 1976, (2) the National Commission on New Technological Uses of Copyright Works (CONTU), (3) the Digital Millennium Copyright Act of 1998, (4) exclusive rights, and (5) Fair Use Contracts. Each is discussed in the following five sections, respectively.

10.5.1 Copyright Act of 1976[44]

The first copyright statute in the United States was adopted by Congress in 1790 and has been revised several times since then. Currently the Copyright Act of 1976, which took effect on January 1, 1978, and has been updated in part several times since then, is the copyright law in the United States.[45] It lists seven categories under the term "work of authorship": (1)literary works, including computer programs; (2) musical works, including accompanying words; (3) dramatic works, including accompanying music; (4) pantomimes and choreographic works; (5) pictorial, graphic, and sculptural works, including photographs, maps, technical drawings, diagrams, and cartoon characters; (6) motion pictures and other audiovisual works; and (7) sound recordings.[46]

In each of these seven categories, what is actually copyrighted is the unique expression or presentation of materials or information, not the facts, scientific data, or raw information underlying those materials. Thus, many works created by compilation, derivation, and joint authors may or may not be protected by U.S. copyright law. To legally use a copyrighted work, a user must obtain permission from the copyright owner.

10.5.2 National Commission on New Technological Uses of Copyright Works (CONTU)

As technology advanced in the 1970s, Congress became increasingly more concerned about the impact of new technological developments on the law's ability to protect copyrighted works. For this reason, in 1974 Congress established the National Commission on New Technological Uses (CONTU) of Copyright Works.

The purpose of CONTU was to study and compile data on:

(1) *the reproduction and use of copyrighted works of authorship;*

  (A) *in conjunction with automatic systems capable of storing, processing, retrieving and transferring information, and*

  (B) *by various forms of machine reproductions, not including reproduction by or at the request of instructors for use in face-to-face teaching activities; and*

(2) *the creation of new works by the application or intervention of such automatic systems or machine reproduction,"*[47] *and*

(3) *the extent to which computer programs should be protected by copyright law.*

To accomplish these goals, the CONTU commission included experts in copyright law, representatives from the publishing and other affected industries, and representatives of the public, including users of the new technologies. CONTU's Final Report, published in 1978[48] led to the 1980 amendments of the Copyright Act, which (1) make it explicit that computer programs, to the extent they embody an author's original creation, are proper subject matter of copyright, (2) apply to all computer uses of copyright programs by the deletion of the present § 117; and (3) ensure that rightful possession of copies of computer programs may use or adopt these copies for their use.[49] This significantly impacted copyrights on the Internet.

10.5.3 1998—Digital Millennium Copyright Act (DMCA)

In the 1980s and 1990s Congress amended and modified the U.S. Copyright Law numerous times, but in 1998, Congress enacted the Digital Millennium Copyright Act (DMCA), an update of the U.S. copyright law to include the impact of digital technology and to comply with various international copyright treaties. Among other things, the DMCA provided three important changes: (1) it prohibited people from circumventing technological controls to access copyrighted works; (2) it limited the liability of service providers such as Internet service providers (ISPs), as discussed in Section 10.8; and (3) it amended the then existing copyright law so that third-party service organizations could load diagnostic software and operating systems onto a computer system for maintenance purposes without infringing the software or systems copyright. However, the Internet Age raises new concerns because it offers both an immediate, inexpensive method for creators to reach a worldwide audience, and a nearly unmanageable opportunity for unauthorized people to use, copy, and manipulate copyrighted material without authorization or payment to the copyright owners. In addition, tracing the chain of infringement is nearly impossible. A sampling of several current issues in this area include, in general, works at risk of such electronic infringement and/or use in seven categories: (1) music, (2) photos, (3) software, (4) text, (5) exclusive rights, (6) fair use, and (7) electronic rights.

1. **Music (Napster, etc.).** The recent legal issues surrounding Napster and MP3 perfectly describe these issues concerning digitized music and its rapid dissemination over the Internet to the forefront of everyone's awareness. While the cases were settled with fines and the issuance of licenses, the issues have not truly been resolved. In this case, technical solutions, such as electronic tags connected to each copy, may be required rather than changes in the law. As in most of these cases, the law cannot react quickly enough to respond to each of the latest technical developments.

2. **Digital photos.** Scanners are inexpensive, readily available, and capable of copying published or original photos into digitized form, which can then be used on Web sites, electronic bulletin boards, or other documents, and transmitted around the world. Photos can also be digitally regenerated and/or cropped to create a new or composite photo which may or may not infringe.

3. **Software.** While copyright law protects computer software, people still share "pirated" copies of software. On a small scale, these infringements are illegal and frustrating to the developers, but relatively small in actual damages. True economic harm to the software developer can result, however, when unauthorized copies of software are placed on the Internet. This most frequently happens with game software. Again, a technical solution may be required, rather than a legal solution.

4. **Text.** Written material can readily be copied, altered, and transmitted, especially over the Internet. For this reason, it is more difficult to attach an electronic tag or otherwise apply a technical solution to textual material. The magnitude of impact, however, could have a serious chilling effect, especially as huge files can be attached easily to an email, and emails can be copied to large electronic distribution lists.

10.5.4 Exclusive Rights

Current copyright law grants five exclusive rights to the owner of a copyrighted work. These include the right to: (1) distribute, (2) reproduce, (3) adapt, (4) perform, and (5) publicly display the protected work. Infringement of the copyright occurs when one or more of these actions occur without the copyright owner's consent.

However, in the electronic world the definitions of these exclusive rights are not clear. For example, courts question whether the transmission of copyrighted material over the Internet constitutes a *public distribution* of the work?

They also ask, does *public display* of a work include its availability on the Internet? If so, what remedies are available to the author if no one can determine who placed the material on the Internet? Does uploading or downloading material from the Internet, a bulletin board service, or a database constitute a "reproduction" of the work? What liability do Web site managers and ISPs have in monitoring such activity or the ownership of material? All of these issues are currently being evaluated but no resolution exists as yet.

10.5.5 Impact of Fair Use on Copyright Contracts

Further, as discussed in Chapter 7, the "fair use" doctrine allows people other than the copyright owner to use copyrighted material in a reasonable manner without the owner's consent, notwithstanding the monopoly granted to the owner. To determine whether fair use of a copyrighted work has been made, courts consider the nature and objects of the selections, the quantity and value of material used, and the extent to which the use may diminish the value of the original work.[50] Thus, applying the concept of *fair use* involves a balancing process in which numerous variables are considered to determine if other interests override the rights of the creators. The U.S. copyright law explicitly identifies four interests: (1) the purpose and character of the use, including its commercial nature; (2) the nature of the copyrighted work; (3) the proportion of the work that was "taken" (the amount and substance of the work used); and (4) the economic impact of the "taking," (the impact on the market).[51] The issues of fair use in the Internet Age, however, have yet to be decided. In the meantime, they have created confusion for at least three types of contracts concerning copyrights, publishing contracts, Web site design, and database ownership contracts, and work-for-hire contracts. The issues concerning each are outlined in the following three sections.

Copyright, Publishing Contracts

Current publishing contracts and copyright agreements between authors and publishers generally grant the publishers the exclusive right to publish the authors' works, but they do not define what "publish" includes. Therefore, it is not clear whether online distribution of a work significantly differs enough from other distribution methods that it requires a separate contract or approval by the author and possibly additional payment to the author. In many cases, publishers have attempted to distribute their authors' works online without separate approval by the authors and often without payments to them for the additional "publication." Some authors have argued that "electronic rights" to their works are separate from the traditional publication agreements and need to be separately addressed in copyright contracts with publishers. Publishers, on the other hand, cite the provisions concerning "collective works" in the Copyright Act. While the issue continues to be addressed, most contracts now specifically address the issue as a separate, negotiated item between authors and publishers.

Web Site Design, Web Content, and Database Ownership

When a company hires a Web page designer and places content on the site, the agreement between the two must be written carefully or the company runs the risk of losing ownership of the site and content to the designer and/or losing the right to make changes or upgrades to the site without infringing the rights of the Web page designer.

Work for Hire

In the Copyright Act,[52] the "work for hire" doctrine has two branches. First, it can be a work prepared by an employee within the scope of his or her employment; or second, it can be a work specially ordered or commissioned for use as a contribution to a collective work. As such, it can be a part of a motion picture or other audiovisual work, a translation, a supplementary work, a compilation, an instructional text, a test, answer material for a test, or an atlas. While copyrightable works created by an employee within the scope of his or her employment are owned by the employer, works created by a nonemployee, such as a Web page designer and other contractors, are owned by the contractor unless a written contract, signed by all parties, expressly states otherwise.

10.6 TRADE SECRETS

A sixth issue of concern with the Internet is trade secrets. Trade Secrets are *information* that must be kept secret to retain their economic value. Some examples of trade secrets include corporate strategies, scientific formulas, recipes, drawings, blueprints, plans, income and expense statements, test records, engineering information, measurements, and statistical models.

Since secrecy is the key factor concerning trade secrets, the ease of access to information over the Internet introduces several unique issues for companies trying to protect their secrets. For example, corporate espionage rises to a new level, when uniform resource locators (URLs), which are part of a firm's Web site, provide significant information to partners and competitors, and the manner in which information is compiled, recorded, and stored impacts its security. In response, Congress enacted the Economic Espionage Act of 1996,[53] and the Uniform Trade Secrets Act,[54] which has served as the model for trade secrets statutes in over 40 states. However, the issues surrounding trade secret management and electronic espionage are not fully resolved.

10.7 DEFAMATION ON THE INTERNET

A seventh issue concerning the Internet includes defamation. Defamation is "an intentionally false communication, either published or publicly spoken, that injures another's reputation or good name, or holds a person up to ridicule, scorn, or contempt in a respectable and considerable part of the community.[55] The intent of defamation is to diminish the esteem, respect, goodwill, or confidence in which a person is held or to create unpleasant feelings or opinions about the person. Defamation includes both slander and libel and may be prosecuted under both criminal and civil law.

The ease of "publishing" information, correct or not, to millions of "listeners" worldwide over the Internet has caused defamation to become an increasing problem. For example, one of the newest types of Web sites on the Internet are "sucks sites," Web sites that use a domain name that includes the name and/or trademark of a company and then contains information critical of that company. Such sites go beyond just publishing negative comments about an individual or company and actively establish a specific Web site with an identifiable name such as *Lucentsucks.com*. A key question in this area is, What remedies does the law provide to the victims of such actions?

In addition to regular tort law, the U.S. Anticybersquatting Consumer Protection Act (ACPA)[56] addresses "sucks" sites in its last section. Also, three cases decided in 2000 serve as good examples of the current law in this new area: (1) *Morrison & Forerster v. Wick,* (2) *Lucent Technologies v. Johnson,* and (3) *Lucent Technologies v. lucentsucks.com*. The guidance provided by each case is outlined in the three sections below as an example of the direction this still evolving area of the law is taking.

10.7.1 *Morrison & Forerster v. Wick,* 94 F. Supp.2d 1125 (D. Colo. 2000)

In *Morrison & Forerster v. Wick,* the defendant, Wick, registered various Web site domain names similar to the law firm of Morrison and Forerster including: *morrisonandforerster.com, morrisonandforester.com,* and *morrisonfoerster.com*. At each site, Wick posted critical remarks about the law firm and provided links to anti-Semitic and racist sites. Wick argued that the First Amendment protected these Web sites as noncommercial use of a mark. However, the District Court of Colorado disagreed, finding that the sites did not meet the standard for parody, and based on Wick's testimony, were meant to "get even" with the law firm.

10.7.2 *Lucent Technologies v. Johnson*, Civ. No. 00-05668 (C.D. Cal. Sept. 12, 2000)

In a related case involving the same parties and domain names, Wick argued, in a motion to dismiss the case, that the court should adopt a *per se* rule that *yourcompanynamesucks.com*, as a group, should be given a safe harbor First Amendment defense. The California federal court declined to do so.

10.7.3 *Lucent Technologies v. lucentsucks.com*, 54 U.S.P.Q.2d 1653 (E.D. Va. 2000)

However, a court in the eastern district of Virginia noted that the word "sucks" has entered the vernacular as a word "loaded with criticism" and that the domain name *lucentsucks.com* therefore, is effective parody. This ruling may be inconsequential, however, because the court dismissed the *in rem* case on the grounds that the plaintiff failed to comply with the "due diligence" notice and service requirements of the Anticybersquatting Consumer Protection Act (ACPA).

Since the falseness of a defamatory action is a key factor, truth may be a defense, but not always, depending on the maliciousness of the act. Unlike in *Morrison & Forerster v. Wick*, defendants have also argued First Amendment Free Speech, but other cases may meet with more success. The First Amendment of the U.S. Constitution states: "Congress shall make no law respecting an establishment of religion, or prohibiting the free exercise thereof; or abridging the freedom of speech, or of the press; or the right of the people peaceably to assemble, and to petition the Government for a redress of grievances."

This focuses on speech and the press because those were the issues the framers of the Constitution were familiar with. However, while the wording of the First Amendment seems clear, it has had to be interpreted and applied to American life throughout our country's history—especially as those rights have collided with other rights, such as privacy and impartial juries—and as new forms of communication have evolved. The debate continues.

10.8 LIABILITY OF ISPS AND COMPUTER SYSTEM OPERATORS

An eighth unresolved issue concerning the Internet is the liability of Web sites and ISPs for defamation. Section 581(1) of the Restatement (Second) of

Torts concerning liability for defamation states that, "one who only delivers or transmits defamatory matter published by a third person is subject to liability if, but only if, he knows or has reason to know of its defamatory character." Thus, in common law, knowledge and control over the information distributed by Internet service providers (ISPs), or computer system operators are the key factors in determining their liability or accountability for the information they transmit or make available. As a result, publishers, distributors, and common carriers have had different levels of liability in the past, but these decisions are less clear in the Internet Age. Thus, the issue of liability is still evolving, as described in the following sections.

10.8.1 Publishers

Publishers, such as newspapers, magazines, and television and radio stations, create, edit, and package information for distribution. As such, they have tremendous control over the content and are held to the highest level of accountability for it. In fact, they can be liable even for material they do not originate, such as letters to the editor and guest contributors. Some exceptions exist for information the publisher could not reasonably verify, including wire service information and misleading advertising published in good faith.

10.8.2 Distributors

Distributors, on the other hand, do not create information, but do distribute it. Distributors include libraries, bookstores, and newsstands. Typically they are not held liable for content unless they knew of problems.

10.8.3 Common Carriers

Common carriers, such as telephone companies and private mail carriers, tend to have the least liability for defamatory information over their networks or other distribution systems because they act simply as conduits for the flow of unedited information from one party to another. They do not create or change the information, and in fact are prohibited by law from doing so. Thus, even if the common carrier knows of the offensiveness of the communication, they may not be able to interfere to stop it.

The issue in the Internet Age, is that the new ISPs and companies providing bulletin board and email services do not fit easily into these traditional categories. Instead, they share attributes with each type and may or may not use electronic screening and editing policies. Thus, at present, each case concerning the following questions is determined on a case-by-case basis.

1. Should the liability of ISPs vary depending on how "interactive," "transactional," or "passive" the Web site is?
2. What liability exists for threats? For following through with a threat?
3. Should ISPs be required to filter information they carry for users?
4. What liability or possible exposure do the users of the Internet have if they simply upload information and messages?

10.8.4 Claims on Web Sites

If an ISP or Web site assures its customers that, as a matter of policy, it does not do certain things, such as: place cookies on their computers or retain any personally identifiable information, it must make certain that third-party advertisers and/or servers also do not do such things. If they do, the site-owning company must state that cookies may be generated by other site participants. Additionally, liability has not yet been determined.

10.8.5 Digital Millennium Copyright Act

Additionally, Congress enacted the Digital Millennium Copyright Act to provide protection against copyright liability when companies that provide chat rooms and other similar services face potential liability if copyright infringing material is posted on the Web site by third parties. While ISPs are definitely covered by the DMCA, it is not yet clear to what extent Web site operators, e-commerce companies, and "brick and mortar" companies are covered. The DMCA requires registering with the Copyright Office and adopting a policy of removing material alleged to be infringing.

10.9 OBSCENITY AND VIOLENCE ON THE INTERNET

A ninth issue affecting the Internet Age is that of obscenity and violence on the Internet. As part of the Telecommunications Act of 1996, Congress included the Communications Decency Act (CDA).[57] In 1997 in the case of *Reno v. American Civil Liberties Union* (*Reno I*), it was found to be unconstitutional by the U.S. Supreme Court because its vagueness violated the First Amendment. This decision created one of the most controversial areas in the law of free speech expression in the Internet Age and generated subsequent laws with which users and ISPs must comply, including the Child Online Protection Act (COPA), the Child Online Privacy Protection Act of 1998 (COPPA), and numerous state laws. The following five sections discuss these statutes and court cases.

10.9.1 1996—The Communications Decency Act (CDA)

The Communications Decency Act stated, in part, that

Whoever in interstate or foreign communications knowingly uses any interactive computer service to display, in a manner available to a person under 18 years of age, any comment, request, suggestion, proposal, image or other communications that, in context, depicts or describes, in terms patently offensive as measured by contemporary community standards, sexual or excretory activities or organs, regardless of whether the user of such service placed the call or initiated the communications, shall be fined (up to $250,000) or imprisoned not more than two years, or both.[58]

The intent of the Communications Decency Act (CDA) was to protect minors from "indecent" and "patently offensive" communications on the Internet unless the communications can be made unavailable to minors. This broad language affects even constitutionally protected free speech. In the classic case in this area, *FCC v. Pacifica Foundation*, the U.S. Supreme Court upheld the ban against indecent material over the airwaves during hours when children are likely to be in the audience, because of the uniquely pervasive presence of broadcast signals and its easy access to children.[59] These two elements are not necessarily present on the Internet and in other modern electronic media such as cable television.[60]

Therefore, numerous groups immediately protested the CDA as being overbroad and a threat to First Amendment freedom of speech, education, and commerce for adults. On February 8, 1996, the day President Clinton signed the CDA into law, the American Civil Liberties Union (ACLU), joined by several other plaintiffs, filed a First Amendment and Due Process case against the CDA.[61] Several days later, a federal judge in Philadelphia granted a temporary restraining order prohibiting enforcement of the CDA, and a second suit was filed by the Citizens Internet Empowerment Coalition, the American Library Association, the American Society of Newspaper Editors, The Society of Professional Journalists, and Microsoft Corporation, plus 16 other groups, stating that language was so broadly written that most teachers of biology, literature and/or art history could be convicted. The three-judge District Court in Pennsylvania made extensive findings of fact concerning (1) the character and dimensions of the Internet, (2) the availability of sexually explicit material on the Internet, and (3) the problems confronting age verification for recipients of Internet communications. The district court also found that the CDA abridged freedom of speech protected by the First Amendment. The United States government immediately appealed the district court's decision to the Supreme Court in *Reno v. American Civil Liberties Union, et al.*

10.9.2 *Reno v. American Civil Liberties Union (Reno II)*[62]

In 1997, the Supreme Court upheld the District Court decision, stating that the Communications Decency Act (CDA) is unconstitutional on its face, because it is so vague and overbroad that it violates the First Amendment. The Supreme Court concluded first that the text of the CDA is unacceptably vague in that it is unclear what speech the CDA prohibits. Second, the harmful effects of the CDA's vagueness are to chill free speech and to create the potential for the arbitrary enforcement of the CDA. Specifically, the vague contours of the CDA's coverage unquestionably silences some speakers whose messages would be entitled to constitutional protection. Third, the CDA's burden on protected speech cannot be justified if it could be avoided by a more carefully drafted statute. To correct this, the U.S. Congress passed the Child Online Protection Act, applying it to both users and ISPs.

10.9.3 Child Online Protection Act (COPA)

In October 1998, Congress passed the Child Online Protection Act (COPA), a second attempt to create legislation to prevent minors from accessing "harmful materials," especially pornography, on the Internet. On October 21, 1998, President Clinton signed COPA into law.[63]

However, even before COPA was signed into law, it was attacked as being unconstitutional. On October 5, 1998, the U.S. Department of Justice sent a letter to Congress expressing numerous concerns about the constitutionality of COPA,[64] many of them the same concerns as those raised against the Communications Secrecy Act.

10.9.4 *American Civil Liberties Union v. Reno (Reno III)*

On October 22, 1998, one day after COPA became law, the American Civil Liberties Union and 16 other coplaintiffs filed the case *American Civil Liberties Union v. Reno (Reno III)*[65] in the eastern district of Pennsylvania, arguing that COPA is unconstitutional, overbroad, and vague. The court agreed and issued a decision on February 1, 1999. In response, on April 2, 1999, Attorney General Reno appealed the district court's ruling.

10.9.5 Child Online Privacy Protection Act (COPPA)

In 1998, as a third try to protect minors from problems on the Internet, Congress passed the Child Online Privacy Protection Act (COPPA), which required, among other things, that an ISP provide certain notice and "opt-out" options to customers. Compliance with these requirements should be included in the carrier's contract with each subscriber, but the issue will continue to be discussed with input from private censorship and international standards and regulations on the topic. Remaining issues include the following:

1. How can the availability and distribution of child pornography on the Internet be avoided, controlled, or eliminated?
2. How can children's access to pornography on the Internet be avoided, controlled, or eliminated?
3. What is considered "obscenity" on the Internet?
4. What First Amendment rights of adults do these controls impact?
5. What compromise solutions, if any, exist?
6. Should sexually explicit information be readily available on the Internet?
7. How should photos, movies, and chat rooms be managed?
8. How effective are "filters" that allow parents to screen out words, Web sites, or images they deem inappropriate? What privacy and access issues do filters raise for adults, particularly when installed in libraries and other public or academic locations?

10.9.6 State Laws

Acknowledging the importance of the obscenity and violence issues on the Internet, the states have also passed numerous laws in this area. The following provides a sample, but not a complete list, of such legislation, challenges to it, and several Web sites to check for updates.

1. New York: N.Y. Penal Law § 235.21(3) challenged by *American Library Assoc., et al. v. Pataki,* 969 F.Supp 160 (S.D.N.Y. 1997), *http://www.aclu.org/court/nycdadec.html.*
2. New Mexico: Section 30-37-1 of the New Mexico Statutes Annotated, challenged in *American Civil Liberties Union, et al. v. Johnson,* 4 F.Supp.2d 1029 (D.N.M. 1998), aff'd. No. 98-2199 (10th Cir. Nov. 2, 1999).
3. Virginia: Virginia Code § 18.2-390 and Library Board policy, challenged in *Mainstream Laudoun v. Board of Trustees of Loudoun County Library,* 2 F. Supp. 2d 783 (E.D. Va. 1998).
4. Michigan: Mich. Comp. Laws § 722.675(1). Challenged in *Cyberspace Communications, Inc., et al. v. Engler,* 55 F.Supp.2d 737 (E.D. Mich. 1999).

Many states require filters to protect children, allowing parents to screen out material they deem inappropriate. These have also been challenged.

10.10 FRAUD ON THE INTERNET

A tenth issue impacting the Internet concerns fraud on the Internet. As with any new technology, criminals have found new ways to use the Internet to commit crimes, including the use of fraud. Some of the more recent fraudulent schemes include (1) chain letters; (2) pyramid schemes; (3) get-rich-quick investment scams; (4) work-at-home scams; (5) travel/vacation fraud; (6) pay-per-call telephone solicitation scams; (7) health-care frauds; (8) Internet auction fraud in which some e-auction sites take cashier's checks or money orders from consumers as payment, but never deliver the goods purchased; (8) "rebate" check scams; (9) Web site design promotion scams; and (10) credit card-cramming scams. Rebate check scams occur when companies such as ISPs or telephone or cable providers mail "rebate" checks in small amounts, usually under $5.00, to consumers. By cashing the checks, the consumers unknowingly agree to become customers of the ISP or telephone or cable provider. The company then begins placing monthly charges on the consumers' telephone bills and makes it very difficult for the consumer to cancel the service or to receive refunds. Internet Web site design promotion scams, also known as "Web cramming" occur when companies, including some phone companies, offer small businesses and nonprofit organizations a "free" Web page or Web site design and then charge them for the service on their monthly bills. Credit card cramming occurs when operators of some Internet Web sites, usually adult-oriented Web sites, charge consumers' credit cards or phone bills for services the consumers did not order or authorize.

To address these new developments, the following areas are being examined by the courts and legislatures.

1. How should the law respond to the increasing numbers of scams and fraudulent criminal activity perpetrated over the Internet?
2. Is current criminal law sufficient to address these crimes, or do unique aspects exist within the Internet that require new laws and/or tracking methods?
3. What unique aspects of Internet fraud exist, if any, that are not addressed by current criminal law?
4. Is current criminal law sufficient to respond to the increasing numbers of scams and fraudulent criminal activity perpetrated over the Internet, or do unique aspects exist that are not addressed by current criminal law?

The rapid increase of these scams and the borderless, rapid access and distribution aspects of the Internet have caused the consumer protection enforcement agencies from over 29 countries to form the International Marketing Supervision Network.[66] The purpose of the network is to share information concerning cross-border scams and to suggest appropriate responses. Members include Australia, Canada, Finland, Germany, Ireland, New Zealand, Norway, the United Kingdom, and the United States, including the United Kingdom's Department of Trade and Industry and Office of Fair Trading and the U.S. FTC. In addition, consumer protection agencies from these countries cooperated in a year-long effort to target the top 10 Internet scams. Their comments and ongoing activities are available at *www.ftc.gov/opa/2000/10/topten.htm.*

CONCLUSION

Even though these ten issues and numerous others like them are currently unresolved, the purpose of this chapter is to present the complexity and importance of the issues in this rapidly evolving era of modern communications and to provide the information needed to understand their resolution. Each issue significantly affects society worldwide and likely will evolve into still other issues. However, all ten, along with the other issues discussed in this book, describe the enormously dynamic and exciting nature of telecommunications law in the Internet Age.

ENDNOTES

[1]*See, for example,* the *1999 Report of Law of Commerce in Cyberspace Committee,* Business Law Section, Washington State Bar Association available at *www.wsba.org/sections/biz/lccc/report/1999;* a letter from CommerceNet to the American Law Institute dated May 11, 2000, copy available at *www.commerce.net/resources/work/pubdocs/2.4-1.pdf;* and documents from the Federal Trade Commission (FTC) and the Federal Reserve Board (FRB), the two key governing and decision-making agencies in this area.

[2]15 USC § 1681(b)(2).

[3]Federal Trade Commission, *www.ftc.gov.*

[4]For example, the Direct Marketing Association, *www.the-dma.org.*

[5]The Uniform Electronic Transactions Act (UETA) was first adopted by California and has since been adopted by several states.

[6]*See, for example,* Step-Saver Data Sys., Inc. v. Wyse Technology, 939 F.2d 91, 100-04 (3rd Cir. 1991) and Arizona Retail Sys. v Software Link, Inc. 831 F. Supp. 759 (D. Ariz. 1993).

[7]*See* Pro-CD, Inc. v. Zeldenberg,, 86 F. 3d 1147 (7th Cir. 1996) and Hill v. Gateway 1000, Inc., 105 F. 3d 1147 (7th Cir. 1997) cert. denied, 118 S. Ct. 47 (1997).

[8]Interagency Working Group on Electronic Commerce, *A Framework for Global Electronic Commerce,* § II.3 (July 1, 1997). A copy of the document available at: *www.whitehouse.gov/WH/NEW/Commerce/read.html.*

[9]Black's Law Dictionary, West Publishing Co., St. Paul, MN.

[10]Pennoyer v. Neff, 95 U.S. 714, 24 L.Ed. 565 (1877).

[11]Burger King Corp. v. Rudzewicz, 471 U.S. 462, 476 (1985).

[12]Superfos Investments Ltd. v. FirstMiss Fertilizer, Inc., 774 F.Supp. 393, 397-98 (E.D. Va. 1991).

[13]World-Wide Volkswagen Corp. v. Woodson, 444 U.S. 286, 297 (1980).

[14]California Software Inc. v. Reliability Research, Inc., 631 F.Supp. 1356, 1363 (C.D. Cal. 1986).

[15]Ticketmaster-New York, Inc. v. Aliota, 26 F. 3d 201,212 (1st Cir. 1994).

[16]Digital Equip. Corp. v. Altavista Tech. Inc., 960 F. Supp. 456, 463 (D. Mass. 1997).

[17]Internet Tax Freedom Act (ITFA), S. 442, 105th Cong., 2nd Sess. (1998) (enacted Oct. 21, 1998).

[18]California Internet Tax Freedom Act (CITFA), Cal. Rev. & Tax Code § 65001 *et seq.* (Part 32), (August 24, 1998).

[19]California Assembly Joint Resolution No. 20, introduced by Assembly Member Lempert.

[20]Proposed legislation AB 1614 introduced in June 1997 by California Board of Equalization member

Johan Klehs (1st district) and authorized by California Assembly Member Lempert.

[21]Introduced September, 1997 by California State Senator Mountjoy and referred on January 6, 1998 to the Senate Revenue and Taxation Committee.

[22]Introduced February 19, 1998 by Senator Vasconcellos (whose district included Silicon Valley) and reviewed in April 1998 by the California Senate Energy, Utilities and Communications Committee.

[23]Introduced February 23, 1998 by California Assembly Member Bowen and reviewed in May 1998 by the California Assembly Judiciary Committee.

[24]California Internet Tax Freedom Act (CITFA), *supra* note 17.

[25]Cal. Rev. & Tax Code § 65004.

[26]National Bella Hess, Inc. v. Dept. of Revenue, 386 U.S. 753 (1967).

[27]The Ohio Table Pad Co., N.Y. Div. Tax App. No. 815122 (N.Y. Administrative Law Judge Arthur Bray, Feb. 5, 1998).

[28]Florida Dep't. of Revenue v. Share Int'l Inc., 676 So. 2d 1362 (Fla. 1996), cert. denied, 117 S.Ct. 685 (1997).

[29]The U.S. Internal Revenue Code (IRC) §§ 861(a) and 865 (1996).

[30]Ford Motor Co. v. Ford Financial Solutions Inc., 55 U.S.P.Q.2d 1271 (N.D. Iowa 2000).

[31]Washington Speakers Bureau, Inc. v. Leading Authorities, Inc., 33 F. Supp. 2d 488, 491, n.4 (E.D. Va. 1999).

[32]The Lanham Act, codified at 15 U.S.C. §§ 1051 *et seq*.

[33]*See* Sandra Edelman, *Cybersquatting Claims Take Center Stage*, Computer & Internet Lawyer, Vol. 18, No.1, Jan. 2001, pp. 1-6.

[34]Ford Motor Co. v. Ford Financial Solutions Inc., 55 U.S.P.Q.2d 1271 (N.D. Iowa 2000).

[35] www.icann.org.

[36]Weber-Stephen Products v. Arnitage Hardware, 54 U.S.P.Q.2d 1766 (N.D. Ill. 2000).

[37]*Id*.

[38]Broadbridge Media v. Hypercd.com, 55 U.S.P.Q.2d 1426 (S.D.N.Y. 2000).

[39]Anticybersquatting Consumer Protection Act (ACPA), codified at 15 USC § 1125(d) (1999).

[40]*See, for example,* United Greeks v. Klein, 2000 WL 554196 (N.D.N.Y. 2000) and Electronic Boutique Holdings v. Zuccarini, Civ. No 00-4055 (E.D. Pa. Oct. 30, 2000).

[41]Cello Holdings v. Lawrence Dahl Companies, 89 F. Supp. 2d 464 (S.D.N.Y. 2000).

[42]U.S. Const. art.1, § 8, cl. 8.

[43] John D. Zelezny, Communications Law: Liberties, Restraints, and the Modern Media, Second Edition, 494-498, Wadsworth Publishing Company, Belmont,CA (1997).

[44]17 U.S.C. §§ 101 *et seq*. (1988).

[45]*Id*.

[46]17 U.S.C. § 102(a) (1988).

[47]Final Report of the National Commission on the New Technological Uses of Copyright Works, (1978), H.R. Rep. No. 1307, 96th Cong., 2d Sess, at 105.

[48]*Id*.

[49]M Kramer Mfg. Co. v. Andrews, 783 F.2d 421, 432 n. 8 (4th Cir. 1986).

[50]Rosemont Enterprises, Inc. v. Random House, Inc., 256 F.Supp. 55, 65-66. (1966).

[51]17 U.S.C. § 107 (1988).

[52]17 U.S.C. § 101 (1988).

[53]18 U.S.C. § 1831, *et seq*.

[54]Uniform Trade Secrets Act, 14 U.L.A. 462 (1990). *See also:* R. Mark Halligan, "The Trade Secrets Home Page" at *www.execpc.com/-mhallign*.

[55]Restatement, (Second), of Torts §§ 559, 563 (1977). *See also,* McGowen v. Prentice, 341 So.2d 55, 57 La.App.; (1976) and Wolfson v. Kirk, 273 So.2d 774, 776 Fla.App. (1973)

[56]Anticybersquatting Consumer Protection Act (ACPA), Codified at 15 USC § 1125(d).

[57]47 U.S.C. 223 *et seq*.

[58]47 U.S.C. 223(a)(1)(b).

[59]Federal Communications Commission v. Pacifica Foundation, 438 U.S. 726, 98 S.Ct. 3026, 57 L.Ed.2d 1073 (1978).

[60]Cruz v. Ferre, 755 F.2d 1415 (11th Cir. 1985).

[61] American Civil Liberties Union v. Reno, 929 F Supp. 824, 830-849 (ED Pa. 1996). [Reno I].

[62] Reno v. American Civil Liberties Union, 521 U.S. 844, 117 S.Ct. 2329 138 L.Ed. 2d. 874 (1997). [Reno II].

[63] Children's Online Protection Act (COPA), codified at 47 U.S.C. § 231.

[64] Letter from L. Anthony Sutin, Acting Assistant Attorney General, United States Dep't. of Justice, to Representative Thomas Bliley, Chairman, U.S. House of Representatives, Committee on Commerce (October 5, 1998). Entire text of letter is available at the ACLU Web site at *www.aclu.org/court/acluvrenoII_doj_letter.html.*

[65] American Civil Liberties Union v. Reno, 31 F. Supp. 2d 473(E.D. Pa. 1999). [Reno III].

[66] *Law Enforcement Officials from Around the World Tackle "Top Ten" Online Scams*, Computer & Internet Lawyer, Vol. 18, No.1, Jan. 2001, p. 31.

APPENDIX A Key Documents and Decisions Concerning Local Number Portability

1. July 2, 1996—In the Matter of Telephone Number Portability, 11 FCC Rcd 8352 (1996) [*First Report and Order on Local Number Portability (LNP)*.
2. Aug. 19, 1997—In the Matter of Application of Ameritech Michigan Pursuant to Section 271 of the Communications Act of 1934, as amended, To Provide In-Region, InterLATA Services In Michigan, 12 FCC Rcd 20543 (1997) [*Ameritech Michigan 271 Order*].
3. Aug.18, 1997—In the Matter of Telephone Number Portability, 12 FCC Rcd 12281 (1997) [*Second Report and Order on LNP, FCC 97-289*].
4. Dec. 24, 1997—In the Matter of Application of BellSouth Corporation, et al. Pursuant to Section 271 of the Communications Act of 1934, as amended, To Provide In-Region, InterLATA Services In South Carolina, 13 FCC Rcd 539 (1997) [*Bell So/Carolina 271 Order*].
5. Feb. 4, 1998—In the Matter of Application by BellSouth Corporation, et al. Pursuant to Section 271 of the Communications Act of 1934, as amended, To Provide In-Region, InterLATA Services In Louisiana, 13 FCC Rcd 6245 (1998) [*First Bell So. Louisiana 271 Order*].
6. May 12, 1998—In the Matter of Telephone Number Portability, 13 FCC Rcd 11701 (1998) [*Third LNP Report and Order*].
7. April 17, 1998—Operating Support Systems (OSS) Performance Measures. *www.fcc.gov.*
8. Aug. 7, 1998—In the Matters of Deployment of Landline Services Offering Advanced Telecommunications Capability; Petition of Bell Atlantic Corporation For Relief from Barriers to Deployment of Advanced Telecommunications Services; Petition of U S WEST Communications, Inc. For Relief from Barriers to Deployment of Advanced Telecommunications Services; Petition of Ameritech Corporation to Remove Barriers

to Investment in Advanced Telecommunications Technology; Petition of the Alliance for Public Technology Requesting Issuance of Notice of Inquiry and Notice of Proposed Rulemaking to Implement Section 706 of the 1996 Telecommunications Act; Petition of the Association for Local Telecommunications Services (ALTS) for a Declaratory Ruling Establishing Conditions Necessary to Promote Deployment of Advanced Telecommunications Capability Under Section 706 of the Telecommunications Act of 1996; Southwestern Bell Telephone Company, Pacific Bell, and Nevada Bell Petition for Relief from Regulation Pursuant to Section 706 of the Telecommunications Act of 1996 and 47 U.S.C. § 160 for ADSL Infrastructure and Service, 13 FCC Rcd 24011 (1998) [*Sec. 706 Memoranda and Opinion*].

9. Oct. 13, 1998—In the Matter of Application of BellSouth Corporation, BellSouth Telecommunications, Inc., and BellSouth Long Distance, Inc., for Provision of In-Region, InterLATA Services in Louisiana, 13 FCC Rcd 20599 (1998) [*Second Order—FCC Bell South/Lousiana*].
10. Oct. 20, 1998—In the Matter of Telephone Number Portability, 13 FCC Rcd 21204 (1998) [*Second Memo and Order on Reconsideration of LNP*].
11. Mar. 31, 1999—In the Matters of Deployment of Landline Services Offering Advanced Telecommunications Capability, 14 FCC Rcd 4761 (1999) [*First Report and Order on Sec. 706*].

APPENDIX B Key Documents and Decisions Concerning Universal Service

1. Sept. 27, 1995—Case 94-C-0095, Proceeding on Motion of the Commission to Examine Issues Relating to the Continuing Provision of Universal Service and to Develop a Regulatory Framework for the Transition to Competition in the Local Exchange Market; Order Instituting Framework for Directory Listing, Carrier Interconnection and Inter-carrier Compensation (Case 94-C-0095) at 12-13 (1995).
2. March 8, 1996—Notice of Proposed Rulemaking (NPRM) & Order Establishing a Federal-State Joint Board on Universal Service, FCC 96-93 (1996).
3. Nov. 24, 1998—Joint Board Makes Universal Service Recommendations to FCC, FCC 98J-7 (1998).
4. July 17, 1998—Order Referring Issues to Joint Board and Extending Nonrural Implementation Date, FCC 98-160 (1998).
5. Dec. 22, 2000—Federal-State Joint Board on Universal Service, FCC 00J-4 (2000).
6. Oct. 4, 2000—Public Notice on Universal Service, FCC 00J-3 (2000).

Updated FCC Orders on universal service are available at *www.fcc.gov/ccb/universal_service/orderssubsequent.html,* and updated FCC reports to Congress are available at *fcc.gov/ccb/universal_service/report.html.*

APPENDIX C Key Documents and Decisions Concerning Access and Reciprocal Compensation

Date	Federal	State	Title
1980	FCC		Second Computer Inquiry [*Computer II*], 77 F.C.C.2d 384 (1980) (initial FCC preemption of state authority over enhanced services).
1982	D.C. Cir.		FCC's preemption of state authority over enhanced services upheld in Computer & Communications Industry Ass'n v. FCC, 693 F.2d 198 (D.C. Cir. 1982), *cert. denied,* 461 U.S. 938 (1983).
1983	FCC		MTS and WATS Market Structure, Third Report and Order, 93 F.C.C.2d 241 (1983), *aff'd in principal part and remanded in part* National Ass'n of Regulatory Utility Commissioners v. FCC, 737 F.2d 1095 (D.C. Cir. 1984), *cert. denied* 469 U.S. 1227 (1985) [contains *FCC 1983 Access Charge Order*]. *See also,* 47 C.F.R. § 69.1 *et seq.* (Part 69).
1983	FCC		*MTS and WATS Market Structure,* 97 F.C.C.2d 682, 715 (1983).
1984	D.C. Cir.		NARUC v. FCC, 737 F.2d 1095, 1137 (D.C.Cir. 1984), *cert. denied* 469 U.S. 1227 (1985). (affirmed FCC's decision not to impose access charges on ISPs).
1985	Congress		MTS and WATS Market Structure: Establishment of a Joint Board, Amendment, 50 Fed. Reg. 939 (1985).
1986			47 C.F.R. §§ 69.601-69.610 (1986).

(continued)

Date	Federal	State	Title
1986	FCC		Amendment of Section 64.702 of the Commission's Rules and Regulations, Phase I Report and Order, 104 F.C.C.2d 958, 1964 (1986) (Third Computer Inquiry) [*Computer III*].
1987	FCC		MTS and WATS Market Structure, Recommended Decision and Order, 2 FCC Rcd 2324 (1987).
1987	FCC		MTS and WATS Market Structure, Report and Order, 2 FCC Rcd 2953 (1987).
1988	FCC		Amendments of Part 69 of the Commission's Rules Relating to Enhanced Service Providers, 3 FCC Rcd 2631, 2633 (1988).
1988	FCC		MTS and WATS Market Structure, Amendment of Part 67 of the Commission's Rules and Establishment of a Joint Board, Report and Order, 3 FCC Rcd 4543 (1988) (Introduced the "subscriber line charge").
1988	FCC		Filing and Review of Open Network Architecture Plans, Phase I Memorandum Opinion and Order, 4 FCC Rcd 1 (1988). [hereinafter ONA Agreements.]
1989	FCC		Amendment of Part 69 of the Commission's Rules Relating to the Common Line Pool Status of Local Exchange Carriers Involved in Mergers or Acquisitions, 4 FCC Rcd 740, at ¶¶ 7-9 (1989).
1990	FCC		Amendment of Part 69 of the Commission's Rules Relating to the Common Line Pool Status of Local Exchange Carriers Involved in Mergers or Acquisitions, 5 FCC Rcd 231, at ¶ 6n.11 (1990).
	FCC		Memorandum Opinion and Order on Reconsideration of ONA, 5 FCC Rcd 3084 (1990).
	FCC		Memorandum Opinion and Order on ONA, 5 FCC Rcd 3104 (1990).
1990	9th Circuit Court		California v. FCC, 905 F.2d 1217, 1223 and 1238 (9th Cir. 1990) (*California I*) (Services involving data processing are "enhanced" services and not subject to FCC regulation. In contrast, common carriers provide "basic" communications services.) (The FCC made a "plausible case" that competitive forces plus technological advances in ONA offered sufficient safeguards against discrimination to justify the policy.)

Date	Federal	State	Title
1990	FCC		Computer I Remand Proceedings, Report and Order, 5 FCC Rcd 7719 (1990).
1992	FCC		Application of Teleport Communications, Memorandum Opinion and Order, 7 FCC Rcd 5986, at ¶ 14 (1992).
1993	FCC		Expanded Interconnection with Local Telephone Co. Facilities, Memorandum Opinion and Order, 8 FCC Rcd 4871, 7359-7363 and ¶ 1 (1993).
1993	9th Circuit Court		California v. FCC, 4 F.3d 1505, 1511 (9th Cir. 1993) (*California II*).
1994	9th Circuit Court		California v. FCC, 39 F.3d 919, 927 (9th Cir. 1994) (*California III*). (The nonstructural safeguards applied by the FCC in *Computer II* were reviewed by the Ninth Circuit in California III. The court decided that the Open Network Architecture (ONA) design and plans did not provide the necessary unbundling of fundamental local exchange transmission facilities. The court required the FCC to review its cost-benefit analysis of the nonstructural safeguards framework and it required that the FCC return to the CEI plan approach and be provided on a service-by-service basis).
1994	FCC		Price Cap Performance Review for Local Exchange Carriers, 9 FCC Rcd 1687, at ¶ 22 (1994).
Sept. 27, 1995	FCC		Proceeding on Motion of the Commission to Examine Issues Relating to the Continuing Provision of Universal Service and to Develop a Regulatory Framework for the Transition to Competition in the Local Exchange Market; Order Instituting Framework for Directory Listing, Carrier Interconnection and Inter-carrier Compensation, Case 94-C-0095 at 12-13, (Sept. 27, 1995).
Jan. 17, 1996		Delaware	DPUC Investigation Into the Unbundling of the Southern New England Telephone Company's Local Telecommunications Network—Reopened, Docket No. 94-10-02, Decision at 72 (Jan. 17, 1996).
Feb. 8, 1996	Congress		Telecommunications Act of 1996, Pub. L. No. 104-104, 110 Stat. 56 (1996), codified at 47 U.S.C. § 252(d)(2)(a).
Aug. 8, 1996	FCC		In the Matter of Implementation of the Local Competition Provisions in the Telecommunications Act of 1996;

(continued)

Date	Federal	State	Title
Aug. 8, 1996 *(continued)*			Interconnection Between Local Exchange Carriers and Commercial Mobile Radio Service Providers, CC Docket No. 96-98: CC Docket No. 95-185, 11 FCC Rcd 15505, 1996 FCC LEXIS 4312, 4 Comm. Reg. (P&F) 1 (released Aug. 8, 1996).
			Local Competition Order, CC Docket No. 96-98 (adopted August 8, 1996) [Telecommunications Act and FCC Interconnection Order, CC Docket No. 96-98 (Aug. 8, 1996)].
	FCC		In the Matter of Access Charge Reform, Price Cap Performance Review for Local Exchange Carriers, Transport Rate Structure and Pricing, and End User Common Carrier Line Charges, First Report and Order, CC Docket Nos. 96-262, 94-1, 91-213, 95-72 at ¶ 345 (many of the characteristics of ISP traffic are shared by other classes of business endusers).
Nov. 4, 1996		Illinois Commerce Commission	Ameritech v. TCG (1996).
Nov. 8, 1996		Wisconsin	Ameritech v. TCG (1996).
Dec. 24, 1996	FCC		Notice of Proposed Rulemaking (NPRM) on Reform of Interstate Access Charges, available at *www.fcc.gov.* CC Docket No. 96-262, FCC 96-488 (released Dec. 24, 1996) (Following request from LECs, FCC agreed to examine whether LECs should be permitted to assess special access charges against the ISPs).
Dec. 24, 1996	FCC		Notice of Inquiry (NOI), CC Docket No. 96-262 FCC 96-488 (released Dec. 24, 1996) [available at *www .fcc.gov.*] (FCC initiated broad study to examine whether FCC should encourage higher-speed data communications to and from consumers and study methods to do so).
Jan. 1997		Washington Utilities & Transportation Board	Final Order Approving Negotiated and Arbitrated Interconnection Agreement, In the Matter of the Petition for Arbitration of an Interconnection Agreement Between MFS Comm. Co., Inc. and US WEST Comm. Inc., pursuant to 47 USC § 252, Docket No. UT-960323 (1997).
Feb. 12, 1997		Bell Atlantic	Report of Bell Atlantic on Internet Traffic, available at *www.ba.com/ea/fcc/report.htm* at § 3 (Feb. 12, 1997).

Date	Federal	State	Title
March 1997	FCC		Kevin Werbach, Counsel for New Technology Policy, Digital Tornado: The Internet and Telecommunications Policy, FCC Office of Plans and Policy (OPP) Working Paper No. 29 (March 1997). Available at *www.fcc.gov /Bureaus/OPP/working_papers/oppwp29pdf.*
May 8, 1997			*FCC Approves Historic Universal Service and Access Charge Reforms,* Communications Daily (May 8, 1997).
May 16, 1997			In the Matter of Access Charge Reform, Price Cap Performance Review for Local Exchange Carriers, Transport Rate Structure and Pricing, End User Common Line Charges, *First Report and Order,* CC Docket Nos. 96-262, 94-1, 91-213, 95-72 at ¶ 345; 12 FCC Rcd 15982 (1997) (Many of the characteristics of ISP traffic are shared by other classes of business endusers).
May 21, 1997	FCC		Price Cap Performance Review for Local Exchange Carriers: Access Charge Reform, (Continued) CC Docket 94-1, 96-262, FCC 97-159 (Released May 21, 1997) (FCC increased the X-factor, the number by which LECs reduce their access charges each year after first increasing them by the rate of inflation. The X-factor is the means by which the FCC passes the LECs' productivity gains to customers).
June 20, 1997		ALTS	Letter to FCC requesting expedited clarification about inclusion of local calls to ISPs within Reciprocal Compensation agreements (CC No. 96-98) (1997).
July 17, 1997		New York Public Service Commission	Order Denying Petition and Instituting Proceeding on Motion to Investigate Reciprocal Compensation Related to Internet Traffic, Case 97-C-1275; Case 93-C-0033; Case 93-C-0103; Case 97-C-0895; Case 97-C-0918; Case 97-C-0979, 1997 N.Y.PUC LEXIS 444 (July 17, 1997).
1997	FCC		In the Matter of Implementation of the Local Competition Provision in the Telecommunications Act of 1996, 11 FCC Rcd 15499, 16013 (¶ 1034) (1997).
July 18, 1997	8th Circuit Court		In the Matter of Implementation of the Local Competition Provision in the Telecommunications Act of 1996, 11 FCC Rcd 15499, 16013 (¶ 1034), *aff'd in part and vacated in part sub nom., Iowa Utilities Bd. v. FCC,*

(continued)

Date	Federal	State	Title
July 18, 1997 *(continued)*			No. 96-3321, et. al. (8th Cir. July 18, 1997) (imposing reciprocal compensation obligation for terminating access for local traffic).
July 18, 1997	8th Circuit Court		Iowa Utilities Bd. v. FCC, 120 F.3d 753 (8th Cir. 1997) (imposing reciprocal compensation obligation for terminating access for local traffic).
Sept. 29, 1997		Minnesota PUC	In the Matter of a Request for Approval of a Reciprocal Compensation Agreement for Extended Area Service Between US WEST Comm., Inc. and GTE-Minnesota, 1997 Minn. PUC LEXIS 159 (Sept. 29, 1997).
Oct. 14, 1997	U.S. District Court, Wisconsin		TCG Milwaukee, Inc. v. Pub. Service Commission of Wisconsin and Wisconsin Bell, Inc. d/b/a Ameritech Wisconsin, 980 F. Supp. 992, 1997 U.S. Dist. LEXIS 16109 (Oct. 15, 1997).
Oct. 27, 1997		Colorado State Court	US WEST Communications, Inc. v. Colorado Public Utilities Commission, et.al., District Court, City and County of Denver, Colorado Case No. 96-CV-2566, (Oct. 27, 1997). (As part of the transition of local telecommunications service from a fully regulated to a competitive environment, the PUC passed several rules including rules concerning (1) interconnection and unbundling [Docket No. 94R-556T], (2) resale of regulated telecommunications service [Docket No. 94R-557T], and (3) universal telephone service [Docket No. 94R-558T], which became effective July 1, 1996. [The rules required, among other things, the establishment of methods of paying for cost-based, unbundled, nondiscriminatory interconnection, resale of services and means of assessing and distributing contributions to Colorado universal service fund, known as the "Colorado High Cost Fund." US WEST requested judicial review of these rules. Intervenors: AT&T, MCI, Colorado Telephone Association (CTA).)
Jan. 6, 1998	U.S. District Court, Washington		U.S. West Communications, Inc. v. MFS Intelenet, Inc., No. C97-222WD, Slip Op. at 8 (W.D. Wash. Jan. 6, 1998). (The Washington Utilities and Transportation Comm. had not acted arbitrarily or capriciously in "deciding not to change the current treatment of ESP call termination from reciprocal compensation to special access fee.")

Date	Federal	State	Title
Jan. 22, 1998		Ohio PUC	ICG v. Ameritech, Case No. 97-1557-TP-CSS, 1998 Ohio PUC LEXIS 53 (Jan. 22, 1998).
Jan. 28, 1998		Michigan Public Service Commission	Brooks Fiber Comm., TCG Detroit, MFS Intelenet of Michigan, and MCI v. Ameritech Michigan, Case No. U-11178, Case No. U-11502, Case No. U-11522, Case No. U-11553, Case No. U-11554, 1998 Mich. PSC LEXIS 47 (Jan. 28, 1998).
Feb. 5, 1998		Texas PUC	PUC Commissioners unanimously ruled that calls to ISPs are local calls and that Southwestern Bell must reimburse with interest Time Warner Comm. for costs incurred. [PUC News Release, Feb. 5, 1998.]
Feb. 6, 1998			In re Implementation of Section 703(e) of the Telecommunications Act of 1996, Report and Order, CC Docket No. 97-151, 13 FCC Rcd 67777 (released Feb. 6, 1998). (FCC established formulas for determining the attachment rents that utilities could charge).
April 10, 1998			In the Matter of the Federal-State Joint Board on Universal Service (FCC Report to Congress on Universal Service and Internet Access Charges), CC Docket No. 96-45, FCC 98-67 (released April 10, 1998). 1. FCC reaffirmed its regulatory distinction between "telecommunications service providers" and "information service providers." 2. Stated that telecom service providers are subject to all of the following, while information service providers a. are not providers of transmission capability and capacity, b. receive access charges paid by long-distance carriers, c. have regulated rates, and d. have universal service obligations. 3. Where information service providers use their own facilities to provide content (information services), the FCC confirmed that a. it does not currently require information service providers to contribute to the universal service

(continued)

Date	Federal	State	Title
April 10, 1998 *(continued)*			fund, but reserves the right to do so in the future and on a continuing case-by-case review, and b. include Internet Service Providers ISPs.
April 21, 1998		Tennessee Regulatory Authority	Brooks Fiber v. BellSouth, Docket No. 98-00118 (1998).
April 23, 1998		Missouri Public Service Commission	Birch Telecom of Missouri, Inc. v. Southwestern Bell Telephone Co., Case No. TO-98-278, 1998 Mo. PSC LEXIS 17 (Apr. 23, 1998).
June 4, 1998		Florida Public Service Commission	Complaints of Worldcom, TCG, Intermedia Comm., Inc., and MCI Metro Access against BellSouth Telecommunications, Inc., Docket No. 971478-TP, Docket No. 980184-TP, Docket No. 980495-TP, Docket No. 980499-TP; Order No. PSC-98-0769-PHO-TP; 1998 Fla. PUC LEXIS 1182; 98 FPSC 6; 124 (June 4, 1998).
June 16, 1998	U.S. District Court, Texas		Southwestern Bell Tel. Co. v. Public Util. Comm'n., No. 98 CA 043, Slip Op. at 14-25 (W.D. Tex. June 16, 1998) (held that calls to an ISP are "local traffic" and therefore eligible for reciprocal compensation).
July 21, 1998	U.S. District Court, Illinois		Ameritech Illinois v. Worldcom, MFS Intelenet, TCG, MCI Metro Access, AT&T Comm., Focal Comm. Corp., and Commissioners of the Illinois Commerce Comm., No. 98 C 1925, 1998 U.S. Dist. LEXIS 11344 (July 21, 1998).
Aug. 6, 1998	FCC		Memorandum Opinion and Order; Notice of Proposed Rulemaking, FCC 98-188 (Aug. 6, 1998).
Aug. 27, 1998		Pennsylvania PUC	Opinion and Order—Investigation of Issuance of Local Telephone Numbers to Internet Service Providers by CLECs, P-00981404 (Aug. 27, 1998).
Sept. 3, 1998		Maryland Pub. Service Commission	MFS Intelenet v. Bell Atlantic, Case No. 8731, Order No. 74557, 1998 Md. PSC LEXIS 43 (Sept. 3, 1998).
1998	U.S. 8th Circuit Court of Appeals		Southwestern Bell Telephone Co. v. FCC, 153 F.3d 523, 537 and 544 (8th Cir. 1998) (upheld the FCC's access charge reform order). [The three-judge panel unanimously decided that the FCC made a reasonable choice from the several available policy alternatives it considered (at 537). It also affirmed the FCC's decision to continue the ISPs' exemption from paying access charges (at 544).]

Date	Federal	State	Title
Sept. 4, 1998	U.S. 5th Circuit Court of Appeals		SBC Communications v. FCC, No. 98-10140 (1998).
Sept. 4-9, 1998		BellSouth	Cornetto, Jon, *BellSouth to Charge for Voice-over-IP Services,* Infoworld Elec., Sept. 9, 1998 at *www.inforworld .com/cgi-bin/displayStory.pl?98098.wntbell.htm.*
Sept. 10, 1998	U.S. 7th Circuit Court of Appeals		Ameritech Illinois v. Worldcom Tech., Inc., et al. (1998).
Sept. 18, 1998	FCC		In the Matter of Pacific Bell Telephone Co., CC Docket No. 98-103 (1998).
Oct. 13, 1998		Delaware Public Service Commission	MCI Telecommunications Corp. v. Bell Atlantic-Delaware, PSC Docket No. 97-323; Order No. 4923, 1998 Del. PSC LEXIS 227 (Oct. 13, 1998).
Oct. 14, 1998		Ohio PUC	Time Warner of Ohio v. Ameritech, Case No. 98-308-TP-CSS, 1998 Ohio PUC LEXIS 484 (Oct. 14, 1998).
Oct. 14, 1998		Ohio PUC	MCImetro v. Ameritech, Case No. 97-1723-TP-CSS, 1998 Ohio PUC LEXIS 485 (Oct. 14, 1998).
Oct. 30, 1998	FCC		Memorandum Opinion and Order In the Matter of GTE Telephone Operating Cos., CC Docket No. 98-79, FCC 98-292 (Oct. 30, 1998) (FCC determined that high-speed Internet services are interstate communications services and thus within the FCC's jurisdiction). *See also, GTE Service is Ruled Interstate, Suggesting Likely Boon for Bells,* Wall St. J., Nov. 2, 1998, at B6.
1999	D.C. Circuit Court		United States Telephone Ass'n (USTA) v. FCC, 1999 U.S. App. LEXIS 9768 (D.C. Cir. 1999).
1999	U.S. Supreme Court		Iowa Utilities Board v. FCC, 525 U.S. 366, 119 S.Ct. 721, 142 L.Ed.2d 834 (1999).
1999	U.S. Supreme Court		Ameritech Corp. v. FCC, 119 S.Ct. 2016, 143 L.Ed.2d 1029, 1999 U.S. LEXIS 3671 (1999).
Aug. 9, 1999	9th Circuit Court		AT&T et al. v. City of Portland, Appeal No. 99-35609 (9th Cir. Aug. 9, 1999). District Court Opinion: *www.techlawjournal.com/courts /portland/19990604op.htm*

(continued)

Date	Federal	State	Title
Aug. 9, 1999 *(continued)*			AT&T Appeal Brief: *www.techlawjournal.com/courts/portland/19990809.htm* Portland's Opposition Brief: *www.techlawjournal.com/courts/portland/19990907port.htm* FCC's Amicus Quriae Brief: *www.techlawjournal.com/courts/portland/19990916fcc.htm*
Aug. 27, 1999	FCC		In the Matter of Access Charge Reform, Price Cap Performance Review for Local Exchange Carriers, Interexchange Carrier Purchases of Switched Access Services Offered by Competitive Local Exchange Carriers and Petition of US West Communications, Inc., for Forbearance from Regulation as a Dominant Carrier in the Phoenix, Arizona MSA, Fifth Report and Order and Further Notice of Proposed Rulemaking, CC Docket Nos. 96-262, 94-1, 98-157 (released Aug. 27, 1999).
Dec. 6, 1999	AT&T		*Letter from AT&T to FCC Chairman William Kennard* (Dec. 6, 1999) available at *www.techlawjournal.com/broadband/19991206let.htm* [In response to some of the criticism raised by open access advocates against AT&T in AT&T v. City of Portland, AT&T agreed to provide access to an ISP (MindSpring Enterprises, Inc.) through AT&T's cable systems. This action established a model for AT&T's possible relationships with nonaffiliated ISPs].

APPENDIX D Legal Instruments Embodying the Results of the Uruguay Round

Volume 1:	The Legal Texts:
	• The Final Act • The Marrakesh Agreement Establishing the WTO • Ministerial Decisions and Declarations
Volumes 2–26:	Tariff Schedules for Trade in Goods:
	Vol. 2: Australia, Brazil, Myanmar Vol. 3: Canada Vol. 4: Sri Lanka, Chile, China Vol. 5: Cuba, India, New Zealand Vol. 6: Norway, Pakistan Vol. 7: South Africa Vol. 8: United States Vol. 9: Indonesia, Dominican Republic, Finland, Nicaragua, Sweden Vol. 10: Uruguay, Austria, Peru, Turkey Vol. 11: Japan Vol. 12: Malaysia Vol. 13: Nigeria, Gabon, Senegal, Madagascar, Cote d'Ivoire, Zimbabwe, Switzerland/ Liechtenstein Vol. 14: Republic of Korea Vol. 15: Iceland, Egypt, Argentina Vol. 16: Poland, Jamaica, Trinidad & Tabago, Romania Vol. 17: Hungary, Singapore, Suriname, the Philippines Vol. 18: Colombia, Mexico, Zambia, Thailand

Vol. 19: European Community and member states

Vol. 20: Morocco, Hong Kong, Tunisia, Bolivia, Costa Rica, Venezuela, El Salvador, Guatemala, Macau

Vol. 21: Namibia

Vol. 22: Paraguay, Czech Republic

Vol. 23: Slovak Republic, Honduras, Antigua & Barbuda, Bahrain, Barbados, Belize

Vol. 24: Brunei Darussalam, Cameroon, Cyprus, Dominica, Fiji, Ghana, Guyana, Kenya, Kuwait, Malta, Mauritius, St. Lucia, St. Vincent & the Grenadines.

Vol. 25: Swaziland

Vol. 26: Benin, Mauritania, Niger, Bangladesh, Congo, Tanzania, Uganda

Volume 27:	Agreements:

- Agriculture
- Application of Sanitary and Phytosanitary Measures
- Textiles and Clothing
- Technical Barriers to Trade
- Trade-related Investment Measures
- Implementation of Article VI (GATT 1994)

Volumes 28–32:	Services Agreement (Vol. 28) and schedules of services commitments and/or MFN exemptions (Vols. 28–30, 32):

Vol. 28: The General Agreement on Trade in Services; Schedules for Algeria, Antigua & Barbuda, Australia, Austria, Bahrain, Bangladesh, Barbados, Belize, Benin, Bolivia, Brazil, Brunei Darussalam, Burkina Faso, Cameroon, Canada, Chile, China, Columbia, Congo, Costa Rica, Cote d'Ivoire, Cuba, Cyprus, Czech Republic, Dominica, Dominican Republic, Egypt, El Salvador, European Community and member states, Fiji

Vol. 29: Finland, France for New Caledonia, Gabon, Ghana, Grenada, Guatemala, Guyana, Honduras, Hong Kong, Hungary, Iceland, India, Indonesia, Israel, Jamaica, Japan, Kenya, Korea (Rep. of), Kuwait, Liechtenstein, Macau, Madagascar, Malaysia, Malta, Mauritius, Morocco, Mexico, Mozambique, Myanmar, Namibia, the Netherlands for Aruba, the Netherlands for the Netherlands Antilles

Vol. 30: New Zealand, Nicaragua, Niger, Nigeria, Norway, Pakistan, Paraguay, Peru, the Philippines, Poland, Romania, Senegal, St. Lucia, St. Vincent & the Grenadines, Senegal, Singapore,

Slovak Republic, South Africa, Sri Lanka, Suriname, Swaziland, Sweden, Switzerland/ Tanzania, Thailand, Trinidad & Tabago, Tunisia, Turkey, Uganda, United States, Uruguay, Venezuela, Zambia, Zimbabwe

Agreements (Vol. 31):

- Intellectual property
- Dispute settlement
- Trade policy review mechanism
- The plurilateral trade agreements: government procurement (including schedules of commitments,) meat, and dairy

Schedules on services submitted after 15 April 1994 (Vol. 32):

Angola, Botswana, Burundi, Central African Republic, Chad, Djibouti, the Gambia, Guinea, Guinea-Bissau, Haiti, Lesotho, Malawi, Maldives, Mali, Mauritania, Rwanda, Sierra Leone, Solomon Islands, Togo, Zaire

Volume 33:	**Tariff Schedules on Goods submitted after 15 April 1994:**
	Angola, Botswana, Burkina Faso, Burundi, Central African Republic, Chad, Djibouti, The Gambia, Guinea, Guinea-Bissau, Haiti, Lesotho, Malawi, Maldives, Mali, Mozambique, Rwanda, Sierra Leone, Solomon Islands, Togo, Zaire
Volume 34:	**Schedules on Goods and Services:**
	Slovenia

APPENDIX E Membership of the World Trade Organization

Upon the accession of Latvia, on February 10, 1999 the number of members of the World Trade Organization reached 134. The following is the list of members and the effective dates of their membership:

Government	*Date of Membership*
Angola	December 1, 1996
Antigua and Barbuda	January 1, 1995
Argentina	January 1, 1995
Australia	January 1, 1995
Austria	January 1, 1995
Bahrain	January 1, 1995
Bangladesh	January 1, 1995
Barbados	January 1, 1995
Belgium	January 1, 1995
Belize	January 1, 1995
Benin	February 22, 1996
Bolivia	September 14, 1995
Botswana	May 31, 1995
Brazil	January 1, 1995
Brunei Darussalam	January 1, 1995
Bulgaria	December 1, 1996
Burkina Faso	June 3, 1995
Burundi	July 23, 1995
Cameroon	December 13, 1995
Canada	January 1, 1995
Central African Republic	May 31, 1995
Chad	October 19, 1996
Chile	January 1, 1995
Colombia	April 30, 1995
Congo	March 27, 1997
Costa Rica	January 1, 1995
Côte d'Ivoire	January 1, 1995
Cuba	April 20, 1995
Cyprus	July 30, 1995
Czech Republic	January 1, 1995
Democratic Republic of the Congo	January 1, 1997
Denmark	January 1, 1995
Djibouti	May 31, 1995
Dominica	January 1, 1995
Dominican Republic	March 9, 1995
Ecuador	January 21, 1996
Egypt	June 30, 1995
El Salvador	May 7, 1995
European Community	January 1, 1995
Fiji	January 14, 1996
Finland	January 1, 1995
France	January 1, 1995
Gabon	January 1, 1995
The Gambia	October 23, 1996
Germany	January 1, 1995

Government	*Date of Membership*
Ghana	January 1, 1995
Greece	January 1, 1995
Grenada	February 22, 1996
Guatemala	July 21, 1995
Guinea	October 25, 1995
Guinea Bissau	May 31, 1995
Guyana	January 1, 1995
Haiti	January 30, 1996
Honduras	January 1, 1995
Hong Kong, China	January 1, 1995
Hungary	January 1, 1995
Iceland	January 1, 1995
India	January 1, 1995
Indonesia	January 1, 1995
Ireland	January 1, 1995
Israel	April 21, 1995
Italy	January 1, 1995
Jamaica	March 9, 1995
Japan	January 1, 1995
Kenya	January 1, 1995
Korea	January 1, 1995
Kuwait	January 1, 1995
Kyrgyz Republic	December 20, 1998
Latvia	February 10, 1999
Lesotho	May 31, 1995
Liechtenstein	September 1, 1995
Luxembourg	January 1, 1995
Macau	January 1, 1995
Madagascar	November 17, 1995
Malawi	May 31, 1995
Malaysia	January 1, 1995
Maldives	May 31, 1995
Mali	May 31, 1995
Malta	January 1, 1995
Mauritania	May 31, 1995
Mauritius	January 1, 1995
Mexico	January 1, 1995
Mongolia	January 29, 1997
Morocco	January 1, 1995
Mozambique	August 26, 1995
Myanmar	January 1, 1995
Namibia	January 1, 1995
The Netherlands, For the Kingdom in Europe and for The Netherlands Antilles	January 1, 1995
New Zealand	January 1, 1995
Nicaragua	September 3, 1995
Niger	December 13, 1996
Nigeria	January 1, 1995
Norway	January 1, 1995
Pakistan	January 1, 1995
Panama	September 6, 1997
Papua New Guinea	June 9, 1996
Paraguay	January 1, 1995
Peru	January 1, 1995
The Philippines	January 1, 1995
Poland	July 1, 1995
Portugal	January 1, 1995
Qatar	January 13, 1996
Romania	January 1, 1995
Rwanda	May 22, 1996
Saint Kitts and Nevis	February 21, 1996
Saint Lucia	January 1, 1995
Saint Vincent and the Grenadines	January 1, 1995
Senegal	January 1, 1995
Sierra Leone	July 23, 1995
Singapore	January 1, 1995
Slovak Republic	January 1, 1995
Slovenia	July 30, 1995
Solomon Islands	July 26, 1996
South Africa	January 1, 1995
Spain	January 1, 1995
Sri Lanka	January 1, 1995
Suriname	January 1, 1995
Swaziland	January 1, 1995
Sweden	January 1, 1995
Switzerland	July 1, 1995
Tanzania	January 1, 1995
Thailand	January 1, 1995
Togo	May 31, 1995
Trinidad and Tobago	March 1, 1995
Tunisia	March 29, 1995

Government	*Date of Membership*
Turkey	March 26, 1995
Uganda	January 1, 1995
United Arab Emirates	April 10, 1996
United Kingdom	January 1, 1995
United States	January 1, 1995
Uruguay	January 1, 1995
Venezuela	January 1, 1995
Zambia	January 1, 1995
Zimbabwe	March 3, 1995

Observer governments:

Albania
Algeria
Andorra
Armenia
Azerbaijan
Belarus
Bhutan
Cambodia
Cape Verde
People's Republic of China
Croatia
Estonia
Ethiopia
Former Yugoslav Republic of Macedonia
Georgia
Holy See (Vatican)
Jordan
Kazakstan
Lao People's Democratic Republic
Lebanon
Lithuania
Moldova
Nepal
Oman, Sultanate of
Russian Federation
Samoa
Saudi Arabia
Seychelles
Sudan
Chinese Taipei
Tonga
Ukraine
Uzbekistan
Vanuatu
Vietnam
Yemen

Note: All observer countries have applied to join the WTO except the Holy See (Vatican) and, for the time being, Ethiopia, Cape Verde, and Bhutan.

International organization observers to General Council (observers in other councils and committees differ):

United Nations (UN)

United Nations Conference on Trade and Development (UNCTAD)

International Monetary Fund (IMF)

World Bank

Food and Agricultural Organization (FAO)

World Intellectual Property Organization (WIPO)

Organization for Economic Co-operation and Development (OECD)

Source: *www.wto.org.*

APPENDIX F Commitments and Most Favored Nation Exemptions

HIGHLIGHTS OF COMMITMENTS AND MOST FAVORED NATION EXEMPTIONS RESULTING FROM THE NEGOTIATIONS ON BASIC TELECOMMUNICATIONS*

TRADE TOPICS:

Antigua & Barbuda
The government of Antigua and Barbuda offered to liberalize its international voice telephony services by 2010; but reserved domestic voice telephony for its exclusive operator. It also agreed to undertake liberalization of other basic telecommunication services, such as data transmission and private leased circuit services by 2012. It opened to full competition the provision of data transmission over closed user groups, Internet, and Internet access services (excluding voice), teleconferencing, and several value-added services. It also allowed the provision of terrestrial-based mobile services (cellular, data, personal communications service (PCS), paging, and trunked radio system) through commercial presence and committed to allow cross-border supply of satellite-based mobile services (including voice, data, PCS, paging, and trunked radio systems) and of fixed satellite services through commercial arrangements with the exclusive operator. The government committed to the Reference Paper on regulatory principles and submitted a most-favored nation (MFN) Exemption List to enable the government to extend to nationals of other Caribbean Community (Caricom) Member countries treatment equal to its own nationals with respect to joint venture requirements.

Argentina
Argentina agreed to a phased-in commitment liberalizing voice telephony (local, long-distance, and international) and provision of other basic telecommunications services supplied on an international basis by November 2000. It offered full competition in basic services other than voice, such as data transmission, and supplied in the national market and leased circuit services (international and national) without phase in. It also offered open competition in mobile telecommunications services such as data, paging, and trunking and committed to duopoly in mobile cellular services, undertaking to allow new entrants, subject to an economic needs test

for the provision of mobile personal communication services. Argentina's improved offer committed to the Reference Paper on regulatory principles and it submitted an MFN Exemption List on telecommunications services involving the supply of fixed satellite services by geostationary satellites.

Australia

Australia offered unrestricted competition in virtually all basic telecommunications services as of July 1997. As part of this, it committed its existing free markets for voice telephone and other basic services on a resale basis. Also by July 1997, Australia offered to end its limits on the number of satellite service providers (set at two in January 1997); on primary suppliers of public mobile cellular telephony; and on facilities–based carriers (both then set at three). It offered: no limits on foreign equity for new carriers; permitted foreign equity up to 11.7% of the government-controlled carrier, TELSTRA; and required majority Australian ownership of the mobile carrier, Vodaphone. Its improved offer removed foreign equity limitation of Optus. Australia also commited to the Reference Paper on regulatory principles.

Bangladesh

Bangladesh licensed two operators, in addition to the government operator, to supply local voice, domestic long distance, and transmission facilities (leased circuit) services. It committed to full competition in voice and data transmission over closed user groups and for Internet access services. Licenses for four suppliers of cellular mobile voice telephone services were granted and the country indicates that it will review the possibility of adding regulatory principles in the future.

Bangladesh submitted an MFN Exemption List to permit the government or the government-run operator to apply differential measures, such as accounting rates, in bilateral agreements with other operators or countries and listed other related points in its final Report to the Group.

Barbados*

Barbados agreed to liberalize basic telecommunications services including voice telephony, data transmission, and private leased circuit services offered to the general public in 2012 when the monopoly exclusivity expires. For nonpublic use, the supply of voice telephony, data transmission, and facsimile services are opened for competition on the basis of facilities leased from the monopoly up until 2012 and on each providers' own facilities thereafter. However, as of 1999, Barbados allows unrestricted supply of terrestrial- and satellite-based mobile services (cellular, data, PCS, and paging), and a variety of value-added services, including Internet access and V-SAT services for nonpublic use. It committed to the Reference Paper on regulatory principles.

Belize

Belize offered a phased-in commitment to allow open competition in trunked radio service and teleconferencing by 2003, as well as paging and various value-added telecommunications services by 2008. It committed to the Reference Paper on regulatory principles.

Bolivia

Bolivia offered to phase in competition in all domestic long distance and international basic services by November 2001. Its local voice telephony is to be provided by 16 exclusive local suppliers while all other basic services for local market are to be liberalized. Bolivia offered full competition without phase in for the local, national, and international supply of all basic services (including voice) to closed user groups and for the supply of mobile services, including cellular services, mobile data services, radio navigation, paging services, PCS, and mobile satellite services. It placed no restrictions on foreign ownership but did include some regulatory principles.

Brazil

Brazil's improved commitment offered to end its monopoly on public telecommunications services, and to submit new commitments and regulatory principles within one year, rather than two, from the date of passage of its new telecommunications law. It committed, with no phase in, to open markets for paging services, many value-added services, and the supply to closed user groups of all basic telecommunications services, including voice. It further committed to establish cellular telephone duopolies in each of several designated markets and to phaseout the 49% (direct and indirect) foreign equity limits on cellular telephony and satellite transport services as of July 1999. It allowed competition in satellite transport services, subject to the requirement that satellites with orbital positions notified by Brazil must be used unless their services are not equivalent to those of satellites notified by other countries. Furthermore, Brazil submitted an MFN Exemption List on the distribution of radio or television programming directly to consumers.

Brunei/Darussalam

Brunei/Darussalam offered to undertake a policy review in 2010 on whether to allow additional suppliers in local public voice telephone services beyond its current monopoly, international public voice telephone services beyond its current duopoly, and in cellular telephone services beyond its current monopoly. It committed to the Reference Paper.

Bulgaria

Bulgaria agreed to phased-in commitments to liberalize *public* voice telephone, telegraph, and telex services on a nonfacilities basis as of 2003 and on a facilities basis as of 2005. It offered full liberalization without phase-in of data transmission, paging, mobile data, nonpublic mobile voice telephone, VSAT, and other nonvoice satellite services. It committed to liberalize digital and analogue cellular telephone subject to a requirement that, until 2003, all international traffic use the network facilities of the monopoly. Bulgaria committed to the Reference Paper on regulatory principles.

Canada

In its commitment made on February 14, 1997, Canada improved certain routing restrictions and foreign equity limits on many services that previously were scheduled to be phased out by the year 2000. It also removed a requirement that Canadian equity holding in mobile satellite systems must equal Canadian usage levels and liberalized the regime for resale-based competition in local telephone services and in most other basic telecommunications services. However, it maintained a few limitations on market access for telephone service in certain cities or provinces and limited foreign equity in all facilities-based suppliers to 20% direct and 46.7% combined direct and indirect foreign ownership. As of October 1998, Canada agreed to remove its restrictions on obtaining licenses to land submarine cables, Teleglobe's monopoly on oversea (non-U.S.) facilities-based service, and to raise its foreign equity limits to the 46.7%. Finally, it advanced the date for the elimination of Telesat's exclusive rights on satellite facilities and earth stations serving the U.S./Canadian market to March 2000. Canada committed to the Reference Paper on regulatory principles.

Chile

Chile offered full competition in the national long distance and international markets for all basic telecommunications services, including mobile and satellite services. However, it made no commitment on the provision of basic telecommunications services in local markets. It revised its commitment on the Reference Paper based on regulatory principles.

Colombia
In its commitment, Colombia offered open competition in facilities-based local voice telephony and data transmission for public use, as well as voice and other basic telecommunications services over closed user groups. Public facilities-based long distance and international voice telephony were reserved to the public operator, while additional operators were to be determined based on an economic needs test. Colombia also committed to the provision of satellite transport capacity exclusively for geostationary satellite systems and a foreign equity limit of 70% for all telecommunications service providers. Paging and facilities-based trunking were to be opened to a limited number of suppliers by July 1997, and cellular telephone service, the regional duopolies, were to be liberalized in September 1999, after which new entrants would be permitted subject to technical constraints. Finally, personal communications services (PCS) were to be opened to a limited number of suppliers by 2000. Colombia improved its commitment to the Reference Paper based on regulatory principles.

Côte d'Ivoire
Côte d'Ivoire reserved monopoly provision of voice telephone service over fixed network infrastructure and telex for 10 years, but will thereafter open both to unrestricted competition. Open market access without phase in is offered for all other basic telecommunications services including data transmission, all mobile networks and services, video transmission services and satellite services, links, capacity, and earth stations. Côte d'Ivoire committed to the Reference Paper on regulatory principles.

Cyprus*
Cyprus stated that while telecommunication services were under monopoly control at the time it submitted its commitment to the WTO, the government had commissioned a study to review telecommunications policy issues and to make changes necessary for the gradual liberalization of its market. Cyprus also indicated that its ministers were expected to decide in 1998 on the specific measures to be adopted.

Czech Republic
The Czech Republic committed to full competition in all segments of voice telephone, leased circuit services, and satellite services after the year 2000. Before then, it agreed to open markets, with no phase in, for voice over *closed* user groups, data transmission, various mobile services (excluding international voice until 2000), video transport, and frame relay. The country's improved offer included the Reference Paper on regulatory principles.

Dominica
Dominica offered full competition for data transmission over *closed* user groups, teleconferencing, several value-added telecommunications services, and Internet access and services (excluding voice). It agreed to allow cross-border supply of satellite-based mobile services (including voice, data, PCS, paging, and trunked radio system) and fixed satellite services through commercial arrangements with the exclusive operator. Dominica committed to the Reference Paper on regulatory principles.

Dominican Republic
The Dominican Republic committed to allow the provision of all basic telecommunications services, including voice, data transmission, private leased circuits, mobile maritime, and air-to-ground, through commercial presence. Its improved offer included the Reference Paper on regulatory principles.

Equador

Equador committed to unrestricted market access for cellular telephony.

El Salvador

El Salvador offered full competition of basic telecommunications services (facilities-based and resale) for all market segments (local, long distance, and international) including, for example, voice telephony, data transmission, private leased circuits, paging, and mobile cellular services. It also committed to the Reference Paper on regulatory principles.

European Community

The European Community committed to complete liberalization of basic telecommunications services (facilities-based and resale) across the European Community for all market segments (local, long distance, and international). This included, for example, satellite networks and services and all mobile and personal communications services and systems. However, certain individual member countries committed to different specific dates and/or criteria for implementation, most on a delayed basis. For example, Portugal advanced the liberalization date for facilities-based services to July 1999, internationally connected mobile and personal communication services to later in 1999, and public voice telephony to 2000. It also undertook an additional commitment to partially remove the foreign equity restriction (then 25%) by 1999, subject to parliamentary approval. Spain advanced the liberalization date for both voice telephone and facilities-based services to December 1998 (issuing one additional nationwide license in January 1998) and removed its 25% foreign equity restriction. Ireland agreed to liberalization of internationally connected mobile and PCS by 1999, and voice telephone and facilities-based services in 2000. Greece committed to full liberalization of public voice telephony and facilities-based services by 2003. Restrictions on foreign equity limits were removed by Belgium to 49% and by France to 20% radio-based services, direct investment only. The European Community committed to the Reference Paper on regulatory principles.

Ghana

Ghana committed to maintain two facilities-based suppliers providing local, long distance, and international public voice telephone services and private leased circuit services. It offered to licence additional suppliers of local voice services to underserved population centers and to undertake a policy review, possibly allowing new entrants to supply voice telephony once the five-year exclusivity of the duopoly operators have expired. Ghana also offered full competition in data transmission, Internet, and Internet access (excluding voice) and teleconferencing. It committed to opening mobile services, (both terrestrial and satellite-based), including mobile data services, fixed satellite services, paging and cellular, with the reservation that cross-border voice services can only be supplied through commercial arrangements with the duopoly operators. Ghana also committed to the Reference Paper on regulatory principles.

Grenada

Grenada offered to phase in liberalization of most basic telecommunication services in all market segments including voice telephony, data transmission, private leased circuits, and terrestrial mobile services by 2006. With this, it committed to full competition in value-added telecom services, trunked radio systems and Internet and Internet access service (excluding voice). Grenada will retain the supply of satellite-based mobile services (including voice, data, PCS) and fixed satellite services through commercial arrangements with the exclusive operator until 2006, but will open the market with no restrictions thereafter. Grenada committed to the Reference Paper on regulatory principles.

Guatemala
Guatemala offered full competition in basic telecommunications services (facilities based and resale) for all market segments (local, long distance, and international) including voice telephony, data transmission, private leased circuits, paging, mobile cellular, and satellite services. It also committed to the Reference Paper on regulatory principles.

Hong Kong
In a February 1998 revision, Hong Kong committed to international simple resale of its facsimile and data transmission services. It already provided access to the local market for many basic telecommunications services including voice and data transmission, as well as mobile radio telephone and mobile data services. For local fixed-network services, four licences were already issued, and issuance of further licences were to be given consideration in June 1998. Furthermore, Hong Kong agreed to permit call back and other alternative international calling services, certain satellite services, virtual private networks, and mobile satellite services and committed to the Reference Paper on regulatory principles.

Hungary
In its February 1998 revision, Hungary indicated that its reservation of land mobile services to three existing suppliers will end by 2004. It also committed to competition in domestic long-distance and international public voice telephone as of 2003 and in local voice service as of 2004. It required 25% Hungarian equity for local and domestic voice services and facilities-based international service, but not for international resale of voice. Services such as paging, data transmission, and leased circuit services are to be fully liberalized without phase in. It committed to the Reference Paper on regulatory principles.

Iceland
Iceland liberalized essentially all basic telecommunications services, including both facilities-based and resale, and data transmission, voice telephone, satellite communications, cellular mobile telephony, and other mobile services. It committed to the Reference Paper on regulatory principles.

India
India committed to review in 1999 further opening of national long-distance service and in 2004 of international services. For fixed networks providing many basic services in the local markets and for long-distance service within each of a number of service areas, India committed to allow one new operator, in addition to MTNL, under licenses to be valid for a period of 10 years. The licenses for these operators will be issued as the need for the additional licenses is determined by the relevant authorities. India limited foreign equity participation to 25%, and committed to the Reference Paper on regulatory principles. With this, India submitted an MFN Exemption List to permit the government or the government-run operator to apply differential measures, such as accounting rates, in bilateral agreements with other operators or countries. It filed related points in its final Report on the Group.

Indonesia
In the final days of the 1997 talks, Indonesia improved its offer by deleting an economic needs test for new entrants in domestic mobile cellular telephone services, personal mobile cellular communication services, and regional and national paging services. Public voice telephony, circuit-switched public data network and teleconferencing services are currently supplied by a number of suppliers with exclu-

sive rights. However, Indonesia committed to a policy review to determine whether to admit additional suppliers upon the expiration of the exclusive rights. Exclusivity rights expire for international service in 2005, for long-distance service in 2006, and for local service in 2011. Indonesia offered competition for packet-switched public data network services, telex, telegraph, and Internet access services, subject to use of networks of PT Indosat and PT Satelindo for international traffic. It further offered competition in domestic mobile cellular telephone services, paging, public payphone services and limited foreign equity to 35% for all services except personal communication services that require joint venture with state-owned company. Indonesia committed to the Reference Paper on regulatory principles.

Israel

Israel committed to three operators of international voice services, removed limits on the number of operators for global satellite systems, and indicated that regulations for opening competition in domestic voice services and network infrastructure will be published when the monopoly rights end in 2002. Israel further committed to reexamine its policy on competition in international voice services by 2001 and to open competitive market access in cellular telephone and paging, voice over closed user groups, international private leased circuit services (excluding voice), and data transmission. It limited foreign equity to 74% on all service providers except for wireless service providers, where it allowed 80%. Israel committed to the Reference Paper on regulatory principles.

Jamaica

Jamaica offered to phase in domestic facilities-based and international voice telephony and other basic telecommunication services by 2013. It will undertake a policy review before submitting additional commitments on voice over closed user groups and voice over Internet currently reserved to exclusive supply until 2013. In addition, international satellite-based mobile telephone and fixed satellite services will continue to be provided through commercial arrangements with the exclusive operator until 2013. Terrestrial cellular mobile telephone and domestic satellite-based mobile telephone services will continue to be provided by an exclusive operator under a 5- to 10-year licence. On the other hand, Jamaica committed to full competition in data transmission, digital mobile data services, personal communication services, paging, teleconferencing, Internet, and Internet access (excluding voice), trunked radio systems, video transport (excluding teleconferencing) as well as several value-added services. It also committed to allow supply of international voice, data, and video transmission services to firms involved in information processing located within free zones. Jamaica further committed to the Reference Paper on regulatory principles.

Japan

In April 1996, Japan agreed to remove longstanding foreign equity limits on Type I carriers and radio-based services, leaving only two companies, KDD and NTT, with foreign equity limits of 20%. It also deleted the reservation concerning international simple resale of voice services. Aside from these company-specific restrictions, Japan committed to open market access in all market segments for basic telecommunications services (facilities-based and resale). It further committed to the Reference Paper on regulatory principles.

Korea

Korea permits full competition in wire-based telephone services and resale of all telecommunications services except voice without phasein International simple voice, resale with phase in, will occur by 2001. Also in 2001, Korea will increase its permitted foreign equity limit in the national supplier (KT) from 20% to 33%. It permitted market access to domestic voice resale in 1999, allowing up to 49% foreign equity participation, which will increase to 100% in 2001. Korea also committed on the Reference Paper on regulatory principles.

Malaysia

Malaysia permits up to 30% foreign shareholding in existing facilities-based public telecommunications operators. Services supplied by the existing operators include voice telephony (wire or wireless), data transmission, private leased circuit services, domestic and international satellite services and satellite links/capacity, satellite earth stations, terrestrial- and satellite-based mobile services and video transport services. In its commitment, Malaysia listed some regulatory principles as additional commitments.

Mauritius

In its improved offer of February 1997, Mauritius committed to opening competition in mobile satellite-based private mobile radio and paging services. It also committed to eliminate existing de facto monopoly and exclusive rights in all basic telecommunications services by 2004 and to introduce regulatory principles in the future.

Mexico

Mexico committed to competition in all market segments of public telecommunications services on both facilities and a resale basis. This includes voice telephone, data transmission, private leased circuit, paging, and certain cellular telephone services. For cellular telephony, Mexico committed to ending the exclusivity of the regional duopolies and now allows more than 49% foreign investment subject to prior authorization. For all other telecommunications services, Mexico raised its foreign equity limitation from 40% to 49% in 1997. It committed to the Reference Paper on regulatory principles.

Morocco

Morocco opened market access for packet-switched data transmission and frame-relay services, but agreed only to phased-in supply of voice telephone services over fixed infrastructure by December 2001. Provision of mobile telephone and mobile data services, personal communication services (PCS), and paging were reserved for an unspecified number of operators yet to be licensed beyond the one mobile telephone operator licensed to that date. Foreign equity participation may also be limited, but the level was unspecified. Morocco's commitment included some regulatory principles.

New Zealand

New Zealand committed to open markets for all basic telecommunication services for all market segments (local, long distance, and international). While a national treatment limitation indicates that no single foreign entity is permitted to hold more than 49.9% of Telecom New Zealand, this does not limit the overall foreign shareholding in that operator. New Zealand committed to the Reference Paper on regulatory principles.

Norway

Norway committed to complete liberalization of all basic telecommunications services. This includes, for example, data transmission, voice telephone, paging and other mobile services, and satellite communications (including voice) in all market segments. Norway also committed to the Reference Paper on regulatory principles.

Pakistan*

Pakistan committed to open markets for data transmission, email, Internet and Intranet, domestic VSAT, trunked radio services, videoconferencing, telemedicine, and tele-education. It allows competition in satellite-based services, including voice telephone and value-added services subject only to restrictions on cross-border supply to preserve monopoly rights on basic and international networks and services until their expiration. Pakistan also moved forward the dates of the phase-in of certain commitments by one year. Now, as of January 2004, it will both end exclusivity on cross-border supply of voice telephony, with no commitment on commercial presence, and commit to full competition in private leased-circuit services, including transmission capacity. Pakistan committed to the Reference Paper on regulatory principles, and submitted an MFN Exemption List to permit the government or the government-run operator to apply differential measures, such as accounting rates, in bilateral agreements with other operators or countries. It included related points in the final Report on the Group.

Papua New Guinea

Papua New Guinea received all telecommunications services for an exclusive service provider until 2002. However, it offered to review and announce the issuance of additional operating licenses by the year 2000. It also committed to the Reference Paper on regulatory principles.

Peru

Peru committed to liberalization of its voice telephone services (domestic, long distance, and international) in 1999. Other basic services were to be liberalized without phase-in for the local market and with phase-in during 1999 for the long-distance and international market segments. Where scarce resources such as frequency availability are involved, licenses will be issued through public tender. Peru further committed to the Reference Paper on regulatory principles.

Philippines

The Philippines offered competition through commercial presence in the following services on a facilities basis for public use by means of all types of technologies except cable television and satellite: voice telephone, data transmission, and cellular mobile telephone services in all market segments (local, long distance, and international). Market access for new entrants is to be determined by meeting the criteria of a public convenience and necessity test with foreign equity limited to 40%. The Philippines commitment included some regulatory principles.

Poland

Poland committed to liberalize international public voice and facilities, telex, and telegraph in 2003. Long-distance public voice service and facilities are also to be liberalized by 2003. Poland offered market access with no phase in for local public voice and facilities (in geographic areas assigned by license), voice over closed user groups in all market segments, and data transmission. It committed to liberalization of domestic telex and telegraph by 2000, committed to permit cellular mobile telephone services and networks subject to use, until 2003, of international monopoly network facilities. It committed to permit mobile satellite services and networks in 2003 subject to foreign equity limitations of 49% for all international and domestic long-distance services and facilities networks and public cellular telephone services. Poland commited to the Reference Paper on regulatory principles.

Romania
Romania offered competition with no phase in for data transmission, telex, telegraph, facsimile, and paging services, and the supply of both VSAT and voice telephone offered to closed user groups. It committed to bind the two existing licenses for digital cellular mobile telephony of public voice telephone services (local, long distance, and international) and leased circuit services are to be liberalized while analogue cellular mobile telephony is to be liberalized by 2002. In 2000, Romania committed to the Reference Paper on regulatory principles.

Senegal
Senegal offered to review policy with respect to licensing additional operators when existing monopoly rights expire between 2003 and 2006 in voice telephony, data transmission, private leased circuit services, and fixed satellite services. The number of operators is limited to three in the following services: paging and trunked radio systems. In 1997, two cellular mobile operators (including mobile data) were licensed with opportunities for additional licenses to provide mobile satellite services. Senegal committed to the Reference Paper on regulatory principles.

Singapore
Singapore committed to phase in competition in facilities-based telecommunication services beginning April 2000 while open markets for cellular telephony, mobile data, paging and trunked radio services. Foreign equity on these facilities-based systems, however, is limited to 49%. It also agreed to the provision of domestic and international resale of public-switched capacity (not including the connection of leased lines to public network) for most basic services, including voice, data, and Integrated Services Digital Network (ISDN). Singapore committed to the Reference Paper on regulatory principles.

Slovak Republic
The Slovak Republic offered competition, not subject to phase in, for voice telephony within closed user groups, data transmission, private leased circuits services with no connection to the public network, and all mobile and personal communication services (excluding analogue cellular voice services), except the mobile supply of international voice. In 2003, it offers competition in public voice services and network infrastructure, private leased circuits with connection to the public network, and international mobile voice services. The Slovak Republic committed to the Reference Paper on regulatory principles.

South Africa
South Africa committed to end its monopoly control of public-switched, facilities-based services, including voice, data transmission, telex, facsimile, private leased circuits, and satellite-based services by the end of 2003. At that time it will introduce at least a second supplier and will review the feasibility of allowing additional suppliers of public-switched services. It also committed to duopoly supply of mobile cellular telephony, but placed no limitations on the number of suppliers of paging, personal radio communication, and trunked radio systems. Foreign investment in telecommunications suppliers is limited to 30%. Between 2000 and 2003, South Africa offered to liberalize resale services. It also committed to the Reference Paper on regulatory principles.

Sri Lanka
Sri Lanka offered a duopoly in international basic voice services after 2000, subject to satisfactory progress by the monopoly on

tariff rebalancing. It limited foreign equity participation up to 35% for a strategic partner in the government-owned operator SLT. For local and domestic long-distance mobile cellular services, four operators were licensed by 2000, and the government agreed to review the number of additional licences to be permitted after that. For provision of basic voice telephony, data transmission, payphones, voice mail, and facsimile by wireless local loop, Sri Lanka committed to two licenses (in addition to SLT), to be guaranteed exclusivity for five years. It committed to five licenses for public payphones and paging services, with the possibility of additional suppliers for each to service subject to economic needs tests. It further committed to six operators of data communication services and is considering additional licenses for GMPCS services supplied through its own gateways. For all suppliers other than SLT, foreign equity is permitted up to 40%, with investments over 40% subject to case-by-case approval. Sri Lanka committed to the Reference Paper on regulatory principles and submitted an MFN Exemption List to permit the government or the government-run operator to apply differential measures, such as accounting rates, in bilateral agreements with other operators or countries. It also filed related points in the final Report on the Group.

Suriname*

Suriname committed to duopoly provision of public voice telephone services, fixed network infrastructure, and fixed satellite services and to maintain existing licenses for mobile telephone and PCS. However, it also committed to determine, by 2003, the circumstances for the licensing of additional operators of these services. It liberalized nonpublic voice services and public and nonpublic data transmission, Internet services (excluding voice), and teleconferencing subject to use of the duopolies' facilities. It fully liberalized provision of public mobile data, paging, and trunked radio systems and committed to the Reference Paper on regulatory principles.

Switzerland*

Switzerland's commitment reflected its new telecommunications law committing to complete liberalization of basic telecommunications services (facilities-based and resale, public and nonpublic) for all market segments (local, long distance, and international) and by any type of technology. It committed to the Reference Paper on regulatory principles.

Thailand

In its revised offer of February 1997, Thailand agreed to review its public local, long-distance, and international voice telecommunications services and to introduce revised commitments for these services and markets in 2006, conditional upon the passage of and consistent with the provisions in its proposed new communications acts. It also committed to introduce in the future some regulatory principles conditional upon the passage and entry into force of its new telecommunications acts.

Trinidad & Tobago

The government of Trinidad and Tobago offered competition in voice telephony, data transmission, telex, telegraph and private leased circuit services for public use after 2010. It also committed to phased in competition in mobile satellite-based services for public use, including mobile telephone services, mobile data, fixed satellite services, and personal communication services, as well as several value-added services. Other mobile services, Internet and Internet access and teleconferencing for private use are unconfirmed offers to be negotiated.

Trinidad and Tobago committed to the Reference Paper on regulatory principles.

Tunisia

Tunisia offered competition in telex and packet-switched data transmission after 1999; in mobile telephone, frame relay, paging, and teleconferencing after 2000; and in local telephone services after 2003. After 2002, foreign participation in the capital of Tunisie Telecom will be allowed up to 10%. For all other services, foreign equity is limited to 49%.

Turkey

In its revised offer that includes some regulatory principles, Turkey committed to end its monopoly's exclusive rights on voice telephony and other basic telecommunications services by 2006 and to open competition in cellular mobile services and paging services. It committed to market access for data transmission services without phase in. Turkey submitted an MFN Exemption List with two entries: one relating to two neighboring countries with respect to fees for transit land connections and the usage of satellite ground stations, and the other to permit the government or the government-run operator to apply differential measures, such as accounting rates, in bilateral agreements with other operators or countries. Some related points were made in the final Report on the Group.

United States

The United States committed to open markets for essentially all basic telecommunications services (facilities-based and resale) for all market segments (local, long distance, and international), including unrestricted access to a common carrier radio licenses for operators that are indirectly foreign owned. This offer included, for example, satellite-based services, cellular telephony, and other mobile services. Limitations on market access include no issuance of radio licenses to operators with more than 20% direct foreign ownership, and Comsat's retention of exclusive rights to links with Intelsat and Inmarsat satellites. The United States committed to the Reference Paper on regulatory principles and submitted an MFN Exemption List on telecommunications services involving the one-way satellite transmission of DTH and Direct Broadcast Satellite (DBS) television services and digital audio services.

Venezuela

Venezuela committed to open markets for facilities-based voice telephone services in all market segments (local, long distance, and international) as of November 2000. With this, it offered full competition in facilities-based telecommunication services such as mobile telephony, data transmission, teleconferencing, and paging without phase in.

*Some governments have submitted new or improved schedules of commitments on basic telecommunications since the conclusion of the 1997 negotiations. Such schedules are not attached to the Fourth Protocol, but have the same legal status as those that are attached. Source: *www.wto.org.*

APPENDIX G Chapters within Title 19 of the U.S. Code Customs Duties

Title 19 of the U.S. Code provides most U.S. trade law. The chapters within Title 19, listed below, provide a good overview of the evolution of U.S. trade law. The later laws generally amend the earlier laws, but many of the earlier laws still apply today. Therefore, it is important to be familiar with all of the trade laws, not just the most recent or only those that directly address telecommunications issues.

Chapter 1. Collection Districts, Ports, and Officers
Chapter 1A. Foreign Trade Zones
Chapter 2. The United States International Trade Commission
Chapter 3. The Tariff and Related Provisions
Chapter 4. Tariff Act of 1930
Chapter 5. Smuggling
Chapter 6. Trade Fair Program
Chapter 7. Trade Expansion Program
Chapter 8. Automotive Products
Chapter 9. Visual and Auditory Materials of Educational, Scientific and Cultural Character
Chapter 10. Customs Service
Chapter 11. Importation of pre-Columbian Monumental or Architectural Sculpture or Murals
Chapter 12. Trade Act of 1974
Chapter 13. Trade Agreements Act of 1979
Chapter 14. Convention on Cultural Property
Chapter 15. Caribbean Basin Economic Recovery

Chapter 16. Wine Trade
Chapter 17. Negotiation and Implementation of Trade Agreements
Chapter 18. Implementation of the Harmonized Tariff Schedule
Chapter 19. Telecommunications Trade
Chapter 20. Trade Preference for the Andean Region
Chapter 21. North American Free Trade Agreement (NAFTA)
Chapter 22. Uruguay Round Trade Agreements

APPENDIX H State Privacy Laws

State	Enacted	Proposed
Alabama, Code of Ala.	13A-8-100, *et. seq.* (1999)	
Alaska	Alaska Stat. 11.46.484.	1999 Bill Tracking AK H.B. 338; (New Bill), 21st Legislature second session, HOUSE BILL 338. Introduced: February 4, 2000, Last-action: February 4, 2000; To House Committee on Judiciary Relates to crimes involving computers, access devices, other technology, and identification documents; relates to the crime of criminal impersonation; relates to crimes committed by the unauthorized access to or use of communications in electronic storage. Subject: Communication and information, electronic and business equipment, computers and information systems, electronic text and mail, privacy and records, records-misc, law and justice, criminal law, fraud, telecommunications crime. 1999 Bill Tracking AK S.B. 245; (New Bill), 21st Legislature—second session, Senate Bill 245. Introduced: February 3, 2000, Last-action: February 3, 2000; To Senate Committee on Judiciary. Relates to crimes involving computers, access devices,

(continued)

State	Enacted	Proposed
Alaska *(continued)*		other technology, and identification documents; relates to the crime of criminal impersonation; relates to crimes committed by the unauthorized access to or use of communications in electronic storage. Subject: communication and information, electronic and business equipment, computers and information systems, electronic text and mail, law and justice, criminal law, ammunition and protective clothing, telecommunications crime. Last-action: February 3, 2000; To Senate Committee on Judiciary.
Arizona	Ariz. Rev. Stat. Ann. 41-1750.	
Arkansas	Ark. Stat. Ann. § 5-41-101, *et. seq.* (1999), Computer-related crimes.	
Colorado	C.R.S. 18-5.5-101, *et. seq.,* (LEXIS 1999). Computer crime "Any person who knowingly uses any computer, computer system, computer network, or any part thereof for the purpose of devising or executing any scheme or artifice to defraud; obtaining money, property, or services by means of false or fraudulent pretenses, representations, or promises; using the property or services of another without authorization; or committing theft commits computer crime."	2000 Bill Tracking CO H.B. 1107, 2nd regular session of the 62nd General Assembly, House Bill 1107, Introduced: January 5, 2000, Last-action: January 21, 2000; To House Committee on Appropriations. Relates to computer crime; relates to drunk driving. Subject: Law and justice, criminal law, communication and information, electronic and business equipment, computers and information systems, fraud, telecommunications crime.
	C.R.S. 18-9-309 (LEXIS 1999): Telecommunications crime "A person commits a class 3 misdemeanor if he or she knowingly: (a) Accesses, uses, manipulates, or damages any telecommunica tions device without the authority of the owner or person who has the lawful possession or use thereof . . ."	2000 Bill Tracking CO H.B. 1111, 2nd regular session of the 62nd General Assembly, House Bill 1111, Introduced: January 5, 2000, Last-action: January 18, 2000; To House Committee on Appropriations. Adequately allows Colorado prosecuting attorneys to prosecute persons using the Internet and other mediums for the forgery of identity documents.

State	Enacted	Proposed
		2000 Bill Text CO H.B. 1107, Colorado 2nd Regular Session of the 62nd General Assembly, Introduced, version-date: January 5, 2000. A Bill for an Act Concerning Substantive Changes for the Strengthening of the Criminal Laws, and Making an Appropriation Therefor. (Note with particularity, Sections 5, 6, and 7).
California	Cal. Penal Code 502 and 11771.	
Connecticut	Computer Crimes Act, Conn. Gen. Stat. 53a-250 *et seq.*	
Delaware	Del. Code Ann. Tit. 11, 932-935 and 8606.	
Florida	Fla. Stat. § 815.01 (1999).	
Georgia	1999 Bill Tracking GA H.B. 213, 145th General Assembly, 1999-00 Regular Session, House Bill 213. Introduced: January 25, 1999, Last-action: April 13, 1999. Enacts Computer Pornography and Child Exploitation Prevention Act; relates to offenses related to minors generally; defines the crime of computer pornography. Status: Signed by governor as Act No. 155.	
Hawaii	Haw. Rev. Stat. 708-890 to 708-896.	
Idaho	Idaho Code § 18-2201, *et. seq.* (1999); computer crime, appears to track model form.	
Illinois	1999 Bill Tracking IL H.B. 249, 91st General Assembly, 1999-00 General Assembly, House Bill 249, Introduced: January 22,	1999 Bill Tracking IL H.B. 788, 91st General Assembly, 1999-00 General Assembly, House Bill 788. Introduced: February 10, 1999, Last-action: May 8,

(continued)

State	Enacted	Proposed
Illinois *(continued)*	1999, Last-action: July 22, 1999; Signed by governor. Public Act No. 91-222, provides that it is a Class 4 felony to disclose the name, address, telephone number, or email address of a person under 17 years of age on an adult obscenity or child pornography Internet site.	1999; Re-referred to Senate Committee on Rules. Amends the Criminal Code of 1961; provides that it is a Class 4 felony for a first offense and a Class 3 felony for a second or subsequent offense for a person to send a message to a minor by telephone, email, the Internet, or online service, that is harmful material, with the intent of arousing, appealing to, or gratifying the lust or passions or sexual desires of the person or of the minor with the intent of seducing a minor. 1999 Bill Tracking IL H.B. 789, 91st General Assembly, 1999-00 General Assembly, House Bill 789. Introduced: February 10, 1999, Last-action: April 26, 1999; To Senate Committee on Rules. Amends the Criminal Code of 1961; creates the offense of knowing dissemination of obscene material to a minor by computer; establishes certain evidence that may be admissible in prosecutions for the offense; establishes an affirmative defense; sets penalty as a Class 4 felony. 1999 Bill Tracking IL H.B. 791, 91st General Assembly, 1999-00 General Assembly, House Bill 791, Introduced: February 10, 1999, Last-action: March 19, 1999; In House. Read third time. Passed House and sent To Senate. Amends the Criminal Code of 1961; creates the offense of facilitating theft of online services. 1999 Bill Tracking IL H.B. 793, 91st General Assembly, 1999-00 General Assembly, House Bill 793. Introduced: February 10, 1999, Last-action: May 8, 1999; Re-referred to Senate Committee on Rules. Prohibits Internet gambling.
Indiana	Burns Ind. Code Ann. § 35-42-4-4. *et seq.* (1999), Child exploitation. "A person who knowingly or intentionally:	

State	Enacted	Proposed
	(1) manages, produces, sponsors, presents, exhibits, photographs, films, or videotapes any performance or incident that includes sexual conduct by a child under eighteen (18) years of age; or (2) disseminates, exhibits to another person, offers to disseminate or exhibit to another person, or sends or brings into Indiana for dissemination or exhibition matter that depicts or describes sexual conduct by a child under eighteen (18) years of age; commits child exploitation, a Class D felony. However, the offense is a Class C felony if it is committed by using a computer network." Burns Ind. Code Ann. § 35-45-2-2 (1999), Harassment. "A person who, with intent to harass, annoy, or alarm another person but with no intent of legitimate communication: . . . (4) uses a computer network (as defined in IC 35-43-2-3 (a)) or other form of electronic communication to: (A) communicate with a person; or (B) transmit an obscene message or indecent or profane words to a person; commits harassment."	
Iowa	Iowa Code 716.A.1 and 2.	
Kansas	K.S.A. § 21-3755 (1998), Computer crime; computer password disclosure; computer trespass. Appears to track model form.	1999 Bill Tracking KS H.B. 2982, 78th Legislature, 2000 Regular Session, House Bill 2982. Introduced: February 11, 2000, Last-action: February 14, 2000; To House Committee on Judiciary. Concerns the theft of computer information; authorizing a civil cause of action therefore.

(continued)

State	Enacted	Proposed
Kansas *(continued)*		1999 Bill Tracking KS S.B. 384, 78th Legislature, 2000 Regular Session, Senate Bill 384, Introduced: December 15, 1999, Last-action: January 10, 2000; To Senate Committee on Judiciary. Relates to indecent solicitation and aggravated indecent solicitation of a child; creates the crime of electronic harassment.
Kentucky	KRS § 434.840, *et seq.* (Michie 1998). Unlawful access to a computer. Appears to track model form.	
Louisiana	La. R.S. 14:73.1 *et seq.* (LEXIS 2000). Computer related crime. Appears to track model form.	
Maine	Me. Rev. Stat. Ann. Tit. 17 431.	
Maryland	Maryland Ann. Code of 1957, art. 27, 146.	
Massachusetts	Mass. Ann. Laws ch. 266, § 33A (1999), Fraudulent obtaining of commercial computer service; penalty.	1999 Bill Tracking MA H.B. 2662, 181st General Court, 1999 Regular Session, House Bill 2662. Introduced: January 6, 1999, Last-action: September 20, 1999; From Joint Committee on Criminal Justice: Accompanied Study Order H 4737. Establishes the crime of luring or abducting of minors by computer communication.
	Mass. Ann. Laws ch. 266, § 120F (1999), Unauthorized accessing of computer systems; penalty; password requirement as notice. "Whoever, without authorization, knowingly accesses a computer system by any means, or after gaining access to a computer system by any means knows that such access is not authorized and fails to terminate such access, shall be punished."	
	Massachusetts law, Employment Records—requires any employer	

State	Enacted	Proposed
	that maintains employee records to provide access for employee to review and correct.	
Michigan	MCL § 752.791 *et. seq.*, Fraudulent access to computers, computer systems and networks, "An Act to prohibit access to computers, computer systems, and computer networks for certain fraudulent purposes; to prohibit intentional and unauthorized access, alteration, damage, and destruction of computers, computer systems, networks, computer software programs, and data; and to prescribe penalties."	1999 Bill Tracking MI H.B. 4869, 90th Legislature, 1999 Regular Session, House Bill 4869. Introduced: September 28, 1999, Last-action: September 28, 1999; To House Committee on Criminal Law and Corrections. Prohibits using the Internet to threaten to kill or injure another person or to threaten to cause property damage.
	MSA § 28.342b, Use of Internet or computer system; prohibited communication.	1999 Bill Tracking MI H.B. 5185, 90th Legislature, 1999 Regular Session, House Bill 5185. Introduced: December 9, 1999, Last-action: February 17, 2000; To Senate Committee on Judiciary. Prohibits access to computers, computer systems, and computer networks for certain fraudulent purposes; prohibits intentional and unauthorized access, alteration, damage, and destruction of computers, computer systems, computer networks, computer software programs, and data; prescribes penalties.
		1999 Bill Tracking MI S.B. 7, 90th Legislature, 1999 Regular Session, Senate Bill 7. Introduced: January 13, 1999, Last-action: June 1, 1999; Signed by governor. Public Act 32. Relates to crimes and computer.
Minnesota	Minn. Stat. 270B.18 and 609.87-89.	
Mississippi	Miss. Code Ann. § 97-45-1, *et. seq.* (2000), Computer crimes. Appears to track model form.	2000 Bill Tracking MS H.B. 617, Creates the tort of stalking; includes the use of elec tronic communication devices; prohibits

(continued)

State	Enacted	Proposed
Mississippi *(continued)*		making threats with electronic communication devices; provides penalties for violations.
	Miss. Code Ann. § 97-5-33 (2000), Exploitation of children; prohibitions: "No person shall, by any means including computer, cause or knowingly permit any child to engage in sexually explicit conduct or in the simulation of sexually explicit conduct for the purpose of producing any visual depiction of such conduct." Miss. Code Ann. § 97-25-54 (2000), Theft of telephone and other communication services prohibited; definitions; manufacture and possession of devices to facilitate theft prohibited; penalties: "It shall be unlawful for any person to use a telecommunication device intending to avoid the payment of any lawful charge for service to the device."	
Missouri	§ 569.095 R.S.Mo. (1999) § 569.095 R.S.Mo., Tampering with computer data "A person commits the crime of tampering with computer data if he knowingly and without authorization or without reasonable grounds to believe that he has such authorization." § 569.097 R.S.Mo. (1999) § 569.097 R.S.Mo., Tampering with computer equipment: "A person commits the crime of tampering with computer equipment if he knowingly and without authorization or without reasonable grounds to believe that he has such authorization." § 569.099 R.S.Mo. (1999) § 569.099 R.S.Mo., Tampering with computer users.	2000 Bill Tracking MO H.B. 1215, Creates a crime of indecent solicitation of a child; includes solicitation by electronic means.

State	Enacted	Proposed
Montana	Mont. Code Anno., § 45-6-311 (1999), Unlawful use of a computer: "A person commits the offense of unlawful use of a computer if the person knowingly or purposely: . . . obtains the use of or alters or destroys a computer, computer system, computer network, or any part thereof as part of a deception for the purpose of obtaining money, property, or computer services from the owner of the computer, computer system, computer network, or part thereof or from any other person."	
Nebraska	Neb. Rev. Stat. 28-1343 and 29-3518 *et seq.*	
Nevada	Nev. Rev. Stat. 242.111.	
New Hampshire	N.H. Rev. Stat. Ann. 638:17.	
New Jersey	N.J. Stat. Ann. 2A:38A-3.	
New Mexico	N.M. Stat. Ann. 15-1-9 and 30-45-5.	
New York	N.Y. Penal Law 156.05.	
North Carolina	N.C. Gen. Stat. § 14-453, *et. seq.* (1999), Computer-related crime.	1999 Bill Tracking NC H.B. 813, General Assembly of NC—Session of 1999, House Bill 813. Introduced: April 1, 1999, Last-action: October 5, 1999; 1999 General Assembly, first session adjourned 07/21/1999; carried over to 1999 General Assembly; Second Session. Criminalizes the use of electric mail or other electronic communications with the intent to harass, threaten, annoy, terrify, defame or embarrass anyone; criminalizes the introduction of any computer virus into electronic mail or electronic communication.
	N.C. Gen. Stat. § 14-196 (1999), Using profane, indecent or	1999 Bill Tracking NC S.B. 907, General Assembly of NC-Session of 1999, Senate

(continued)

State	Enacted	Proposed
North Carolina *(continued)*	threatening language to any person over telephone; annoying or harassing by repeated telephoning or making false statements over telephone, "... shall include communications made or received by way of a ... computer modem."	Bill 907. Introduced: April 14, 1999, Last-action: October 5, 1999; 1999 General Assembly—First Session Adjourned 07/21/1999; not carried over to 1999 General Assembly; Second Session. Prohibits the dissemination of obscenities by computer transmission.
North Dakota	General criminal law.	
Ohio	Ohio Rev. Code Ann., 2913.04.	
Oklahoma	Okla. Stat. Ann. Tit. 21, 1953-1958.	
Oregon	ORS § 164.377 (1997), Computer crime: "Any person commits computer crime who knowingly accesses, attempts to access or uses, or attempts to use, any computer, ... for the purpose of" committing fraud.	
	ORS § 164.125 (1997), Theft of services: "'services' includes ... use of telephone, computer and cable television systems."	
Pennsylvania	18 Pa.C.S. § 3933, Unlawful use of computer: "A person commits an offense if he: accesses, alters, damages or destroys any computer, with the intent to interrupt the normal functioning of an organiza tion or to devise or execute any scheme or artifice to defraud or deceive or control."	1999 Bill Tracking PA H.B. 1535, 183rd General Assembly, 1999-00 Regular Session, House Bill 1535. Introduced: May 13, 1999, Last-action: May 17, 1999; To House Committee on Judiciary. Regulates criminal conduct concerning computer activities. Provides for definition of crime of computer interference, involving unauthorized access to computer to damage program; gain control of money, property or data; use computer services; or to defraud, deceive or extort.
	18 Pa.C.S. § 910, Manufacture, distribution or possession of devices for theft of telecommunications services.	1999 Bill Tracking PA H.B. 1974, 183rd General Assembly, 1999-00 Regular Session, House Bill 1974. Introduced: October 18, 1999, Last-action: October 19, 1999; To House Committee on Judiciary. Amends the Crimes and Offenses Code.

State	Enacted	Proposed
		Provides for the misdemeanor offense of gambling by computer.
	18 Pa.C.S. § 6312, Sexual abuse of children: "Any person who causes or knowingly permits a child under the age of 18 years to engage in a prohibited sexual act or in the simulation of such act is guilty of a felony of the second degree if such person knows, has reason to know or intends that such act may be photographed, videotaped, depicted on computer. . . ."	
Rhode Island	R.I. Gen. Laws § 11-52-1, *et. seq.* (1999), computer crime.	1999 Bill Tracking RI H.B. 7680/ 1999 Bill Tracking RI S.B. 2689, 1999–2000 Legislative Session, House Bill 7680. Introduced: February 3, 2000; Last-action: February 3, 2000; To House Committee on Judiciary. Relates to cyberstalking.
	1999 Bill Tracking RI H.B. 5644, 1999-2000 Legislative Session, House Bill 5644. Introduced: February 2, 1999, Last-action: July 6, 1999; Became law without governor's signature. Expands the definition of "computer crime" to include the crime of "computer trespassing" and electronic mail and internet activities.	1999 Bill Tracking RI S.B. 2749, 1999-2000 Legislative Session, Senate Bill 2749. Introduced: February 10, 2000, Last-action: February 10, 2000; To Senate Committee on Judiciary. Provides for making use of the Internet for immoral or illegal purposes involving children a felony.
South Carolina	S.C. Code Ann. 16-16-20.	
South Dakota	S.D. Codified Laws § 43-43B-1 (2000), Unlawful uses of computer "A person is guilty of unlawful use of a computer if he: (1) Knowingly obtains the use of, or accesses, (2) Knowingly alters or destroys computer programs or data,	

(continued)

State	Enacted	Proposed
South Dakota *(continued)*	(3) Knowingly obtains use of, alters, accesses or destroys a computer system, or any part thereof, as part of a deception for the purpose of obtaining money, property or services, (4) Knowingly uses or discloses to another or attempts to use or disclose to another the numbers, codes, passwords or other means of access to a computer, ..."	
Tennessee	Tenn. Code Ann. § 39-14-601, *et. seq* (1999), Computer offenses.	
Texas	Tex. Penal Code § 33.01, *et. seq.* (1999), Computer crimes.	
Utah	Utah Code Ann. 53-5-214.	
Vermont	13 V.S.A. § 4101, *et. seq.* (2000), Computer crimes. Appears to track model form.	1999 Bill Tracking VT H.B. 100, 65th Biennial Session, House Bill 100. Introduced: January 27, 1999, Last-action: July 14, 1999; 65th Biennial Session Adjourned, 05/15/1999; carried over to adjourned session of 1999–2000 biennium. Prohibits the sending of commercial electronic mail that uses a third party's Internet domain name without the third party's permission.
Virginia	Computer Crimes Act, Va Code Ann. 18.2-152.1 to 152.14.	
Washington	Rev. Code Wash. (ARCW) § 9A.52.110, *et. seq.,* Computer trespass in the first degree. Rev. Code Wash. (ARCW) § 9.26A.100, *et. seq.,* Fraud in obtaining telecommunications service "... through ... the commission of computer trespass."	

State	Enacted	Proposed
West Virginia	W. Va. Code 61-3C-5.	
Wisconsin	Wis. Stat. 943.70.	
Wyoming	Wyo. Stat. § 6-3-501, *et. seq.* (1999), Computer crimes. Wyo. Stat. § 6-4-303 (1999), Sexual exploitation of children: "A person is guilty of sexual exploitation of a child if, for any purpose, he knowingly: . . . (iii) Manufactures, generates, creates, receives, distributes, reproduces, delivers or possesses with the intent to deliver, including through digital or electronic means, whether or not by computer, any child pornography."	

APPENDIX I Resources to Detect and Delete Cookies

1. **Albert's Ambry** (*www.alberts.com*). Albert's offers free downloads, shareware, and resale of anticookie software, including APK Cookie Killing Engine 98++ SR-4 by AlecStaar Systems, Inc., Cookie Crusher v2.11, and Cookie Cutter PC v2.53.
2. **Anonymizer.com** (*www.anonymizer.com* and *www.Anonominity.com*). Both Anonymizer.com and Anonominity.com. Web sites offer "proxy" servers that provide intermediaries or "clean portals" so that users can conduct anonymous Web searches, and thus protect their own privacy. The site also sells Webroot's Window Washer software.
3. **Blackboard Software, Inc.** (*www.blackboardsoftware.com/*). Blackboard's Internet Privacy v4.6 (12/22/99) permanently deletes history, cache, newsgroups, email trash, recently visited URLs and chat rooms, and cookies created from using the Internet. It provides support for Netscape 4, and Microsoft Internet Explorer 4/5, AOL4/5, ICQ, Microsoft Outlook/Express and StarOffice.
4. **Computer Incident Advisory Capability (Department of Energy)** (*ciac.llnl.gov/ciac/bulletins/i-034.shtml*). The site provides a thorough explanation of cookie technology, information on how to avoid cookies, and links to other cookie-related sites.
5. **Cookie Central** (*www.cookiecentral.com/*). Cookie Central provides a broad base of information about cookies including a consumer-friendly explanation about cookies and how to prevent them. It also provides a number of links to anticookie software, and a host of cookie software package downloads. The software packages includes Cookie Web Kit, Cookie Pal, ZDNet Cookie Master 2, Cookie Crusher 1.5 *The Limit Software,* Luckman's Anonymous Cookie for Internet Privacy, Buzof, PGP-cookie.cutter for Windows NT, IEClean (16-bit) Version: 3.01 for Internet Explorer, and NSClean (32-bit) Version: 4.05 for Netscape.

6. **Cookie Cutter PC by AyeCor** (*www.ayecor.com/html/fcc.html*). The Cookie Cutter site provides software that searches your hard drives for cookies, displays the cookie files, and gives you the option to delete these files automatically, on start up, at intervals, or piecemeal.
7. **Cookie Monster 1.5.1** (*www.geocities.com/Paris/1778/monster.html*). Cookie Monster deletes (Macintosh's) MagicCookie and cookies.txt files each time it is launched.
8. **CookieMaster.** CookieMaster is a ZDNET software product available free, which allows Internet users to chose which cookies to erase from their hard drives and which to keep.
9. **Privacy Software Corporation's NSClean Privacy Software** (*www.wizvax.net/kevinmca/*). This site provides anticookie software for Netscape (NSClean), Internet Explorer (IEClean), and Macintosh (MacClean). These software packages allow Internet users to use an alias in place of their user IDs or to purge cookies on their computers. Demos of the products can be purchased through such sites as: *www.axxis.com/altus/products/*.
10. **Junkbuster.com** (*www.junkbusters.com/ht/en/cookies.html*). The Junkbuster site provides a detailed explanation of what cookies are, and information on enabling your browser to screen for and block cookies. The Internet Junkbuster 2.0 program for Windows and UNIX users removes cookies and banner ads, and blocks Spam ads. It also provides information on online privacy, junk email, Web ads, and related information in 15 languages. Junkbusters Declare (*www.junkbusters.com/ht/en/declare.html*) provides a collection of form letters for consumers to use to request that marketers remove the consumers' names and data from commercial lists. This addresses junk email, junk "snail mail," junk faxes, and telemarketing calls. The Junkbuster site also provides links to many other Web sites for customers to request removal of their names and information from marketing lists.
11. **Luckman's Anonymous Cookie for Internet Privacy** (*www.luckman.com*). This free resource from Luckman Interactive, Inc. erases cookies.
12. **Net Angels.** Net Angels offers "proxy" servers that provide intermediaries through which to make anonymous Web searches and thus protect one's privacy.
13. **Netscape Navigator 3.0** and **Microsoft's Internet Explorer 3.0.** Both of these packages allow users to decide whether to accept a cookie or not. However, these programs present the option to users when the cookie arrives at their computer. While users appreciated the option, many found the continuous window warnings of cookie arrivals in Navigator 3.0 to be annoying. Thus Navigator 4.0 offered filtering that allows users to automatically reject all cookies.

14. **PGPcookie cutter** (*www.pgp.com*). PGPcookie cutter created by Pretty Good Privacy, Inc.
15. **PrivNet.** Privnet prevents cookies and unwanted email, known as "spam," from being deposited in the user's computer.
16. **Research Central** (*www.softouch.on.ca/rc/cookies.htm*). Research Central provides information about cookies and how to get rid of them.
17. **Snooper 2.0** (*www.softouch.on.ca/rc/cookies.htm*). Snooper 2.0 provides a demonstration of the type of information that can be detected and stored with cookies.
18. **Webroot Software Inc.** (*www.webroot.com*). Webroot's Window Washer 3.0 cleans the browser's cache of cookies affecting the system's recycle bin, Windows temporary folders, recently opened document list, Window registry streams, Windows find history, Windows run history, and CHK scan disk files, for IE, Netscape, NeoPlanet, AOL, and Opera.
19. **Webwasher.com.** This site provides forms and other information for consumers to use to request removal of their names and information from marketing lists.
20. **ZDnet** (*www.ZDnet.com*). ZDnet provides downloads of various anti-cookie software packages including PC Magazine's CookieCop v1.1, Cookie Cutter PC (32-bit) v2.61, Cookie Cutter PC (16-bit) v2.4, Cookie Terminator v1.0, Cookie Pal v1.5e, Cookie Cutter v1.0, and Cookie Web Kit v2. It also provides a reference article entitled "Protect Yourself from Cookie Collectors: Take the Distribution of Cookies into Your Own Hands with This Web Cookie Explainer," by Leena Pendharkar, at (*www.zdnet.com/zdhelp/stories/main/0,5594,2420690,00.html*).

APPENDIX J State Laws Concerning Encryption, Key Escrow, and Digital Signatures

State	Enacted	Proposed
Alabama	Electronic Tax Filing Act, Code of Alabama § 40-30-1, *et. seq.* (1997).	2000 Bill Text AL H.B. 390. Introduced: February 8, 2000. An act to adopt the Uniform Electronic Transactions Act to apply to any electronic record or signature created, sent, communicated, or received on, and after January 1, 2001. 2000 Bill Text AL S.B. 312. Introduced: February 10, 2000. An act to adopt the Uniform Electronic Transactions Act to apply to any electronic record or signature created, sent, communicated, or received on and after January 1, 2001.
Alaska	Alaska Stat. § 09.25.500, *et. seq.* (2000), electronic signatures act.	
Arizona	A.R.S. § 41-132 (2000) A.R.S. § 41-132, electronic and digital signatures act.	2000 Bill Text AZ H.B. 2069. Prefiled: January 3, 2000. An act relating to electronic transactions.
Arkansas	Ark. Stat. Ann. § 25-31-101, *et. seq.* (1999), electronic records and signatures act. Ark. Stat. Ann. § 25-27-101, *et. seq.* (1999), Arkansas Information Network. See particularly, Ark. Stat. Ann.	

(continued)

State	Enacted	Proposed
Arkansas *(continued)*	§ 25-27-104 (1999), "(a) The Information Network of Arkansas shall have the following duties: (1) To develop and implement an electronic gateway system to provide electronic access to members of the public to public information and to develop, implement, and promote the use of electronic commerce and digital signature applications within the state in cooperation with the Department of Information Systems."	
California	Cal Civ Code § 1633, (1999), validity of contract with broker-dealer resulting from customer's electronically transmitted application "an application by a prospective customer to enter into a brokerage agreement with a broker-dealer, which application is transmitted electronically and is accompanied by the prospective customer's electronic signature or digital signature as described in subdivisions (d), (e), (f), and (g), shall be deemed, upon acceptance by the broker-dealer, to be a fully executed, valid, enforceable, and irrevocable written contract, unless grounds exist which would render any other contract invalid, unenforceable, or revocable.	1999 Bill Tracking CA A.B. 374. Introduced: February 11, 1999. Requires the Insurance Commissioner, in consultation and agreement with the Chief Information Officer and the Secretary of the State, to adopt rules creating minimal acceptable standards regarding the use in the insurance industry of digital signatures and public-key infrastructures.
	Cal Gov Code § 16.5, (1999), use of digital signature: "(a) In any written communication with a public entity, as defined	1999 Bill Tracking CA S.B. 1124. Introduced: February 26, 1999. Provides that an application by a prospective customer to enter into a brokerage agreement

State	Enacted	Proposed
	in Section 811.2, in which a signature is required or used, any party to the communication may affix a signature by use of a digital signature that complies with the requirements of this section. The use of a digital signature shall have the same force and effect as the use of a manual signature if. . . ."	with a broker-dealer, as defined, shall be deemed to be a valid contract if the customer to the broker-dealer transmits the application electronically, is accompanied by the prospective customer's electronic or digital signature, and is accepted by the broker-dealer.
Colorado	C.R.S. 24-71-101, *et. seq* (1999), electronic Signatures.	2000 Bill Text CO H.B. 1329, concerning the adoption of the Uniform Electronic Transactions Act.
Connecticut	Conn. Gen. Stat. § 19a-25a (1999), regulations regarding electronic signatures for medical records. 1999 Ct. ALS 155; 1999 Ct. P.A. 155; 1999 Ct. HB 6592, PUBLIC ACT NO. 99-155, An Act Concerning Electronic Records and Signatures.	
Delaware		1999 Bill Tracking DE H.B. 492. Introduced: March 16, 2000. Creates enforceable electronic contracts and validates electronic signatures.
District of Columbia	None found.	
Florida	Fla. Stat. § 282.70, *et. seq.* (1999), Electronic Signature Act. See particularly, Fla. Stat. § 282.73 (1999), force and effect of electronic signature "Unless otherwise provided by law, an electronic signature may be used to sign a writing and shall have the same force and effect as a written signature."	1999 Bill Text FL S.B. 1224. Prefiled: February 16, 1999. An act authorizing filing of tangible personal property tax returns in a form initiated through electronic data interchange.

(continued)

State	Enacted	Proposed
Florida *(continued)*	Fla. Stat. § 199.052 (1999), annual tax returns; payment of annual tax, "(b) A taxpayer may choose to file an annual intangible personal property tax return in a form initiated through an electronic data interchange using an advanced encrypted transmission by means of the Internet or other suitable transmission."	
Georgia	O.C.G.A. § 10-12-1, *et. seq.* (1999), Electronic Records and Signatures Act. See particularly, O.C.G.A. § 10-12-4 (1999), legal effect. 1999 Bill Text GA S.B. 62. Enacted: April 19, 1999. An act to amend Chapter 12 of Title 10 of the Official Code of Georgia Annotated, the Georgia Electronic Records and Signatures Act, so as to change the provisions relating to definitions, to provide for the legal effect, validity, and admissibility of electronic records, electronic signatures, and secure electronic signatures.	1999 Bill Tracking GA S.B. 403, relates to filing of [court] documents by electronic means; provides for the electronic signature and verification of such pleadings and to provide for methods of service and docketing of such pleadings.
Hawaii	Hawaii Revised Statutes Annotated Title 14 § 231-8.5, electronic filing of tax returns.	1999 Bill Tracking HI S.B. 1434. Introduced: January 27, 1999. Relates to information technology; establishes legal framework for using digital signatures as a means of authenticating computer-based information.
Idaho	Idaho Code § 67-2351 (1999), Electronic Signature and Filing Act. See particularly Idaho Code § 67-2354 (1999), electronic signatures, "In any communication filed with or issued by any public agency, in which a	2000 Bill Text ID S.B. 1334. Enacted: April 14, 2000. An act amending Title 28, Idaho Code, by the addition of a new Chapter 50, to establish the Uniform Electronic Transactions Act.

State	Enacted	Proposed
	signature is required or used, any signing party may affix a signature by use of an electronic signature that complies with the requirements of this section."	
Illinois	5 ILCS 175/5-105, *et. seq.* (1999), Electronic Commerce Security Act. See particularly,	1999 Bill Tracking IL H.B. 3420. Introduced: January 20, 2000. Authorizes the Secretary of State . . . to accept electronic signatures and utilize an electronic payment system for required fees and taxes.
	5 ILCS 175/10-110 (1999), secure electronic signature, "(a) If, through the use of a qualified security procedure, it can be verified that an electronic signature is the signature of a specific person, then such electronic signature shall be considered to be a secure electronic signature at the time of verification, if . . ."; 5 ILCS 175/15-101 (1999), Effect of Digital Signature.	
	205 ILCS 705/1, *et. seq.* (1999), Financial Institutions Digital Signatures Act. See particularly, 205 ILCS 705/10 (1999), electronic documents; digital signatures, "(a) . . . shall have the same force and effect as one comprised, recorded, or created on paper or other tangible form by writing, typing, printing, or similar means. . . ."	
Indiana	Burns Ind. Code Ann. § 5-24-1-1, *et. seq.* (1999), Electronic Digital Signature Act. See particularly, Burns Ind. Code Ann. § 5-24-3-1 (1999), effectiveness of digital signature.	

(continued)

State	Enacted	Proposed
Indiana *(continued)*	2000 Bill Text IN H.B. 1395. Enacted: March 15, 2000, an act to amend the Indiana Code concerning commercial law "(D) If a law requires a signature, or provides consequences in the absence of a signature, the law is satisfied with respect to an electronic record if the electronic record includes an electronic signature."	
Iowa	Iowa Code § 48A.13 (1999), electronic signatures on voter registration records, "Electronic signatures shall be accepted. However, before the use of electronic signatures is accepted on voter registration forms, the state voter registration commission shall prescribe by rule the technological requirements for guaranteeing the security and integrity of electronic signatures. . . ." Iowa Code § 155A.27 (1999), requirements for prescription, "The name, address, and written or electronic signature of the practitioner issuing the prescription." 1999 Bill Tracking IA H.B. 624. Enacted: May 19, 1999, relates to electronic commerce security; effectuates electronic communications by means of reliable electronic records; promotes interoperability; pertains to secure digital signatures, algorithms, standards, legal proceedings.	1999 Bill Text IA H.B. 2205, Passed First House, Amended: March 8, 2000. An act relating to electronic commerce by establishing requirements for electronic transactions and electronic records and providing penalties.
Kansas	K.S.A. § 60-2616, Digital Signature Act, "(c) A digital signature may be accepted as a substitute for, and, if	1999 Bill Text KS H.B. 2879. Version-date: March 16, 2000. An act concerning electronic transactions; enacting the uniform electronic transactions act.

State	Enacted	Proposed
	accepted, shall have the same force and effect as any other form of signature."	
		1999 Bill Text KS S.B. 559. Version-date: February 21, 2000. An act concerning electronic transfers; repealing K.S.A. 1999 Supp. 60-2616.
Kentucky	KRS § 369.010, *et. seq.* (Michie 1998), use of electronic signature. See particularly, KRS § 369.030 (Michie 1998), Use of electronic record or electronic signature. 2000 Ky. HB 571, adopts the Uniform Electronic Transactions Act drafted by the National Conference of Commissioners on Uniform State Laws, and making changes incidental thereto, "A new section of KRS Chapter 369 is created to read as follows: . . . " 2000 Bill Tracking KY H.B. 939. Signed by Governor: April 10, 2000. States the desire of the General Assembly to automate the filing of campaign finance information and further disclose that information to the public; provides definitions for electronic reporting, security procedure, electronic signature, filer and filer-side software; provides for electronic filing of campaign finance reports; makes provisions contingent upon funding.	
Louisiana	La. R.S. 12:2, business corporation law, filing methods, "A. (1) The secretary of state may	

(continued)

State	Enacted	Proposed
Louisiana *(continued)*	accept any filing, authorized by this Title, by electronic or facsimile transmission. All electronic filings authorized by this Title shall include an electronic or digital signature. . . ." La. R.S. 22:2.1, insurance code, public records; forms and methods; electronic signatures and filings, "D. Subject to such guidelines and limitations as may be promulgated by the commissioner, electronic signatures are hereby authorized. . . ." La. R.S. 40:2144, Hospital Records and Retention Act, "E. (1) A hospital record or hospital chart may be kept in any written, photographic, microfilm, or other similar method or may be kept by any magnetic, electronic, optical, or similar form of data compilation. . . ."	
Maine	11 M.R.S. § 3-1401 (1999), Signature, "(2) A signature may be made: (a) Manually or by means of a device or machine. . . ."	1999 Bill Tracking ME H.B. 1451. Introduced: March 30, 1999. Provides for validity and admissibility . . . of electronic records and also allows for an electronic signature to have the same legal force and effect as a manual signature. 1999 Bill Text ME S.B. 995. Introduced: February 15, 2000. An Act to Implement the Recommendations of the Blue Ribbon Commission to Establish a Comprehensive Internet Policy.
Maryland	Md. STATE GOVERNMENT Code Ann. § 8-504, Digital Signature Pilot Program, "(2) . . . to allow for the use of a digital signature in any communication in which a	2000 Bill Tracking MD S.B. 3. Introduced: January 4, 2000. Adopts the Maryland Uniform Electronic Transactions Act; limits the application of the act to transactions where the parties have agreed to conduct transactions electronically; provides that an

State	Enacted	Proposed
	signature is required or used within the agency or between the agency and another agency or governmental entity. (3) The use of a digital signature under this section shall have the same force and effect as the use of a manual signature."	electronic record or signature may not be denied legal effect or enforceability solely based on its electronic form; provides that the requirements of laws that specify a record or signature be in written form are met by an electronic record or electronic signature. 2000 Bill Text MD H.B. 18. Enrolled: April 7, 2000. An act concerning commercial law—The Maryland Uniform Electronic Transactions Act.
Massachusetts		1999 Bill Text MA H.B. 4494, An act protecting the privacy of medical records. "Section 6. (a) For the purposes of this chapter, the requirement of informed consent shall be deemed to have been satisfied if. . . . For purposes of this section documentation of informed consent may be satisfied by the use of electronic signatures . . . in lieu of paper based documentation."
Michigan		1999 Bill Tracking MI H.B. 4406; (NEW BILL). Introduced: March 10, 1999. Creates the Digital Signature Act. 1999 Bill Text MI H.B. 5537. Introduced: March 22, 2000. A bill to authorize and provide the terms and conditions under which information and signatures can be transmitted, received, and stored by electronic means.
Minnesota	Minn. Stat. § 325K.01, *et. seq.* (1999), Electronic Authentication Act. See particularly, Minn. Stat. § 325K.13 (1999), control of private key, "By accepting a certificate issued by a licensed certification authority, the subscriber identified in the	1999 Bill Text MN H.B. 3109. Version-date: April 3, 2000. Enacting the Uniform Electronic Transactions Act adopted by the National Conference of Commissioners on Uniform State Laws; proposing coding for new law as Minnesota Statutes, Chapter 325L.

(continued)

State	Enacted	Proposed
Minnesota *(continued)*	certificate assumes a duty to exercise reasonable care to retain control of the private-key and prevent its disclosure to a person not authorized to create the subscriber's digital signature. . . ." Minn. Stat. § 325K.19 (1999), satisfaction of signature requirements; Minn. Stat. § 325K.21 (1999), digitally signed document is written.	
	Minn. Stat. § 221.173 (1999), Carriers, Chapter 221 Motor Carriers; Pipeline Carriers, Motor Carriers, 221.173, electronic signature "(a) The commissioner may accept in lieu of a required document completed on paper, an electronically transmitted document authenticated by an electronic signature."	1999 Bill Tracking MN H.B. 3066. Introduced: February. Provides for technical amendments to provisions regarding digital signatures.
		1999 Bill Text MN H.B. 1312, Version-date: March 22, 1999. A bill for specifying the consequences of accepting certain digital signatures.
		1999 Bill Tracking MN S.B. 820. Introduced: February 18, 1999. Specifies the consequences of accepting certain digital signatures.
Mississippi	Miss. Code Ann. § 25-63-1, *et. seq.* (2000), Digital Signature Act. See particularly, Miss. Code Ann. § 25-63-9 (2000), force and effect of digital signature.	2000 Bill Tracking MS H.B. 1164. Introduced: February 21, 2000. Provides that an electronic medical record or medical order containing an electronic signature is considered to be signed.
		2000 Bill Tracking MS S.B. 2864. Introduced: February 18, 2000. Provides for electronic signatures on medical records (died in committee).
Missouri	None found.	

State	Enacted	Proposed
Montana	Mont. Code Anno., § 2-20-101, *et. seq.* (1999), electronic transactions with state government. See particularly, Mont. Code Anno., § 2-20-102 (1999), legislative intent "(3) [To] facilitate and promote electronic commerce by eliminating uncertainties over what constitutes verified electronic records and electronic signatures in transactions with state agencies and local government units. . . ."	
Nebraska	R.R.S. Neb. § 86-1701 (2000), digital signatures authorized, "(1) In any written communication in which a signature is required or used, any party to the communication may affix a signature by use of a digital signature that complies with the requirements of this section. The use of a digital signature shall have the same force and effect as the use of a manual signature if and only if it embodies all of the following attributes. . . ."	1999 Bill Text NE L.B. 929. Introduced: January 5, 2000. An act to adopt the Uniform Electronic Transactions Act.
	1999 Neb. ALS 628; 1999 Neb. Laws 628; 1999 Neb. LB 628, an Act relating to records; to amend sections 84-712, 84-712.01, 84-712.03, and 86-1701, Reissue Revised Statutes of Nebraska; to provide requirements for the provision of copies of public records; to change provisions relating to public records and digital signatures; to provide for electronic signatures; and to repeal the original sections.	1999 Bill Tracking NE L.B. 1080. Introduced: January 7, 2000. Relates to digital signatures; provides for use of electronic communications by state government.

(continued)

State	Enacted	Proposed
Nevada	Nev. Rev. Stat. Ann. § 720.010, *et. seq.* (2000), electronic transfers, digital signatures. See particularly, Nev. Rev. Stat. Ann. § 720.160 (2000), agreement to use digital signature: satisfaction of statute or rule of law requiring signature or writing. Nev. Rev. Stat. Ann. § 80.003 (2000), business associations, securities, commodities, foreign corporations. "Signed" means to have executed or adopted a name, word or mark, including, without limitation, a digital signature. Nev. Rev. Stat. Ann. § 171.103, court clerk may accept complaint filed electronically.	
New Hampshire	RSA 294-D:1 *et. seq.,* New Hampshire Digital Signature Act. RSA 382-A:9-410, facsimile and electronic signatures. RSA 506:8, *et. seq.,* prevention of frauds and perjuries, digital signatures.	
New Jersey		2000 Bill Text NJ S.B. 1183. Introduced: March 27, 2000. Creates the Uniform Electronic Transactions Act, supplementing Title 12A of the New Jersey Statutes.
New Mexico	New Mexico Statutes Annotated §14-15-1 *et. seq.* (2000), Electronic Authentication of Documents Act.	
New York	NY CLS State Technology Law § 101, *et. seq.* (1999), Electronic Signatures and Records Act. See particularly, NY CLS State Technology Law § 104, (1999), Use of electronic signatures.	1999 Bill Tracking NY A.B. 2566. Introduced: January 25, 1999. Enacts the Digital Signatures Act; establishes licensing and regulation of the certification of electronic communications; implements an infrastructure in which computer users use certification authorities, online databases

State	Enacted	Proposed
		called repositories and public-key encryption technology in order to sign electronic documents in a legally binding fashion.
		1999 Bill Tracking NY S.B. 2155. Introduced: February 3, 1999. Relates to Electronic Signatures Act to provide for the certification of electronic signatures that are used in lieu of written signature; grants electronic signatures the same force and effect as written signatures.
		1999 Bill Text NY S.B. 2155. Version-date: June 13, 1999. An act to amend the general obligations law, in relation to certification of electronic signatures.
North Carolina	N.C. Gen. Stat. § 66-58.1 (1999) N.C. Gen. Stat. § 66-58.1 *et. seq.,* electronic commerce in government. See particularly, N.C. Gen. Stat. § 66-58.4 (1999) N.C. Gen. Stat. § 66-58.4, use of electronic signatures; N.C. Gen. Stat. § 66-58.5 (1999) N.C. Gen. Stat. § 66-58.5, validity of electronic signatures.	1999 Bill Tracking NC S.B. 808. Introduced: April 12, 1999. Appropriates funds for a digital signature pilot project.
	N.C. Gen. Stat. § 90-412 (1999), electronic medical records "(b) Notwithstanding any other provision of law, any health care provider or facility licensed, certified, or registered . . . may permit authorized individuals to authenticate orders and other medical record entries . . . by electronic or digital signature in lieu of a signature in ink. . . ."	1999 Bill Text NC H.B. 957. Version-date: June 23, 1999. An act to provide that an electronic or facsimile signature of a physician providing medical certification of death is acceptable.
North Dakota	N.D. Cent. Code, § 31-08-01.2 (2000), medical records authentication. "If appropriate safeguards	

(continued)

State	Enacted	Proposed
North Dakota *(continued)*	have been taken to limit access to medical records in an electronic data storage system, a medical record in an electronic data storage system may be authenticated by an electronic signature or a computer-generated signature code."	
Ohio		1999 Bill Tracking OH H.B. 488. Introduced: October 27, 1999. Enacts the Electronic Records and Signatures Act by providing for regulation of electronic signatures, including digital signatures, and electronic records.
Oklahoma	15 Okl. St. § 962 (1999)15 Okl. St. § 96215 Okl. St. § 962, Electronic Records and Signature Act of 1998.	1999 Bill Text OK H.B. 2508. Version-date: March 1, 2000. An Act relating to electronic commerce; amending Sections 3, 4, 5, 6, 7, 9, and 11, Chapter 308, O.S.L. 1998 (15 O.S. Supp. 1999, Sections 960, 961, 962, 963, 964, 966 and 968), which relate to the Electronic Records and Signature Act of 1998; changing the name of the Electronic Records and Signature Act to the Uniform Electronic Transaction Act.
	74 Okl. St. § 5060.50, *et. seq.* (1999) Electronic Commerce, § 5060.50, electronic commerce pilot program. See particularly, 74 Okl. St. § 5060.51 (1999)74 Okl. St. § 5060.5174 Okl. St. § 5060.51, electronic signature certification authority technology.	1999 Bill Text OK S.B. 249. Version-date: March 8, 1999. An Act relating to (among other provisions) electronic chattel paper.
	1999 Bill Text OK H.B. 1411. Enacted: May 24, 1999. An Act relating to technology; amending Section 1, Chapter 308, O.S.L. 1998 (74 O.S. Supp. 1998, Section 5060.50), which relates to the electronic commerce pilot program; extending the reporting date for the pilot program; authorizing the Office of State Finance to	1999 Bill Text OK S.B. 1337. Version-date: March 13, 2000. Oklahoma Uniform Computer Information Transactions Act.

State	Enacted	Proposed
	implement and issue electronic signature certification authority,	
Oregon	ORS § 192.825 (1997), Electronic Signature Act. See particularly, ORS § 192.840 (1997), force and effect of electronic signature. ORS § 709.335 (1997), regulation of trust business, electronic and digital signatures. "A trust company may conduct transactions using electronic and digital signatures, may be an authentication authority and may issue certificates for the purpose of verifying digital signatures."	
Pennsylvania	1999 Pa. ALS 69; 1999 Pa. Laws 69, an act Regulating Electronic Transactions. 1999 Bill Tracking PA S.B. 555, signed by governor. Act No. 69 of 1999. Creates the Electronic Transactions Act. Provides for the use of electronic signatures and records.	1999 Bill Tracking PA S.B. 519; (NEW BILL). Introduced: March 9, 1999. Relates to digital signatures.
Rhode Island	R.I. Gen. Laws § 42-127-1, *et. seq.* (2000), Electronic Signatures and Records Act. See particularly, R.I. Gen. Laws § 42-127-4 (2000), electronic signatures; R.I. Gen. Laws § 42-127-5 (2000), electronic records.	1999 Bill Tracking RI H.B. 7694. Introduced February 3, 2000. Pertains to Electronic Signatures and Records Act; provides for the repeal of such act and replacing its provisions with the Uniform Electronic Transaction Act. 1999 Bill Tracking RI S.B. 2569. Introduced: February 9, 2000. Relates to the Uniform Electronic Transactions Act, which will provide for the validity of and govern the use of electronic signatures and records.
South Carolina	S.C. Code Ann. § 26-5-50 (1998), Electronic Commerce Act.	

(continued)

State	Enacted	Proposed
South Carolina *(continued)*	"(B) The Secretary of State is authorized to develop, implement, and facilitate the use of model procedures for the use of electronic records, electronic signatures, and security procedures for all other purposes, including private commercial transactions and contracts. The Secretary of State is also authorized to promulgate methods, means, and standards for secure electronic transactions including administration by the Secretary of State and/or the licensing of third parties to serve in such capacity.	
South Dakota	2000 S.D. SB 193, an act to establish certain requirements and procedures regarding electronic transactions. 2000 Bill Tracking SD S.B. 193. Introduced: January 24, 2000. Establishes certain requirements and procedures regarding electronic transactions.	
Tennessee	None found.	
Texas	Tex. Bus. & Com. Code § 2.108 (1999), digital signature, "(a) A written electronic communication sent from within or received in this state in connection with a transaction governed by this chapter is considered signed if a digital signature is transmitted with the communication. . . ." Tex. Bus. & Com. Code § 2A.110 (1999), digital signature, "(a) A written electronic communication sent from within or received in this state in connection with a transaction governed by this	

State	Enacted	Proposed
	chapter is considered signed if a digital signature is transmitted with the communication. . . ."	
Utah	Utah Code Ann. § 46-1-1 (1999), Notarization and authorization of documents and digital signatures.	2000 Bill Tracking UT S.B. 76, 2000 General Session, Senate Bill 76. Introduced: December 14, 1999. Last-action: February 21, 2000. Signed by governor. Relates to digital signatures; amends the provisions mandating that the Division of Corporations and Commercial Code be a certification authority.
	Utah Code Ann. § 46-3-101, *et. seq.* (1999), Utah Digital Signatures Act.	
	2000 Bill Tracking UT H.B. 163, 2000. Signed by governor. Relates to commerce, trade and the criminal code; creates the crime of identification number fraud; defines identification number to include . . . digital signatures and private keys in addition to any other number or information that can be used to access personal and financial information.	
	2000 Bill Tracking UT S.B. 77, 2000. Signed by governor. Relates to the duties of the chief information officer; prescribes responsibility for coordination of the development of electronic authentication methods and technology to facilitate electronic transactions between government and citizens and business.	
Vermont	9 V.S.A. § 4208 (2000). Registration by qualification. "(9) Electronic securities registration depository. The	1999 Bill Tracking VT H.B. 609. Introduced: January 5, 2000. Uniform Electronic Transactions Act.

(continued)

State	Enacted	Proposed
Vermont *(continued)*	commissioner may, as an alternative method of registering securities under this chapter, or in conjunction with the pertinent provisions of this chapter, register securities and record the registration of securities by means of or through the facilities of an electronic securities registration depository which is maintained on a multi-state basis in cooperation with and for the use of state securities agencies or administrators. In the event of conflict between this provision and other pertinent provisions of this chapter, the commissioner may elect that this provision prevail. . . ."	
Virginia	Va. Code Ann. § 18.2-152.15 (1999). Encryption used in criminal activity. Va. Code Ann. §§ 59.1-467 to 469, electronic signatures. 2000 Bill Tracking VA H.B. 499. Signed by governor. Adopts the Uniform Electronic Transactions Act promulgated by the National Conference of Commissioners on Uniform State Laws; includes the exemption for the court filings, the incorporated sections; makes technical amendments throughout the Virginia Code to conform to the provisions of the new chapter; provides rules and procedures for using electronic records and electronic signatures in both commercial and governmental transactions.	
Washington	Rev. Code Wash. (ARCW) § 19.34.010. *et. seq.,*	1999 Bill Tracking WA H.B. 2174. Introduced: February 17, 1999. Promotes

State	Enacted	Proposed
	Electronic Authentication Act. 1999 Bill Tracking WA S.B. 5962. Signed by Governor. Promoting electronic commerce through digital signatures.	electronic commerce through digital signatures.
West Virginia	W. Va. Code § 39-5-1 (2000). Electronic Signatures Authorization Act. See particularly, W. Va. Code § 39-5-3 (2000), acceptance of electronic signatures generally.	2000 Bill Tracking WV H.B. 4493. Introduced: February 10, 2000. Uniforms the Electronic Transactions Act.
Wisconsin	Wis. Stat. § 137.04 through 137.06, electronic signatures.	1999 Bill Tracking WI A.B. 267. 1999 Introduced: April 9, 1999. Relates to the use and regulation of electronic signatures.
Wyoming	Wyo. Stat. § 9-2-2501 (2000), Wyoming on-line government commission; duties; electronic transaction of business.	

Index